For information about Vols. 1–23, please contact your bookseller or Springer-Verlag.

Vol. 24: R. F. Snipes, Statistical Mechanical Theory of the Electrolytic Transport of Nonelectrolytes. V, 210 pages. 1973.

Vol. 25: Constructive Quantum Field Theory. The 1973 "Ettore Majorana" International School of Mathematical Physics. Edited by G. Velo and A. Wightman. III, 331 pages. 1973.

Vol. 26: A. Hubert, Theorie der Domänenwände in geordneten Medien. XII, 377 Seiten. 1974.

Vol. 27: R. K. Zeytounian, Notes sur les Ecoulements Rotationnels de Fluides Parfaits. XIII, 407 pages. 1974.

Vol. 28: Lectures in Statistical Physics. Edited by W. C. Schieve and J. S. Turner. V, 342 pages. 1974.

Vol. 29: Foundations of Quantum Mechanics and Ordered Linear Spaces. Advanced Study Institute, Marburg 1973. Edited by A. Hartkämper and H. Neumann. VI, 355 pages. 1974.

Vol. 30: Polarization Nuclear Physics. Proceedings 1973. Edited by D. Fick. IX, 292 pages. 1974.

Vol. 31: Transport Phenomena. Sitges International Schools of Statistical Mechanics, June 1974. Edited by G. Kirczenow and J. Marro. XIV, 517 pages. 1974.

Vol. 32: Particles, Quantum Fields and Statistical Mechanics. Proceedings 1973. Edited by M. Alexanian and A. Zepeda. V, 132 pages. 1975.

Vol. 33: Classical and Quantum Mechanical Aspects of Heavy Ion Collisions. Proceedings 1974. Edited by H. L. Harney, P. Braun-Munzinger, and C. K. Gelbke. VII, 311 pages. 1975.

Vol. 34: One-Dimensional Conductors GPS Summer School Proceedings, 1974. Edited by H. G. Schuster. VII, 371 pages. 1975.

Vol. 35: Proceedings of the Fourth International Conference on Numerical Methods in Fluid Dynamics, 1974. Edited by R. D. Richtmyer. V, 457 pages. 1975.

Vol. 36: R. Gatignol, Théorie Cinétique des Gaz à Répartition Discrète de Vitesses. II, 219 pages. 1975.

Vol. 37: Trends in Elementary Particle Theory. Proceedings 1974. Edited by H. Rollnik and K. Dietz. V, 472 pages. 1975.

Vol. 38: Dynamical Systems, Theory and Applications. Proceedings 1974. Edited by J. Moser. VI, 624 pages. 1975.

Vol. 39: International Symposium on Mathematical Problems in Theoretical Physics. Proceedings 1975. Edited by H. Araki. XII, 562 pages. 1975.

Vol. 40: Effective Interactions and Operators in Nuclei. Proceedings 1975. Edited by B. R. Barrett. XII, 339 pages. 1975.

Vol. 41: Progress in Numerical Fluid Dynamics. Proceedings 1974. Edited by H. J. Wirz. V, 471 pages. 1975.

Vol. 42: H II Regions and Related Topics. Proceedings 1975. Edited by D. Downes and T. L. Wilson. XII, 488 pages. 1975.

Vol. 43: Laser Spectroscopy. Proceedings 1975. Edited by S. Haroche, J. C. Pebay-Peyroula, T. W. Hänsch, and S. E. Harris. X, 466 pages. 1975.

Vol. 44: R. A. Breuer, Gravitational Perturbation Theory and Synchrotron Radiation. VI, 196 pages. 1975.

Vol. 45: Dynamical Concepts on Scaling Violation and the New Resonances in e^+e^- Annihilation. Edited by B. Humpert. VII, 248 pages. 1976.

Vol. 46: E. J. Flaherty, Hermitian and Kählerian Geometry in Relativity. VIII, 365 pages. 1976.

Vol. 47: Padé Approximants Method and Its Applications to Mechanics. Edited by H. Cabannes. XV, 267 pages. 1976.

Vol. 48: Interplanetary Dust and Zodiacal Light. Proceedings 1975. Edited by H. Elsässer and H. Fechtig. XII, 496 pages. 1976.

Vol. 49: W. G. Harter and C. W. Patterson, A Unitary Calculus for Electronic Orbitals. XII, 144 pages. 1976.

Vol. 50: Group Theoretical Methods in Physics. 4th International Colloquium. Nijmegen 1975. Edited by A. Janner, T. Janssen, and M. Boon. XIII, 629 pages. 1976.

Vol. 51: W. Nörenberg und H. A. Weidenmüller. Introduction to the Theory of Heavy-Ion Collisions. IX, 273 pages. 1976.

Vol. 52: M. Mladjenović, Development of Magnetic β-Ray Spectroscopy. X, 282 pages. 1976.

Vol. 53: D. J. Simms and N. M. J. Woodhouse, Lectures on Geometric Quantization. V, 166 pages. 1976.

Vol. 54: Critical Phenomena. Sitges International School on Statistical Mechanics, June 1976. Edited by J. Brey and R. B. Jones. XI, 383 pages. 1976.

Vol. 55: Nuclear Optical Model Potential. Proceedings 1976. Edited by S. Boffi and G. Passatore. VI, 221 pages. 1976.

Vol. 56: Current Induced Reactions. International Summer Institute, Hamburg 1975. Edited by J. G. Körner, G. Kramer, and D. Schildknecht. V, 553 pages. 1976.

Vol. 57: Physics of Highly Excited States in Solids. Proceedings 1975. Edited by M. Ueta and Y. Nishina. IX, 391 pages. 1976.

Vol. 58: Computing Methods in Applied Sciences. Proceedings 1975. Edited by R. Glowinski and J. L. Lions. VIII, 593 pages. 1976.

Vol. 59: Proceedings of the Fifth International Conference on Numerical Methods in Fluid Dynamics. 1976. Edited by A. I. van de Vooren and P. J. Zandbergen. VII, 459 pages. 1976.

Vol. 60: C. Gruber, A. Hintermann, and D. Merlini, Group Analysis of Classical Lattice Systems. XIV, 326 pages. 1977.

Vol. 61: International School on Electro and Photonuclear Reactions I. Edited by C. Schaerf. VIII, 650 pages. 1977.

Vol. 62: International School on Electro and Photonuclear Reactions II. Edited by C. Schaerf. VIII, 301 pages. 1977.

Vol. 63: V. K. Dobrev et al., Harmonic Analysis on the n-Dimensional Lorentz Group and Its Application to Conformal Quantum Field Theory. X, 280 pages. 1977.

Vol. 64: Waves on Water of Variable Depth. Edited by D. G. Provis and R. Radok. 231 pages. 1977.

Vol. 65: Organic Conductors and Semiconductors. Proceedings 1976. Edited by L. Pál, G. Grüner, A. Jánossy and J. Sólyom. 654 pages. 1977.

Vol. 66: A. H. Völkel, Fields, Particles and Currents. VI, 354 pages. 1977.

Vol. 67: W. Drechsler and M. E. Mayer, Fiber Bundle Techniques in Gauge Theories. X, 248 pages. 1977.

Lecture Notes in Physics

Edited by J. Ehlers, München, K. Hepp, Zürich
R. Kippenhahn, München, H. A. Weidenmüller, Heidelberg
and J. Zittartz, Köln
Managing Editor: W. Beiglböck, Heidelberg

106

Feynman Path Integrals

Proceedings of the International Colloquium
Held in Marseille, May 1978

Edited by
S. Albeverio, Ph. Combe, R. Høegh-Krohn,
G. Rideau, M. Sirugue-Collin, M. Sirugue and R. Stora

Springer-Verlag
Berlin Heidelberg GmbH 1979

Editors

S. Albeverio
Fakultät für Mathematik
der Universität Bielefeld
D-4800 Bielefeld 1

R. Høegh-Krohn
Matematisk Institut
Universite
N-Blindern-Oslo 3

R. Stora
C.N.R.S. - Luminy - Case 907
Centre de Physique Théorique
F-13288 Marseille Cedex 2

and

C.E.R.N.
Division Théorique
CH-1211 Genève 23

Ph. Combe
M. Sirugue-Collin
M. Sirugue
C.N.R.S. - Luminy - Case 907
Centre de Physique Théorique
F-13288 Marseille Cedex 2

G. Rideau
Université de Paris VII
Laboratoire de Physique
Théorique et Mathématique
Tour 33–43
2, place Jussieu
F-75221 Paris Cedex 05

ISBN 978-3-540-09532-3 ISBN 978-3-540-35039-2 (eBook)
DOI 10.1007/978-3-540-35039-2

Originally published by Springer-Verlag Berlin Heidelberg New York in 1979

2153/3140-543210

- FOREWORD -

The circle of ideas which generated the so called Feynman path integrals is contained in work by Dirac in the thirties and Feynman in the forties. Especially in the latter work the representation of transition amplitudes in terms of heuristic integrals on "path space" - or their approximations - was developed into an alternative general formulation of quantum dynamics, equivalent to the previous formulations (Heisenberg, Schrödinger) and the more recent versions (Schwinger) in certain cases (e.g. non relativistic quantum mechanics with Lagrangians at most quadratic in the velocities). Mathematically, the heuristic integrals introduced by Feynman being integrals over spaces of functions, have their natural place in the realm of functional integration, a discipline which has its classical roots in work connected with the calculus of variations (Volterra) and had a first great impact in analysis through Wiener's introduction (1923) of Wiener integrals (on a space of continuous functions) to handle the problem of Brownian motion and heat diffusion. Whereas Wiener integrals, and more generally the integrals introduced later on in connection with the study of stochastic processes, are integrals with respect to positive finite measures, the heuristic expression for Feynman path integrals is in terms of a complex formal density which does not define a measure. Thus the mathematical definition of the objects understood under the name of Feynman path integrals posed genuine new problems. Such problems have been attacked by different methods since the fifties.

It is important to realize (and we hope this will be conveyed also by these Proceedings) that the ideas connected with Feynman path integrals have a vast range of implications, and that different mathematical realizations of these are possible, according to the type of applications one is aiming to. We should hope that all present points of views are well represented in these Proceedings. Part of the contributions deal with various mathematical definitions of Feynman path integrals and the development of the mathematical tools needed to make Feynman integrals work. The contributions concerned with these mathematical problems in view of applications to non relativistic quantum mechanics as well as scalar quantum field theory ("commutative Feynman path integrals") are gathered in Section I.

In order that quantization by Feynman path integrals might be a true alternative formulation of quantum dynamics, the problems given by the existence of half integer-valued spin particles should also be handled. Mathematical approaches to "non commutative Feynman integrals" are gathered in Section II.

Since their inception the Feynman path integrals have prompted connections (Kac, Dynkin, Gelfand, Minlos, Yaglom, Nelson, Symanzik) with integrals associated with stochastic processes, on the basis of the natural observation that one can get solutions of Schrödinger equations by analytic continuation from those of the corresponding diffusion equations. Such connections have received several new impulses in the last fifteen years, through a successful Euclidean approach to quantum fields, and through the discovery of the connection between stochastic mechanics and quantum mechanics. These aspects are represented in Section III of these Proceedings.

One beautiful feature of Feynman path integrals is that they contain in a simple direct way the relation of quantum dynamics with classical dynamics. Namely the detailed quantum behaviour can be obtained from the detailed classical behaviour by a method of stationary phase (the classical behaviour being dictated by the fundamental variational principles) following the original proposals of Dirac, Feynman, Pauli and Schwinger. In this way one gets connections of the Feynman path integrals with the theory of oscillatory integrals and Fourier integral operators (aspects of these relations are illustrated also in Section I . Conversely, Feynman path integrals, in as much as they embody a "quantization procedure" can be expected to have close relations, worth while to be studied in details, with other quantization procedures. In particular it is natural to try to understand better the connections with such quantization procedures as the "geometric quantization" and the method of "Poisson brackets", both of which are being geometric in nature, closely related with the symplectic and differential geometric features of the above mentioned methods of stationary phase. (Incidentally, there are here connections also with problems of group representations (Kirillov, Kostant, Auslander, e.g.)). Some of these "geometrical aspects" are presented in Section IV.

Since Feynman path integrals give a new definition of quantum dynamics, it is interesting to apply them, even on a heuristic level, to domains where the usual approaches to quantization meet difficulties, for instance general relativity. In particular a definition of Feynman path integrals on curved spaces is needed. C. De Witt discusses the possibility of such a definition in analogy with the work done for the Wiener integrals on manifolds. Another domain where both the usual and the Feynman quantization procedures meet difficulties is afforded by Lagrangians involving higher than quadratic terms in the velocities. Such problems are discussed in Section V.

Another domain where ideas connected with Feynman path integrals have played an important role, in laying down lines of research, is concerned with gauge fields. On the basis of the general heuristic principle (justified mathematically in several cases, as mentioned above) according to which Feynman path integrals contain the classical dynamics and give rise to asymptotic expansions around the classical solutions, an intensive study of classical gauge fields has been made in the last few years. This area is represented is Section VI where both some classical and some quantum situations are studied with the hope to come closer eventually to a construction of quantized gauge fields.

The study of the asymptotic series themselves that are obtained by a formal expansion around the classical limit is an interesting object of study in general, not only for gauge fields. This is discussed in Section VII.

We hope that the above remarks might help in making understandable to the reader that the diversity of aspects and approaches which is reflected in these Proceedings is just a sign of the richness of the approach to quantization given by Feynman path integrals. The same diversity should also make clear that the subject of Feynman path integrals should not be considered as a closed one, on the contrary, much work is needed, on the conceptual mathematical and physical level, in order to bring to fruition all the beautiful potentialities contained in those ideas. Also the presence of heuristic suggestions in some of the contributions should act as a stimulus to new mathematical efforts to give mathematical form to the different parts of the building.

We have collected in these Proceedings the invited lectures, ordered along the lines of above "themes", as well as some contributed communications.

We would like to express our gratitude to Professor Mohammed Mebkhout, Directeur de l'UER Pluridisciplinaire de Luminy, and Professor Claude Mesliand, Président de l'Université de Provence, for their kind interest in this Colloquium.

We are very grateful to the secretaries of the Centre de Physique Théorique and to Mrs. G. Niard for their assistancy during the meeting. Our special thanks are due to Maryse Cohen-Solal whose experience has been of constant invaluable help throughout all stages. We particularly thank her for her patience and skill in the preparation of these Proceedings.

We also gratefully acknowledge the financial support of the Université de Paris VII, of the Université de Provence and the Université d'Aix-Marseille II, as well as the Centre de Physique Théorique de Marseille.

Marseille, November 1978

S. Albeverio, Ph. Combe, R. Høegh-Krohn,
G. Rideau, M. Sirugue-Collin, M. Sirugue, R. Stora

with gauge fields, on the basis of the general geometric principles [illegible] [illegible] contain the classical [illegible] and [illegible] around the classical solutions, [illegible] progressive study of classical [illegible] has been made in the last few years. This area is represented as Section II [illegible] to some extent contributing to a construction of quantized gauge fields.

The study of the [illegible] [illegible] is discussed in Section [illegible].

We hope that the above remarks might help in making understandable the [illegible] diversity of aspects and approaches which is reflected [illegible]

[illegible] corresponding [illegible] to the different parts of the [illegible]

[illegible]

[illegible] during the meeting. Our special thanks [illegible] whose [illegible] constant [illegible] the preparation of these proceedings.

We also gratefully acknowledge the financial support of the Université de Paris VII, of the Université de Provence and the Université d'Aix-Marseille II, as well as the Centre de Physique Théorique de Marseille.

Marseille, November 1978

S. Albeverio, Ph. Combe, R. Høegh-Krohn, G. Rideau, M. Sirugue-Collin, M. Sirugue, R. Stora

- CONTENTS -

- PARTICIPANTS -

T. AABERGE	Université de Neuchâtel
S. ALBEVERIO	Universität Bielefeld ; CPT,CNRS Marseille
E. AMAR	Université de Paris VII
H. BACRY	Université d'Aix-Marseille II, Luminy CPT, CNRS Marseille
J. BERTRAND	Université de Paris VII
Ph. BLANCHARD	Universität Bielefeld
O. BRATTELI	CPT, CNRS Marseille
G. BURDET	Université de Dijon
R. CARMONA	Université d'Aix-Marseille II, Luminy
P. COLLET	Université de Genève
Ph. COMBE	Université d'Aix-Marseille II, Luminy CPT, CNRS Marseille
I. DANA	Technion, Haifa
C. DEWITT-MORETTE	University of Texas at Austin
P. DUCLOS	Centre Universitaire de Toulon CPT, CNRS Marseille
Ch. DUVAL	Université d'Aix-Marseille II, Luminy CPT, CNRS Marseille
J. ELHADAD	Université de Provence ; CPT,CNRS Marseille
K.D. ELWORTHY	University of Warwick
M. FARIA	Université de Provence ; CPT, CNRS Marseille
R. FERRARI	Università di Pisa
R. FIGARI	Università di Napoli
H.M. FRIED	Brown University, Providence
W. GARCZYNSKI	University of Wroclaw
L. GARRIDO	University of Barcelona
F. GHABOUSSI-FARHAD	Universität Hamburg
P. GHEZ	Centre Universitaire de Toulon CPT, CNRS Marseille
M. GINOCCHIO	Université de Paris VII
N. GIOVANNINI	Université de Genève
A. GROSSMANN	CPT, CNRS Marseille
F. GUERRA	Università di Salerno
H. HAHN	Techn. Universität Braunschweig
R. HØEGH-KROHN	CPT, CNRS Marseille, and Oslo University
J.-Cl. HOUARD	Université de Paris VII
G. IMMIRZI	Università di Napoli
B. IOCHUM	Université de Provence ; CPT, CNRS Marseille

M. IRAC	Université de Paris VII
C. ITZYKSON	CEN Saclay
A. JAKUBIEC	University of Warsaw
W. KARWOWSKI	University of Wroclaw
K. KELLER	Universität Dortmund
W. KERLER	Philipps Universität, Marburg
A. KHAKPOUR	University of Teheran
P. KREE	Université de Paris VI
A. LAMBERT	Université d'Aix-Marseille II, Luminy CPT, CNRS Marseille
F. LANGOUCHE	University of Leuwen, Heverlee
J. LERAY	Collège de France, Paris
H. LESCHKE	Universität Dortmund
A. LICHNEROWICZ	Collège de France, Paris
A. MAHESHWARI	University of Mysore
M. MANOLESSOU-GRAMMATICOU	Ecole Polytechnique, Palaiseau
Ch. MARTIN	Université de Dijon
M.M. MIZRAHI	Center for Naval Analyses, Arlington
R.P. MONDAINI	ICTP, Trieste
J. NUYTS	Université de Mons ; CPT, CNRS Marseille
E. ONOFRI	Università di Parma
L.C. O'RAIFEARTAIGH	Dublin Institute for Advanced Studies
C. PALMIERI	Università di Napoli
M. PERRIN	Université de Dijon
O. PIGUET	Universität Karlsrühe
J.L. RICHARD	CPT, CNRS Marseille
G. RIDEAU	Université de Paris VII
J.M. RIVERA	Advanced School of Physics, Trieste
R. RODRIGUEZ	Université d'Aix-Marseille II, Luminy CPT, CNRS Marseille
D. ROEKAERTS	University of Leuwen, Heverlee
A. ROUET	CPT, CNRS Marseille
S. SAKAKIBARA	RWTH Aachen
M. SAN MIGUEL RUIBAZ	University of Barcelona
W. SCHNEIDER	Brown Boveri Research Center, Baden-Dättwil
R. SEILER	Freie Universität, Berlin
D.J. SIMMS	Trinity College, Dublin
M. SIRUGUE	CPT, CNRS Marseille
M. SIRUGUE-COLLIN	Université de Provence ; CPT, CNRS Marseille

A. SLAVNOV	CEN Saclay
J.M. SOURIAU	Université de Provence; CPT, CNRS Marseille
R. STORA	CPT, CNRS Marseille
N. SZABO	Université de Genève
M. TALON	Université de Montpellier
J. TARSKI	Universität Clausthal
A. TRUMAN	Heriot Watt University, Edinburgh
J.W.F. VALLE	Syracuse University
A. VOROS	CEN Saclay
D.N. WILLIAMS	University of Michigan, Ann Arbor
A. WULFSOHN	Open University, Milton Keynes
K. YAJIMA	ETH Zürich
K. YOSHIDA	Università di Salerno

- SECTION I -

FEYNMAN PATH INTEGRALS AND THE CORRESPONDING METHOD OF STATIONARY PHASE

by

S. ALBEVERIO and R. HØEGH-KROHN

CNRS - CPT Marseille, Université d'Aix-Marseille II, UER de Luminy
Fakultät für Mathematik, Universität Bielefeld
Matematisk Institutt, Universitetet i Oslo

ABSTRACT

We give a review of our work concerning the mathematical definition of Feynman path integrals as particular cases of oscillatory integrals on infinite dimensional spaces, to which the finite dimensional theory (in particular the stationary phase method) is extended. Applications are given to quantum mechanics and quantum field theory.

- CONTENTS-

1. Introduction

Feynman path integrals were introduced by Feynman [1] in his general formulation of quantum dynamics, see e.g., for history and references, the introductions of [2] , [3] . Feynman's physical ideas are connected with previous work by Dirac [4] on the Lagrangian formalism in quantum mechanics. The Feynman path integrals were introduced as heuristic tools, Kac [5] showed however that for the solutions of the diffusion equation, with a potential term (instead of Schrödinger'sequation) the heuristic idea has actually a mathematical expression in terms of functional integrals of Wiener type (integration with respect to Wiener's measure [6] ; such integrals , as developed by Wiener, Lévy, Cameron, Martin and others, were at that time already a well developed subject). Soon after Kac's work integration of multiplicative functionals of Brownian motion (corresponding, for the diffusion instead of the Schrödinger equation, to the averages considered by Feynman) was developed to a great extent, by work of Doob, Dynkin and others. The mathematical theory of the Feynman integrals themselves had a slower development, due to inherent difficulties (the linear functionals given formally by Feynman path integrals cannot be represented as complex integrals with respect to some complex measure [7]) and also due to the broad range of applications one wanted to be able to extract from them. For a mathematical definition of Feynman path integrals to be a useful concept, it should be close enough to the concept of an integral to possess important properties (transformation, continuity, Fubini - iteration e.g.) that permit to have a useful tool, suitable for some basic applications. E.g. very often the question of allowing for locally singular potentials in the definition of the integral is discussed. This question is best solved by using (following Kac) the indirect definition by Wiener integrals together with analytic continuation (positive measures can easily allow for (positive) singularities in the additive functionals), see e.g. [8]-[10]. However the question of being able to handle the classical limit of the Feynman path integrals for the solutions of the Schrödinger equation (and related ones, e.g. the S-matrix), obtaining (according to classical suggestions by Dirac, Feynman, Pauli, Schwinger, Maslov and others) expansions around the classical limit of quantum mechanics, is on the other hand best treated by taking a direct definition of Feynman path integrals as complex linear functionals, and using a corresponding stationary phase method [11],[12],[45]. Moreover, the same oscillatory nature of such integrals, makes the treatment of lower unbounded potentials possible on the same footing (whereas for the methods using Wiener's measure, lower unbounded potentials are not suitable, due to the "blowing up" of the corresponding multiplicative functionals).

We shall here give an overall view on an approach to the mathematical definition of Feynman path integrals which leads naturally to the kind of applications mentioned, in above sense complementary to the ones obtained by using Wiener's measure. This approach has its origin in work by several people, in particular K. Ito [14] and C. De Witt-Morette [15] , and was developed particularly in [2] , [11] . We urge the reader to look at these references also for relations to other work and to extensions. Let us however stress the extensions by A. Truman [13],[18] and Ph. Combe, G.Rideau, R.Rodriguez, M.Sirugue-Collin [19] of our work as well as a different, independent approach developed by P. Krée [16].

As to the content of this work, let us go shortly through the different sections.

In Section 2, we expose shortly the definition of Feynman path integrals as oscillatory integrals on a Hilbert space, in the case where the phase function is the sum of a positive quadratic form and a bounded continuous function. This case has applications to quantum mechanics (of finitely many degrees of freedom with bounded interactions), applications which are described shortly in Section 3.

In Section 4, the definition of Feynman path integrals in the case where the phase function is the sum of a non necessarily positive (or negative) definite quadratic form and a bounded continuous function is shortly given. This case has applications to the quantum mechanics of anharmonic oscillators (finitely and infinitely many degrees of freedom, the latter including the case of relativistic quantum fields), and some of these applications are mentioned in Section 5. In particular we give indications how the solutions of the "Schrödinger equation for the quantum fields" can be constructed, a line which seems to be very promising, see [22] for more details. (Note that some of the results concerning the case of finitely many degrees of freedom have also been obtained using different definitions, e.g. [18] , [19] , however our approach is basically directed towards natural extensions to the infinite dimensional case, and it does not seem that our results in this case have been derived using these other definitions). In Section 6 we give a short exposition of our results concerning the asymptotic expansion of Feynman path integrals, defined in our way, in powers of the small parameter h (which is in fact Planck's constant divided by 2π). These results represent the first extension, for the particular phase functions of interest for applications to quantum physics, of the theory of oscillatory integrals and Fourier integral operators from the finite dimensional case to the infinite dimensional one. In fact we extend the methods used in the finite dimensional situation (which have their beginnings in work of last century (Stones and Kelvin) and have received impressive development in the last decade, through the work of Maslov, Hörmander, Duistermaat, Arnold, Leray, Bernshtein and others, see Sect. 6 for references). In a sense, our infinite dimensional situation permits a "uniform-in-dimension treatment of degeneracies" (as opposite to the increasing number of parameters required in the finite dimensional situation, according to the type of singularity). As in the finite dimensional case we first treat the case of one and only one stationary point, in the regular case, and then show how essentially analyticity of the potential is enough to permit the treatment of the case of several critical points, the integral being split in an infinite dimensional part with one and only one stationary point, a regular one, and a finite dimensional part. The case of degeneracy is also treated by a similar method. We find then, applying the results to quantum mechanics, asymptotic expansions in powers of h for, e.g., solutions of the Schrödinger equation with a bounded continuous potential. The case of perturbed anharmonic oscillators is also treated, by an application of a general theory (developed by ourselves in collaboration with Ph. Blanchard [12]) for oscillatory integrals where the phase function has a quadratic form non necessarily positive or negative semi-definite. The latter results permit in particular the proof of a trace formula ("Poisson type formula") for the corrected propagator of the time dependent Schrödinger equation for an anharmonic oscillator on $\mathbb{R}^d$ (ζ -function for the anharmonic oscillators). Further results along these lines will be given elsewhere [12], [45]).

Finally, a little note concerning references. In this lecture, our referencing has its origin in more or less random association of ideas. We feel partly excused for doing so because we have given some more systematic referencing elsewhere ([2] , [3] , [11]). We apologize however for any omission or distortion that might have occurred ; it was certainly not our intention to do so. Also it should be clear that we are reviewing here one line of approach, and we mention only some somewhat directly related work. However there are problems (like e.g. the case of velocity-dependent forces and the non commutative case) which are completely left out here. Fortunately there are other contributions in these Proceedings concerned with these other problems and we urge the interested reader to consult them.

2. The normalized integral in a Hilbert space

Let H be a real separable Hilbert space, considered as a measurable Hilbert space with the σ-algebra generated by its open subsets. Let $|\ |$ be the norm in H. We want to define a normalized integral on H of the following form

$$\text{"}\tilde{\int_H} e^{\frac{i}{2}|\gamma|^2} f(\gamma)\, d\gamma\text{"}, \tag{2.1}$$

where i is the imaginary unit, f belongs to a suitable class of "integrable functions" on H and $\sim$ above the integral reminds us to the normalization. Normalization should mean that

$$\tilde{\int_H} e^{\frac{i}{2}|\gamma|^2} d\gamma = 1. \tag{2.2}$$

To see what this implies let us look first to the case where H is finite dimensional i.e. $H = \mathbb{R}^d$ = d-dimensional Euclidean space. In this case one has

$$(2\pi i)^{-d/2} \int_{\mathbb{R}^d} e^{\frac{i}{2}|\gamma|^2} d\gamma = 1 \tag{2.3}$$

with $\quad i^{-d/2} = e^{-\frac{\pi}{4} id}$. $\quad$ Hence from (2.3), (2.2) we see that

$$\tilde{\int_{\mathbb{R}^d}} e^{\frac{i}{2}|\gamma|^2} d\gamma = (2\pi i)^{-d/2} \int_{\mathbb{R}^d} e^{\frac{i}{2}|\gamma|^2} d\gamma. \tag{2.4}$$

We also want that the normalized integral (2.1) be translation invariant, which is formally expressed by the writing $d\gamma$ as a "Lebesgue measure"; in fact when $H = \mathbb{R}^d$ the translation invariance of the Lebesgue measure $d\gamma$ implies the translation invariance of the usual integral

$$\int_{\mathbb{R}^d} e^{\frac{i}{2}|\gamma|^2} f(\gamma)\, d\gamma.$$

The condition for translation invariance is

$$\tilde{\int_H} e^{\frac{i}{2}|\gamma+\alpha|^2} d\gamma = 1 \tag{2.5}$$

for all $\alpha \in H$. Let now $(\ ,\ )$ be the scalar product in H. Then $|\gamma+\alpha|^2 = |\gamma|^2 + 2(\gamma,\alpha) + |\alpha|^2$, hence (2.5) is equivalent with

$$\tilde{\int_H} e^{\frac{i}{2}|\gamma|^2} e^{i(\gamma,\alpha)} e^{\frac{i}{2}|\alpha|^2} d\gamma = 1, \tag{2.6}$$

i.e., observing that $e^{\frac{i}{2}|\alpha|^2}$ is independent of the integration variable γ, we get, with the obvious postulate that the normalized integral be $\mathbb{C}$-linear:

$$\tilde{\int_H} e^{\frac{i}{2}|\gamma|^2} e^{i(\gamma,\alpha)} d\gamma = e^{-\frac{i}{2}|\alpha|^2} . \tag{2.7}$$

Set now

$$f_\alpha(\cdot) \equiv e^{i(\cdot,\alpha)} \tag{2.8}$$

and

$$I(f_\alpha) \equiv \tilde{\int_H} e^{\frac{i}{2}|\gamma|^2} f_\alpha(\gamma)\, d\gamma . \tag{2.9}$$

Then from (2.7) - (2.9) we get

$$I(f_\alpha) = e^{-\frac{i}{2}|\alpha|^2} . \tag{2.10}$$

(2.10) gives us the evaluation of the functional $I(\cdot)$ for the value of the argument f_α. Extending now naturally the definition of $I(\cdot)$ to linear combinations of functions (2.8), where α varies over H, and observing that when α is the zero vector in H we have $f_\alpha(\cdot) = 1$ and $I(1) = 1$, by (2.10) (which is in this case precisely the normalization condition (2.2)), we see that $I(\cdot)$ becomes a normalized linear functional on the linear vector space generated by the functions (2.8) (which is actually identical with the algebra generated by such functions). We have then

$$\tilde{\int_H} e^{\frac{i}{2}|\gamma|^2} \Sigma c_n f_{\alpha_n}(\gamma)\, d\gamma = I(\sum_n c_n f_{\alpha_n}) = \sum_n c_n e^{-\frac{i}{2}|\alpha_n|^2} = \int_H e^{-\frac{i}{2}|\alpha|^2} \sum_n c_n \delta(\alpha-\alpha_n), \tag{2.11}$$

where $\delta(\cdot)$ is the measure concentrated at the origin in H. Observing that $f(\gamma) \equiv \sum_n c_n f_{\alpha_n}(\gamma)$ is the Fourier transform on H of the measure $d\mu_f(\gamma) \equiv$ $\equiv \sum_n c_n \delta(\alpha-\alpha_n)$ i.e.

$$f(\gamma) = \int_H e^{i(\gamma,\alpha)} d\mu_f(\alpha) \tag{2.12}$$

we can rewrite (2.11) in the form

$$\tilde{\int_H} e^{\frac{i}{2}|\gamma|^2} f(\gamma)\, d\gamma = I(f) = \int_H e^{-\frac{i}{2}|\alpha|^2} d\mu_f(\alpha) . \tag{2.13}$$

We now observe that the right hand side of (2.13) is well defined also when μ_f is replaced by any bounded complex measure μ on the measurable space H, $e^{-\frac{i}{2}|\alpha|^2}$ being a bounded continuous function. Since any such measure has a

Fourier transform on H, namely a bounded uniformly continuous function $f_{(\mu)}$ such that

$$f_{(\mu)}(\gamma) = \int_H e^{i(\gamma,\alpha)} d\mu(\alpha), \tag{2.14}$$

it is natural to extend the definition of $I(f)$ to all such functions f. We then have

$$I(f) = \int_H e^{-\frac{i}{2}|\alpha|^2} d\mu(\alpha) \tag{2.15}$$

for all functions f on H which are Fourier transforms of bounded complex measures μ on H, i.e. such that

$$f(\gamma) = \int_H e^{i(\gamma,\alpha)} d\mu(\alpha). \tag{2.16}$$

Because of (2.13) it is natural to use for $I(f)$ also the notation

$$I(f) \equiv \tilde{\int_H} e^{\frac{i}{2}|\gamma|^2} f(\gamma)\, d\gamma, \tag{2.17}$$

and to take this, together with (2.15), as the definition of the normalized integral (2.1). We have then that the normalized integral is a linear normalized complex-valued functional defined at least for all functions f in the linear vector space $F(H)$ of Fourier transforms of bounded complex measures on H. The structure of $F(H)$ is well studied. It is very easy to see [2] that $F(H)$ is a Banach space equipped with the norm $\|f\|_o = \|\mu\|$, where $\|\mu\|$ is the total variation norm for the corresponding measure μ. In fact $F(H)$ is a Banach function algebra with respect to the pointwise multiplication of functions. $F(H)$ is bijectively isometric with $M(H)$, where $M(H)$ is the space of bounded complex measures on H equipped with the total variation norm, the bijection being given by the Fourier transform (2.16). An intrinsic characterization of $F(H)$ is e.g. as the complex linear hull of those positive definite functions f which are continuous in the so-called Minlos-Sazonov-Gross topology i.e. those continuous positive definite functions f on H for which for any $\varepsilon > 0$ there exists a nuclear operator (positive (symmetric) trace class) N_ε such that $\mathrm{Re}\,(f(0) - f(\gamma)) < \varepsilon$ whenever $(\gamma, N_\varepsilon \gamma) \le 1$. We call $F(H)$ the space of (Fresnel) integrable functions and we call $I(f)$ the (Fresnel) integral of f. $I(f)$ is the normalized integral of $e^{\frac{i}{2}|\cdot|^2} f(\cdot)$, as illustrated above. The properties of this integral are studied in [2]. We recall here shortly those properties which will be useful in the following exposition.

P. 1) $I(\cdot)$ is a complex linear normalized (i.e. such that $I(1) = 1$) bounded functional on the Banach algebra $F(H)$. Thus in particular

$$|I(\prod_{j=1}^{n} f_j)| \leq \prod_{j=1}^{n} \|f_j\|_o , \tag{2.18}$$

where $|\ |$ on the left hand side is the absolute value in the field $\mathbb{C}$ of complex numbers and we recall that $\|\ \|_o$ is the norm in $F(H)$.

P. 2) When H is finite dimensional i.e. $H = \mathbb{R}^d$ then

$$I(f) = \widetilde{\int_{\mathbb{R}^d}} e^{\frac{i}{2}|\gamma|^2} f(\gamma)\, d\gamma = (2\pi i)^{-d/2} \int_{\mathbb{R}^d} e^{\frac{i}{2}|\gamma|^2} f(\gamma)\, d\gamma$$

for all f in $F(\mathbb{R}^d)$ for which the (improper) Riemann or Lebesgue integral

$$\int_{\mathbb{R}^d} e^{\frac{i}{2}|\gamma|^2} f(\gamma)\, d\gamma \quad \text{exists (e.g. all } f \in F(\mathbb{R}^d) \cap L^1(\mathbb{R}^d)).$$

Another useful property is the linearity of $I(f)$ under smooth partitions $\{\varphi_k\}$ of the unit i.e.

$$I(f) = \widetilde{\int_{\mathbb{R}^d}} e^{\frac{i}{2}|\gamma|^2} f(\gamma)\, d\gamma = \sum_k \widetilde{\int_{\mathbb{R}^d}} e^{\frac{i}{2}|\gamma|^2} f(\gamma)\, \varphi_k(\gamma)\, d\gamma$$

$$= \sum_k I(f\varphi_k),$$

with φ_k of compact support (e.g. $\varphi_k(\gamma) = \psi_k(\gamma) - \psi_{k-1}(\gamma)$, $k = 1, 2, 3 \ldots$, $\psi_o(\gamma) \equiv 0$, $\psi_k(\gamma) \equiv \psi(\gamma/_k)$, $\psi(\gamma)$ being a C^∞ function equal to 0 for $|\gamma| \geq 2$ and equal to 1 for $|\gamma| \leq 1$).

P. 3) Let $F\,F(H)$ be the subalgebra of $F(H)$ consisting of functions f in $F(H)$ with the property that $f(\gamma) = f(P_f\gamma)$ for some individual finite dimensional (i.e. with finite rank) projection on H. I.e. the functions on $F\,F(H)$ are the cylinder or tame functions on H. Let for $f \in F\,F(H)$ be $I_{P_fH}(\cdot)$ the normalized integral on the finite dimensional space $P_f H$. Then $I(f) = I_{P_f H}(\tilde{f})$, where $\tilde{f}$ is the function on $P_f H$ such that $\tilde{f}(P_f\gamma) = (f \circ P_f)(\gamma)$. In particular for $f \in F\,F(H)$

$$I(f) = (2\pi i)^{-\dim(P_fH)} \int_{\mathbb{R}^d} e^{\frac{i}{2}|\gamma|^2} \tilde{f}(\gamma)\, d\gamma ,$$

whenever the Riemann or Lebesgue integral on the right hand side exists. [1])

P. 4) The translation invariance of the normalized integral has been built in from the beginning in our presentation above. One has thus

$$\widetilde{\int_H} e^{\frac{i}{2}|\gamma+\alpha|^2} f(\gamma+\alpha)\, d\gamma = \widetilde{\int_H} e^{\frac{i}{2}|\gamma|^2} f(\gamma)\, d\gamma ,$$

for all $\alpha \in H$. Moreover for any bounded operator on H with everywhere defined bounded inverse we have

$$\tilde{\int_H} e^{\frac{i}{2}|\gamma|^2} f(\gamma)\, d\gamma = \tilde{\int_H} e^{\frac{i}{2}|T\gamma|^2} f(T\gamma)\, d\gamma .$$

In particular one has the Euclidean invariance of the normalized integral. 2)

P. 5) For the normalized integral a "Fubini Theorem" concerning iterated integration holds, namely if $H = H_1 \oplus H_2$ (where $\oplus$ means direct sum) then

$$\tilde{\int_H} e^{\frac{i}{2}|\gamma|^2} f(\gamma)\, d\gamma = \tilde{\int_{H_2}} e^{\frac{i}{2}|\gamma_2|^2} \left(\tilde{\int_{H_1}} e^{\frac{i}{2}|\gamma_1|^2} f(\gamma_1, \gamma_2)\, d\gamma_1 \right) d\gamma_2$$

$$= \tilde{\int_{H_1}} e^{\frac{i}{2}|\gamma_1|^2} \left(\tilde{\int_{H_2}} e^{\frac{i}{2}|\gamma_2|^2} f(\gamma_1, \gamma_2)\, d\gamma_2 \right) d\gamma_1 .$$

P. 6) Other properties of the normalized integral follow from relation with integrals defined in different ways. E.g. $I(\cdot)$ is the restriction to $F(H)$ of an integral defined by Ito [14], a fact which yields e.g. the approximation

$$I(f) = \lim_{\substack{n \to \infty \\ n \in \mathbb{N}}} \left(\prod_{j=1}^{\infty} (1 - in\lambda_j)^{1/2} \right) E_{nA} \left(e^{\frac{i}{2}|\cdot|^2} f(\cdot) \right),$$

E_{nA} being the expectation with respect to the Gauss measure of mean zero and covariance nA, with A a bounded symmetric strictly positive trace class operator with eigenvalues λ_j, $j = 1, 2, \ldots$. Further relations and extensions are either contained or can be obtained from the work [40]. Fortunately part of this work will be illustrated directly by some of the authors in these Proceedings, and we refer to their contributions for more information. More indirect relations are obtained by connecting the normalized integral $I(f)$ with Wiener integrals associated with H , i.e. integrals with respect to the normal distribution given by H. Such integrals being linear functionals given by probability measures extend to larger classes of functions, however the connection with $I(f)$ is obtained by analytic continuation with respect to a suitable parameter, and for this continuation again rather strong restrictions are needed (the stronger the more properties of $I(f)$ one wants to preserve, e.g. in order to be able to make the theory of asymptotic expansions of oscillatory integrals, see Sect. 6 below).

For work in which an integral of the type $I(f)$ is obtained by methods involving such an analytic continuation procedure see e.g. the references in [10] of [2].

Finally for relations of the normalized integral I(f) with limits of finite dimensional integrals, possibly combined with Lie-Trotter approximations of the latter, see the references in footnote 1 to this Section, as well as e.g. the references in [11] of [2].

3. Some applications to non relativistic quantum mechanics

3.1 The Schrödinger equation (finite time)

We first consider the case of a Schrödinger operator of the form $-\Delta + V(x)$, where Δ is the Laplacian on $\mathbb{R}^d$ and $V(x)$ is a function in $F(\mathbb{R}^d)$. According to Feynman's heuristic considerations (see Section 1) we expect that the solution of the corresponding Schrödinger equation

$$\frac{i\partial}{\partial\tau}\psi(\tau,x) = (-\Delta + V(x))\,\psi(\tau,x) \tag{3.1}$$

on the time interval $[0,t]$, with initial condition $\psi(0,x) = \varphi(x)$, (3.2)
be given by a suitable normalized integral

$$\widetilde{\int\limits_{\tilde{\gamma}(t)=x}} e^{\frac{i}{2} S_t(\tilde{\gamma})} \varphi(\tilde{\gamma}(0))\, d\tilde{\gamma} \tag{3.3}$$

where $S_t(\tilde{\gamma})$ is the action correspondent to the given problem. In our problem the action is

$$S_t(\tilde{\gamma}) = \int_0^t \dot{\tilde{\gamma}}^2(\tau)\, d\tau - \int_0^t V(\tilde{\gamma}(\tau))\, d\tau\ , \tag{3.4}$$

and the integration should be in a suitable space of paths $\tilde{\gamma}$ ending at the instant t in x. In order to be able to actually compute integrals of the form (3.3) it is natural to try to exploit the special feature of the phase function $S_t(\tilde{\gamma})$, namely of consisting of a positive quadratic part plus the term involving V, considered as a "perturbation". [1]) The most simple way then of taking advantage of the quadratic part as a "known" part is to introduce as an integration space a linear space of paths γ for which

$$\int_0^t \dot{\gamma}^2(\tau)\, d\tau$$

is finite i.e. a subset of all absolutely continuous paths. [2]) For simplicity it is natural to restrict the paths to those building a Hilbert space H in which $\int_0^t \dot{\gamma}^2(\tau)\, d\tau$ is the norm and for which Fourier transforms are particularly easily computed, which is the case when the Fourier transform of $e^{\frac{i}{2}\int_0^t \dot{\gamma}^2(\tau)d\tau}$ evaluated at $\Gamma(s,\cdot)$ is equal to $e^{-\frac{i}{2}\Gamma(s,s)}$, where $\Gamma(s,\tau) \equiv t - \max(s,\tau)$ is the kernel of $-\frac{d^2}{d\tau^2}$ with zero boundary condition e.g. at $\tau = t$ and zero derivative condition at $\tau = 0$. This then selects H to be the Hilbert space of absolutely continuous paths on $[0, t]$ of finite kinetic energy and being at the origin at time t (note that the space has to be at least as big as to contain $\Gamma(s, \cdot)$).[3] Then we have for paths $\gamma \in H$ that the expression (3.4) reads, for paths $\tilde{\gamma} = \gamma + x$ (with the correct property for (3.4) of ending at time t in x)

$$S_t(\tilde{\gamma}) = S_t(\gamma + x) = |\gamma|^2 - \int_0^t V(\gamma(\tau) + x)\, d\tau ,$$

where $|\ |$ is the norm in H. Thus the candidate for (3.3) becomes

$$\widetilde{\int_H} e^{\frac{i}{2}|\gamma|^2} e^{-i\int_0^t V(\gamma(\tau)+x)\,d\tau} \varphi(\gamma(0) + x)\, d\gamma. \qquad (3.5)$$

That (3.5) is well defined and yields indeed the wanted result is the content of the results below, which we shall split, for later convenience, in 2 Theorems (cfr. also Section 3 in [2]).

Theorem 3.1 Assume $\varphi \in F(\mathbb{R}^d)$, $V \in F(\mathbb{R}^d)$ and define for $t \geq 0$

$$(T_t\varphi)(x) \equiv \widetilde{\int_H} e^{\frac{i}{2}\int_0^t \dot{\gamma}^2(\tau)\,d\tau} e^{-i\int_0^t V(\gamma(\tau)+x)\,d\tau} \varphi(\gamma(0)+x)\, d\gamma .$$

Then T_t is a continuous bounded semigroup on $F(\mathbb{R}^d)$ with $\|T_t\varphi\|_0 \leq e^{t\|V\|_0}\|\varphi\|_0$, where $\|\ \|_0$ is the norm on $F(\mathbb{R}^d)$. [5]

Proof: Let $\gamma \in H$ and write

$$\gamma(\tau) = \gamma_1(\tau) + \gamma_2(\tau) \qquad (3.6)$$

where $\gamma_2(\tau)$ is constant on the subinterval $[0,t_1]$, $0 < t_1 < t$ and $\gamma_2(\tau) = \gamma(\tau)$ on $[t_1, t]$, and where $\gamma_1(\tau) = 0$ on $[t_1, t]$, $\gamma_1(\tau) = \gamma_2(\tau) - \gamma_2(t_1)$ on $[0,t_1]$. Clearly (3.6) gives a splitting of H in the direct sum $H = H_1 \oplus H_2$, where H_1 consists of the paths γ_1 and H_2 consists of the paths γ_2. By the property P5) of the normalized integral (Section 2, "Fubini Theorem") we have

$$(T_t\varphi)(x) \equiv \widetilde{\int_H} e^{\frac{i}{2}\int_0^t \dot{\gamma}^2(\tau)\,d\tau} e^{-i\int_0^t V(\gamma(\tau)+x)\,d\tau} \varphi(\gamma(0)+x)\,d\gamma =$$

$$= \widetilde{\int_{H_2}} e^{i\int_{t_1}^t \dot{\gamma}_2^2(\tau)\,d\tau} e^{-i\int_{t_1}^t V(\gamma_2(\tau)+x)\,d\tau} \Big(\qquad (3.7)$$

$$\widetilde{\int_{H_1}} e^{i\int_0^{t_1} \dot{\gamma}_1^{\,2}(\tau)\,d\tau} e^{-i\int_0^{t_1} V(\gamma_1(\tau)+\gamma_2(t_1)+x)\,d\tau}$$

$$\varphi(\gamma_1(0) + \gamma_2(t_1) + x)\, d\gamma_2)\, d\gamma_1 = (T_{t-t_1}(T_{t_1}\varphi))(x),$$

which proves that T_t is a semigroup. The continuity of T_t and the bound on its norm follow from the explicit computation of the normalized integral defining $T_t\,\varphi$, which is done by using the property P1) (Section 2) of the normalized integral i.e. that the Fresnel integral is a bounded linear functional on the Banach algebra $F(H)$, so that

$$(T_t\varphi)(x) = \widetilde{\int_H} e^{\frac{i}{2}\int_0^t \dot\gamma^2(\tau)\,d\tau}\, e^{-i\int_0^t V(\gamma(\tau)+x)\,d\tau}\, \varphi(\gamma(0)+x)\, d\tau =$$

$$= \sum_{n=0}^{\infty} \frac{(-i)^n}{n!} \int_0^t \dots \int_0^t \prod_{j=i}^{n} \Big(\widetilde{\int_H} e^{\frac{i}{2}\int_0^t \dot\gamma^2(\tau)\,d\tau}\, V(\gamma(\tau_j)+x)\, d\tau_j\Big)\, \varphi(\gamma(0)+x)\, d\gamma$$

$$= \sum_{n=0}^{\infty} \frac{(-i)^n}{n!} \int_0^t \dots \int_0^t \int_{\mathbb{R}^d} \dots \int_{\mathbb{R}^d} e^{i\sum_{j,k=0}^{n} \alpha_j \Gamma(\tau_j,\tau_k)\alpha_k}\, e^{i\sum_{j=0}^{n}\alpha_j x} \prod_{j=1}^{n} (d\mu_V(\alpha_j)\, d\tau_j)\, d\mu_\varphi(\alpha_0), \quad (3.8)$$

where μ_V, μ_φ are the measures on $\mathbb{R}^d$ of which V and φ are Fouriertransforms. In fact (3.8) yields the bound $\|T_t\varphi\|_0 \le \sum_{n=0}^{\infty} \frac{1}{n!} t^n \|\mu_V\|\,\|\mu_\varphi\| = e^{t\|V\|_0}\, \|\varphi\|_0$. ∎

Theorem 3.2 The solution of the Schrödinger equation (3.1), (3.2) with V and φ in $F(\mathbb{R}^d)$ is given by $\psi(t,x) = (T_t\varphi)(x)$, where T_t is given in Theorem 3.1 i.e. we have

$$\psi(t,x) = \widetilde{\int_H} e^{\frac{i}{2}\int_0^t \dot\gamma^2(\tau)\,d\tau}\, e^{-i\int_0^t V(\gamma(\tau)+x)\,d\tau}\, \varphi(\gamma(0)+x)\, d\gamma .$$

In the case where V is real the semigroup T_t, $t \ge 0$ is a semigroup mapping isometrically and bijectively $F(\mathbb{R}^d) \cap L^2(\mathbb{R}^d)$ into itself, hence extends naturally to an unitary group U_t on $L^2(\mathbb{R}^d)$. In fact $U_t = e^{-itH}$, $t \in \mathbb{R}$, where H is the self-adjoint operator $-\Delta + V$ in $L^2(\mathbb{R}^d)$.

Proof: It suffices to compare (3.8) with the computation of the convergent Dyson series for $e^{-itH}\varphi$ (see [2]). ∎

Remark 1.

The expression for the solution of Schrödinger's equation is the realization of the formal Feynman expression (3.3), where the Feynman path $\tilde{\gamma}$ ending at time t in x is written as $\tilde{\gamma} = \gamma + x$, where γ is a path in H i.e. ending at time t in 0. Actually the particular translation $\gamma \to \gamma + x$ could be replaced, by using the translation invariance of the normalized integral (property P4) of Section 2), by any translation $\gamma \to \gamma + x + \gamma_o$ with $\gamma_o \in H$. The formulae in Theorems 3.1, 3.2 would then take a different aspect (but would of course be equivalent to the previous ones). In this sense the formula (3.3) can be understood as a short notation for a family of equivalent formulae, e.g. for the one given in Theorem 3.2.

Remark 2.

The expression (3.8) for the solution of the Schrödinger equation (3.1), (3.2) can be put directly in relation with the one obtained from Varadhan and Maslov-Chebotarev's way of expressing the Feynman integrals as integrals with respect to complex Poisson measures, see [41] and Maslov - Chebotarev's contribution to the present Proceedings.

Let now $G_t(x, y) = (e^{-itH})(x, y)$ be the Green's function for (3.1), (3.2) such that the solution of Schrödinger's equation (3.1), (3.2) is given by

$$\psi(t, x) = \int_{\mathbb{R}^d} G_t(x, y)\varphi(y)\, dy. \tag{3.9}$$

Using P5) of Section 2) ("Fubini Theorem" for the normalized integral) together with the splitting $H = H_o \oplus H_o^{\perp}$, with

$$H_o \equiv \{\gamma \in H \mid \gamma(0) = 0\},\ H_o^{\perp} = \{y\eta \mid y \in \mathbb{R}^d,\ \eta(\tau) \equiv 1 - \frac{\tau}{t}\}$$

we get the following (see [11]):

Theorem 3.3 The Green's function for Schrödinger equation (3.1), (3.2) with $\varphi, V \in F(\mathbb{R}^d)$ is given by

$$G_t(x, y) = G_t^o(x, y) \widetilde{\int_{H_o}} e^{\frac{i}{2}\int_0^t \dot{\gamma}^2(\tau)\, d\tau} e^{-i\int_0^t V(\gamma(\tau)+(y-x)\eta+x)d\tau} d\gamma,$$

where $G_t^o(x, y) \equiv (2\pi i t)^{-d/2} e^{i\frac{|x-y|^2}{2t}}$ is the Green's function in the case $V = 0$. ∎

Remark The expression in Theorem 3.3 is a mathematical way of expressing the formal Feynman integral

$$(2\pi i t)^{-d/2} \int_{\substack{\tilde{\gamma}(0) = y \\ \tilde{\gamma}(t) = x}} e^{i S_t(\tilde{\gamma})} d\tilde{\gamma}$$

for the Green's function $G_t(x,y)$.

Remark. The results of Theorems 3.1, 3.2, 3.3 extend immediately to the case of Schrödinger operators describing N particles interacting by a sum of ν-body potentials, $\nu = 1, 2, \dots,$ where translation invariance is allowed. All is needed is that the ν-body potentials be Fourier transforms of bounded complex measures.

3.2 The wave operators and the scattering matrix

Let us consider again a Schrödinger operator of the form $-\Delta+V(x)$, $x\in\mathbb{R}^d$, with $V\in F(\mathbb{R}^d)$ and real. Let $H_0 = -\Delta$ and $H = H_0 + V$, as self-adjoint operators on $L^2(\mathbb{R}^d)$. Consider $e^{-itH}\, e^{itH_0}$, for $t \in \mathbb{R}$. The following is proven in [2] (Sect. 3) by a direct computation:

Theorem 3.4

Let $V \in F(\mathbb{R}^d)$, $\varphi \in F(\mathbb{R}^d) \cap L^2(\mathbb{R}^d)$, then, for all $x \in \mathbb{R}^d$,

$$(e^{-itH}\, e^{itH_0}\, \varphi)(x) = \int_{\mathbb{R}^d} e^{i\beta\cdot x}\, W_t(x,\beta)\, d\mu_\varphi(\beta),$$

where $\beta\cdot x$ is the scalar product in $\mathbb{R}^d$, μ_φ is the measure on $\mathbb{R}^d$ such that V is the Fourier transform of μ_φ, and the quantity $W_t(x,\beta)$ is given by the following normalized integral:

$$W_t(x,\beta) \equiv \widetilde{\int_{H_-}} e^{\frac{i}{2}\int_{-\infty}^{0} \dot{\gamma}^2(\tau)d\tau}\; e^{-i\int_{-t}^{0} V(\gamma(\tau) + \beta\tau + x)\, d\tau}\, d\gamma ,$$

where H_- is the Hilbert space of absolutely continuous paths γ on the time interval $(-\infty, 0]$, with values in $\mathbb{R}^d$, of finite kinetic energy $\int_{-\infty}^{0} \dot{\gamma}^2(\tau)\, d\tau$ ($\dot{\gamma}$ is defined a.e., γ being absolutely continuous) equal to the norm in H_- and ending at time 0 at the origin.

For potentials such that the wave operators $W_\pm$ exist as strong limits in $L^2(\mathbb{R}^d)$ of $e^{-itH}\, e^{itH_0}$ as $t \to \mp\infty$ we have :

$$(W_+\varphi)(x) = \int_{\mathbb{R}^d} e^{i\underline{v}\cdot x}\, W_+(x,\underline{v})\, d\mu_\varphi(\underline{v}),$$

with

$$W_+(x,\underline{v}) = \widetilde{\int_{H_-}} e^{\frac{i}{2}\int_{-\infty}^{0} \dot{\gamma}^2(\tau)d\tau}\; e^{-i\int_{-\infty}^{0} V(\gamma(\tau) + \underline{v}\tau + x)\, d\tau}\, d\gamma ,$$

where the integral is an "improper Fresnel integral" defined as the limit for

$t \to -\infty$ of the Fresnel integral of the Fresnel integrable function

$$e^{-i\int_{-\infty}^{0} V(\gamma(\tau)+v_-\tau+x)\,d\tau}.$$

Similarly

$$(W_-\varphi)(x) = \int_{\mathbb{R}^d} e^{-i v_+ \cdot x} \, W_-^*(x, v_+)\, d\mu_\varphi(v_+)$$

with

$$W_-^*(x, v_+) = \tilde{\int_{H_+}} e^{\frac{i}{2}\int_0^\infty \dot{\gamma}^2(\tau)\,d\tau} \, e^{-i\int_0^\infty V(\gamma(\tau)+v_+\tau+x)\,d\tau} \, d\gamma ,$$

where H_+ is defined as H_- with the time interval $(-\infty, 0]$ replaced by $[0, +\infty)$.

<u>Remark 1:</u> In the case where V is gentle [4]) (such that the perturbation series in powers of V for the wave operators converges) the improper Fresnel integrals can also be defined by the termwise limit as $t \to \pm\infty$ of the convergent perturbation series obtained by expanding in powers of V and interchanging the sum and the Fresnel integral.

<u>Remark 2:</u> $W_+(x, v_-)$ has e.g. the interpretation of the wave function at time 0 of a particle which had prescribed velocity v_- at time $-\infty$.

<u>Remark 3:</u> Analogously as before we can write the formula for $W_+(x, v_-)$ (and similarly for $W_-^*(x, v_+)$) symbolically as

$$W_+(x, v_-) = \tilde{\int}_{\substack{\tilde{\gamma}(0)=0 \\ \tilde{\gamma}(t)\to\gamma_0 \\ t\to-\infty}} e^{i(S_-(\tilde{\gamma}) - S_0(\gamma_0))} \, d\tilde{\gamma}$$

where $S_-(\tilde{\gamma})$ is the total action along the path $\tilde{\gamma}$ which is in 0 at time 0 and with asymptote given by the rectilinear path γ_0 through x at time 0 and given by the tangent vector v_- at $t = -\infty$ (the path a classical particle coming in with velocity v_- at time $t = -\infty$, on a line through x, would follow if moving without any force) and $S_0(\gamma_0)$ is the free action along the rectilinear path γ_0 (note that both $S_-(\tilde{\gamma})$ and $S_0(\gamma_0)$ are infinite, but their difference is finite, for $\tilde{\gamma}(\tau) = \gamma(\tau) + v_-\tau + x$, $\gamma \in H_-$.

The above "Feynman integral expression" has been first discussed in [2].

For physics the most interesting quantity is the scattering operator $S = W_-^* W_+$. In [2] the following result has been proven for the first time, using the

Fubini theorem for the Fresnel integrals:

Theorem 3.5

Let $V \in F(\mathbb{R}^d)$ and such that the wave operators W_+, W_- exist. Then, for all $f \in L^2(\mathbb{R}^d)$

$$(S\tilde{f})(v_+) = \int S(v_+, v_-)\, \tilde{f}(v_-)\, dv_-$$

with

$$S(v_+, v_-) = \int e^{i[S(\tilde{\gamma}) - S_o(\gamma_o)]}\, d\tilde{\gamma}\ ,$$

$$\lim_{t \to \pm\infty} \frac{\tilde{\gamma}(t)}{t} = v_{\pm}$$

$$\gamma_o(\tau) = \begin{cases} y + v_-\tau & \tau \leq 0 \\ y + v_+\tau & \tau \geq 0 \end{cases},$$

where the symbolic integral stands for

$$\int_{\mathbb{R}^d} W_-^*(x, v_-)\, e^{-(v_+ - v_-).x}\, W_+(x, v_+)\, dx =$$

$$= \int_{\mathbb{R}^d} dx\, e^{i(v_+ - v_-).x} \tilde{\int}_{H_- \oplus H_+} e^{i(S_- + S_+)(\gamma)}\, d\gamma\ ,$$

with $S(\tilde{\gamma})$ = total action along $\tilde{\gamma}$, $S_o(\gamma_o)$ = free action along γ_o, S_- as in Remark 3 following Theorem 3.4 and S_+ defined as S_- with the interval $(-\infty, 0]$ replaced by $[0, \infty)$ (and v_+ replacing v_-).

Remark: Formulae similar to those of above type, published first in [2], have been suggested recently in the context of quantum field theory (quantum S-matrix for solitons) by L.D.Faddeev and V.E.Korepin, see e.g. [30].

§ 4. The normalized integral with respect to a quadratic form

In order to treat on the same footing quantum mechanical systems and quantum field theoretical systems in which a quadratic term in the coordinates arises in the action (and in the Hamiltonian) it is useful to introduce the concept of an integral normalized with respect to a (non necessarily positive) quadratic form. We shall later discuss the relation of this concept with other ones discussed in the literature.

Let us think as an example to a system in $\mathbb{R}^d$ with classical action for the time interval [0,t] :

$$S_t(\gamma) = S_t^o(\gamma) - \int_o^t V(\gamma(\tau))\, d\tau\ ,$$

where $S_t^o(\gamma) \equiv \frac{1}{2}\int_0^t \dot\gamma^2(\tau)\,d\tau - \frac{1}{2}\int_0^t (\gamma(\tau), A^2\gamma(\tau))\,d\tau$ and where A is a $d \times d$ matrix. Since quadratic functions are not Fourier transform of bounded complex measures, it is natural to let the quadratic form $S_t^o(\gamma)$ take the role of the kinetic energy term in the previous definition of normalized integral (Feynman path integrals for the case under consideration). Mathematically it turns out that this can be done as a particular case of the following situation.

Let H be a real separable Hilbert space. Let B be a densely defined symmetric operator in H, then $(\gamma, B\gamma)_H$ is a real quadratic form on D(B). We would like to define, somewhat in analogy with § 2, a "normalized integral"

$$\tilde{\int_H} e^{\frac{i}{2}(\gamma, B\gamma)_H} f(\gamma)\,d\gamma, \tag{4.1}$$

where $(,)_H$ is the scalar product in H.

The normalization should be such that

$$\tilde{\int_H} e^{\frac{i}{2}(\gamma, B\gamma)_H} d\gamma = 1. \tag{4.2}$$

From the case $H = \mathbb{R}^d$ we see that it is natural to assume that B is non degenerate, namely in order to define (4.1) in analogy with (2.13), i.e. by a "Parseval relation" involving the Fourier transforms of $e^{\frac{i}{2}(\gamma, B\gamma)}$ and $f(\gamma)$. Let us look first at the situation where B is strictly positive and bounded everywhere in H. Then, with $(\gamma,\gamma)_B \equiv (\gamma, B\gamma)_H$ and H_B the closure of H in the $(,)_B$-norm, we have by (2.13)

$$\tilde{\int_{H_B}} e^{\frac{i}{2}(\gamma,\gamma)_B} f(\gamma)d\gamma = \int_{H_B} e^{-\frac{i}{2}(\alpha,\alpha)_B} d\mu_f^B(\alpha), \tag{4.3}$$

for any $f \in F(H_B)$, where μ_f^B is the measure whose Fourier transform in H_B is f i.e.

$$f(\gamma) = \int_H e^{i(\gamma,\alpha)_B} d\mu_f^B(\alpha) = \int_H e^{i(\gamma,\alpha)_H} d\mu_f(\alpha) \tag{4.4}$$

with

$$d\mu_f(\alpha) = d\mu_f^B(B^{-1}\alpha). \tag{4.5}$$

From (4.3) - (4.5) we get

$$\tilde{\int_{H_B}} e^{\frac{i}{2}(\gamma,\gamma)_B} f(\gamma)\,d\gamma = \int_{H_B} e^{-\frac{i}{2}(\alpha, B\alpha)_H} d\mu_f^B(\alpha) \tag{4.6}$$

$$= \int_{R(B)} e^{-\frac{i}{2}(B^{-1}\beta,\beta)_H} d\mu_f(\beta) = \int_{R(B)} e^{-\frac{i}{2}\Delta(\beta,\beta)} d\mu_f(\beta),$$

with Δ the bilinear form such that $\Delta(\beta,\beta) = (B^{-1}\beta, \beta)_H$. R(B) means the range of B. Hence the normalized integral on H_B is given by a "Parseval relation" involving an integral of $e^{-\frac{i}{2}\Delta(\beta,\beta)}$ against a bounded complex measure.

This suggests to define in general (4.1) by

$$\int_D e^{-\frac{i}{2}\Delta(\alpha,\alpha)} d\mu_f(\alpha), \tag{4.7}$$

where D is a suitable linear subset of H and Δ is a suitable "inverse" of B. By the above considerations we see, that it is natural to require that Δ be a symmetric bilinear form on D x D, such that $\Delta(\alpha, B\beta) = (\alpha,\beta)_H$ $\forall \alpha \in D$, $\beta \in D(B)$, assuming that the range of B is contained in D. For applications to the case of quantum fields, e.g., it is useful to allow Δ to be complex-valued and we see that the simplest assumption to define (4.7) is to require that the integration be taken on some nice space, e.g. D a separable Banach space densely contained in H and with norm $| \ |_D$ stronger than the one, $| \ |_H$, in H (i.e. $|\cdot|_H \leq a\,|\cdot|_D$ for some a), and with Δ continuous on D x D, with $\mathrm{Im}\,\Delta(\gamma,\gamma) \leq 0$. The definition of (4.1) by (4.7) clearly depends on the choice of Δ, hence it is useful to write

$$I_\Delta(f) \equiv \int_H^\Delta e^{\frac{i}{2}(\gamma,B\gamma)_H} f(\gamma)\,d\gamma$$

instead of (4.1). Hence we come to the definition: Let (H, D, B, Δ) be a "Fresnel fourtuple" as described above, i.e. H a real separable Hilbert space, D a real separable Banach space densely contained in H and with norm dominating the one in H, B a symmetric densely defined operator (not necessarily bounded) with range containing D and Δ a complex symmetric bilinear form with non positive imaginary part and satisfying $\Delta(\alpha, B\beta) = (\alpha,\beta)_H$ $\forall \alpha \in D$, $\forall \beta \in D(B)$. Let $F(D^*)$ be the Banach algebra of functions f on the topological dual D^* of D which are Fourier transforms of bounded complex measures μ_f on D i.e. $f(\gamma) = \int e^{i<\gamma,\alpha>} d\mu_f(\alpha)$, where $<,>$ is the dualization between D and D^*. Multiplication of functions is, as in $F(H)$, the pointwise one, and the norm $\|f\|_0$ in $F(D^*)$ is defined as the total variation $\|\mu_f\|$ in the corresponding Banach algebra $M(D^*)$ of bounded complex measures $M(D^*)$ on D^*. In [2] $I_\Delta(f)$ was called <u>normalized</u> (or <u>Fresnel</u>) <u>integral with respect</u> to <u>the quadratic form Δ.</u> This integral enjoys properties related to those of the normalized integral defined in Section 2, e.g.

P 1) $f \to I_\Delta(f)$ is a complex-valued, linear, bounded, continuous normalized functional on the Banach function algebra $F(D^*)$. One has

$$|I_\Delta(\prod_{j=1}^n f_j)| \leq \prod_{j=1}^n \|f_j\|_0 , \quad \text{with } \|f\|_0 = \|\mu_f\| =$$

total variation on D of μ_f .

P 2) When B = 1 then $I_\Delta(f) = I(f) = \tilde{\int}_H e^{\frac{i}{2}(\gamma,\gamma)_H} f(\gamma)\, d\gamma$.

More generally when B is strictly positive and with range dense in D then

$$I_\Delta(f) = \tilde{\int}_{H_B} e^{\frac{i}{2}(\gamma, B\gamma)_B} f(\gamma)\, d\gamma .$$

In particular for $B \equiv 1$

$$I_\Delta(f) = I(f) = \tilde{\int}_H e^{\frac{i}{2}(\gamma,\gamma)_H} f(\gamma)\, d\gamma .$$

Thus the normalized integral with respect to a quadratic form reduces to the normalized integral when the quadratic form is the unit one on H .

P 3) If B^{-1} is bounded and everywhere defined then $D = H$ with equivalent norms and $\Delta(\alpha, \beta) = (\alpha, B^{-1}\beta)\ \forall\, \alpha, \beta \in H$. Moreover B is self-adjoint, one has a direct sum decomposition $H = H_+ \oplus H_-$ of H and one has $B = B_+ \ominus B_-$ with B_+ strictly positive, where $B_\pm$ are the parts of B in H_+ respc. H_- . One has then the "Fubini theorem"

$$\tilde{\int}_H^{\Delta} e^{\frac{i}{2}(\gamma, B\gamma)} f(\gamma) d\gamma = \tilde{\int}_{H_+} e^{\frac{i}{2}|\gamma_1|^2_{H_+}} [\tilde{\int}_{H_-} e^{\frac{i}{2}|\gamma_2|^2_{H_-}} \overline{f(\gamma_1, \gamma_2)}\, d\gamma_2]^- \, d\gamma_1$$

where - and [$]^-$ mean complex conjugates.

P 4) If both B and B^{-1} are bounded everywhere on H then P3) applies with $H = D$, $\Delta = B^{-1}$ and in this case

$$I_\Delta(f) = \tilde{\int}_{H_B} e^{\frac{i}{2}(\gamma,\gamma)_B} f(\gamma) d\gamma = \int_H e^{-\frac{i}{2}(\alpha, B\alpha)_H} d\mu_f^B(\alpha) .$$

In this case one also has nice transformations properties of the integral and other types of "Fubini theorems". In fact $I_\Delta(f)$ has in this case an extension to the concept of a normalized integral on a Banach space B instead of an Hilbert space H_B, with a symmetric bounded non degenerate bilinear form $B(\alpha, \beta)$ instead of $(\alpha, B\beta)$, see [2].

P 5) If $\dim H < \infty$ i.e. $H = \mathbb{R}^d$ then

$$\int_H^{\Delta} e^{\frac{i}{2}(\gamma, B\gamma)} f(\gamma) d\gamma = (2\pi)^{-d/2} \,|\mathrm{Det}\, B|\, e^{-i\frac{\pi}{4}\,\mathrm{sign}\, B}$$

$$\int_{\mathbb{R}^d} e^{\frac{i}{2}(\gamma, B\gamma)} f(\gamma)\, d\gamma ,$$

where Det B and sign B are the determinant resp. the signature of B.

P 6) Invariance under translations by elements in D(B) is easily seen:

$$\int_H^{\Delta} e^{\frac{i}{2}(\gamma+\alpha,\, B(\gamma+\alpha))_H} f(\gamma+\alpha)\, d\gamma = \int_H^{\Delta} e^{\frac{i}{2}(\gamma, B\gamma)_H} f(\gamma)\, d\gamma.$$

§ 5. Some applications of the normalized integral with respect to a quadratic form.

5.1 The Schrödinger equation for the anharmonic oscillator

Consider the operator $-\Delta + \frac{1}{2}(x, A^2 x)_d + V(x)$, $x \in \mathbb{R}^d$, where Δ is the Laplacian on $\mathbb{R}^d$, $(,)_d$ is the scalar product in $\mathbb{R}^d$ and A^2 is a strictly positive (definite) matrix in $\mathbb{R}^d$.[1]) Let λ_j be the eigenvalues of A. The following result is proven in [2] (Theor. 5.1.):

Theorem 5.1 Let $t \neq (k+1)\frac{\pi}{\lambda_j}$, $j = 1, 2, \ldots$, for all integers k. Let $V, \varphi \in F(\mathbb{R}^d)$. Then

$$\psi(t,x) = c_t \widetilde{\int_{\tilde{\gamma}(t)=x}} e^{i S_t(\tilde{\gamma})} \varphi(\tilde{\gamma}(0))\, d\tilde{\gamma}$$

solves the Schrödinger equation

$$\begin{cases} i\partial_t \psi(t,x) = \left(-\Delta + \frac{1}{2}(x, A^2 x)_d + V(x)\right)\psi(t,x) \\ \psi(0,x) = \varphi(x) . \end{cases}$$

The path integral should be understood in the following sense: Let $H = D = \{\gamma : [0,t] \to \mathbb{R}^d,\ \gamma$ absolutely continuous and such that $\int_0^t \dot{\gamma}^2(\tau)\, d\tau < \infty$ and $\gamma(t) = 0\}$; $(\gamma, B\gamma) = 2 S_o^t(\gamma) \equiv \int_0^t \dot{\gamma}^2(\tau) d\tau - \int_0^t (\gamma(\tau), A^2\gamma(\tau))_d\, d\tau$; $\Delta = B^{-1}$, $\tilde{S}_t(\gamma) \equiv S_o^t(\tilde{\gamma}) - \int_0^t V(\tilde{\gamma}(\tau))\, d\tau \equiv$ total action along the absolutely continuous path $\tilde{\gamma}$ of finite kinetic energy $\int_0^t \dot{\tilde{\gamma}}^2(\tau)\, d\tau$ ending at time t in x; then

$$\widetilde{\int_{\tilde{\gamma}(t)=x}} e^{i S_t(\tilde{\gamma})} \varphi(\tilde{\gamma}(0))\, d\tilde{\gamma} \equiv \int_H^{\Delta} e^{\frac{i}{2}(\gamma, B\gamma)} f(\gamma)\, d\gamma ,$$

with $f(\gamma) \equiv e^{-\int_0^t V(\gamma(\tau)+\beta(\tau))\,d\tau}\ \varphi(\gamma(0)+\beta(0)),\ e^{-\frac{i}{2}(x,\,\mathrm{Atg}\,\mathrm{At}x)_d}$

where $\beta(\tau) \equiv \frac{\cos A\tau}{\cos At}\ x$.

Also $c_t = [\,\mathrm{Det}\,(\cos At)\,]^{-1/2}$.

Remark $\beta(\tau)$ solves the equation $\ddot{\beta} + A^2 \beta = 0$, $\beta(t) = x$, $\dot{\beta}(0) = 0$ i.e. is so chosen that $(\beta, B\beta) = -(x, \mathrm{AtgAt}x)_d$. Other formulae giving the path integral can be obtained by using the translation invariance of the integral, e.g. replacing β by $\beta + \tilde{\beta}$ with $\tilde{\beta}$ absolutely continuous, of finite kinetic energy and such that $\tilde{\beta}(t) = 0$.

The following result, proven in [2], permits the direct computation of expectation values:

Theorem 5.2 Let $\varphi_1, \varphi_2 \in F(\mathbb{R}^d) \cap L^2(\mathbb{R}^d)$ and let $V \in F(\mathbb{R}^d)$. Let $H = -\Delta + \frac{1}{2}(x, A^2 x)_d + V(x)$ as a self-adjoint operator in $L^2(\mathbb{R}^d)$. Then for all t s.t. the matrix $\sin At$ is non singular one has

$$(\varphi_1, e^{-itH} \varphi_2)_{L^2(\mathbb{R}^d)} = \tilde{c}(t) \int_H^{\Delta} e^{iS_t(\gamma)}\ \overline{\varphi_1}(\gamma(t))\, \varphi_2(\gamma(0))\, d\gamma\ ,$$

with $\tilde{c}(t) \equiv [\,\mathrm{Det}\,(\frac{i}{2\pi} A \sin At)\,]^{-1/2}$, $H = D = \{\gamma$ absolutely continuous from $[0, t]$ in $\mathbb{R}^d$ and such that

$$\int_0^t [\,\dot{\gamma}^2(\tau) + \gamma^2(\tau)\,]\, d\tau < \infty \ \} ;$$

and B, Δ and $S_t(\gamma)$ as in Theorem 5.1.

5.2. Feynman path integrals for relativistic quantum fields and infinitely many oscillators

Let $K = L^2_{\mathbb{R}}(\mathbb{R}^s)$ be the real separable Hilbert space which is going to play for relativistic fields an analogous role as the configuration space $\mathbb{R}^d$ played (see Sects. 3, 5.1) for non relativistic quantum mechanics. Let $A^2 \equiv -\Delta + m^2$, where Δ is the Laplacian as an operator in $L^2_{\mathbb{R}}(\mathbb{R}^s)$ and m is a fixed constant (which is going to be interpreted as the mass of the free relativistic field).

The action for a free relativistic scalar field $\gamma(t, \vec{x})$, $t \in \mathbb{R}$ (time), $\vec{x} \in \mathbb{R}^s$ (space) is

$$S_0(\gamma) \equiv \frac{1}{2} \int_{\mathbb{R}^{s+1}} [\,\dot{\gamma}^2(\tau, \vec{x}) - (\nabla\gamma)^2(\tau, \vec{x}) - m^2 \gamma^2(\tau, \vec{x})\,]\, d\tau\, d\vec{x} \quad (5.1)$$

where $\dot{\gamma} \equiv \frac{\partial}{\partial\tau} \gamma$ and $\nabla\gamma$ is the gradient of γ. We shall define, analogously as in Theorem 5.1, $H = H_1^{\mathbb{R}} (\mathbb{R}^{s+1}) \equiv$ real Sobolev space over $\mathbb{R}^{s+1}$ consisting of generalized functions γ with finite Sobolev-norm $\int [\dot{\gamma}^2 + (\nabla\gamma)^2]\, d\tau\, d\vec{x}$. Moreover, for $\gamma \in H$, $(\gamma, B\gamma) \equiv 2 S_0(\gamma)$, for all $\gamma \in D(B)$, where the domain $D(B)$ of B consists of all paths γ in H which have compact support. Let D be the set of all paths in H whose Fourier transform $\hat{\gamma}$ is in $C^1(\mathbb{R}^{s+1})$ and such that $\|\gamma\|_D \equiv$

$$\left(|\gamma|_H + \sup_{\substack{j=0,..,s \\ p\in\mathbb{R}^{s+1}}} \left| \frac{\partial\hat{\gamma}}{\partial p_j} \right| \right) < \infty .$$

One verifies then that the range of B, $R(B)$, has the form $\frac{p_0^2 - \omega(\vec{p})^2}{p^2+1} \hat{\gamma}(p)$, with $\gamma \in D(B)$, $\omega(\vec{p}) \equiv \sqrt{\vec{p}^{\,2} + m^2}$ and the notation $p = (p_0, \vec{p})$, where p_0 resp. $\vec{p}$ are the conjugate variables to t resp. $\vec{x}$. One verifies $R(B) \subset D$ and defining then, for $\gamma_1, \gamma_2 \in D$, $\Delta_c(\gamma_1, \gamma_2) =$

$$\int_{\mathbb{R}^s} P \int_{\mathbb{R}} dp^0\, \overline{\hat{\gamma}_1(p)}\, (p_0^2 - \omega(\vec{p})^2)^{-1}\, \hat{\gamma}_2(p)\, (p^2+1)^2\, dp$$

$$- i\pi \int_{\mathbb{R}^{s+1}} c(\vec{p})\, \overline{\hat{\gamma}_1(p)}\, \delta(p_0^2 - \omega(\vec{p})^2)\, \hat{\gamma}_2(p)\, (p^2+1)^2\, dp, \tag{5.2}$$

with P the principal value distribution and $c(\vec{p})$ an arbitrary non negative measurable function, we see that([2]) (H, D, B, Δ) is a Fresnel fourtuple.

Hence the normalized integral $I_{\Delta_c}(\cdot)$ with respect to the quadratic form Δ_c is defined and we have for all $g \in F(D^*)$

$$I_{\Delta_c}(g) = \int_H^{\Delta_c} e^{\frac{i}{2} S_0(\gamma)}\, g(\gamma)\, d\gamma . \tag{5.3}$$

In particular, for $f_j \in D(A^{-1/2}) \subset L^2(\mathbb{R}^s)$, we have that

$$I_{\Delta_c}\left(\prod_{j=1}^n e^{i\gamma_{t_j}(f_j)} \right) \equiv \int_H^{\Delta_c} e^{\frac{i}{2} S_0} \prod_{j=1}^n \left(e^{i\gamma_{t_j}(f_j)} \right) d\gamma, \tag{5.4}$$

with $\gamma_t(f) \equiv \int_{\mathbb{R}^s} f(\vec{x})\, \gamma(t, \vec{x})\, d\vec{x}$, is well defined and in fact it is equal to

$$e^{-\frac{i}{2} \sum_{j,k}^{n} \int \overline{\hat{f}_j(\vec{p})}\, G_c(t_j - t_k, \vec{p})\, \hat{f}_k(\vec{p})\, d\vec{p}}\,, \tag{5.5}$$

with

$$G_c(t,\vec{p}) \equiv -\frac{1}{2A}\,(\sin|t|A + ic(\vec{p})\cos tA), \qquad (5.6)$$

where A is understood here as the operator of multiplication by $\omega(\vec{p}) = (\vec{p}^2 + m^2)^{1/2}$ on functions of $\vec{p}$.

For functions $c(\cdot) \geq 1$ and only for such functions the integrals (5.4) define ([2]) states on the Weyl algebra associated with $K^c \equiv L^2(\mathbb{R}^s)$ (the square integrable complex-valued functions on $\mathbb{R}^s$). In fact one has, for $c(\cdot) \geq 1$,

$$\int_H^{\Delta} e^{iS_0(\gamma)} \prod_{j=1}^{n} e^{i\gamma_{t_j}(f_j)}\, d\gamma = \omega_c\left(\prod_{j=1}^{n} e^{i\Phi_{t_j}(f_j)}\right) \qquad (5.7)$$

with $e^{i\Phi_t(f)} \equiv \alpha_t(e^{i\Phi(f)}) \equiv e_W\left(e^{itA}\frac{f}{\sqrt{2A}}\right)$,

where $\Phi(f) \equiv e_W\left(\frac{f}{\sqrt{2A}}\right)$ is the time zero field in the representation of the Weyl algebra $\{z\in K^c, e_W(z)\}$ over K^c given by ω_c and α_t is the corresponding time automorphism induced by $e_W(z) \rightarrow e_W(e^{itA} z)$.

For $c(\vec{p}) \equiv 1$ we have that Δ_c is the Feynman propagator and correspondingly the state ω_c is the one given by the free Fock vacuum.

Other choices of $c(\cdot)$ s. t. $c(\cdot) \geq 1$ yield other states on the Weyl algebra associated with K^c, e.g. $c(\vec{p}) \equiv \mathrm{ctgh}\left(\frac{\beta}{2}\omega(\vec{p})\right)$, where $\beta \in [0,\infty]$ is a constant, yields the Gibbs states introduced by Høegh-Krohn [31].

One can extend above results to the following situation ([2], Sect. 8, p. 90 - 104). Let K be a real separable Hilbert space and let A^2 be a positive self-adjoint operator, such that 0 is not an eigenvalue. Define now (H, D, B, Δ) in a correspondent way as above, i.e. $H \equiv \{$ paths from $\mathbb{R}$ into K such that $\int [\dot{\gamma}^2(\tau) + (\gamma(\tau), A^2\gamma(\tau))_K]\, d\tau < \infty\}$ and define $(\gamma, B\gamma) = \int [\dot{\gamma}^2(\tau) - (\gamma(\tau), A^2\gamma(\tau))_K]\, d\tau$, with $D(B) \equiv \{\gamma(t,\omega)$, supp γ compact$\}$, where $\gamma(t,\omega)$ is the representation of $\gamma \in H$ given by the isomorphism $K \cong L^2_{\mathbb{R}}(d\nu)$, where $d\nu$ is the spectral measure of A. Moreover take D to be, correspondingly, the space of all paths in H, with values in $L^2(d\nu) \cong K$, whose Fourier transform $\hat{\gamma}(p,w)$, with respect to t, is in $C^1(\mathbb{R})$, as a function of p, and in $C(\mathbb{R})$ as a function of ω, with norm $|\gamma|_D \equiv |\gamma|_H + \sup_{\omega,p} |\frac{\partial\hat{\gamma}}{\partial p}(p,\omega)|$. Define Δ_c by (5.2), with p replaced by (p_0, ω) and dp replaced by $dp_0\, d\nu(\omega)$, $p^2 + 1$ by $p^2 + \omega^2$ and obviously $\int_{\mathbb{R}^s} P \int_{\mathbb{R}}$ by $\int d\nu(\omega)\, P\int_{\mathbb{R}}$. Then we have again that (H, D, B, Δ) is a Fresnel fourtuple and in particular the normalized integral with respect to Δ_c is well defined. Moreover in [2], Sect. 8 we proved the following

Theorem 5.3

All quasi-free positive states on the Weyl-algebra $(z \in K^c, e_w(z))$ over K^c which are invariant under the one-parameter group of $*$-automorphisms α_t of the Weyl-algebra describing the (free) time evolution (i.e. $\alpha_t(e_w(z)) \equiv e_w(e^{itA}z))$ for a self-adjoint A on K^c are of the form $\omega_c(e_w(z)) = e^{-\frac{1}{2}(z, c^w z)}$, where (,) is here the scalar product in $K \oplus K$, with c any $d\nu$-measurable function larger or equal 1, where $d\nu$ is the spectral measure of A. These states are given in terms of normalized integrals with respect to Δ_c by

$$\omega_c\left(\prod_{j=1}^{n} e^{i(u, \Phi(t_j))_K}\right) = \int_H^{\Delta_c} e^{i\int_{-\infty}^{\infty}[\dot{\gamma}^2 - (\gamma, A^2\gamma)_K]\, d\tau}$$

$$\left(\prod_{j=1}^{n} e^{i(u_j, \gamma(t_j))_K}\right) d\gamma$$

with $$e^{i(u, \Phi(t))_K} \equiv \alpha_t\left(e^{i(u,\Phi)_K}\right) \equiv e_w\left(e^{itA}\frac{u}{\sqrt{2A}}\right),$$

for any $u \in K$, $\Phi \in K$.

In [2], § 9, pp. 111-114 and [2], pp. 185-192, we also found a representation of quantities yielding the Schwinger functions of scalar fields interacting by trigonometric interactions (with space and ultraviolet cut-off) in terms of normalized integrals. Let namely χ be a C^∞-function on $\mathbb{R}^s$ with compact support in the ball $|\vec{x}| \leq 1$ and with $\int_{\mathbb{R}^s}\chi(\vec{x})\, d\vec{x} = 1$, so that $\chi_\varepsilon(\vec{x}) \equiv \varepsilon^{-s}\chi(\vec{x}/\varepsilon)$ converges to the δ-distribution as $\varepsilon \downarrow 0$. Let $v \in F_{\mathbb{R}}(\mathbb{R})$ and define, for any bounded open set $\Lambda \subset \mathbb{R}^s$,

$$V^\varepsilon_\Lambda \equiv \int_\Lambda \left(\int_{\mathbb{R}} e_W\left(\alpha\,(2A)^{-1/2}\chi_\varepsilon^{\vec{x}}\right) d\mu_v(\alpha)\right) d\vec{x},$$

where μ_v is the symmetric measure on $\mathbb{R}$ of which v is the Fourier transform, $A = \sqrt{-\Delta+m^2}$, $\chi_\varepsilon^{\vec{x}}(\vec{x}) \equiv \chi_\varepsilon(\vec{x}-\vec{y})$. The integral exists as strong Riemann integral in any representation of the Weyl-algebra associated with $K = L^2_{\mathbb{R}}(\mathbb{R})$, induced by a state which is space translation invariant and V^ε_Λ represents there a space cut-off and regularized (ultraviolet cut-off) trigonometric interaction.[2]) If the state is also invariant under the free time automorphism, induced by a unitary group e^{-itH_0}, then we have that $H_0 + V^\varepsilon_\Lambda = H$ is a self-adjoint operator, inducing the automorphism α_t on the algebra of bounded operators in the representation space for the Weyl-algebra. With the notation of (5.7) we have the following Theorem (cfr. [2] and [3]).

Theorem 5.4

For any $f_1, \dots, f_n$ in $L^2(\omega(\vec{p})^{-2}\, d\vec{p})$, $\vec{p} \in \mathbb{R}^s$ and any $t_1 \leq t_2 \leq \dots \leq t_n$

$$\omega_c \left(\prod_{j=1}^{n} e^{i\Phi_{t_j}(f_j)} \right) = \int_{H_1(\mathbb{R}^s)}^{\Delta_c} e^{iS_o(\gamma)} \, e^{-i\int_{t_1}^{t_n} (\int_\Lambda v((\gamma_\varepsilon(t\vec{x}))d\vec{x})dt} \prod_{j=1}^{n} (e^{i\gamma t_j (f_j)}) \, d\gamma ,$$

with $\gamma_\varepsilon \equiv \gamma * \chi_\varepsilon$, $*$ meaning convolution.

In the case $c \equiv 1$, where ω_c is the state given by the free Fock vacuum, the left hand side is equal to

$$(\Omega_o , \prod_{j=1}^{n} e^{-it_j H_\Lambda^\varepsilon} e^{i\Phi(f_j)} e^{it_j H_\Lambda^\varepsilon} \Omega_o),$$

where

$$H^\varepsilon_\Lambda = H_o + \int_\Lambda v \, (\xi_\varepsilon (\vec{x})) \, d\vec{x} ,$$

H_o being the kinetic energy operator in Fock space and $\xi_\varepsilon (\vec{x}) \equiv (\xi * \chi_\varepsilon) (\vec{x})$, $\xi (\vec{x})$ being the free time zero field in the Fock space. This permits then to obtain, in particular, the time ordered expectation values in the Fock vacuum for the space cut-off and ultraviolet cut-off trigonometric interactions (e.g. the so called Sine-Gordon interaction) by means of normalized integrals (yielding thus a definition of "Feynman history integrals" for these models).

In the next section we shall see how we can remove the cut-off for other quantities associated with such models.

5.3 Solution of the Schrödinger equation for local relativistic fields

In this section we describe shortly some new results which will appear in full elsewhere [22]. We have seen in Sect. 3.1, Theorem 3.1 that the solution of the Schrödinger equation $i \frac{\partial}{\partial \tau} \psi = - \frac{1}{2} \Delta \psi + V \psi$ on $\mathbb{R}^d$ with initial condition $\psi (0,x) = \varphi (x)$ on the time interval $[0,t]$ is given, whenever $\varphi, V \in F(\mathbb{R}^d)$, by the path integral

$$\psi (t,x) = \tilde{\int}_{\gamma(t)=x} e^{\frac{i}{2} \int_0^t \dot{\gamma}(\tau)^2 d\tau} \, e^{-i\int_0^t V(\gamma(\tau))\, d\tau} \varphi(\gamma(0))d\gamma, \quad (5.8)$$

defined as equal to

$$\tilde{\int}_H e^{\frac{i}{2} \int_0^t \dot{\gamma}(\tau)^2 d\tau} \, e^{-i\int_0^t V(\gamma(\tau)+x)\, d\tau} \varphi(\gamma(0)+x)\, d\gamma, \quad (5.9)$$

where now the integral is the normalized integral on the Hilbert space H of paths from $[0,t]$ into $\mathbb{R}^d$ with weak square integrable derivative and vanishing at time 0. Using the invariance of the normalized integral under the translation

$\tau \to \tau - t$ in the index set, we can also write (5.9) in the form

$$\widetilde{\int\limits_{\gamma(0)=0}} e^{\frac{i}{2}\int_{-t}^{0} \dot\gamma(\tau)^2 d\tau} e^{-i\int_{-t}^{0} V(\gamma(\tau)+x)\, d\tau} \varphi(\gamma(-t)+x)\, d\gamma \quad (5.10)$$

where now the integral is the normalized integral on the Hilbert space $H_{[-t,0]}$ of all paths in $[-t,0]$, vanishing at time 0 and with square integrable weak derivatives. But $H_{[-t,0]}$ is naturally embedded in $H_{(-\infty,0]}$ as the subspace of functions which are constant, and equal to $\gamma(-t)$, on the interval $(-\infty,-t]$. Since the linear functional $\gamma \to \gamma(\tau)$, $\tau \in (-t,0]$ on $H_{(-\infty,0]}$ is represented by an element in the subspace $H_{[-t,0]}$ we get by the Fubini theorem (P5) in Sect.2)

$$\psi(t,x) = \widetilde{\int\limits_{\gamma(0)=0}} e^{\frac{i}{2}\int_{-\infty}^{0} \dot\gamma(\tau)^2 d\tau} e^{i\int_{-t}^{0} V(\gamma(\tau)+x) d\tau} \varphi(\gamma(-t)+x)\, d\gamma \quad (5.11)$$

Let now consider the Schrödinger equation for fields on $\mathbb{R}^s$, with s odd (e.g. $s = 3$, the physically interesting case). Let $\chi \geq 0$ be in $C^\infty(\mathbb{R}^s)$ with supp χ contained in $|\vec{x}| \leq 1$, $\vec{x} \in \mathbb{R}^s$ and $\int_{\mathbb{R}^d} \chi(\vec{x})\, d\vec{x} = 1$. Let H be the space of paths $\gamma(t,\vec{x})$ with values in $S(\mathbb{R}^s)$ and vanishing at $t = 0$ and let $(\gamma, B\gamma) = 2S(\gamma)$, with $S(\gamma) \equiv \frac{1}{2}\int_{-\infty}^{0}\int_{\mathbb{R}^s} \dot\gamma^2(\tau,\vec{x}) - \nabla\gamma(\tau,x)^2 - m^2\gamma(\tau,x)^2) d\tau\, d\vec{x}$. Let Δ_R be the inverse of $\Box - m^2 \equiv \frac{\partial^2}{\partial\tau^2} - \Delta_s - m^2$ with support on $\tau < 0$ ("retarded propagator"). Let $v \in F(\mathbb{R})$, $\eta \in S(\mathbb{R}^s)$ and let $\psi_\eta(\tau,\vec{x})$ be the solution of $(\Box - m^2)\psi_\eta(\tau,\vec{x}) = 0$ in $\tau < 0$ with $\psi_\eta(0,\vec{x}) = \eta(\vec{x})$ and $\partial_0\psi_\eta(0,\vec{x}) = 0$. We can show that for any φ in the Fresnel space $F(S'(\mathbb{R}^s))$ the normalized integral (similar to the one in Theorem 5.4)

$$\int\limits_{\gamma(0,\vec{x})=0}^{\Delta_R} e^{\frac{i}{2}\int_{-\infty}^{0}\int_{\mathbb{R}^s} [\dot\gamma^2 - (\nabla\gamma)^2 - m^2\gamma^2]\, d\vec{x}\, d\tau}$$

$$e^{-i\int_{-t}^{0}\int_\Lambda v(\gamma_\varepsilon(\tau,\vec{x}) + \psi_\eta(\tau,\vec{x}))\, d\vec{x}\, d\tau} \varphi(\gamma(-t,\cdot) + \psi_\eta(-t,\cdot))\, d\gamma \quad (5.12)$$

is well defined. Clearly (5.12) is the complete analogue for the case where the configuration space $\mathbb{R}^d$ is replaced by an infinite dimensional space $(S'(\mathbb{R}^s))$, of the normalized integral (5.11) giving the semigroup (Theorem 3.1) and the solution of the Schrödinger equation for quantum mechanics. It is therefore natural to interpret the quantity $(U_t^{\Lambda,\varepsilon}\varphi)(\eta)$ defined by the right hand side of (5.12) as the solution of the relativistic Schrödinger equation[3] for quantum fields with a space and momentum cut-off interaction V_Λ^ε (equal to the one already discussed in Sect. 5.2) i.e.

$$V_\Lambda^\varepsilon \equiv \int_\Lambda v(\xi_\varepsilon(\vec{x}))\, d\vec{x} \quad , \quad (5.13)$$

$\xi_\varepsilon(\vec{x})$ being the time zero ultraviolet cut-off field in Fock space. In fact we compute from the definition (5.12) that the normalized integral is equal to

$$(U_t^{\Lambda,\varepsilon}\varphi)(\eta) = \sum_{n=0}^{\infty} \frac{(-i)^n}{n!} \int_{-t}^{0}\int_{\Lambda}\cdots\int_{-t}^{0}\int_{\Lambda} e^{-\frac{i}{2}\sum_{i,j}\alpha_i\, {}^{\varepsilon}\Delta_R^{\varepsilon}(\tau_i-\tau_j,\vec{x}_i-\vec{x}_j)\alpha_j}$$

$$e^{-i\sum_j \alpha_j \psi_\eta(\tau_j,\vec{x}_j)}\; e^{-i\sum_j \alpha_j \int \Delta_R^{\varepsilon}(\tau_j+t,\vec{x}_j-\vec{x})\beta(\vec{x})\,d\vec{x}} \tag{5.14}$$

$$e^{i\int\beta(\vec{x})\psi_\eta(-t,\vec{x})\,d\vec{x}}\; d\mu_\varphi(\beta)\prod_{j=1}^{n} d\mu_v(\alpha_j)\,d\tau_j\,d\vec{x}_j\,,$$

where μ_φ resp. μ_v are the bounded complex measures on $S'(\mathbb{R}^s)$ resp. $\mathbb{R}$ whose Fourier transforms are φ resp. v (i.e. $\varphi(\eta) = \int e^{i<\eta,\beta>}\,d\mu_\varphi(\beta)$, $v(s) = \int_{\mathbb{R}} e^{i\alpha s}\,d\mu_v(\alpha)$, where $<,>$ is the dualization between $S(\mathbb{R}^s)$ and $S'(\mathbb{R}^s)$) and ${}^{\varepsilon}\Delta_R^{\varepsilon}(t,\cdot) \equiv$

$$\chi_\varepsilon * \Delta_R(t,\cdot) * \chi_\varepsilon,\quad \Delta_R^{\varepsilon}(t,\cdot) \equiv \Delta_R(t,\cdot) * \chi_\varepsilon\,.$$

This formula is the analogue of the formula (3.8) in Sect. 3 and we see in the same way using the reality of Δ_R, that $U_t^{\Lambda,\varepsilon}\varphi \in F(S'(\mathbb{R}^s))$, hence $U_t^{\Lambda,\varepsilon}$ maps the Fresnel space $F(S'(\mathbb{R}^s))$ into itself and one has

$$\| U_t^{\Lambda,\varepsilon}\varphi\|_o \le e^{t|\Lambda|\,\|v\|_o}\,\|\varphi\|_o\,, \tag{5.15}$$

where we recall that $\|\ \|_o$ is the Banach space norm of the Fresnel space $F(S'(\mathbb{R}^s))$ (i.e. $\|f\|_o = \|\mu_f\|$, where $\|\mu_f\|$ is the total variation of the bounded complex measure μ_f whose Fourier transform is f). Hence we see that $U_t^{\Lambda,\varepsilon}\varphi$ is a bounded map of $F(S'(\mathbb{R}^s))$ into itself, the bound being uniform in ε. We shall now see that the limit as $\varepsilon \to 0$ exists. We have from the support properties of Δ_R that

$$\sum_{i,j}\alpha_i\,{}^{\varepsilon}\Delta^{\varepsilon}(\tau_i-\tau_j,\vec{x}_i-\vec{x}_j)\alpha_j = \sum_{i\neq j}\alpha_i\,{}^{\varepsilon}\Delta_R^{\varepsilon}(\tau_i-\tau_j,\vec{x}_i-\vec{x}_j)\alpha_j\,,$$

since ${}^{\varepsilon}\Delta_R^{\varepsilon}(0,\vec{x}) = 0$. Moreover ${}^{\varepsilon}\Delta_R^{\varepsilon}(\tau,\vec{x})$ converges pointwise to $\Delta_R(\tau,\vec{x})$ outside the cone $|\vec{x}| = |\tau|$. Hence by the Lebesgue Theorem on dominated convergence we have that (5.14) converges as $\varepsilon \downarrow 0$ and by taking the limits of the terms in (5.14) we get the following

Theorem 5.5

The solution $(U_t^{\Lambda,\varepsilon}\varphi)(\cdot)$ of the Schrödinger equation for relativistic fields in an odd number of space dimensions s with space cut-off and momentum cut-off interaction (5.13) with density v in $F(\mathbb{R})$ and with initial condition $\varphi \in F(S'(\mathbb{R}^s))$ (i.e. equal to the Fourier transform of a measure on $S'(\mathbb{R}^s)$) is defined by the normalized integral, with $\eta \in S(\mathbb{R}^s)$,

$$(U_t^{\Lambda,\varepsilon}\varphi)(\eta) = \int_{\gamma(o,\vec{x})=o}^{\Delta_R} e^{\frac{i}{2}\int_{-\infty}^{o}\int_{\mathbb{R}^s}[\dot{\gamma}^2-(\nabla\gamma)^2-m^2\gamma^2]d\vec{x}\,d\tau}\; e^{-i\int_{-t}^{o}\int_\Lambda v(\gamma_\varepsilon(\tau,\vec{x})+\psi_\eta(\tau,\vec{x}))d\vec{x}d\tau}$$

$$\varphi(\gamma(-t,o)+\psi_\eta(-t,o))\,d\gamma.$$

$U_t^{\Lambda,\varepsilon}$ is also given by (5.13) and is a bounded operator on $F(S'(\mathbb{R}^s))$, uniformly bounded in ε and in fact satisfying

$$\| U_t^{\Lambda,\varepsilon}\varphi\|_o \le e^{t|\Lambda|\,\|v\|_o}\,\|\varphi\|_o ,$$

where $\| \ \|_o$ is the norm on the Fresnel space of Fourier transforms of measures (given by the total variation of the correspondent measure). Moreover for each $\eta \in S(\mathbb{R}^s)$ we have that $(U_t^{\Lambda,\varepsilon}\varphi)(\eta)$ converges as $\varepsilon \downarrow 0$ and the limit $(U_t^{\Lambda}\varphi)(\eta)$ is given by

$$(U_t^{\Lambda}\varphi)(\eta) = \sum_{n=o}^{\infty}\frac{(-i)^n}{n!}\int_{-t}^{o}\int_{\Lambda}\cdots\int_{-t}^{o}\int_{\Lambda} e^{-\frac{i}{2}\sum_{i\neq j}^{n}\alpha_i\,\Delta_R(\tau_i-\tau_j,\vec{x}_i-\vec{x}_j)\,\alpha_j}$$

$$e^{-i\sum_{j=1}^{n}\alpha_j\,\psi_\eta(\tau_j,\vec{x}_j)}\qquad e^{-i\sum_{j=1}^{n}\alpha_j\int\Delta_R(\tau_j+t,\vec{x}_j-\vec{x})\,\beta(\vec{x})\,d\vec{x}}$$

$$e^{i\int\beta(\vec{x})\,\psi_\eta(-t,\vec{x})\,d\vec{x}}\qquad d\mu_\varphi(\beta)\prod_{j=1}^{n}(d\mu_v(\alpha_j)\,d\tau_j\,d\vec{x}_j),$$

with $\psi_\eta(\tau,\vec{x}) \equiv \int_{\mathbb{R}^s}\partial_\tau\,\Delta_R(\tau,\vec{x}-\vec{y})\,\eta(\vec{y})\,d\vec{y}$.

<u>Remark:</u> From the construction it is natural to call $(U_t^{\Lambda}\varphi)(\eta)$ the solution of the Schrödinger equation for quantized relativistic local fields with space cut-off Λ.

Using now that $\Delta_R(\tau,\vec{x})$ has a bounded support when $-t < \tau < 0$ together with methods of statistical mechanics (similarly as in [43]) (see [22] for details) we then get the following

<u>Theorem 5.6</u>

For interactions given in a finite space region $\Lambda \subset \mathbb{R}^s$, s odd, by

$$\int_\Lambda v(\xi_\varepsilon(\vec{x}))\,d\vec{x}$$

with $v(s) = \int_{\mathbb{R}} e^{is\alpha}\,d\mu_v(\alpha)$ there exists a constant C such that for

$\int d\,|\mu_\nu| < C$ and for arbitrary $\varphi \in F\,(S'\,(\mathbb{R}^s))$ we have that

$$\lim_{\Lambda \uparrow \mathbb{R}^s} \frac{(U_t^{\Lambda}\,\varphi)\,(\eta)}{(U_t^{\Lambda}\,1)\,(\eta)} = (U_t\,\varphi)\,(\eta)$$

exists for all $\eta \in S\,(\mathbb{R}^s)$ as Λ converges to $\mathbb{R}^s$ through the filter of bounded measurable subsets.

Remark By the construction it is natural to call $U_t\,\varphi$ the solution of the Schrödinger equation for relativistic local fields with trigonometric interactions (in fact any interaction given by a density which is the Fourier transform of a bounded complex measure).

For further results using $U_t\,\varphi$ see [22].

§ 6. The asymptotic expansion of oscillatory integrals and the expansions of quantum mechanics around the classical limit

According to a beautiful idea of Dirac and Feynman [1],[4] from the quantum dynamics, as expressed by Feynman path integrals, the detailed approach to the classical limit (when Planck's constant h goes to zero) should come out by a "stationary phase method in infinite dimensions". This should intuitively be so since the quantum mechanical quantities are represented by oscillatory path integrals

$$\int e^{\frac{i}{\hbar} S_t\,(\gamma)}\; g(\gamma)\,d\gamma\;, \tag{6.1}$$

where $S_t\,(\gamma)$ is the classical action along the path γ and as $h \to 0$ only the behaviour in a neighborhood of the stationary points of the phase function $S_t\,(\gamma)$ should give a non vanishing contributions. However, by Hamilton's principle, these stationary points are precisely the "classical paths", i.e. the solutions of the classical Newton equations of motion. In order to transform these heuristic suggestions into mathematical results we should study oscillatory integrals of the type

$$\int_H e^{\frac{i}{\hbar}\Phi\,(\gamma)}\; g\,(\gamma)\,d\gamma\;, \tag{6.2}$$

where H is some infinite dimensional space , Φ is a real-valued "phase function" on H and g is a complex-valued "amplitude function" on H. For applications to quantum mechanics we are mainly interested in phase functions of the form $\Phi\,(\gamma) = (\gamma,\,B\gamma) - W\,(\gamma)$, where $(\gamma,\,B\gamma)$ is a quadratic form (corresponding to the kinetic energy or, in the case of an anharmonic oscillator, the kinetic energy minus the potential energy of an harmonic oscillator), and $W(\gamma)$ is a term (which will be assumed bounded) coming from the potential. Such oscillatory integrals are indeed precisely those for which we were able to give a mathematical definition (cfr. Sections 2),4)). We shall now describe shortly the methods by which we were able to develop a method of stationary phase and a theory of asymptotic expansions of oscillatory integrals in infinite dimensions (extending for our phase functions the corresponding finite dimensional theory of Maslov [32], Hörmander [34], Duistermaat [35], Arnold [36], Leray [37].

We shall first consider the case of integrals of the form (6.2) where H is a real separable Hilbert space, $\Phi(\gamma) = (\gamma, B\gamma) - W(\gamma)$, with B strictly positive. We can as well take $B = 1$, changing if necessary the scalar product in H. So we want to study first normalized integrals of the form

$$I_h \equiv \widetilde{\int_H} e^{\frac{i}{h}(\gamma,\gamma)} e^{-\frac{i}{h}W(\gamma)} g(\gamma)\, d\gamma, \tag{6.3}$$

(γ,γ) being the scalar product in H and $W, g \in \mathcal{F}(H)$, W real-valued. In particular we want to study I_h as a function of h, especially in a neighborhood of $h = 0$. We have given the detailed study in [11], we shall here only sketch the results and some of the methods, and we refer to the original reference for more details. Being inspired by the corresponding theory when H is finite dimensional, we first look at the stationary points of the phase function

$$\Phi(\gamma) = (\gamma,\gamma) - W(\gamma), \tag{6.4}$$

since, according to the experience in the finite dimensional case the relevant contributions (as $h \to 0$) come from such points. Assume thus Φ has a Fréchet derivative (in order to be able to discuss stationary points!) $(d\Phi)(\gamma_0)$, at each point $\gamma_0 \in H$, so that, identifying, by the natural duality of H, $d\Phi(\gamma_0)$ with an element in H, we have

$$\Phi(\gamma) = \Phi(\gamma_0) + (d\Phi(\gamma_0), \gamma - \gamma_0) + O(|\gamma - \gamma_0|^2), \tag{6.5}$$

where we recall that $(\, , \,)$ is the scalar product in H, $|\cdot|$ is the norm in H and $O(\lambda)$ is a real-valued function such that $O(\lambda)/\lambda$ is bounded as $\lambda \to 0$.

In order that $d\Phi(\gamma_0)$ exists for all $\gamma_0 \in H$ it is sufficient that

$$\int |\alpha| \, d|\mu_W|(\alpha) < \infty, \tag{6.6}$$

where μ_W is the complex (symmetric) measure on H whose Fourier transform is W. In such a case one can thus consider the stationary points. Now the simplest case is certainly the one where there exists one and only one stationary point. To get this case we observe that under the further assumption on μ_W that

$$\int |\alpha|^2 \, d|\mu_W|(\alpha) < \infty \tag{6.7}$$

we have that $W(\gamma)$ is a twice continuously differentiably function of γ and that $\gamma \to d^2 W(\gamma)$ is a strongly continuous map from H into the symmetric trace class operators on H, with trace norm

$$\| d^2 W(\gamma) \|_1 \leq \int |\alpha|^2 \, d|\mu_W|(\alpha) . \tag{6.8}$$

If now $\int |\alpha|^2 \, d|\mu_W|(\alpha) < 1$ then one sees easily that

$$\| d W(\gamma) - d W(\gamma') \| \leq |\gamma - \gamma'| \tag{6.9}$$

for all $\gamma, \gamma' \in H$, i.e. $d W(\gamma)$ is a contraction, which then yields, by the contraction mapping principle, that $d\Phi(\gamma) = 0$ i.e. $\gamma = d W(\gamma)$ for one and only one value $\gamma = \gamma_o$ of γ.

Let us also note that under the assumption $\int |\alpha|^2 d|\mu_W|(\alpha) < 1$ one has that the unique stationary point γ_o given by $d\Phi(\gamma_o) = 0$ is non degenerated (i.e. regular) in the sense that $d^2\Phi(\gamma_o)\, x = 0$, $x \in H$ implies $x = 0$. In fact $d^2\Phi(\gamma_o) = 1 - d^2 W(\gamma_o)$ and $d^2\Phi(\gamma_o)\, x = 0$ is equivalent with $x = d^2 W(\gamma_o)\, x$, which has the only solution $x = 0$, since $|d^2 W(\gamma_o) x| \leq \int |\alpha|^2 d|\mu_W|(\alpha) |x| < |x|$, for all $x \neq 0$.

Thus we see that under the assumption $\int |\alpha|^2 d|\mu_W|(\alpha) < 1$ one has one and only one stationary point of the phase and this point is non degenerated. In the following the asymptotic expansion in powers of h in such a situation will be considered first. Later on we shall consider the case where there are possibly more stationary points and or there is degeneracy.

6.1. Asymptotic expansions in the case where there exists one and only stationary point and this is regular.

The simplest case is the one in which the stationary point of the phase $\Phi(\gamma)$ is at the origin i.e. dW exists and $d\Phi(0) = 0$ which is satisfied if $dW(0) = 0$. The following result has been proven in [11].

Theorem 6.1 Assume $W, g \in F(H)$ and such that, for some $\lambda > 0$,

$\int e^{\sqrt{2}\lambda|\alpha|} \, d|\mu_W|(\alpha) < \lambda^2$ and $\int e^{\sqrt{2}\lambda|\alpha|} \, d|\mu_g|(\alpha) < \infty$,

and moreover $dW(0) = \int \alpha \, d\mu_W(\alpha) = 0$.

Then the phase function $\Phi(\gamma) = \frac{1}{2}|\gamma|^2 - W(\gamma)$ in the oscillatory integral

$$I(h) = \widetilde{\int_H} e^{\frac{i}{2h}|\gamma|^2} \, e^{-\frac{i}{h}W(\gamma)} \, g(\gamma)\, d\gamma$$

has one and only stationary point, this point is $\gamma = 0$ and is non degenerated. Moreover $e^{\frac{i}{\hbar}W(0)} I(h)$ is bounded for $h \in \mathbb{R}$, analytic in $\operatorname{Im} h < 0$, C^∞ in h on the real axis and has the asymptotic expansion at $h = 0$

$$e^{\frac{i}{\hbar}W(0)} I(h) = \sum_{m=o}^{\infty} \frac{h^m}{m!} I_m$$

with $I_m = (-i)^m \sum_{n=o}^{\infty} (-\frac{1}{2})^h \frac{1}{n!(n+m)!} \sum_{j=1}^{n} (\frac{1}{i}\nabla_{\gamma_j} + \frac{1}{i}\nabla_\gamma)^{2(m+n)}$

$$\tilde{W}(\gamma_1) \ldots \tilde{W}(\gamma_n)\, g(\gamma) |_{\gamma_1 = \ldots = \gamma_n = \gamma = 0}, \quad \tilde{W} \equiv W - W(0).$$

The following estimate on the remainder holds

$$|R_N| \equiv | e^{\frac{i}{\hbar}W(0)} I(h) - \sum_{m=o}^{N} \frac{h^m}{m!} I_m | \leq$$

$$\leq |h|^{N+1} \frac{1}{\lambda^{2(N+1)}} (N+1)! \frac{\int e^{\sqrt{2}\lambda|\beta|} d|\mu_g|(\beta)}{(1 - \frac{1}{\lambda^2}\int e^{\sqrt{2}\lambda|\beta|} d|\mu_W|(\alpha))^{N+2}}$$

Remark: The leading term I_0 can also be written $\frac{g(0)}{|1 - d^2W(0)|^{1/2}}$, where $|1 - d^2W(0)|^{1/2}$ is the square root of the Fedholm determinant of the operator $1 - d^2W(0)$.

For the proof one first observes that by scaling one can reduce oneself to the case $\lambda = 1$. Moreover an expansion in powers of h of $e^{\frac{i}{\hbar}W(\gamma)}$ under the normalized integral in $I(h)$ and an interchange of the operations of summation and taking the normalized integral (possible because of the continuity of the normalized integral and the fact that the space of integrable functions is a Banach algebra) we get that $I(h)$ is equal to

$$\sum_{n=o}^{\infty} \frac{(-i)^n}{n!} \frac{1}{\hbar} \tilde{\int_H} e^{\frac{i}{2\hbar}|\gamma|^2} W(\gamma)^n g(\gamma)\, d\gamma. \tag{6.10}$$

The integrals on the right hand side are then computed, using the definition of the normalized integral, and one verifies that multiplication by $e^{\frac{i}{\hbar}W(0)}$ gives that the derivatives up to order n vanish, hence $e^{\frac{i}{\hbar}W(0)}$ times (6.10) is indeed an expansion in positive powers of h, with coefficient of h^m given by $\frac{1}{m!} I_m$.

We shall now describe the results when there is one and only one stationary point not at the origin.

Theorem 6.2. Assume W, g are as in Theorem 6.1 but $dW(0) \neq 0$. Then the same result of Theorem 6.1 holds, with the difference that the stationary point is no more in $\gamma = 0$ and in the expression for I_m the derivatives have to be taken at the stationary point γ_c instead of $\gamma = 0$ and $e^{\frac{i}{\hbar}W(0)}$ has to be replaced everywhere by $e^{\frac{i}{\hbar}(W(\gamma_c) - \frac{1}{2}(\gamma_c)^2)}$ and $\tilde{W}$ by $W - (W(\gamma_c) - \frac{1}{2}|\gamma_c|^2)$.

For the proof the formal idea is to consider the translated integral

$$\tilde{\int_H} e^{\frac{i}{2\hbar}\Phi(\gamma+\gamma_c)} g(\gamma+\gamma_c)\, d\gamma = e^{\frac{i}{\hbar}|\gamma_c|^2} \tilde{\int_H} e^{\frac{i}{\hbar}|\gamma|^2} e^{2\frac{i}{\hbar}(\gamma_c,\gamma)} e^{-\frac{i}{\hbar}W(\gamma+\gamma_c)} g(\gamma+\gamma_c)\, d\gamma ,$$

and to expand $W(\gamma+\gamma_c)$ around $\gamma = 0$, whereby then one is reduced to the case of Theorem 6.1. However in the expansions terms not in $F(H)$ arise, which are therefore not integrable with respect to the normalized integral. However this difficulty can be circumvented noticing that in the finite dimensional case the expansion can indeed be done and all terms are well defined as improper Riemann integrals. In fact in the corresponding finite dimensional case one has the result of Theor. 6.2 and the general case is then obtained by proving that for any sequence of finite dimensional projection P_n in H converging strongly to 1 one has that if Φ has one and only one stationary point γ_c then $\Phi_n(\gamma) \equiv \frac{1}{2}|\gamma|^2 - (W \circ P_n)(\gamma)$ has also one and only one stationary point γ_n and $\gamma_n \to \gamma_c$ as $n \to \infty$.

6.2. Asymptotic expansions in the case where there exist several critical points

In the case where H is finite dimensional the treatment of phase functions with more than one stationary point, with no accumulation points, is done by a decomposition χ_n of the unit for g, the decomposition of the unit being such that $g\chi_n$ has support enclosing only one stationary point, and then summing over the contributions coming from the individual stationary points, according to Theorem 6.2. This method cannot be carried over immediately to the case where H is infinite dimensional, because for Theor. 6.2 one needs $\int e^{\sqrt{2}\lambda|\beta|}\, d|\mu_g|(\beta) < \infty$, which makes g entire hence not with compact support. This problem has been overcome in [11] under the assumptions

$$\int e^{\sqrt{2}\lambda|\beta|}\, d|\mu_W|(\alpha) < \infty \quad \text{and} \quad \int d|\mu_W|(\alpha) < \lambda^2 \tag{6.11}$$

Note that these assumptions are much weaker than the ones in Theor. 6.2 in as much as the smallnes condition of th.6.1 on W and its derivative is now replaced by a condition implying essentially only smallness of the values of W. In fact the condition for having just one stationary point is no more satisfied, and in general one will have several stationary points. In fact one can show using the Fubini Theorem for infinite dimensional oscillatory integrals (P5) in Sect. 2) that the integral $I(h)$ can be split into an oscillatory integral over an infinite dimensional subspace $H - P_n H$ with a partial phase function $\Phi_n(\gamma) = \frac{1}{2}|P_n\gamma|^2 - W(P_n\gamma)$, having one and only one stationary point (and this is regular) and an oscillatory integral over a finite dimensional subspace $P_n H$, possibly with infinitely many stationary points. The first normalized integral is then treated by Theor. 6.2 and the oscillatory finite dimensional integral is controlled by the finite dimensional theory (e.g. [32] - [39]). In fact under the assumption (6.11) one has, for some sequence of finite dimensional projections P_n converging strongly to 1, that

$$\int e^{\sqrt{2}\,\lambda\,|\alpha - P_n\alpha|}\, d|\mu_W|(\alpha) \to \int d|\mu_W|(\alpha) < \lambda^2,$$

hence $\int_{H-P_nH} e^{\sqrt{2}\,\lambda\,|\alpha - P_n\alpha|}\, d|\mu_W|(\alpha) < \lambda^2$, for n sufficiently large.

Thus in $H - P_n H$ one can apply Theor. 6.2, and for this reason one splits $H = H_1 \oplus H_2$, $H_2 \equiv P_n H$, $H_1 \equiv (1-P_n)H$, and accordingly $\gamma = \gamma_1 \oplus \gamma_2$, $W(\gamma) = W(\gamma_1, \gamma_2)$, $\mu_W(\alpha) = \mu_W(\alpha_1, \alpha_2)$,

$$W(\gamma_1, \gamma_2) = \int_{H_1} e^{i(\alpha_1, \gamma_1)_1}\, d\mu_W^{\gamma_2}(\alpha_1),\quad d\mu_W^{\gamma_2}(\alpha_1) \equiv \int_{H_2} e^{i(\alpha_1, \gamma_2)_2}\, d\mu_W(\alpha).$$

Then one has

$$\int_{H_1} e^{\sqrt{2}\,\lambda\,|\alpha_1|}\, d|\mu_W^{\gamma_2}|(\alpha_1) \leq \int_{H_1 \oplus H_2} e^{\sqrt{2}\,\lambda\,|(1-P_n)\alpha|}\, d|\mu_W|(\alpha) < \lambda^2, \tag{6.12}$$

for n sufficiently big.

The stationary points of the total phase $\Phi(\gamma) = \frac{1}{2}|\gamma|^2 - W(\gamma)$ are given by $d_\gamma \Phi(\gamma) = 0$ i.e., in terms of the variables γ_1, γ_2, by

$$d_1 W(\gamma_1, \gamma_2) = \gamma_1 \tag{6.13}$$

and

$$d_2 W(\gamma_1\ \gamma_2) = \gamma_2\ , \tag{6.14}$$

where $d_j \equiv d_{\gamma_j}$.

Consider now the phase $\Phi(\cdot;\gamma_2)$ as a function on the Hilbert space H_1, of finite codimension, for fixed $\gamma_2 \in H_2$. Its stationary points are given by (6.13).

Now $\mu_W^{\gamma_2}$ is the measure whose Fourier transform is $W(\cdot\,, \gamma_2)$, hence the fact that (6.12) holds for n sufficiently big implies, by the first part of

Theorem 6.1, that (6.13) has one and only one solution (depending on γ_2), say $\gamma_1 = b(\gamma_2)$, and this solution is regular. Thus

$$d_1 W(b(\gamma_2), \gamma_2) = b(\gamma_2), \tag{6.15}$$

where $b(\gamma_2)$ is uniquely given by γ_2, in fact $\gamma_2 \to b(\gamma_2)$ is a real analytic mapping from H_2 into H_1 (one has $d\,b(\gamma_2) = (1 - d_1^{\,2} W(b(\gamma_2), \gamma_2))^{-1} d_1 d_2 W(b(\gamma_2), \gamma_2)$ with $d_1^{\,2} W(b(\gamma_2), \gamma_2)$ of trace class in H_1, namely with trace norm bounded by $\int |\alpha_1|^2 \, d\,|\mu_W^{\gamma_2}|\,(\alpha_1) < 1$).

Consider now (6.14), with γ_1 replaced by the solution $b(\gamma_2)$ of (6.13), i.e.

$$d_2 W(b(\gamma_2), \gamma_2) = \gamma_2 . \tag{6.16}$$

Because of the analyticity of b as a map from H_2 to H_1 and the analyticity of $(x, y) \to W(x, y)$, due to the assumptions on W, we have that the points γ_2 which solve (6.16) do not accumulate i.e. are isolated in H_2 i.e. they form a discrete set $\gamma_2^{(j)}$, $j = 1, 2, \ldots$ in H_2 .

We shall now mention shortly how one gets the asymptotic expansion of the oscillatory integral $I(h)$ under the above assumptions, implying that the total phase has singular points $\{ (b(\gamma_2^{(j)}), \gamma_2^{(j)}), j = 1, 2, \ldots \} \equiv S_{cr}$. Using the Fubini Theorem (P5 in Sect. 2) we have, using the direct splitting $H = H_1 \oplus H_2$ described above

$$I(h) = \widetilde{\int_{H_2}} e^{\frac{i}{2h}|\gamma_2|^2} \left(\widetilde{\int_{H_1}} e^{\frac{i}{2h}|\gamma_1|^2} e^{-\frac{i}{h}W(\gamma_1, \gamma_2)} g(\gamma_1, \gamma_2)\, d\gamma_1 \right) d\gamma_2 \tag{6.17}$$

hence

$$I(h) = \widetilde{\int_{H_2}} e^{\frac{i}{2h}|\gamma_2|^2} I_1(h, \gamma_2)\, d\gamma_2 , \tag{6.18}$$

with $\qquad I_1(h, \gamma_2) \equiv \widetilde{\int_{H_1}} e^{\frac{i}{2h}|\gamma_1|^2} e^{-\frac{i}{h}W(\gamma_1, \gamma_2)} g(\gamma_1, \gamma_2)\, d\gamma_1 .$

By the assumptions on W and our choice of H_2 we have that the phase function in the oscillatory integral $I_1(h, \gamma_2)$ has one and only one stationary point $b(\gamma_2)$, non degenerated, for every fixed $\gamma_2 \in H_2$. Set now

$$\hat{I}_1(h, \gamma_2) \equiv e^{-\frac{i}{h}\left(\frac{1}{2}b(\gamma_2)^2 - W(b(\gamma_2), \gamma_2)\right)} I_1(h, \gamma_2) \tag{6.19}$$

then by Theor. 6.1 we have that $\hat{I}_1$ is as a function of h, C^∞ on the real axis, for any fixed $\gamma_2 \in H$ and has the asymptotic expansion in a neighborhood of $h = 0$

$$\hat{I}_1(h,\gamma_2) = \sum_{m=0}^{N} \frac{h^m}{m!} \hat{I}_1^{(m)}(0,\gamma_2) + R_N(h,\gamma_2), \tag{6.20}$$

where $\hat{I}_1^{(m)}$ is the m-th derivative of $\hat{I}_1$ with respect to h and R_N is the remainder i.e. $R_N = 0(|h|^{N+1})$.

Inserting now (6.20) into (6.18) we get

$$I(h) = \sum_{m=0}^{N} \frac{h^m}{m!} \widetilde{\int_{H_2}} e^{\frac{i}{h}\Phi_2(\gamma_2)} \hat{I}_1^{(m)}(0,\gamma_2)\, d\gamma_2 + $$

$$+ \widetilde{\int_{H_2}} e^{\frac{i}{h}\Phi_2(\gamma_2)} R_N(h,\gamma_2)\, d\gamma_2, \tag{6.21}$$

where $\Phi_2(\gamma_2) \equiv \frac{1}{2}(\gamma_2)^2 + \frac{1}{2} b(\gamma_2)^2 - W(b(\gamma_2),\gamma_2).$ (6.22)

By the property P2 in Section 2 we can insert under the normalized integrals over the finite dimensional space H_2 a decomposition of the unit by smooth functions φ_i of compact supports, pairwise disjoint, and such that each smooth function φ_i has only one point $(\gamma_2^{(j)})$ out of S_{cr} in its support. Then, using also that the normalized integral of smooth functions on a finite dimensional space is essentially the same as the Riemann integral (P3 in Section 2), we get from (6.21)

$$I(h) = (2\pi i)^{-\frac{n}{2}} \sum_{m=0}^{N} \frac{h^m}{m!} c_m + \hat{R}_N(h), \tag{6.23}$$

with

$$c_m = \sum_j \int_{\mathbf{R}^n} e^{\frac{i}{h}\Phi_2(x)} \hat{I}_1^{(m)}(0,x)\, \varphi_j(x)\, dx \tag{6.24}$$

$$\hat{R}_N(h) = (2\pi i)^{-\frac{n}{2}} \sum_j \int_{\mathbf{R}^n} e^{\frac{i}{h}\Phi_2(x)} R_N(h,x)\, d\varphi_j(x) dx \tag{6.25}$$

where $n \equiv \dim H_2$ and the sum is over all j numbering the points in the discrete sets S_{cr}. Clearly for the convergence of the series in (6.24) and (6.25), in the case where S_{cr} has infinitely many points it is sufficient e.g. that

$$\sum_j \int_{\operatorname{supp}\varphi_j} |\hat{I}_1^{(m)}(0,x)|\, dx \quad \text{and} \quad \sum_j \int_{\operatorname{supp}\varphi_j} |R_N(h,x)|\, dx$$

converge. For each finite dimensional integral in (6,24), (6.25) one can now apply the theory of oscillatory integrals on finite dimensional spaces. If the stationary points of the total phases Φ_2 in (6.21) and $\Phi_2(x) + \psi(x,h)$ in (6.25), where $\psi(x)$ is the amplitude function to $R_N(h,x)$ (i.e. $R_N(h,x) = e^{i\psi(x,h)}$

$|R_N(h,x)|$), are regular then each integral in (6.24), (6.25) is C^∞ in h on the real axis. It is shown in [11] that this is indeed the case when the phase Φ of the original oscillatory integral I (h) is non degenerated (in the sense that $d^2\Phi = 1 - d^2 W(\gamma)$ does not have 0 as eigenvalue). For more details we refer to [11] and we shall here limit ourselves to the following summary of the mentioned results in the case of more than one stationary point:

Theorem 6.3 : Consider the oscillatory integral

$$I(h) = \widetilde{\int_H} e^{\frac{i}{h}\Phi(\gamma)} g(\gamma)\, d\gamma$$

with $\Phi(\gamma) = \frac{1}{2}|\gamma|^2 - W(\gamma)$, $W, g \in F(H) \cap C^\infty(H)$. Assume that there exists a splitting $H = H_1 \oplus H_2$, with $\dim H_2 < \infty$ such that there exists a $\lambda > 0$ with $\|\mu_W\| < \lambda^2$ and $\int e^{\sqrt{2}\lambda|\alpha_1|}\, d|\nu|(\alpha_1,\alpha_2) < \infty$, where ν stands for μ_W or μ_g. Then if $\dim H_2 = 0$ the critical set S_{cr} is discrete.

For $0 < \dim H_2 < \infty$ one has that if $(\operatorname{supp} g) \cap S_{cr}$ is a finite set consisting of m points and $(1-d^2W)(\gamma_c^{(j)})$ is non degenerated for each $\gamma_c^{(j)} \in (\operatorname{supp} g) \cap S_{cr}$, then

$$I(h) = \sum_{j=1}^{m} e^{\frac{i}{h}\left(\frac{1}{2}|\gamma_c^{(j)}|^2 - W(\gamma_c^{(j)})\right)} I_j^*(h),$$

where $I_j^*(h)$ is C^∞ in h, hence has an asymptotic expansion in powers of h, obtained from (6.19) and the asymptotic expansion in powers of h of the m finite dimensional integrals with regular stationary points in (6.20). The expansion only involves W and g and their derivatives at $\gamma_c^{(j)}$. The leading term, i.e. the term of order h^0, is given by

$$I_j^*(0) = \frac{e^{i\frac{\pi}{2}n_j}}{|\operatorname{Det}(1-d^2W(\gamma_c^{(j)}))|^{\frac{1}{2}}}\, g(\gamma_c^{(j)}),$$

where n_j is the number of negative eigenvalues of $1 - d^2W(\gamma_c^{(j)})$.

6.3 The case of degenerate stationary points and the case of oscillatory integrals depending on parameters.

In the study of oscillatory integrals over finite dimensional spaces H the occurrence of degenerate stationary points is treated by considering families of oscillatory integrals depending on parameters. The reason is that the behaviour as $h \to 0$ of oscillatory integrals with degenerate stationary point is dictated by the structure of the degeneracy and this structure is best described by having additional parameters at disposal, (and changing variables in such a way that the phase function is brought onto some standard form). In fact by Morse theorem the functions $\Phi(x)$ which have only non degenerate critical points form an open and dense set (in the Whitney topology) in C^∞ with complement of codimension 1. Hence degenerate critical points are unstable in the sense that they disappear under C^∞ - small perturbations. On the other hand by Thom's transversality theorem there is a countable intersection of dense open sets of phase functions $y \to \Phi(\cdot, y) \in C^\infty(H)$, depending parametrically on $y \in \mathbb{R}^k$, such that the functions induced in the jet bundle over H intersect the singular manifolds (i.e. the ones carrying the critical points) in this jet bundle transversally, hence intersect indeed only singular manifolds of codimension at most k, and the intersections are stable. In other words, the study of oscillatory integrals depending on k parameters is the suitable instrument for studying degenerated critical points, with degeneracy of codimension k.

It is then natural to try to study the case of degenerated critical points in the infinite dimensional case along similar lines, i.e. by considering oscillatory integrals depending parametrically on the points in some finite dimensional space $\mathbb{R}^k$ i.e. integrals of the form

$$I(h, y) = \widetilde{\int_H} e^{\frac{i}{\hbar}\Phi(\gamma, y)} g(\gamma, y)\, d\gamma, \tag{6.22}$$

with $y \in \mathbb{R}^k$. It turns out that for these integrals a corresponding theory to the finite dimensional one can be developed, and we shall here summarize our results on this topic, see [11] for more details. We assume that Φ is of the form

$$\Phi(\gamma, y) = \frac{1}{2}|\gamma|^2 - W(\gamma, y), \tag{6.23}$$

which reduces to the case considered before in the case of absence of parameters y. The stationary points of the phase in (6.22) are given by $d_\gamma \Phi = 0$, and they are obviously points (γ, y) in $H \times \mathbb{R}^k$.

In the finite dimensional case one possibility to treat oscillatory integrals depending on parameters is to bring the phase function into some standard form $\tilde{\Phi}$ (in such a way that Φ and $\tilde{\Phi}$ are equivalent as unfoldings of functions on H).

Then the theory of oscillatory integrals with phase functions in standard form can be used [36], especially if one is interested in the behaviour of $I(h, y)$ as a function of y.[2]) However another possibility is to study integrals of the form (6.22) by integrating them against suitable functions of the form $\varphi(y) \equiv e^{-\frac{i}{\hbar}\psi(y)}\chi(y)$, with ψ real valued. It turns out that in the applications to quantum mechanics this second possibility arises quite naturally (in fact φ is here provided automatically by the initial conditions!). So let us consider integrals of the form

$$I(h,\psi) \equiv (2\pi i h)^{-k/2} \int_{Y=\mathbb{R}^k} I(h,y)\; e^{-\frac{i}{h}\psi(y)} \chi(y)\, dy, \tag{6.23}$$

with $\psi \in C^\infty(\mathbb{R}^k, \mathbb{R})$, $\chi \in C_0^\infty(\mathbb{R}^k)$ and $I(h,y)$ given by (6.22). We have, from (6.22), (6.23):

$$I(h,\psi) = \tilde{\int_Y} \left(\tilde{\int_H} e^{\frac{i}{h}\tilde{\Phi}(\gamma,y)} g(\gamma,y)\, d\gamma \right) \chi(y)\, dy \tag{6.24}$$

with

$$\tilde{\Phi}(\gamma,y) \equiv \Phi(\gamma,y) - \psi(y). \tag{6.25}$$

The idea behind the introduction of ψ is to use ψ in order to make the total phase non degenerated. The discussion from this point on takes a nice geometrical character. In fact the stationary points of $\tilde{\Phi}$ are given by

$$d\tilde{\Phi}(\gamma, y) = 0 \tag{6.26}$$

i.e.

$$\left\{ \begin{array}{l} d_1 \tilde{\Phi} = 0 \qquad (6.27) \\ d_2 \tilde{\Phi} = 0 , \qquad (6.28) \end{array} \right.$$

where $d_1 \equiv d_\gamma$, $d_2 \equiv d_y$.

Note that (6.27) and (6.28) are equivalent respectively to

$$d_1 \Phi = 0 \tag{6.29}$$

$$d_2 \Phi = d_2 \psi . \tag{6.30}$$

Let S_W be the singular locus in $H \oplus Y$ i.e., by definition, the set of solutions (γ,y) of (6.29). The points in S_W are thus precisely the stationary points for the integral over H.

The condition for the total phase $\tilde{\Phi}$ to have only non degenerated critical points is that $d^2\tilde{\Phi}$ does not have 0 as an eigenvalue i.e.

$$\left(\left(\begin{array}{cc} d_1^2 \Phi & d_1 d_2 \Phi \\ d_2 d_1 \Phi - d_2^2 \psi & d_2^2 (\Phi - \psi) \end{array} \right)\right) (\gamma, y) \tag{6.31}$$

does not have 0 as an eigenvalue (to an eigenvector in $H \oplus Y$), for any solution (γ,y) of (6.29), (6.30).

Suppose now

$$W(\gamma,y) = \int_H e^{i(\alpha,\gamma)_H} d\mu_W^y(\alpha), \tag{6.32}$$

with

$$\int e^{\sqrt{2}\,\lambda|\alpha|} d|\mu_W^y|(\alpha) < \infty \quad \text{and} \quad \|\mu_W^y\| < \lambda^2, \tag{6.33}$$

for some $\lambda > 0$, where $(\alpha, \gamma)_H$ is the scalar product in H. The following geometrical facts are established in [11] :

1) $d^2 \tilde{\Phi}$ has a bounded inverse, as an operator in $H \oplus Y$ iff it is surjective. In turn $d^2 \tilde{\Phi}$ surjective implies that $d\, d_1 \tilde{\Phi} = (1 - d_{11} W, - d_{12} W)$ is surjective as a map from $H \oplus Y$ to H.

2) Dim Ker $d\, d_1 \tilde{\Phi}$ = dim Y.

3) S_W is a smooth manifold in $H \oplus Y$ and its tangent space is Ker $d\, d_1 \tilde{\Phi}$ (the proof of this fact uses the Fredholm alternative for $1 - d_{11} W$ and analytic perturbation theory).

4) $d\, d_1 \tilde{\Phi}$ is injective (this is in fact equivalent with S_W being smooth).

5) Let J be the mapping : $(\gamma, y) \to (y, - d_2\, W(\gamma, y)) \in T^* Y$. Then $d J \restriction T S_W$ is injective (where T^* resp. T mean cotangent resp. tangent bundle), $J \restriction S_W$ is an immersion, call it Λ_W, of S_W in $T^* Y$. Λ_W is a smooth k-dimensional manifold in $T^* Y$ and is a Lagrange manifold (thus a Lagrange immersion of S_W in $T^* Y$).

6) The condition for the non degeneracy of $\tilde{\Phi}$ at a critical point is precisely (e.g. from (6.29), (6.30)) that the Lagrange manifolds Λ_W and (graph $d\psi$) intersect transversally.

We shall now describe shortly some of the results concerning the asymptotic expansion of $I(h, \psi)$ in powers of h, for other results and details we refer to [11], Sect.4. Assume 6), then under the stated assumptions (6.33) on μ_W we have by Theor. 6.3

$$I(h, \psi) = \sum_P [\, | \operatorname{Det} d^2 \tilde{\Phi} |^{-1/2}\, e^{i \frac{\pi}{2} n (d^2 \tilde{\Phi})}\, e^{\frac{i}{h} \Phi}\, g \chi\,] (P) + O(h) \qquad (6.34)$$

where the sum is over the points $P \in \mathcal{P} \equiv \{(\gamma, y) \in J^{-1} (\Lambda_W \cap (\text{graph } d\psi), y \in \operatorname{supp} \chi\}$ (this set is discrete if the Lagrange immersion $S_W \to \Lambda_W \subset T^* Y$ is proper).

It is now interesting to examine how the quantities Det $d^2 \tilde{\Phi}$ and $n(d^2 \tilde{\Phi})$ depend on ψ. Φ itself is independent of ψ and can be looked upon as a function on Λ_W. Det $d^2 \tilde{\Phi}$ depends on ψ, but only through the tangent plane to (graph $d\psi$) at the points in $\mathcal{P}$. One has $|\operatorname{Det} d^2 \tilde{\Phi}| = \operatorname{Vol} \psi / \operatorname{Vol}_W$, where Vol_W is formally $\delta(d_1 \Phi)$ i.e. the volume in $H \oplus Y \restriction S_W$ (identified with Λ_W), in particular is thus independent of ψ, and Vol_ψ is the volume deformation by the map $T_{(\gamma,y)} S_W \to T_y^* Y$. The dependence of $n(d^2 \tilde{\Phi})$ on ψ is the following: $n(d^2 \tilde{\Phi})$ is the number of negative eigenvalues of the compact operator $1 - d^2 \tilde{\Phi}$ (with the eigenvalue 0 counted with multiplicity $\frac{1}{2}$) and one has $n(d^2 \tilde{\Phi}) =$ $= n(1 - d_{11} W) + n(A_\psi)$, where therefore $n(1 - d_{11} W)$ is independent of ψ and the symmetric map A_ψ is defined by $A_\psi \zeta = \Delta y \in T_y^* Y$, for $\zeta \in T_y^* Y$, where Δy is given by the unique solution of

$$d^2 \tilde{\Phi} \binom{\Delta\gamma}{\Delta y} = \binom{0}{\zeta}, \qquad \Delta\gamma \in H, \quad \Delta y \in T_y Y. \tag{6.35}$$

For different choices of ψ, let us say ψ_1, ψ_2, such that $P_1 = P_2$ (P_i defined as P with ψ replaced by ψ_i), so that the P in (6.34) are the same for ψ_1 and ψ_2, we have

$$n(d^2 \tilde{\Phi}_1) - n(d^2 \tilde{\Phi}_2) = n(A_{\psi_1}) - n(A_{\psi_2}) \tag{6.36}$$

and this is, in the case where there is only one point $P = (y_o, \zeta_o)$, equal to $2\sigma(T^x_{y_o} Y, T_{(y_o,\xi_o)}\Lambda_W ; L_{\psi_1}, L_{\psi_2})$, where L_{ψ_1}, $i = 1,2$, are the tangent spaces of the Lagrange manifolds $(y, d\psi_i(y))$ and σ is Hörmander's invariant (Ref. [34], Sect. 3.3.). One has $\sigma = \langle\gamma, \alpha_E\rangle$, where γ is a closed curve in the set of Lagrange planes in the symplectic space $T^x Y$ which consists of an arc from L_{ψ_1} to L_{ψ_2} of Lagrange planes transversal to $T^x_{y_o} Y$ followed by an arc from L_{ψ_2} to L_{ψ_1} of Lagrange planes transversal to $T_{(y_o,\zeta_o)}\Lambda_W$, and α_E is the cohomology class discussed by Keller, Maslov, Arnold, Hörmander, Leray (the "Maslov index").

In order to discuss naturally the transformation properties of the leading term of the oscillatory integral it is convenient to assume that, χg being a density in Y, χ and g are 1/2-densities in Y (in the sense that under change of coordinates $y \to y'$ in Y they get multiplied by $|dy/dy'|^{1/2}$). Observing also that the assumptions (6.33) can be relaxed in the sense of Theorem 6.3, we can summarize the results conveniently in the following form:

Theorem 6.4 Assume that, for each $y \in Y$, W, g satisfy the conditions

$$\int_H e^{\sqrt{2\lambda}|\beta|} \, d|\mu_W^y|(\beta,\gamma) < \infty, \quad \int_H e^{\sqrt{2\lambda}|\beta|} \, d|\mu_g^y|(\beta,\gamma) < \infty, \quad \int d|\mu_W^y| < \lambda^2, \quad \text{for some}$$

splitting $H = H_1 \oplus H_2$, $\dim H_2 < \infty$, some $\lambda > 0$. Suppose ψ is so chosen that $(y, d\psi(y))$ and Λ_W intersect transversally, the Lagrange immersion $S_W \to \Lambda_W$ is proper and, for any $y \in Y$, the phase function $\frac{1}{2}\gamma^2 - W(\gamma,y), \gamma \in H$, is non degenerate (i.e. $(1 - d_{11} W, - d_{12} W)$ is surjective in $H \oplus Y \to H$). Then the leading term in the asymptotic expansion of $|\mathrm{Vol}_\psi|^{1/2} I(h,\psi)$ (where $I(h,\psi)$ is given by (6.24) and Vol_ψ is the volume deformation by the map $T_{(\gamma,y)} S_W \to T_y{}^*Y$, where S_W is the singular locus in $H \oplus Y$) is an element of ($\frac{1}{2}$ densities on $T_{(y_o,\zeta_o)}\Lambda_W$) $\otimes$ (fiber at (y_o,ζ_o) of Maslov canonical line bundle on Λ_W). The leading term is given by

$$\sum_P \left[|\mathrm{Vol}_W|^{\frac{1}{2}} e^{i\frac{\pi}{2} n(1-d_{11}W)} e^{\frac{i}{2}\pi n(A_\psi)} e^{\frac{i}{h}\Phi} g\chi \, e^{-\frac{i}{h}\psi} \right]_P$$

where the sum and evaluation $[\ ]_P$ is over the discrete set of points P in S_W which are in $P \equiv \{(\gamma,y) \in J^{-1}(\Lambda_W \cap \text{graph } d\psi), y \in \text{supp}\,\chi\}$. Φ is given uniquely (up to an additive constant) by the Lagrange immersion Λ_W, and in fact if Φ is considered locally as a function on Λ_W, $d\Phi = \zeta\, dy$. Moreover changes of ψ_i bring about changes given by (6.36), at a common point of intersection P.

The question how does $I(h,\psi)$ change when W or g are changed is answered in [11] by observing that under the assumptions in Theorem 6.4 above the oscillatory integrals with functions $W_i\ g_i$, $i = 1, 2$ are equivalent (can be transformed into each other by smooth y-dependent transformation) (modulo oscillatory factors of the form $e^{\frac{i}{\hbar}C}$, where C is a constant, independent of h) if and only if the germs of the corresponding phase functions $\frac{1}{2}|\gamma|^2 - W_i(\gamma, y)$ at the points in the singular locus define the same germs of Lagrange immersions Λ_{W_i} in T^*Y.

6.4
Applications to the classical limit of quantum mechanics

We saw above (Theorem 3.3) that the Green's function for the time-dependent Schrödinger equation on $\mathbb{R}^d$ in the time interval [0,t], with potential V, is given by

$$G_t(x,y) = G_t^0(x,y) \tilde{\int_{H_0}} e^{\frac{i}{2h}\gamma^2} e^{-\frac{i}{\hbar}W(\gamma,y)}\, d\gamma, \tag{6.37}$$

with $\gamma^2 \equiv \int_0^t \dot{\gamma}^2\, d\tau$,

$$W(\gamma,y) \equiv \int_0^t V(\gamma(\tau) + (y-x)(1-\tfrac{\tau}{t}) + x)\, d\tau,$$

where H_0 is the Hilbert space of continuous paths starting at time 0 in 0 and ending at time t in 0, with finite kinetic energy (in the sense of distributions). $G_t^0(x,y)$ is the free Green's function i.e. the kernel of $e^{i(t/h)\Delta}$ i.e.

$$G_t^0(x,y) = (2\pi iht)^{-d/2}\, e^{i(x-y)^2/2ht}.$$

The solution of the time-dependent Schrödinger equation with initial value $\varphi(y) \equiv e^{-\frac{i}{\hbar}\psi(y)}\tilde{\psi}(y)$ is then given by

$$\psi(t,x) = \int_{\mathbb{R}^d} G_t(x,y)\, \varphi(y)\, dy =$$

$$= (2\pi ih)^{-d/2} \int_{\mathbb{R}^d = Y} \left(\int_{H_0} e^{\frac{i}{\hbar}\tilde{\Phi}(\gamma,y)} g(\gamma,y)\, d\gamma \right) \chi(y)\, dy, \tag{6.38}$$

with

$$\tilde{\Phi}(\gamma,y) \equiv \frac{1}{2}\gamma^2 - W(\gamma,y) - \psi(y)$$

$$g(\gamma,y)\,\psi(y) \equiv (2\pi iht)^{d/2}\, \tilde{\chi}(y)\, e^{\frac{i}{2ht}(y-x)^2} \tag{6.39}$$

We see that the integral (6.38) is of the form (6.24), with $k = d$, $H = H_0$. It is then natural to apply the results discussed in the preceding subsection concerning integrals of the form (6.24) to the present case. This is done in details in Ref. [11], Sect. 5.

The results are as follows. Assume the Schrödinger potential V is in $F(\mathbb{R}^d)$ and the measure μ_V of which it is the Fourier transform satisfies $\int e^{|\beta|\varepsilon} d|\mu_V|(\beta) < \infty$ for some $\varepsilon > \infty$. We shall see that this assumption (implying of course analyticity of V, but no smallness assumption) is actually enough to permit the splitting of the Hilbert space H_0 in a part on which the phase function has one and only one stationary point and a finite dimensional part, so that the result of Theorem 6.3 can be applied (as was assumed for the results (Theorem 6.4) of the precedings subsection). In fact the decomposition $\gamma \in H_0$, $\gamma = \gamma_1 + \gamma_2$ with γ_2 piecewise linear and such that $\gamma_2(\frac{kt}{n}) = \gamma(\frac{kt}{n})$, $k = 0, 1, \ldots, n$ and γ_1 defined by $\gamma_1 \equiv \gamma - \gamma_2$, is a direct decomposition (since

$$\int \dot{\gamma}_1 \dot{\gamma}_2 \, d\tau = \sum_{k=0}^{n-1} c_k \int_{k\frac{t}{n}}^{(k+1)t/n} \dot{\gamma}_1 \, d\tau = 0)$$

and one has with $H_1 \equiv$ (all paths of the form γ_1), $H_2 \equiv$ (all paths of the form γ_2), $H = H_1 \oplus H_2$ with $\dim H_2 = (n-1)d \equiv$ dimension of the space of possible coordinates of the points $\gamma(\frac{kt}{n})$, $k = 1, .., n-1$. In order to satisfy the assumptions of Theorem 6.3 we have to verify that there exists $\lambda > 0$ and a choice of n such that

$$\int_{H_1} e^{\sqrt{2}\,\lambda\,|\beta|} \, d|\mu_W|\,(\beta,\delta) < \lambda^2 \tag{6.40}$$

for all $\delta \in H_2$, where $|\beta|$ is the norm of $\beta \in H_1$ in H_0. However μ_W is given by the following

$$W(\gamma_1,\gamma_2,y) \equiv \int_0^t V(\gamma(\tau) + (y-x)(1-\tfrac{\tau}{t}) + x)\, d\tau = \int_{H_1 \oplus H_2} e^{i(\alpha,\gamma)_{H_0}} d\mu_W(x)$$

$$= \int_{H_1 \oplus H_2} e^{i(\alpha,\gamma)} \int_0^t \int_{\mathbb{R}^d} \delta_{\varepsilon\gamma_\tau}(\alpha)\, e^{i[(y-x)(1-\frac{\tau}{t})+x]\varepsilon} d\mu_V(\varepsilon) d\tau, \tag{6.41}$$

where $\varepsilon\gamma_\tau$ is the element in H_0 such that $(\gamma, \varepsilon\gamma_\tau) = \varepsilon\,\gamma(\tau)$, for all $\gamma \in H_0$, i.e. $\gamma_\tau(s) = (1-\frac{\tau}{t})\,s$, $0 \leq s \leq \tau \leq t$. Note that $\varepsilon \in \mathbb{R}^d$ and $\gamma_\tau(\tau') \in \mathbb{R}$, for each $\tau, \tau' \in [0,t]$. For any $\alpha \in H_0$ we have thus

$$d\mu_W(\alpha) = \int_0^t \int_{\mathbb{R}^d} \delta_{\varepsilon\gamma_\tau}(\alpha)\; e^{i[(y-x)(1-\frac{\tau}{t})+x]\varepsilon} d\mu_V(\varepsilon) d\tau \tag{6.42}$$

hence the condition (6.40) involves estimating

$$\int_0^t \int_{\mathbb{R}^d} e^{\sqrt{2}\,\lambda |\varepsilon\gamma_\tau^{(1)}|_{H_1}} d|\mu_V|\,(\varepsilon)\, d\tau, \tag{6.43}$$

where $\varepsilon\,\gamma_\tau^{(1)}$ is the component of $\varepsilon\gamma_\tau$ in H_1 and $|\ |_{H_1}$ is the norm in H_1. Now on the interval $\frac{kt}{n} \leq \tau < (k+1)\frac{t}{n}$ we have, using that $\gamma_\tau^{(1)}(kt/n) = 0$,

$$|\varepsilon\cdot\gamma_\tau^{(1)}|_{H_1} = |\varepsilon|\;|\gamma_\tau^{(1)}(\tau)| = |\varepsilon|\left|\int_{k\frac{t}{n}}^{\tau} \frac{d\gamma_\tau^{(1)}(s)}{ds}\, ds\right| \leq$$

$$\leq |\varepsilon| \int_{k\frac{t}{n}}^{\tau} \left| \frac{d\gamma_\tau^{(1)}(s)}{ds} \right| ds \leq \sqrt{\frac{t}{n}}\, |\varepsilon|\, |\gamma_\tau^{(1)}\, \chi\, [k\tfrac{t}{n}, \tau\,]|_{H_0} \leq \sqrt{\frac{t}{n}}\, |\varepsilon|(1-\tfrac{\tau}{t}) \leq \sqrt{\frac{t}{n}}|\varepsilon| . \quad (6.44)$$

(6.43) is therefore estimated as less or equal to

$$t \int_{\mathbb{R}^d} e^{\sqrt{2}\,\lambda\, \sqrt{\frac{t}{n}}\, |\varepsilon|}\, d|\mu_V|\,(\varepsilon) \tag{6.45}$$

and this is less or equal λ^2, provided for fixed $t \in \mathbb{R}$ one choses $\lambda^2 / t > 1$ and n so big that

$$\int_{\mathbb{R}^d} e^{\sqrt{2}\,\lambda\, \sqrt{\frac{t}{n}}\, |\varepsilon|}\, d|\mu_V|\,(\varepsilon) < \frac{\lambda^2}{t} . \tag{6.46}$$

Hence we see that, for such a choice of λ, n, the assumption (6.40) is satisfied, provided $\int_{\mathbb{R}^d} e^{|\varepsilon|\eta}\, d_{|\mu_V|}(\varepsilon) < \infty$ for some $\eta > 0$. Hence we can apply Theorem 6.4 to obtain an expansion of (6.38) in powers of h. In order to describe it we have to identify the different quantities occurring in the general case with particular quantities occurring in our special case. This goes as follows. We take $H = H_0$ and $Y = \mathbb{R}^d$. By the definition of the singular locus, we have

$$S_W = \{ (\gamma, y) \in H_0 \oplus \mathbb{R}^d \mid \gamma = d_1 W(\gamma, y) \} .$$

But the paths satisfying $\gamma = d_1 W(\gamma, y)$ are precisely those for which γ is a stationary point of the phase function $\frac{1}{2}|\gamma|^2 - W(\gamma, y)$, as a function of γ, i.e. due to the definition $|\gamma|^2 = \int_0^t \dot{\gamma}(\tau)^2\, d\tau$ and $W(\gamma, y) = \int_0^t V(\gamma(\tau) + (y-x)(1 - \frac{\tau}{t}) + x)\, d\tau$, are those for which $\tilde{\gamma} \equiv \gamma + (y-x)(1 - \frac{\tau}{t}) + x$ satisfies Newton's equation

$$\frac{d^2}{d\tau^2} \tilde{\gamma} = - \nabla V(\tilde{\gamma}) \tag{6.47}$$

with $\gamma(t) = 0 = \gamma(0)$ i.e. $\tilde{\gamma}(t) = x$, $\tilde{\gamma}(0) = y$.
The mapping J is given by $(\gamma, y) \to (y, -d_2 W(\gamma, y))$. But on S_W we have $d_2 W(\gamma, y) = -d_y S_{t,x;0,y}(\tilde{\gamma})$, $S_{t,x;0,y}(\tilde{\gamma})$ being the classical action

$$S_{t,x;0,y}(\tilde{\gamma}) = \frac{1}{2} \int_0^t \dot{\tilde{\gamma}}^2(\tau)\, d\tau - \int_0^t V[\tilde{\gamma}(\tau)]\, d\tau \tag{6.48}$$

along the path $\tilde{\gamma}$ that solves Newton's equation (6.47). Hence J is given on S_W by

$$J(\gamma, y) = (y, d_y S_{t,x;0,y}(\tilde{\gamma})) . \tag{6.49}$$

But from (6.48)

$$d_y S_{t,x;0,y}(\tilde{\gamma}) = \int_0^t \frac{1}{2} \frac{d\tilde{\gamma}}{d\tau} \frac{d}{d\tau} (d_y \tilde{\gamma}(\tau) - 2\nabla V(\tilde{\gamma})\, d_y \tilde{\gamma}(\tau))$$

hence

$$d_y S_{t,x;0,y}(\tilde{\gamma}) = -\frac{1}{2} \int_0^t [\ddot{\tilde{\gamma}} + 2\nabla V(\tilde{\gamma})]\, d_y \tilde{\gamma}(\tau)\, d\tau - \dot{\tilde{\gamma}}(0) d_y \tilde{\gamma}(0) \qquad (6.50)$$

and using (6.47) and $\tilde{\gamma}(0) = y$ we then get from (6.50)

$$d_y S_{t,x;0,y}(\tilde{\gamma}) = - \dot{\tilde{\gamma}}(0). \qquad (6.51)$$

Thus the image Λ_W of S_W under J is given by $(y, -\dot{\tilde{\gamma}}(0))$
Now $\dot{\tilde{\gamma}}(0)$ is the momentum of a classical particle that starts at y at time 0 and ends at x at time t. Let $p_{cl}^x(y) \equiv -\dot{\tilde{\gamma}}(0)$: this is the function of y described by the momentum, as a function of y, of a Newtonian particle being at y at time 0 and in x at time t, for x fixed. The interpretation of the conditions on ψ in Theorem 6.4 is then as follows. The transversality of the intersection of the Lagrange manifolds $(y, d_y \psi)$ and Λ_W is simply the condition that $\psi''(y) \neq p_{cl}^{x'}(y)$, where $'$ means derivative with respect to y. The points P in Theorem 6.4 coincide here with the points in the intersection of $(y, d\psi(y))$ and $\Lambda_W = (y, p_{cl}^x(y))$ i.e. are the points y such that a Newton particle with initial (time 0) momentum $p_{cl}^x(y) = \psi'(y)$ in y arrives at x at time 0. The result of Theorem 6.4 says that if the set of points P is discrete (which is the case e.g. under the assumptions of [32]; these include V analytic, which follows from our assumption

$$\int e^{|\varepsilon| n}\, d|\mu_V|(\varepsilon) < \infty \quad \text{for some } n > 0 \text{)} \qquad (6.52)$$

and if $\det\left(\left(\frac{\partial \tilde{\gamma}_k}{\partial y_l}(P,t)\right)\right) \neq 0$, where $k, l = 1, \ldots, d$ are the components of vectors in $\mathbb{R}^d$ (this expresses the non degeneracy required for Theorem 6.4), then one has an asymptotic expansion of the solution $\psi(t,x)$ of the time-dependent Schrödinger's equation on $[0,t]$ in $\mathbb{R}^d$ in powers of h. Namely we have the following

<u>Theorem 6.5</u>. Assume $\tilde{\chi} \in F(\mathbb{R}^d) \cap C_0^\infty(\mathbb{R}^d)$, $\psi \in C^\infty(\mathbb{R}^d)$ and $V \in F(\mathbb{R}^d)$ with corresponding measure μ_V satisfying $\int e^{|\varepsilon| n}\, d_{|\mu_V|}(\varepsilon) < \infty$ for some $n > 0$. The solution in x of the time dependent Schrödinger equation on $[0,t]$ in $\mathbb{R}^d$ with initial condition $\psi(0,y) = e^{-\frac{i}{\hbar}\psi(y)} \tilde{\chi}(x)$ is given by

$$\sum_P \frac{1}{(2\pi i h)^{d/2}} e^{\frac{i}{2h}(x-y)^2} \frac{1}{\left|\det \frac{\partial \tilde{\gamma}_k}{\partial y_l}(P,t)\right|^{1/2}} e^{-\frac{i}{2}\pi n(\tilde{\gamma},y)}$$

$$e^{\frac{i}{h} S(\tilde{\gamma}, y_0)} \tilde{\chi}(y) + O(h),$$

where the functions under the sum are evaluated at the points P with coordinate y in $\mathbb{R}^d$ such that a Newtonian particle starting at y with momentum $\psi'(y)$ arrives at x at time t (and follows the path $\tilde{\gamma}$). It is assumed that x, t, ψ are such that $\psi''(y) \neq p_{cl}^{x\,\prime}(y)$, where $p_{cl}^{x}(y)$ is the momentum of a particle that moving according to Newton's equations goes from $(0, y)$ to (t, x) in time t. $n(\tilde{\gamma}, y)$ is the Maslov index of the classical path $\tilde{\gamma}$, starting at y at time 0 with momentum $\psi'(y)$ and ending at x at time t.

Remark : One has $\psi'(y) = p_{cl}^{x}(y)$ for y the coordinate of P, but otherwise in general $p_{cl}^{x}(y) \neq \psi'(y)$.

Remark : For conditions s.t. there are finitely many P see e.g. [32] .

6.5. Asymptotic expansion for the case of normalized integrals with respect to a quadratic form, and a trace formula for anharmonic oscillators.

We shall now shortly indicate the type of results obtained in [12] for the asymptotic expansion of the normalized integrals with respect to a non degenerate, not necessarily positive - or negative - definite quadratic form described in Section 4 (with applications given in Section 5). These are thus oscillatory integrals of the form

$$I(h) \equiv \int_H^{\Delta} e^{\frac{i}{2h}(\gamma, B\gamma)} e^{-\frac{i}{h}W(\gamma)} g(\gamma)\, d\gamma , \tag{6.53}$$

where $\mathrm{Im}\, h \leq 0$ and where (H, D, B, Δ) is a Fresnel fourtuple in the sense of Section 4 and $W, g \in F(D^*)$. The expansion of $I(h)$ around $h = 0$ is obtained along similar lines as the one for $B \equiv 1$ described in Sections 6.1 - 6.3 above. The following theorem is proven in [12] , along similar lines as Theorem 6.1 above :

Theorem 6.6. Assume the complex measure μ_W on D of which W is the Fourier transform satisfies $\int_D d\mu_W(\alpha) = \int_D \Delta(\gamma, \alpha)\, d\mu_W(\alpha) = 0$ for all $\gamma \in D$

and moreover

$$\int_D e^{\sqrt{2C}\,N(\alpha)}\, d|\mu_W|\ (\alpha) < 1$$

for some seminorm $N(\cdot)$ such that $|\Delta\ (\alpha,\alpha)| \leq C\ N(\alpha)^2$ for all $\alpha \in \text{supp } d\mu_W \cap \text{supp } d\mu_g$, where μ_g is the measure on D of which g is the Fourier transform of. Suppose in addition $\int_D e^{\sqrt{2C}\,N(\alpha)}\, d|\mu_g|\ (\alpha) < \infty$. Then $I(h)$, as defined by (6.53), is analytic on the half-plane $\text{Im } h < 0$ and it is a C^∞-function of h on the real axis $\text{Im } h = 0$.

Moreover we have the asymptotic expansion in powers of h

$$I\ (h) = \sum_{m=0}^{N} h^m \left(-\frac{i}{2}\right)^m \sum_{n=0}^{\infty} \left(-\frac{1}{2}\right)^n \frac{1}{n!\ (n+m)\ !}$$

$$\left\{ \int_D \ldots \int_D \left[\Delta\left(\sum_{j=1}^{n} \alpha_j + \beta\ ,\ \sum_{k=1}^{n} \alpha_k + \beta\right)\right]^{n+m} \prod_{j=1}^{n} d\mu_W\ (\alpha_j)\ d\mu_g\ (\beta) \right\} + R_N$$

for any arbitrary integer N, with the remainder R_N satisfying

$$|R_N| \leq |h|^{N+1}\ (N+1)\ !\ \left[1 - \int_D e^{\sqrt{2C}\,N(\alpha)}\ d\,|\mu_W|\ (\alpha)\right]^{-N-2}$$

$$\int_D e^{\sqrt{2C}\,N(\alpha)}\ d\,|\mu_g|\ (\alpha)\ .$$

<u>Remark</u> : The expansion can also be written replacing the term in { } by

$$\left\{ \Delta\left(\sum_{j=1}^{n} \nabla_{\gamma_j} + \nabla_\gamma,\ \sum_{k=1}^{n} \nabla_{\gamma_k} + \nabla_\gamma\right)\right]^{n+m} W(\gamma_1)\ldots W(\gamma_n)\, g\,(\gamma) \right\} \tag{6.54}$$

evaluated at $\gamma_1 = \gamma_2 = \ldots = \gamma_n = \gamma = 0$. In the case where $\text{supp } d\mu_W \subset$ (range B) each term $\Delta\ (\cdot,\cdot)$ in the theorem and in (6.54) can be replaced by $(\cdot\,,\ B^{-1}\ \cdot)$. Moreover if B^{-1} is a bounded operator on H then $H = D$, $\Delta = B^{-1}$ and C, $N(\gamma)$ can be taken to be $\|B^{-1}\|$ resp. $|\gamma|$, where $\|\ \|$ and $|\ |$ are the operator-resp. the norm in H. In this case, the point $\gamma = 0$ is the unique stationary point of the phase function

$$\Phi\ (\gamma) \equiv \frac{1}{2}\ (\gamma,\ B\gamma) - W\,(\gamma)\ ,$$

and this point is regular in the sense that $d^2\Phi(0)\gamma \neq 0$ for all $\gamma \in D(B)$, $\gamma \neq 0$. $\gamma = 0$ is the fixed point of the mapping $\gamma \to B^{-1}\, dW(\gamma)$. In fact the results in the case where B^{-1} is bounded extend also to the case where the stationary point is arbitrary and to the case where there are more stationary points. In fact all results proven for the case $B = 1$ extend to this more general case.

Remark : In [12] we also discuss the case of oscillatory integrals where B is non necessarily real. This gives an answer to questions raised in recent work on steepest descent methods in quantum mechanics and quantum field theory (e.g. [44]). The methods of [12] are then used in [45], in collaboration with Ph.Blanchard, to derive a trace formula for the trace of the Green function for the Schrödinger equation of a anharmonic oscillator in terms of normalized integrals over an Hilbert space of periodic paths. The asymptotic expansions in powers of h of such integrals (by the methods of [11],[12]), yields then a "Poisson formula" for the trace, whereby the eigenvalues of the Schrödinger operator are put in relation with the periodic orbits of the corresponding classical anharmonic oscillator.

ACKNOWLEDGEMENTS

It is a pleasure to acknowledge stimulating discussions with Professors Ph. Blanchard, Ph. Combe, J.M. Combes, A. Grossmann, P. Holm, E. Mourre, G. Rideau, R. Rodriguez, R. Seiler, M. Sirugue, M. Sirugue-Collin. We are also very grateful to Mrs. Burghardt for her great patience with a difficult manuscript and for her nice typing. We are very indebted to the CNRS-CPT and the Université d'Aix-Marseille II for their kind hospitality.

- FOOTNOTES -

§2

1) This property can be used to extend the definition of the normalized or Fresnel integral to classes of functions larger than $F(H)$, in particular containing polynomial functions. See e.g. [14] , [21] , [19] , [18] .

Unfortunately, for applications to quantum mechanical problems, the functions to be integrated being of the form $f(\gamma) = \exp(W(\gamma))$, these extensions are not as powerful as might be expected from the finite dimensional situation. In fact our space $F(H)$ for W is difficult to be extended preserving the discussion of the asymptotic expansion in powers of h as $h \to 0$, when $|\gamma|^2$ is replaced by $\frac{1}{h}|\gamma|^2$ (see below).

2) Transformation properties under mappings of the form $\gamma \to T\gamma = A\gamma + \alpha$ with $\mathrm{Tr}(A^*A - 1)^{\beta/2} < \infty$ for some $\beta < 1$ are known ([14] ; see also e.g. [21] , [18] , [23]).

§3

1) Another possibility is to invert the roles, i.e. take the term involving V as the "known part" and consider in a sense the quadratic term as the perturbation. This second point of view has been pursued by Varadhan [25] and Chebotarev and Maslov [41] (see also the contribution to these Proceedings).

2) Wiener measure on paths $w(\tau)$ in the interval $[0,t]$ is formally given by $e^{-\frac{1}{2}\int_0^t \dot{w}(\tau)^2 d\tau} dw(\tau)$ and it is well known that the kinetic energy $\frac{1}{2}\int_0^t \dot{w}(\tau)^2\, d\tau$ is Wiener a.s. infinite. This is taken advantage in the theory of integration with respect to Wiener measure by allowing e.g. for singular positive potentials. However for "oscillatory integrals" of the form (3.3), which are not integrals with respect to measures, one cannot take advantage of support properties of measures. The reason for choosing a linear space of paths of finite kinetic energy is that it permits to compute easily the measure of which

$$e^{-\frac{i}{2}\int_0^t \dot{\gamma}(\tau)^2 d\tau} \;,\quad e^{-i\int_0^t V(\gamma(\tau))\, d\tau}$$

and $\varphi(\gamma)$ are Fourier transforms. See below. H can of course also be interpreted as the Sobolev space H_2^1 on the interval $[0,t]$. See also [13].

3) (H,Γ) is a reproducing kernel Hilbert space.

4) E.g. $V \in L^1(\mathbb{R}^d) \cap L^\infty(\mathbb{R}^d)$, or, for $d = 3$,
$V \in L_{3/2}(\mathbb{R}^3)$, with sufficiently small norms. See e.g. [42].

5) One can also always add a linear term ax+b to the potential ($e^{i(ax+b)} \in F(\mathbb{R}^d)$!)

§5

1) We could as well consider the case of operators of the form

$$\sum_{j=1}^{N} \frac{1}{2m_j} \Delta_{\vec{x}_j} + \frac{1}{2} \sum_{j,n} \vec{x}_j A_{jn}^2 \vec{x}_n + V(\vec{x}_1,\dots,\vec{x}_N)$$

with $A = ((A_{jn}))$, $j, n = 1, \dots, d$; $\vec{x}_j \in \mathbb{R}^s$, V being a superposition of $1, 2, \dots, \nu$-body potentials, also allowed to be translation invariant.

2) Such interactions were first studied in [43].

3) Formally $\frac{d}{dt}(U_t^{\Lambda,\varepsilon}\Phi)(\xi(\vec{x})) = i[-\int_{\mathbb{R}^s} \frac{\delta^2}{\delta\xi(\vec{x})^2} d\vec{x} + \sum_i \int_{\mathbb{R}^s} (\frac{\partial\xi}{\partial x_i}(\vec{x}))^2 d\vec{x} +$
$+ m^2 \int_{\mathbb{R}^s} \xi(\vec{x})^2 d\vec{x} + \int_\Lambda v(\xi_\varepsilon(\vec{x}))\, d\vec{x}]\; U_t^{\Lambda,\varepsilon}\Phi(\xi(\vec{x}))$.

§ 6

1) $C^\infty(H \times \mathbb{R}^k)$ being a Baire space in the Whitney topology, such a set is in particular dense in $C^\infty(H \times \mathbb{R}^k)$.

2) See [12].

3) For other discussions of the classical limit $h \to 0$ of quantum mechanics using the path integrals, see A. Truman's and C. Dewitt-Morette's contributions to these proceedings.

- REFERENCES -

[1] R.P. FEYNMAN, Space-Time Approach to Non-Relativistic Quantum Mechanics, Rev. Mod. Phys. 20, 367-387 (1948).

[2] S. ALBEVERIO, R. HØEGH-KROHN, Mathematical Theory of Feynman Path Integrals, Lecture Notes in Mathematics, 523 , Springer , Berlin (1976).

[3] S. ALBEVERIO, Mathematical Theory of Feynman Path Integrals, Acta Univ. Wratisl. 368 (1976) , XIIth Winter School of Theoretical Physics in Karpacz (1975).

[4] P.A.M. DIRAC, The Lagrangian in Quantum Mechanics, Phys. Zeitschr. d. Sovyetunion 3 , n° 1, 64-72 (1933).
See also the references in [2] , [3] .

[5] M. KAC, On a Distribution of Certain Wiener Functionals, Trans. Am. Math. Soc. 65 , 1-13 (1949).

[6] N. WIENER, Differential Space, J. Math. and Phys. 58, 131 - 174 (1923).

[7] R.H. CAMERON, A Family of Integrals Serving to Connect the Wiener and Feynman Integrals, J. Math. and Phys. 39 , 126-141 (1960).

[8] J. FELDMAN, On the Schrödinger and Heat Equations, Trans. Am. Math. Soc. 10, 251-264 (1963).

[9] D.G. BABBITT, A Summation Procedure for Certain Feynman Integrals, J. Math. Phys. 4 , 36-41 (1963).

[10] E. NELSON, Feynman Integrals and the Schrödinger Equation, J.Math.Phys. 5, 332-343 (1964).
See also e.g. the references under [10] , in Ref. [2] .

[11] S. ALBEVERIO, R. HØEGH-KROHN, Oscillatory Integrals and the Method of Stationary Phase in Infinitely Many Dimensions, with Applications to the Classical Limit of Quantum Mechanics I , Inventiones Mathem. 40, 59-106 (1977).

[12] S. ALBEVERIO, Ph. BLANCHARD, R. HØEGH-KROHN, Oscillatory Integrals and the Method of Stationary Phase in Infinitely Many Dimensions II. (in preparation)

Some results of [11] were also shortly described in [3] . For a different approach to the classical limit, using an extended definition of Feynman path integrals, see

[13] A. TRUMAN, Feynman Path Integrals and Quantum Mechanics as $h \to 0$, J. Math.Phys. 17, 1852-1862 (1977),

and these Proceedings.

[14] K. ITO, Generalized Uniform Measures in the Hilbertian Metric Space with their Application to the Feynman Path Integral, Proc. Fifth Berkeley Symposium on Mathematical Statistics and Probability, Univ. California Press, Berkeley, vol. II, part 1, 145-161 (1967).

[15] C. DEWITT-MORETTE, Feynman's Path Integral. Definition without limiting Procedure, Commun.math.Phys. 28, 47-67 (1972) ; I. Linear and Affine Techniques, II. The Feynman-Green Function, Commun.math.Phys. 37, 63-81 (1974).

[16] P. KREE, Théorie des distributions et calculs différentiels sur un espace de Banach, Séminaire P. Lelong, 15e année, Paris 1974-75.

See also P. Krée's contribution to these Proceedings.

For another approach using finitely additive complex measures, see e.g.

[17] Yu. L. DALETSKII, Continuous Integrals Connected with Certain Differential Equations, Dokl. Ak. Nauk., 137, 268 (1961),
D.N. DUDIN, Generalized measures or distributions on Hilbert space, Trans. Mosc. M. Soc. 28, 133 - 157 (1973) (transl.).

[18] A.TRUMAN, The classical action in nonrelativistic quantum mechanics, J.Math.Phys. 18, 1499-1509 (1977). See also A.Truman in these Proceedings.

[19] Ph. COMBE, G. RIDEAU, R. RODRIGUEZ, M. SIRUGUE-COLLIN, On the Cylindrical Approximation of the Feynman Path Integral, Rep. Math. Phys. 13, n° 2, 279-294 (1978).

For further references connected with this line of work, see also e.g.

[20] I.M. GELFAND, A.M. YAGLOM, Integration in Functional Spaces, J.Math.Phys. 1, 48-69 (1960).

L. STREIT, An Introduction to Theories of Integration over Function Spaces, Acta Phys. Austr. Suppl. 2, 2-20 (1966).

and e.g.

[21] J. TARSKI, Definitions and Selected Applications of Feynman-Type Integrals, pp. 169-180, in "Functional Integration and its Applications", A.M. Arthurs Edit., Oxford (1975).

See also these Proceedings.

[22] S. ALBEVERIO, R. HØEGH-KROHN, The Schrödinger Equation for the Relativistic Quantum Fields, in preparation.

[23] K. BROCK, On the Feynman Integral, Aarhus University, Mathemat. Inst. Various Publ. Series, n° 26 (Oct. 1976).

[24] R. H. CAMERON, D.A. STORVICK, An Operator Valued Function Space Integral and a Related Integral Equation, J. Math. and Mech. 18, 517-552 (1968).

[25] S.R.S. VARADHAN, unpublished.

[26] K. GAWEDZKI, Construction of quantum mechanical dynamics by means of path integrals in phase space, Rep. Math. Phys. 6, 327 - 342 (1974)

[27] W. GARCZYNSKI, Quantum Stochastic Processes and the Feynman Path Integral for a Single Spinless Particle, Repts. Math. Phys. 4 , 21-46 (1973).

[28] G.N. GESTRIN, On Feynman Integral, Izd. Kark. Univ. 12, 69-81 (1970).

[29] G.W. JOHNSON, D.L. SKOUG, A Banach Algebra of Feynman Integrable Functionals with Applications to an Integral Equation formally equivalent to Schrödinger's Equation, J. Funct. Anal. 12, 129-152 (1973).

[30] L.D. FADDEEV, P.P. KULISH, Quantization of Particle-Like Solutions in Field Theory, pp. 270-278, "Mathematical Problems in Theoretical Physics", Proceedings, Rome 1977, Edts. G. Dell'Antonio, S. Doplicher, G. Jona-Lasinio, Lecture Notes in Physics 80, Springer, Berlin (1978).

[31] R. HØEGH-KROHN, Relativistic Quantum Statistical Mechanics in Two-Dimensional Space-Time, Commun.math. Phys. 38, 195-224 (1974).

[32] V.P. MASLOV, The Quasi-Classical Asymptotic Solutions of some Problems in Mathematical Physics, I, J. Comp. Math. 1 , 123-141 (1961)(transl.); II, J. Comp. Math. 1 , 744-778 (1961) (transl.).

See also e.g.

[33] V.P. MASLOV, Théorie des perturbations et Méthodes asymptotiques, Dunod, Paris (1972) (transl.).

[34] L. HÖRMANDER, Fourier Integral Operators I, Acta Math. 127, 79-183 (1971).

[35] J.J. DUISTERMAAT, Oscillatory Integrals, Lagrange Inversions and Unfolding of Singularities, Comm. Pure Appl. Math. 27, 207-281 (1974).

[36] V.I. ARNOLD, Remarks on the Stationary Phase and Coxeter Numbers, Russ. Math.Surv. 28, 19-48 (1973).

[37] J. LERAY, Solutions asymptotiques et groupe symplectique. In Fourier Integral Operators and Partial Differential Equations, pp. 73-97, in Lecture Notes in Mathematics 459, Springer, Berlin (1975).

See also e.g. [38] , [39].

[38] I.N. BERNSHTEIN, Modules over a Ring of Differential Operators. Study of the Fundamental Solution of Equations with Constant Coefficients, Funct. Anal. and its Appl. 5 (2) , 89-101 (1971).

B. MALGRANGE, Integrales asymptotiques et monodromie, Ann. Scient. Ec. Norm. Sup. 4e S., 7, 405-430 (1974).

[39] V. GUILLEMIN, S. STERNBERG, Geometric Asymptotics, Am. Math. Soc., Providence (1978).

[40] See e.g. [7] - [10] ,[13] - [29].

[41] V.P. MASLOV, A.M. CHEBOTAREV, Generalized Measures and Feynman Path Integrals, Teor. i Mat. Fyz. 28, 3, 291-306 (1976) (russ.).

[42] R. HØEGH-KROHN, Partly Gentle Perturbations with Application to Perturbation by Annihilation. Creation Operators, Proc. Nat. Ac. Sci. 58 , 2187-2192 (1967).

[43] S. ALBEVERIO, R. HØEGH-KROHN, Uniqueness of the Physical Vacuum and the Wightman Functions in the Infinite Volume Limit for some Non Polyno-

mial Interactions, Commun.math.Phys. 30, 171-200 (1973).

[44] J. R. KLAUDER, Continuous Representations and Path Integrals, reviseted, Lecture Notes for the NATO Advanced Study Institute on Path Integrals and their Application, Antwerp, Belgium, July 17-30, 1977.

[45] S.ALBEVERIO, Ph.BLANCHARD, R.HØEGH-KROHN, The Poisson formula and the ζ-function for the Schrödinger operators, in preparation.

PROCESSUS DE SAUTS ET LEURS APPLICATIONS DANS LA MECANIQUE QUANTIQUE

A.M. CHEBOTAREV, V.P. MASLOV
Institut des Constructions Electroniques de Moscou
MOSCOU, URSS

Dans cet exposé nous allons décrire les applications de la théorie des processus de sauts au problème fondamental de la mécanique quantique nonrelativiste. Il s'agit du problème de Cauchy pour l'équation de Schrödinger. Nous allons illustrer sur l'exemple de l'équation de la chaleur la méthode proposée pour la représentation des solutions sous forme de moyenne des fonctionnelles des processus de sauts.

On sait que la solution de l'équation :

$$\partial u/\partial t \; = \; 1/2\Delta u + V(x)u$$

peut être écrite sous forme d'une intégrale fonctionnelle par rapport à une mesure w concentrée sur les trajectoires continues possédant la propriété de Lévy. La mesure w est absolument continue par rapport à la mesure W de Wiener et le potentiel $V(x)$ est responsable de la modification de la mesure w :

$$w(A) = \int_A \exp\{-\int_0^t V(x_s)\,ds\}\, W(dx_s)$$

Si le potentiel $V(x)$ est transformé de Fourier de la mesure complexe $\tilde{V}(dp)$ à variation bornée, on peut proposer un autre traitement de base [2]. Dans ce cas la solution de l'équation de la chaleur peut être écrite sous forme d'une intégrale fonctionnelle par rapport à une mesure m sur les trajectoires à sauts. La mesure m est une transformation absolument continue de la probabilité Π engendrée par la solution fondamentale de l'équation de Kolmogorov :

$$\partial Q(p,t)/\partial t = (2\pi)^{n/2} \int \{Q(u+p,t) - Q(p,t)\}\, |\tilde{V}|(du)$$

où $|\tilde{V}|(du)$ est la variation de la mesure complexe $\tilde{V}(du)$. Dans ce cas l'opérateur de Laplace Δ engendre une transformation absolument continue de la mesure Π :

$$m(B) = \int_B \exp\{-\int_0^t H(p_s)\,ds\}\, \Pi(dp_s)$$

où $H(p) = |p|^2/2$ et $H(-i\nabla) = -\Delta/2$.

La possibilité de deux telles représentations de la solution de l'équation de la chaleur équivaut à une sorte d'égalité de Parseval.

La restriction essentielle propre à la seconde représentation est une condition sur la variation de la mesure $\tilde{V}$ [2],[4],[7]-[12]. Cette condition exclut le cas d'un potentiel croissant et illimité et ne nous permet pas de dépasser le cadre d'une théorie de perturbations non stationnaires. De ce point de vue, la méthode proposée est un analogue non stationnaire du schéma de von Neumann [5].

D'autre part, la transformation absolument continue de la mesure m engendrée par un hamiltonien $H(p)$ nous donne non seulement la solution de l'équation de la chaleur mais celle de l'équation de Schrödinger :

$$\{-i\,\partial/\partial t + H(-i\nabla) + V(x)\}\Psi(x,t) = 0$$

où $H(p)$ est une fonction continue réelle.

La représentation de la fonction Ψ sous forme d'une moyenne de fonctionnelles de processus de sauts permet le calcul par la méthode de Monte-Carlo de la solution de l'équation de Schrödinger et l'évaluation de la dispersion [2] . Il importe que la méthode de Monte-Carlo soit stable par rapport à la norme $C(R^n)$, tandis que la construction des schémas aux différences stables par rapport à cette norme est un problème compliqué en mécanique quantique.

La méthode proposée peut être généralisée et appliquée également pour les équations quasilinéaires de type de Hartree :

$$i\,\partial\Psi/\partial t = \{H(-i\nabla) + V(x) + \langle\Psi,\hat{L}(x)\Psi\rangle\}\Psi$$

où $H(p)$ est une fonction continue réelle, $V(x)$ est la transformée de Fourier d'une mesure complexe à variation bornée et $\hat{L}(x)$ est un opérateur auto-adjoint pseudodifférentiel qui dépend du paramètre $x \in R^n$. $\langle\Psi,\hat{L}(x)\Psi\rangle$ est la moyenne de l'opérateur $\hat{L}(x)$ dans l'état Ψ . Dans ce cas la solution Ψ est égale à une moyenne de fonctionnelles particulières de prodessus de sauts qui ont une composante ramifiée. Nous allons examiner cette représentation dans cet exposé.

1. L'INTEGRALE DE FEYNMAN ET LA THEORIE D'INTEGRATION DE LEBESGUE.

La résolution du problème de Cauchy pour l'équation de Schrödinger

$$-ih\,\partial\Psi/\partial t - h^2/2m\ \Delta\Psi + V(x)\Psi = 0$$
$$\Psi(x,0) = \Psi_0(x) \in C_0^\infty(R^n) \tag{1.1}$$

peut être présentée formellement sous la forme bien connue de l'intégrale fonctionnelle de Feynman :

$$\Psi(x,t) = \int_{x_t=x} Dx_s\, \Psi_0(x_0) \exp i/h \int_0^t \{m|\dot{x}_s|^2/2 - V(x_s)\}ds =$$
$$= \lim_{N\to\infty} (m/2\pi itN^{-1}h)^{nN/2} \int\ldots\int \Psi_0(x_0) \prod_{k=0}^{N-1} dx_k \tag{1.2}$$
$$\times \exp i/h \{m(x_{k+1} - x_k)^2/2tN^{-1} - V(x_k)tN^{-1}\}$$

où $x_N = x$. L'existence de la limite (1.2) est démontrée pour une large classe de potentiels [6] mais la représentation de la solution sous forme d'intégrale sur les trajectoires n'a pas de fondement mathématique. Ces dernières années, on utilise la transformation de Fourier pour étudier l'intégrale fonctionnelle (1.2) [1], [3], [4], [7]-[9]. Nous allons décrire ici la méthode simple de [7], [8].

Après des transformations simples, la formule (1.2) peut être écrite :

$$\Psi(x,t) = \lim_{N\to\infty} \int\ldots\int (2\pi h)^{-nN} \prod_{k=1}^{N} dp_k \exp\{-i|p_k|^2 tN^{-1}/2mh\}$$
$$\times \exp\{ip_N x/h\} \times \int\ldots\int \prod_{k=1}^{N-1} dx_k \exp\{-i/h[V(x_k)tN^{-1} - p_k(x_{k+1} - x_k)]\}\Psi_0(x_T) = \tag{1.3}$$
$$= \int Dp_s \int_{x_t=x} Dx_s \exp i/h\{S(x_s,p_s) + p_t x\}\Psi_0(x_0),$$
$$\text{ou}\quad S(x_s,p_s) = \int_0^t p_s dx_s - |p_s|^2/2mds - V(x_s)ds.$$

La fonction $S(x_s, p_s)$ dans (1.3) est une action le long de la trajectoire sous forme hamiltonienne. C'est pourquoi l'intégrale fonctionnelle (1.3) s'appelle "la forme hamiltonienne de l'intégrale de Feynman".

Soit A l'algèbre de trajectoires à sauts qui est engendrée par les tribus

$$a_{s_1\ldots s_k} = \{p_r : p_r = p_{s_i},\ s_i \leqslant r < s_{i+1} ;\ p_0 = 0 ;$$
$$(p_{s_1} - p_{s_1-0}, \ldots, p_{s_k} - p_{s_k-0}) \in B;\ B \in B(R^{nk})\}$$
$$k = 1,2,\ldots$$

où $B(R^{nk})$ est la σ-algèbre borélienne de R^{nk}. Le fait remarquable est l'existence d'une mesure additive complexe m définie comme limite pour $N \to \infty$ d'une suite d'intégrales par rapport aux coordonnées $x_1 \dots x_N$:

$$m(a) = \lim_{j\to\infty} \lim_{N\to\infty} (1/2\pi h)^{nN} \int \dots \int \prod_{l=1}^{N} dp_l \, \chi_a^j(p_{s,N}) \int \dots \int \prod_{k=1}^{N} dx_k \times$$
$$\times \exp\{[V(x_k)\, tN^{-1} - p_k(x_{k+1} - x_k)]/ih\} \Psi_0(x_1),$$

où a est un élément de l'algèbre A, $\{\chi_a^j\}$ est une suite de fonctions bornées continues qui tend vers la fonction caractéristique de a et $p_{s,N}$ est une trajectoire à sauts qui produit les sauts $p_2 - p_1$, $p_3 - p_2, \dots,$ $p_N - p_{N-1}$ aux moments tN^{-1}, $2tN^{-1}$, $3tN^{-1}$, ..., $t - tN^{-1}$.

La fonction m admet un prolongement unique à la σ-algèbre A engendrée par l'algèbre A [7]. La fonctionnelle :

$$Q(p_s) = \exp\{-i \int_0^t |p_s|^2/2m \, ds\} \tilde{\Psi}_0(p_t) \tag{1.5}$$

où $\tilde{\psi}_0(p)$ est la transformée de Fourier de $\psi_0(x)$, est A-mesurable et bornée. Ainsi, après intégration relativement aux coordonnées d'espace Dx_s en (1.3) on obtient l'intégrale de Lebesgue :

$$\Psi(x,t) = (1/2\pi h)^{n/2} \int_{R^n} \exp\{ipx/h\} \, dp \int_{p_0 = p} m(dp_s) \, Q(p_s) \tag{1.6}$$

que nous appelons "p-représentation de l'intégrale de Feynman".

<u>Théorème 1.1.</u> Soit $\psi_0(p)$ une fonction intégrable, soit $V(x)$ une transformée de Fourier d'une mesure complexe à variation bornée et soit $H(p)$ une fonction continue réelle. Alors la solution du problème (1.1) peut être écrite sous forme d'intégrale de Lebesgue (1.6).

La fonction

$$\exp\{-ih^{-1} \int_0^t V(x_s)\, ds\}$$

est une fonction caractéristique de la mesure complexe m.

On peut trouver ces résultats et leurs démonstrations dans les publications [1], [4], [7]-[9].

Et maintenant nous écrirons l'intégrale (1.6) sous la forme d'une moyenne de fonctionnelles (1.5) et le lien sera établi de cette façon avec la théorie des probabilités.

2. LE PROCESSUS DE SAUTS CORRESPONDANT AU POTENTIEL V(x) .

Supposons qu'il existe une mesure complexe $\tilde{V}(dp)$ à variation bornée qui satisfait la condition suivante :

$$V(x) = (2\pi)^{-n/2} \int \exp\{ipx\} \tilde{V}(dp)$$

Puisque la mesure $\tilde{V}$ est absolument continue par rapport à la mesure positive bornée $|\tilde{V}|$, on a

$$\tilde{V}(B) = \int_B f(p) |\tilde{V}| (dp) , \qquad B \in B(R^n)$$

où $f(p)$ est une fonction $|\tilde{V}|$ -intégrable. La mesure $\tilde{V}$ est également absolument continue par rapport à la mesure positive bornée Q :

$$Q(B) = (2\pi)^{-n/2} \int_B |f(p)| \; |\tilde{V}| (dp)$$

d'où on a

$$d\tilde{V}/dQ = (2\pi)^{n/2} f(p)/|f(p)| = (2\pi)^{n/2} \exp i\Phi(p) ,$$
$$\Phi(p) = \arg f(p) \pmod{2\pi}.$$

Cela signifie que la mesure $\tilde{V}$ peut être écrite sous la forme suivante

$$\tilde{V}(B) = (2\pi)^{n/2} \int_B e^{i\Phi(p)} Q(dp) \tag{2.1}$$

où $\Phi(p)$ est une fonction réelle et $B(R^n)$ -mesurable. On dit que Φ est la phase de la mesure $\tilde{V}$.

D'après la définition des opérateurs pseudodifférentiels [10] on déduit de (2.1) une représentation intégrale de l'opérateur $V(i\nabla_p)$:

$$V(ih\nabla_p) \Psi(p) = \int_{R^n} e^{i\Phi(u)} Q(du) f(p+hu) \tag{2.2}$$

Maintenant nous donnerons la définition du processus de sauts correspondant à la mesure Q .

Soit $n(B,\Delta)$ une mesure de Poisson sur $R^n \times [0, T]$ telle que $M\, n(B,\Delta) = \Delta\, h^{-1} Q(B)$. Pour cette mesure nous posons

$$q_{sp}(t) = p + h \int_s^t u\, n(du, ds), \qquad u, p \in R^n \tag{2.3}$$

avec les notations :

$$q_{op}(t) = q_p(t), \quad q_{so}(t) = q_s(t), \quad q_{oo}(t) = q(t)$$

Le processus $q_{sp}(t)$ est un processus de sauts markovien tel que :

(i) La probabilité conditionnelle pour que le saut appartienne à B, lorsqu'il y a saut, est égale à :

$$P(B) = Q(h^{-1} B) / Q(R^n);$$

(ii) La probabilité qu'il y ait k sauts au cours du temps Δ est égale à :

$$p_k = (\Delta h^{-1} c)^k / k!\, \exp\{-\Delta c/h\}$$

où $c = Q(R^n)$.

Voici alors l'énoncé du théorème général de cette section :

<u>Théorème 2.1.</u> Supposons satisfaites les conditions du théorème 1.1. Soient Φ la phase de la mesure $\tilde{V}$ et q_{sp} le processus de sauts (2.3). Alors, on a la formule suivante pour la solution du problème (1.1) :

$$\Psi(x,t) = (2\pi h)^{-n/2} \int e^{ipx/h}\, \tilde{\Psi}(p,t)\, dp,$$

$$\tilde{\Psi}(p,t) = \exp(cth^{-1})\, M_t \exp\Big\{-ih^{-1} \int_0^t H(q_{sp}(t))\, ds + i \int_0^t \int_{R^n} (\Phi(u) - \pi/2)\, n(du, ds)\Big\} \times \tilde{\Psi}_0(q_p(t)) \tag{2.4}$$

Dans l'exposé [2] nous avons utilisé cette formule pour le calcul de $\tilde{\Psi}(p,t)$ par la méthode de Monte-Carlo.

Nous donnerons ici une démonstration du théorème 2.1 lorsque :

$$\sup_p |H(p)| < \infty .$$

Comme le processus $q_{sp}(t)$ est un processus markovien on a

$$\tilde{\Psi}(p,t+\Delta) = e^{c\Delta h^{-1}} M_{t+\Delta} \exp\{-ih^{-1}\int_t^{t+\Delta} H(q_{sp}(t+\Delta))ds +$$
$$+ i\int_t^{t+\Delta}\int_{R^n} (\Phi(u) - \pi/2)\, n(du, ds)\}\, \tilde{\Psi}(p + q_t(t+\Delta), t).$$

La quantité $q_t(t+\Delta)$ est la somme des sauts de q_{sp} au cours du temps Δ : s'il n'y a pas de sauts, alors $q_t(t+\Delta) = 0$; si le processus q_{sp} produit un saut de valeur u , alors $q_t(t+\Delta) = u$. La probabilité pour que les trajectoires aient plus qu'un saut est égale à $0(\Delta^2)$.

Désignons par $M^{\Delta,k}$ la moyenne conditionnelle sous la condition que le processus q_{sp} produise k sauts au cours du temps Δ . Alors on obtient :

$$M_{t+\Delta} = (1 - c\Delta h^{-1})\, M^{\Delta,0} + c\Delta h^{-1} M^{\Delta,1} + O(\Delta^2)$$

Si le processus $q_{sp}(t)$ produit k sauts $hu_1, \ldots, hu_k$ au cours du temps Δ on a

$$\int_t^{t+\Delta}\int_{R^n} (\Phi(u) - \pi/2)\, n(du, ds) = \sum_{j=1}^{k} (\Phi(u_j) - \pi/2)$$

en accord avec la définition de la mesure de Poisson n . Donc

$$\tilde{\Psi}(p,t+\Delta) = e^{c\Delta h^{-1}} (1 - c\Delta h^{-1})\, e^{-ih^{-1}H(p)\Delta}\, \tilde{\Psi}(p,t) +$$
$$+ c\Delta h^{-1} \int Q(du)/Q(R^n)\; \tilde{\Psi}(p + hu,t) \exp i(\Phi(u) - \pi/2) + O(\Delta^2)$$

En utilisant (2.2) on obtient :

$$\tilde{\Psi}(p,t+\Delta) = \tilde{\Psi}(p,t) + (ih)^{-1}\, \Delta\{ H(p) + V(ih\nabla_p)\}\, \tilde{\Psi}(p,t) + O(\Delta^2)$$

et on en déduit l'équation de Schrödinger (1.1) dans la représentation -p :

$$\{-ih\, \partial/\partial t + H(p) + V(ih\nabla_p)\}\, \tilde{\Psi}(p,t) = 0$$

Si $H(p)$ est une fonction continue croissante réelle, la démonstration de ce théorème repose sur le fait suivant. La mesure des trajectoires des processus $q_{sp}(t)$ décroît quand la valeur maximale de $q_{sp}(t)$ augmente :

$$\lim_{M\to\infty} P\{\sup_s |q_{sp}(t)| \geq M\} = 0$$

Donc dans la moyenne (2.4) on peut exclure la contribution des trajectoires trop grandes, car alors la valeur de la moyenne change arbitrairement peu. Sur les trajectoires qui restent, la fonction $H(p)$ est bornée, et la démonstration précédente est valable.

Appliquons le théorème sur la commutativité de l'application d'un opérateur borné et de l'intégration par rapport à un paramètre pour établir la formule donnant la solution du problème (1.1).

Soit $\tilde{\Psi}_0(p) \in C(R^n) \cap L_2(R^n)$. D'après la définition du processus $q_{sp}(t)$ on a : $q_{sp} = q_s + p$. Donc la fonctionnelle

$$\exp\{(ih)^{-1}\int_0^t H(q_s(t)+p)ds\}\tilde{\Psi}_0(q(t)+p)$$

est une fonction du paramètre $p \in R^n$, qui appartient à $L_2(R^n)$. Alors on a

$$\|\tilde{\Psi}(\ ,t)\|_{L_2} \leqslant \exp\{cth^{-1}\}\|\tilde{\Psi}_0\|_{L_2}$$

La transformation de Fourier est un opérateur unitaire dans l'espace $L_2(R^n)$. Donc le théorème sur la commutativité de l'application d'un opérateur borné et de l'intégration par rapport à un paramètre donne, pour la solution de (1.1), la formule :

$$\Psi(x,t) = \exp(cth^{-1})\, M_t\{\exp i\int_0^t\int_{R^n}(\Phi(u)-\pi/2)\, n(du,ds)\times$$
$$\times(2\pi h)^{-n/2}\int_{R^n} dp\, e^{ixp/h}\exp[-ih^{-1}\int_0^t H(q_s(t)+p)ds]\,\tilde{\Psi}_0(q(t)+p)\}$$

Dans ce cas, $\Psi(\ ,t) \in L_2(R^n)$. La formule (2.5) est aussi valable si $\tilde{\Psi}_0(p)dp$ est une mesure complexe à variation bornée. Alors $\Psi(\ ,t) \in C(R^n)$.

3. EQUATION DE LA CHALEUR.

Nous allons appliquer cette méthode au problème de Cauchy pour l'équation de la chaleur :

$$\begin{aligned}&\{-\partial/\partial t + \Delta/2 + V(x)\}R(x,t) = 0\\ &R(x,o) = R_o(x) \in C_o(R^n)\end{aligned} \qquad (3.1)$$

où $V(x)$ est la transformée de Fourier d'une mesure complexe à variation bornée. Soit $q_{sp}(t)$ un processus de sauts (2.3) et soit $h = 1$. Si, dans (2.5) et (1.1), $H(p) \longrightarrow p^2/2i$, $V(x) \longrightarrow V(x)\, e^{i\pi/2}$, $h \to 1.$, alors le problème (1.1) devient le problème (3.1) et la fonction (2.5) se transforme en sa solution. Si la fonction $R_0(x)$ est une fonction réelle, alors $R(x, t)$ est réelle aussi. Donc la formule (2.5) nous donne :

$$R(x, t) = \exp(-ct)\, M_t \cos\Big(\int_0^t \int_{R^n} \Phi(u)\, n\,(du, ds) \Big) \times$$
$$\times (2\pi t)^{-n/2} \int_{R^n} R_0(y)dy\, \exp\Big\{-1/2t \,\Big| \int_0^t q_s(t)ds - x + y \Big|^2 \Big\} \tag{3.2}$$

4. L'EQUATION DU TYPE DE HARTREE ET LES PROCESSUS DE SAUTS RAMIFIES.

Dans cette section, nous allons décrire un processus de sauts ramifié et une fonctionnelle de ce processus pour lesquels la moyenne est égale à la transformée de Fourier de la solution de l'équation :

$$\Big\{-i\partial/\partial t + H(-i\nabla_x) + V(x) + \int \Psi^*(q, t)\, L\,(x, -i\overset{1}{\nabla}_q, \overset{2}{q})\, \Psi(q, t)\, dq \Big\}\, \Psi(q, t) = 0 \tag{4.1}$$

Nous appellerons cette équation "équation du type de Hartree".

Supposons que $H(p)$ est une fonction continue réelle, $V(x)$ est la transformée de Fourier d'une mesure à variation bornée. Alors $V(x)$ admet la représentation intégrale :

$$V(x) = (2\pi)^{+n/2} \int_{R^n} e^{i(\Phi_c(p) - px)}\, Q(dp)$$

où $\Phi_c(p)$ est une fonction mesurable réelle et $Q(dp)$ une mesure positive bornée. Supposons qu'il existe une mesure positive bornée $D(dp, dv, du)$ sur $R^n \times R^n \times R^n$ telle que

$$L(x, -i\overset{1}{\nabla}_q, \overset{2}{q})\, \Psi(q) = (2\pi)^{-n} \int_{R^n} e^{iqv}\, L(x, dv, q) \int_{R^n} e^{-iq'v}\, \Psi(q')dq'$$

$$L(x, B, q) = (2\pi)^{n/2} \int_{R^n} \int_B \int_{R^n} e^{i[\Phi_R(p, v, u) - px + uq]}\, D(dp, dv, du)$$

où $\Phi_R(p, v, u)$ est une fonction réelle et mesurable. Alors le terme quasilinéaire de (4.1) a dans la représentation-p la forme intégrale :

$$\langle \Psi, L(i\overset{1}{\nabla}_{p'}, -i\nabla_q, \overset{2}{q})\Psi\rangle \tilde{\Psi}(p) = \int_{R^{3n}} e^{i\Phi_R(p', p+v, u-v)}\, D(dp', dv+p, du-v)\, \tilde{\Psi}^*(p+u)\tilde{\Psi}(p+v) \times \tag{4.2}$$
$$\times\, \tilde{\Psi}(p + p')$$

Décrivons maintenant le processus de sauts markovien ramifié, qui est associé à l'équation (4.1).

Supposons que les temps Δ entre les sauts des branches du processus soient distribués exponentiellement et que l'exposant dépende de la position de la trajectoire dans l'espace $P(\Delta > T) = \exp - c(p)T$ où p est une position de la branche avant le saut et

$$c(p) = \int Q(du) + \int\int\int D(dq, dv + p, du - v).$$

Supposons que

$$\sup \; c(p) < \infty \tag{4.3}$$

Quand la trajectoire du processus $q_{sp}(t)$ produit un saut au point p , il peut rester non-ramifié avec la probabilité $P_c(p) = c^{-1}(p) \int Q(du)$ et il peut aussi se ramifier avec la probabilité $P_R(p) = 1 - P_c(p)$. Si la trajectoire a un saut sans ramification, la probabilité pour que la valeur du saut appartienne à $B_0 \in B(R^n)$ est égale à $P(B_0) = \int_{B_0} Q(du)/\int Q(dv)$. Si la trajectoire a un saut avec ramification, elle ramifie sur trois branches. Les probabilités de leur premier saut dépendent du point de départ p de la trajectoire dans l'espace des impulsions :

$$P(B_1, B_2, B_3 \,|p) = \int_{B_1}\int_{B_2}\int_{B_3} D(dq, dv+p, du-v) \,/\, \int\int\int D(dq, dv+p, du-v) \tag{4.4}$$

Après ramification, les branches du processus ont des sauts et des ramifications indépendantes. Supposons ensuite que le processus ramifié $q_{sp}(p)$ commence au moment $s = t$ au point $p \in R^n$ et continue quand $0 < s < t$. Ce processus de sauts ramifié markovien s'appelle le processus associé à l'équation (4.1).

Il est évident que chaque trajectoire de $q_{sp}(t)$ peut être représentée comme un arbre. Chaque branche $b^l_{sp}(t)$ non ramifiée qui va vers le sommet de l'arbre est une composante de la trajectoire $q_{sp}(t)$. La somme de toutes les composantes donne l'arbre :

$$q_{sp}(t) = \bigcup_l b^l_{sp}(t).$$

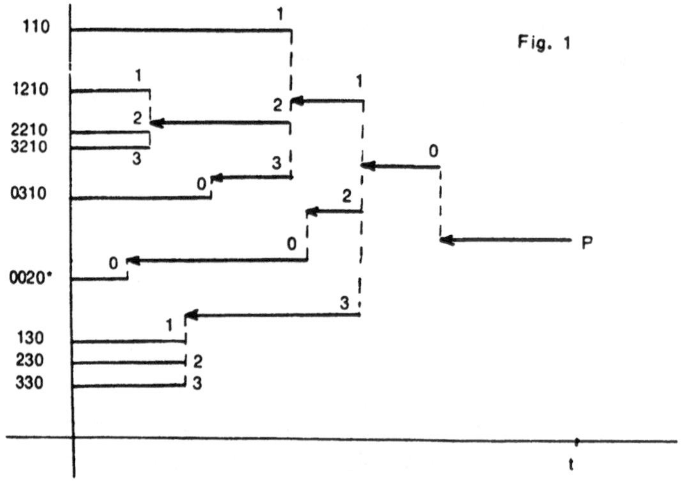

Fig. 1

Il est commode de numéroter la composante à N sauts par la suite de nombres entiers $\{i_1, \ldots, i_N\}$ où $i_k \in \{0, 1, 2, 3\}$. Par exemple, la composante représentée sur la figure 1 a l'indice $\{0020\}$. Nous allons utiliser cette suite comme indice de la composante : b_{0020}. Nous allons utiliser ces notations aussi pour les coordonnées des composantes de la trajectoire ramifiée. Par exemple p_{0020}, p_{020}, p_{20}, p_0, p sont les coordonnées de la composante b_{0020} au temps s :

$$b_{0020} = \{p,\ s_0 < s \leqslant t;\ p_0,\ s_{20} < s \leqslant s_0;\ \ldots;\ p_{0020},\ 0 \leqslant s \leqslant s_{0020}\}$$

et u_0, u_{20}, u_{020}, u_{0020} sont les valeurs des sauts de la composante b_{0020}, aux moments s_0, s_{20}, s_{020}, s_{0020}. Pour terminer les définitions, donnons la définition de la fonctionnelle de la trajectoire ramifiée $\Phi[q_{sp}(t)]$ telle que la moyenne $M\Phi[q_{sp}(t)]$ soit égale à la solution de l'équation (4.1) en représentation -p.

Soit $\tilde{\psi}_0(p) \in C(R^n)$. Supposons que la trajectoire $q_{sp}(t)$ ait N sauts $u_{i_N}, \ldots, u_{i_N \ldots i_1}$ aux moments $s_{i_1}, \ldots, s_{i_N \ldots i_1}$ et soit $p_{i_N \ldots i_1}$ la position de la composante $b_{i_N \ldots i_1}$ au moment $s = 0$. Notons par $\Delta_{i_k \ldots i_1}$ la différence $s_{i_{k-1} \ldots i_1} - s_{i_k \ldots i_1}$ et posons $\Delta_{i_1} = t - s_{i_1}$, $\tilde{\psi}_{i_N \ldots i_1} = \tilde{\psi}_0(p_{i_N \ldots i_1})$. Définissons les valeurs $\tilde{\psi}_{i_{N-1} \ldots i_1}, \ldots, \tilde{\psi}_{i_k, \ldots i_1}, \ldots, \tilde{\psi}_{i_1}$, $\tilde{\psi}$ par récurrence :

$$\tilde{\Psi}_{i_k \ldots i_1} = \exp\{[c(p_{i_k \ldots i_1}) - iH(p_{i_k \ldots i_1})]\,\Delta_{i_k \ldots i_1}\} \times$$
$$\times \begin{cases} \tilde{\Psi}_{0,\, i_k \ldots i_1} \exp i\{\Phi_c(u_{0,\, i_k \ldots i_1}) - \pi/2\}, & \text{si } i_{k+1} = 0 \\ \tilde{\Psi}_{1,\, i_k \ldots i_1}\, \tilde{\Psi}_{2,\, i_k \ldots i_1}\, \tilde{\Psi}^*_{3,\, i_k \ldots i_1} \times & \text{si } i_{k+1} \neq 0 \\ \times \exp i\{\Phi_R(u_{1, i_k \ldots i_1},\ u_{2, i_k \ldots i_1} + p_{i_k \ldots i_1},\ u_{3, i_k \ldots i_1} - u_{2, i_k \ldots i_1}) - \pi/2\} & \end{cases} \qquad (4.5)$$

La quantité $\tilde{\Psi}$ s'exprime à partir de $\tilde{\Psi}_0$ ou $\tilde{\Psi}_1$, $\tilde{\Psi}_2$, $\tilde{\Psi}_3^*$ conformément à (4.5). En examinant toutes les composantes de la trajectoire $q_{sp}(t)$ nous poserons :

$$\Phi[q_{sp}(t)] = \exp\{[c(p) - iH(p)]\Delta_{i_1}\}\tilde{\Psi}$$

La fonctionnelle Φ s'appelle "l'amplitude" de la trajectoire $q_{sp}(t)$. Par exemple, l'amplitude de la trajectoire représentée sur la Figure 2 est égale à :

$$\Phi[q_{sp}(t)] = -i\tilde{\Psi}_0(p_{01})\exp[\{c(p_{01}) - iH(p_{01})\}\Delta_{01} + \{c(p_1) - iH(p_1)\}\Delta_1 + i\Phi_c(u_{01})] \times$$
$$\times\, i^{-1}\tilde{\Psi}_0(p_{02})\exp[\{c(p_{02}) - iH(p_{02})\}\Delta_{02} + \{c(p_2) - iH(p_2)\}\Delta_2 + i\Phi_c(u_{02})] \times$$
$$\times\, i\tilde{\Psi}_0^*(p_{03})\exp[\{c(p_{03}) + iH(p_{03})\}\Delta_{03} + \{c(p_3) + iH(p_3)\}\Delta_3 - i\Phi_c(u_{03})] \times$$
$$\times\, i^{-1}\exp[\{c(p) - iH(p)\}\Delta_1 + i\Phi_R(u_1, u_2 + p, u_3 - u_2)]$$

Supposons maintenant que la moyenne $\tilde{\Psi}(p,t) = M_t\Phi[q_{sp}(t)]$ existe et que $\sup|H(p)| < \infty$ et déduisons l'équation vérifiée par la fonction $\tilde{\Psi}(p,t)$.

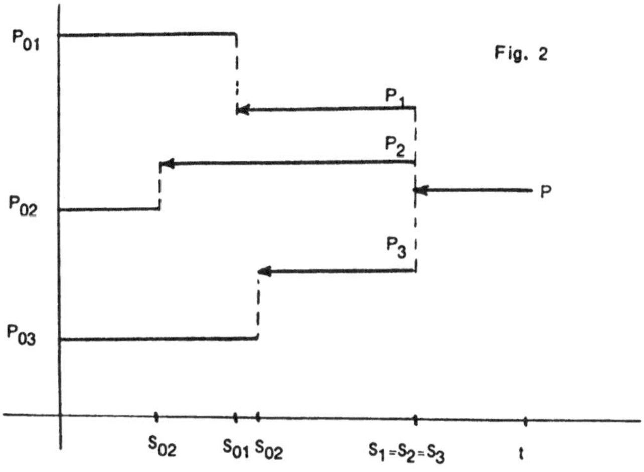

Fig. 2

Le processus $q_{sp}(t)$ est un processus markovien, donc on a

$$\tilde{\Psi}(p, t + \Delta) = M_{t+\Delta}\, M_t\, \Phi[q_{sp}(t + \Delta)]$$

La probabilité pour que le processus $q_{sp}(t+\Delta)$ ne produise pas de sauts au cours du temps Δ est égale à $c(p)\Delta + o(\Delta)$, et la probabilité de produire plus d'un saut est égale à $o(\Delta)$. En adoptant les notations de [2], on obtient :

$$\tilde{\Psi}(p,t+\Delta) = (1 - c(p)\Delta)\, M^{\Delta,0}\, M_t\, \Phi[q_{sp}(t+\Delta)] + c(p)\Delta M^{\Delta,1}\, M_t\, \Phi[q_{sp}(t+\Delta)] + o(\Delta) \qquad (4.6)$$

S'il n'y a pas de sauts au cours du temps Δ , on a, d'après la définition de l'amplitude Φ :

$$\Phi[q_{sp}(t+\Delta)] = \exp \Delta\, \{c(p) - iH(p)\}\, \Phi[q_{sp}(t)]$$

Donc le premier terme dans (4.6) nous donne

$$(1 - c(p)\Delta)\, M^{\Delta,0}\, M_t \Phi = (1 - i\Delta H(p))\, \tilde{\Psi}(p,t) + o(\Delta)\,.$$

Notons par $M_C^{\Delta,1}$ et $M_R^{\Delta,1}$ respectivement les moyennes conditionnelles relativement aux trajectoires qui ont un saut sans ramification et avec ramification, lorsqu'il y a un saut. Alors :

$$M^{\Delta,1} = P_c(p)\, M_c^{\Delta,1} + P_R(p)\, M_R^{\Delta,1}\,.$$

Quand il n'y a pas de ramification au cours du temps Δ et que la trajectoire a un saut u , la définition de l'amplitude nous donne :

$$\Phi[q_{sp}(t+\Delta)] = -i \exp\{i\Phi_c(u)\}\, \Phi[q_{s,p+u}(t)] + o(\Delta)$$

et quand la trajectoire a un saut ramifié, dont les composantes ont les valeurs dr , dv , du , nous obtenons :

$$\Phi[q_{sp}(t+\Delta)] = -i \exp\{\, i\Phi_R(r,\, v+p,\, u-v)\} \times$$
$$\times\, \Phi[q_{s,p+r}(t)]\, \Phi[q_{s,p+v}(t)]\, \Phi^*[q_{s,p+u}(t)]$$

Alors

$$\begin{aligned} M^{\Delta,1} M_t\, \Phi[q_{sp}(t+\Delta)] &= i^{-1} P_c(p) \times \int Q(du)\, e^{i\Phi_c(u)}\, \tilde{\Psi}(p+u,\, t) \,/\, \int Q(R^n) \\ &+ i^{-1} P_R(p) \,/\, \int D(dr, dv+p, du-v) \int D(dr,\, dv+p,\, du-v) \times M_t\{\Phi[q_{s,p+r}(t)] \times \\ &\times \Phi[q_{s,p+v}(t)] \cdot \Phi^*[q_{s,p+u}(t)]\}\, e^{i\Phi_R(r,\, p+v,\, u-v)} + o(\Delta) \end{aligned} \qquad (4.7)$$

Puisque les branches $q_{s,p+v}(t)$, $q_{s,p+u}(t)$ en (4.7) sont indépendantes, la moyenne du produit est égale au produit des moyennes. Les formules (4.7) et (4.6) nous donnent donc :

$$\tilde{\Psi}(p,\ t+\Delta) = \tilde{\Psi}(p,\ t) - iH(p)\Delta\ \tilde{\Psi}(p,\ t) - i\Delta V(i\nabla_p)\ \tilde{\Psi}(p,\ t) -$$

$$- i\Delta < \Psi,\ \hat{L}(i\nabla_p)\ \Psi > \tilde{\Psi}(p,t) + o(\Delta) \tag{4.8}$$

En passant à la limite quand $\Delta \to 0$ en (4.8), on obtient l'équation (4.1) en représentation -p :

$$\{ - i\partial/\partial t + H(p) + V(i\nabla_p) + < \Psi,\ \hat{L}(i\nabla_p)\ \Psi > \}\ \tilde{\Psi}(p,\ t) = 0$$

et $$\lim_{t \downarrow 0} \tilde{\Psi}(p,t) = \tilde{\Psi}_0(p) \tag{4.9}$$

Notre hypothèse sur l'existence de la moyenne $M_t \Phi$ peut être justifiée quand le temps t est suffisamment petit. Dans [12] on établit la proposition suivante :

Proposition 4.1. Soit a_0 la norme $C(R^n)$ de $\tilde{\Psi}_0$. Alors la moyenne $M_t \Phi$ existe si $0 \leqslant t < c^{-1}(p)\ \ \ln(1+a_0^2)/a_0^2$.

Notre supposition que la fonction $H(p)$ est bornée a été utilisée ici pour simplifier la démonstration (4.9). Dans [12] , on trouve le théorème suivant :

Théorème 4.1. Soit $H(p)$ une fonction continue réelle, soient $V(x)$ et $L(x,dv,q)$ ayant les propriétés (4.2-4.3) et $\tilde{\Psi}_0(p) \in C(R^n)$. Soit $q_{sp}(t)$ le processus de sauts ramifié associé à l'équation (4.1) et soit $\Phi\,[q_{sp}(t)]$ l'amplitude de la trajectoire $q_{sp}(t)$. Alors, la moyenne $M_t \Phi\,[q_{sp}(t)] = \tilde{\Psi}(p,t)$ existe et la fonction $\tilde{\Psi}(p,t)$ est solution du problème (4.9) si t est suffisamment petit.

BIBLIOGRAPHIE

[1] *Itô K.*, Generalized Uniform Complex Measures in the Hilbertian Metric Space with their Application to Feynman Integral, Proc. 5th Berkeley Symp. Math. Stat. & Prob. Berkeley & Los Angeles, Univ. of California Press, 1967, *2*, 1, 145-161.

[2] *Maslov V.P., Chebotarev A.M.*, On Monte-Carlo Calculation of the Feynman Path Integrals in the p-Representation, Proc. Second IMACS Int. Symp. Comp. Meth. for Partial Diff. Eq., Lehigh Univ., Bethlehem, Pa., USA, June 22-24, 1977.

[3] *Chebotarev A.M.*, T-Mappings and Functional Integrals, Soviet Math. Dokl., *16*, 6, 1975, 1536-1540.

[4] *Albeverio S.*, Hoegh-Krohn R., Mathematical Theory of Feynman Path Integrals Lect. Math., 1976.

[5] *von Neuman J.*, Various Techniques Used in Connection with Random Rigits. Monte-Carlo Method, Nath. Bur. Stand. Math. Ser. *12*, 1951, 36-38.

[6] *Nelson E.*, Feynman Integrals and the Schrodinger Equation, J. Math. Phys., *5*, 3 1964, 332-343.

[7] *Maslov V.P., Chebotarev A.M.*, The Definition of Feynman's Functional Integral in the p-Representation, Soviet Math. Dokl., *17*,4, 1976, 975-976.

[8] *Маслов В.П., Чеботарев А.М.*, Обобщенная мера в контитуальном интеграле Фейнмана Теоретическая и математическая физика *28*, 3, 1976, 291-307.

[9] *Tarski J.*, Definitions and selected applications Feynman-type integrals, in Functional integration and its applications, Oxford Univ. Press, London, 1975, 169-180.

[10] *Maslov V.P.*, Operational Methods, Moscow, "Mir", 1976.

[11] *Maslov V.P., Chebotarev A.M.*, Representation of the Solution of an Equation of Hartree Type in the Form of a T-Mapping, Soviet Math. Dokl., *16*, 3, 1975, 730-734.

[12] *Маслов В.П., Чеботарев А.М.*, Скачкообразные процессы и их применение в квантовой механике, ВИНИТИ, Итоги Науки, вып. 15, 1978.

THE POLYGONAL PATH FORMULATION OF THE FEYNMAN PATH INTEGRAL

Aubrey Truman, Department of Mathematics,
Heriot-Watt University, Edinburgh, Scotland.

1. Introduction

We give a resumé of the results obtained to date on the polygonal path formulation of the Feynman integral - the Feynman map $\mathcal{F}$. This formulation of the Feynman integral first arose in connection with the earlier definition of the Feynman integral, the Fresnel integral $\mathcal{F}_{AH}$ due to S. Albeverio and R. Høegh-Krohn [1], [32]-[35], [36]. The Albeverio and Høegh-Krohn definition makes great use of the properties of Fourier transform on path-space. The idea for this can be found in the earlier distribution theoretic work of C. DeWitt Morette as well as in the work of P. Kree [8]-[10], [21].

The idea of using polygonal paths to define a path integral is to be found in R. P. Feynman's own writings on this subject and in the mathematical papers of R. H. Cameron [4], [5], [14]. Here we show that in Euclidean space nonrelativistic quantum mechanics the piecewise linear polygonal paths provide a neat, practicable and rigorous definition of a Feynman integral with non-trivial physical applications. There are indications that this may also be true for nonrelativistic quantum mechanics in curved space [13].

In Euclidean space nonrelativistic quantum mechanics it is possible to subsume the path-space Fourier transform definition of the Feynman integral by the polygonal path formulation. This depends upon the fact that for nonrelativistic quantum mechanics the Hilbert path-space H has a reproducing kernel simply related to the piecewise linear polygonal paths. This leads to the result which we discuss first that the Feynman map $\mathcal{F}$ is an extension of $\mathcal{F}_{AH}$ [37].

The potential applicability of our definition of the Feynman integral will, of course, depend upon the class of functionals integrable with respect to $\mathcal{F}$. We show in this paper that associated with our definition of the Feynman integral there is a wide class of integrable functionals. This is achieved by proving a Cameron-Martin formula for $\mathcal{F}$ paralleling the Cameron-Martin formula for the Wiener integral E [39]. We include here some new results for discontinuous kernels similar to the original Cameron-Martin result [6].

Loosely speaking, our Cameron-Martin formula is valid for linear changes of integration variables $(1+K) : H \to H$, where K can be, a not necessarily trace-class, Hilbert-Schmidt operator. This enables us to integrate a wide class of functionals on path-space. Included in this class of integrable functionals is just the kind of exponential quadratic functionals which required Albeverio and Høegh-Krohn to introduce their second definition of the Feynman integral - the Fresnel integral relative to a nonsingular quadratic form [1]. Although this second definition may be necessary for applications to quantum field theory, it seems to us to be unnecessarily complicated for nonrelativistic quantum mechanics.

We go on to discuss the applications of the Feynman map $\mathcal{F}$ to nonrelativistic quantum theory. Most of these applications stem from the Feynman-Itô formula. This Feynman-Itô formula expresses the wavefunction solution of the Cauchy problem for the Schrödinger equation as a Feynman integral [15], [18]. Here we indicate how to prove the Feynman-Itô formula for anharmonic oscillator potentials in one dimension,

$V = Ax^2 + Bx + C + \int e^{i\alpha x} d\mu(\alpha)$, $A > 0$, μ being a measure of bounded absolute variation on $\mathbb{R}^1$. The proof here depends upon the above Cameron-Martin formula and some of the ideas of Albeverio and Høegh-Krohn [1]. A similar result is valid in n dimensions.

The Feynman-Itô formula for the Feynman map $\mathcal{F}$ is, in fact, exactly equivalent to the Feynman-Dirac conjecture, expressing the quantum mechanical amplitude as 'a sum over paths γ' of $\exp\{iS[\gamma]/\hbar\}$, $S[\cdot]$ being the classical action and $\hbar$ being Planck's constant divided by 2π. Thus we establish here the validity of the exact Feynman-Dirac conjecture for the above anharmonic oscillator potentials [11],[12],[15],[37].

The last part of our paper deals with the problem of obtaining the classical mechanical limit of quantum mechanics when $\hbar$ tends to zero [2], [36], [38]. To this end we introduce our quasiclassical representation (qcr) for the wavefunction solution of the Cauchy problem for the Schrödinger equation. Here we only discuss the 1-dimensional Euclidean space qcr [36]. However, following some suggestions of C. DeWitt Morette and the author and the qcr method, K.D. Elworthy has established a formal analogue of the qcr for the Schrödinger equation for a system where configuration space is an n-dimensional Riemannian manifold. This is discussed in C. DeWitt Morette's paper in these proceedings. Here C. DeWitt Morette defines the Feynman integral by prodistributions but the Feynman map definition would do equally well. (Indeed, since it can be *proved* that for the pure imaginary time Schrödinger equation the Feynman map definition converges to the correct Feynman-Kac formula, there are some reasons to prefer the Feynman map definition.) [13], [22].

We show in this paper how our qcr and Cameron-Martin formula lead to the results of V. P. Maslov on the classical mechanical limit of quantum mechanics and the Euclidean space semiclassical expansion of C. DeWitt Morette [23]-[25], [10]. In the former case we obtain the Maslov indices in a simple explicit way without using the principle of stationary phase [2], [23]-[25]. In the latter case the qcr leads unambiguously to the Feynman-Green's function as covariance of the semiclassical expansion. In the original paper of C. DeWitt Morette the inspired choice of the Feynman-Green's function as covariance was made by choosing, in a somewhat arbitrary way, a certain linear transformation [10], [22].

We conclude this paper by giving a simple (possibly new) derivation of the Bohr-Sommerfeld-Maslov quantisation for a particle in a potential well and show how to obtain the corresponding WKB eigenfunctions [17], [42]. This demonstrates just how completely the Feynman path integral meets the traditional requirements of nonrelativistic quantum mechanics.

To minimise typographical difficulties we restrict our attention to the quantum mechanics of a single spinless nonrelativistic quantum particle in $\mathbb{R}^1$. It is not difficult to generalise our results to a finite number of spinless nonrelativistic quantum particles in $\mathbb{R}^n$. Lack of space prevents us from giving all the details of the proofs here. We hope we have given sufficient detail to enable a competent student to complete the arguments.

I am grateful to John Lewis, Ken Brown, Cécile DeWitt Morette and David Elworthy for helpful conversations. I would like to thank the organisers of the conference for inviting me to present this paper and for all their kind hospitality. In addition I must thank the Science Research Council for the forthcoming award of a Senior Visiting Fellowship which provided some of the stimulus for the above work.

2. The Hilbert Path-Space H

In this section we give a concrete realisation of the space of paths for a single spinless non-relativistic quantum particle constrained to move on the real line. We think of these paths as the paths which the quantum mechanical particle might actually describe in a physical experiment. We emphasise here the simple mathematical properties of the path-space required in our subsequent formulation of the Feynman integral.

Definition

H is the space of real-valued continuous functions $\gamma : [0,t] \to \mathbb{R}$ with weak derivative $\frac{d\gamma}{d\tau} \in L^2(0,t)$, γ being normalised so that $\gamma(t) = 0$. H is endowed with the inner product (,).

$$(\gamma,\gamma') = \int_0^t \frac{d\gamma}{d\tau}\frac{d\gamma'}{d\tau}\, d\tau. \tag{2.1}$$

It is not difficult to show that any $\gamma \in H$ can be written as

$$\gamma(\tau) = \alpha_o(\tau-t) + \sum_1^\infty \frac{t\alpha_n}{2\pi n} \sin\left(\frac{2\pi n\tau}{t}\right) + \sum_1^\infty \frac{t\beta_n}{2\pi n}\left[1 - \cos\left(\frac{2\pi n\tau}{t}\right)\right], \quad \tau \in [0,t], \tag{2.2}$$

where $\alpha_o, \alpha_n, \beta_n \in \mathbb{R}$ are the usual Fourier coefficients of $\frac{d\gamma}{d\tau} \in L^2(0,t)$ with $\sum_1^\infty (\alpha_n^2 + \beta_n^2) < \infty$.

Then H is a real separable Hilbert space with reproducing kernel $G(\sigma,\tau) = (t - \sigma^{\vee}\tau)$, where $\sigma^{\vee}\tau = \sup\{\sigma,\tau\}$. The reproducing property reads

$$(G(\sigma,\cdot), \gamma(\cdot)) = \gamma(\sigma), \qquad \forall\, \gamma \in H, \forall\, \sigma \in [0,t]. \tag{2.3}$$

Here $G(\sigma,\tau) = (t - \sigma^{\vee}\tau)$ is simply the Green's function of the Sturm-Liouville differential operator $\left(-\frac{d^2}{d\tau^2}\right)$ with boundary conditions $\frac{dG(\sigma,\tau=0)}{d\tau} = 0$, $G(\sigma,\tau=t) = 0$.

The weak derivative of G is given by

$$\frac{dG}{d\tau}(\sigma,\tau) = -\theta(\tau - \sigma), \tag{2.4}$$

θ being the Heaviside function. The reproducing property now follows by integrating the a.e. convergent Fourier series for $\frac{d\gamma}{d\tau}$ term by term on the subinterval $[\sigma,t]$ [31].

We now introduce the important linear map $P_n : H \to H$ by

$$(P_n\gamma)(\tau) = \sum_{j=0}^{n-1} \left[G\left(\overline{\frac{j+1t}{n}}, \tau\right) - G\left(\frac{jt}{n},\tau\right)\right] \left[\gamma_{j+1} - \gamma_j\right] \frac{n}{t}, \tag{2.5}$$

where $\gamma_j = \gamma\left(\frac{jt}{n}\right)$, $j = 0,1,2,..,n$. From the reproducing property

$$(\gamma',P_n\gamma) = \sum_{j=0}^{n-1} (\gamma'_{j+1} - \gamma'_j)(\gamma_{j+1} - \gamma_j)\frac{n}{t} = (P_n\gamma',\gamma). \tag{2.6}$$

Thus $P_n^* = P_n$.

Substitution of $G(\sigma,\tau) = (t - \sigma^{\vee}\tau)$ gives

$$(P_n\gamma)(\tau) = \gamma_j + \left(\tau - \frac{jt}{n}\right)\left(\gamma_{j+1}-\gamma_j\right)\frac{n}{t}, \qquad \frac{jt}{n} \leqslant \tau < \frac{(j+1)t}{n}. \tag{2.7}$$

Hence, $(P_n\gamma)$ is just the piecewise linear polygonal approximation to γ. Evidently $P_n^2 = P_n$.

$\{(P_n\gamma+X) \mid n = 1,2,\ldots, \gamma \in H, X \in \mathbb{R}\}$ is just the set of piecewise linear polygonal paths of Feynman, satisfying $(P_n\gamma+X)(\tau=t) = X$. The next lemma tells how numerous the the polygonal paths $(P_n\gamma)$ are in H.

<u>Lemma 1</u>

$P_n : H \to H$ is a projection. If $1 : H \to H$ denotes the identity then $P_n \overset{s}{\to} 1$ as $n\to\infty$ i.e. P_n tends to 1 in the strong operator topology as n tends to infinity.

<u>Proof</u>

We have already proved the first part of the lemma. It remains to prove that if

$V = \{\gamma \in H \mid \|P_n \gamma - \gamma\| \to 0 \text{ as } n\to\infty\}$ then $V = H$. This is done by proving that V is a closed subspace of H containing its basis functions.

First V is a closed subspace of H. V is a subspace because P_n is linear. Let $\{\gamma_m \mid m = 1,2,\ldots\} \subset V$ with $\|\gamma_m - \gamma\| \to 0$, as $m \to \infty$, for some $\gamma \in H$. We show that necessarily $\gamma \in V$.

Observe that

$$\|P_n\gamma - \gamma\| \leqslant \|P_n(\gamma-\gamma_m)\| + \|\gamma - \gamma_m\| + \|P_n\gamma_m - \gamma_m\|. \tag{2.8}$$

Hence,

$$\|P_n\gamma - \gamma\| \leqslant 2\|\gamma - \gamma_m\| + \|P_n\gamma_m - \gamma_m\|. \tag{2.9}$$

Given $\varepsilon > 0, \exists\, N_\varepsilon$ such that $\|\gamma - \gamma_m\| < \varepsilon/4$ when $m = N_\varepsilon$. Also, $\exists\, N(m,\varepsilon)$ such that $\|P_n\gamma_m - \gamma_m\| < \varepsilon/2$, $n \geqslant N(m,\varepsilon)$. From above inequality then, for $n \geqslant N(N_\varepsilon,\varepsilon)$, $\|P_n\gamma-\gamma\| < \varepsilon$. Thus, $\gamma \in V$.

It is now a simple matter to show that the basis functions of H are in V [36], [39]. □

3. The Feynman Maps $\mathcal{F}^s$

We now introduce the family of Feynman maps $\mathcal{F}^s$. The label s here is a complex variable with $\operatorname{im} s \leqslant 0$. The Feynman map most important for physical applications corresponds to $s = 1$. We shall often write $\mathcal{F}^1$ as $\mathcal{F}$ [3], [39].

Let $f : H \to \mathbb{C}$ be a complex-valued functional. We define the complex Gaussian $e_s : H \to \mathbb{C}$ by

$$e_s[\gamma] = \exp\left\{\frac{i}{2s}\|\gamma\|^2\right\}, \qquad s \neq 0,\ \operatorname{im} s \leqslant 0, \tag{3.1}$$

where $\|\gamma\|^2 = (\gamma,\gamma)$. We now give the definition of $\mathcal{F}^s[f]$.

Definition

We denote by $\mathcal{F}^s_n$

$$\mathcal{F}^s_n[f] = \int_{P_nH} (fe_s)\circ P_n \, d^n\gamma \left[\int_{P_nH} (e_s\circ P_n)\, d^n\gamma\right]^{-1}, \tag{3.2}$$

where o denotes composition and the integration over P_nH is effected by integrating from $-\infty$ to $+\infty$ each of the variables $\Delta\gamma_j = (\gamma_{j+1} - \gamma_j)$, $\gamma_j = \gamma\left(\frac{jt}{n}\right)$, $j=0,1,2,\ldots,n$.

The normalisation has been chosen so that, for the functional 1, $\mathcal{F}^s_n[1] = 1$. We define the Feynman map $\mathcal{F}^s$ by the limit, when it exists,

$$\mathcal{F}^s[f] = \lim_{n\to\infty} \mathcal{F}^s_n[f]. \tag{3.3}$$

We shall write $f \in \mathfrak{F}^s(P_\infty H)$ iff above limit exists.

$\mathfrak{F}^s(P_\infty H)$ is the class of integrable functionals. To see how large is the space $\mathfrak{F}^s(P_\infty H)$ we require another definition due originally to Albeverio and Høegh-Krohn [1].

Definition

$\mathfrak{F}(H)$ is the space of functionals each of which is the Fourier transform of a measure in $M(H)$, the space of complex-valued measures of bounded absolute variation on H. Thus, $f \in \mathfrak{F}(H)$ can be written

$$f[\gamma] = \int_H \exp\{i(\gamma',\gamma)\}\, d\mu_f(\gamma'), \tag{3.4}$$

$\mu_f \in M(H)$, with $\|\mu_f\| = \int_H d|\mu_f| < \infty$. $\|\ \|$ is a norm on the Banach algebra $M(H)$, closed under addition and convolution of measures $*$, $(\mu * \nu)(A) = \int_H \mu(A-\gamma)d\nu(\gamma)$, Borel $A \subset H$. The functionals in $\mathfrak{F}(H)$ are the simplest type of integrable functionals as our first theorem shows.

Theorem 1

When $\operatorname{im} s \leqslant 0$, $\mathfrak{F}(H) \subset \mathfrak{F}^s(P_\infty H)$ and if $f \in \mathfrak{F}(H)$ is given by $f[\gamma] = \int_H \exp\{i(\gamma',\gamma)\}d\mu_f(\gamma')$, $\mu_f \in M(H)$,

$$\mathfrak{F}^s[f] = \int_H \exp\left\{\frac{-is}{2}\|\gamma\|^2\right\} d\mu_f(\gamma). \tag{3.5}$$

Hence, for $\operatorname{im} s \leqslant 0$, we have

$$|\mathfrak{F}^s[f]| \leqslant \int_H d|\mu_f| \overset{\text{def}}{=} \|f\|_o. \tag{3.6}$$

$\|\ \|_o$ is a norm on the Banach algebra $\mathfrak{F}(H)$. $\mathfrak{F}^s : \mathfrak{F}(H) \to \mathbb{C}$ is a continuous linear map, $\operatorname{im} s \leqslant 0$.

Proof

When $\operatorname{im} s < 0$, interchanging orders of integration by Fubini's theorem in above definition, gives

$$\mathfrak{F}^s_n[f] = \int_H \exp\left\{\frac{-is}{2}(\gamma,P_n\gamma)\right\} d\mu_f(\gamma) \tag{3.7}$$

and the first part of the theorem follows from Lemma 1. When im s = 0, interchanging the orders of integration is slightly more delicate. We refer the reader to Ref [39] for details.

The proof of the remainder of this theorem borrows from Albeverio and Høegh-Krohn [1]. Observe that from Fubini's theorem

$$\int_H \exp\{i(\gamma',\gamma)\}\, d(\mu * \nu)(\gamma') = \int_H \exp\{i(\gamma',\gamma)\}\, d\mu(\gamma') \int_H \exp\{i(\gamma'',\gamma)\}\, d\nu(\gamma''). \quad (3.8)$$

This last observation and the separability of H imply that $\mathcal{F}(H)$, equipped with $\| \ \|_o$, is, under pointwise addition and multiplication, the isometric isomorphic image of the Banach algebra M(H). Thus, $\mathcal{F}(H)$ is a Banach algebra closed under addition, multiplication and composition with entire functions. The continuity of $\mathcal{F}^s$ follows trivially from above.□

Our second lemma tells us about the translation invariance of $\mathcal{F}^s$.

Lemma 2

Let $f \in \mathcal{F}(H)$ be a complex valued functional. Define $f_a : H \to \mathbb{C}$, each $a \in H$, by

$$f_a[\gamma] = f[\gamma + a] \quad (3.9)$$

and define $(e_s^a f_a) : H \to \mathbb{C}$ by

$$[e_s^a f_a][\gamma] = \exp\left\{\frac{i}{s}(a,\gamma)\right\} f_a[\gamma]. \quad (3.10)$$

Then

$$\mathcal{F}^s[e_s^a f_a] = \exp\left\{-\frac{i}{2s}\|a\|^2\right\} \mathcal{F}^s[f]. \quad (3.11)$$

Proof

From equation (3.2) and a simple translation of integration variables we obtain

$$\mathcal{F}_n^s[e_s^a f_a] = \exp\left\{-\frac{i}{2s}(a,P_n a)\right\} \int_H \exp\left\{\frac{-is}{2}(\gamma,P_n\gamma)\right\} \exp\left\{i(\gamma,\overline{1-P_n}\,a)\right\} d\mu_f(\gamma). \quad (3.12)$$

Letting $n \to \infty$, the result follows from Lemma 1 and the dominated convergence theorem for μ_f [39]. □

Theorem 1 shows that $\mathcal{F}^1$ is an extension of $\mathcal{F}_{AH}$ the Albeverio and Høegh-Krohn definition of the Feynman integral. We now seek to obtain more information about $\mathcal{F}^s(P_\infty H)$. This is done in the next section by developing the analogy between the Feynman integral $\mathcal{F}$ and the Wiener integral E.

4. The Wiener Integral and Cameron-Martin Formula

Let E denote the Wiener integral associated with Wiener measure μ_w with supp $\mu_w \subset C_o(0,t)$, the space of continuous functions on $(0,t)$ vanishing at time t. The connection between $\mathcal{F}$ and E is the content of the next theorem.

Theorem 2

Let $f \in \mathcal{F}(H)$ then $\mathcal{F}^s[f]$ is a regular analytic function of s in $\operatorname{im} s < 0$, continuous in $\operatorname{im} s \leqslant 0$. If $f \in \mathcal{F}(H)$ is a continuous functional $f : C_o(0,t) \to \mathbb{C}$, then

$$\mathcal{F}[f] = \lim_{\varepsilon \to 0+} \mathcal{F}^{1-i\varepsilon}[f], \quad E[f] = \lim_{\varepsilon \to 0} \mathcal{F}^{-i+\varepsilon}[f]. \tag{4.1}$$

Interpolation gives

$$|\mathcal{F}^{se^{-i\alpha}}[f]| \leqslant \|f\|_o^{1-\frac{2\alpha}{\pi}} \|f\|_\infty^{\frac{2\alpha}{\pi}}, \quad 0 \leqslant \alpha \leqslant \frac{\pi}{2}, \; s > 0, \tag{4.2}$$

where $\|f\|_\infty = \sup_{\gamma \in C_o(0,t)} |f[\gamma]|$.

Proof

Since $\mathcal{F}^s[f] = \int_H \exp\left\{\frac{-is}{2}\|\gamma\|^2\right\} d\mu_f(\gamma)$, the first part of the theorem follows from the dominated convergence theorem and Morera's theorem. The proof of the second part of the theorem partly imitates Nelson's construction of Wiener measure [27].

Denote by $\dot{\mathbb{R}}$ the 1-point compactification of $\mathbb{R}$. Then let the compact Hausdorff $\Gamma = \underset{0 \leqslant \tau < t}{\times} \dot{\mathbb{R}}$, endowed with (weak) product topology. Γ is the obvious compact model for the path-space. We think of Γ as the space of paths $\gamma : [0,t] \to \mathbb{R} \cup \{\infty\}$. Let $C(\Gamma)$ denote the space of continuous functionals defined on Γ. The complex Stone-Weierstrass theorem implies that $\mathcal{F}_o = C(\Gamma) \cap \mathcal{F}(H)$ is dense in $C(\Gamma)$. $\mathcal{F}^{-i}$ is a positive and, therefore, continuous linear functional defined on $\mathcal{F}_o$. By the Riesz-Markov theorem corresponding to the unique continuous extension of $\mathcal{F}^{-i}$ there is a regular Borel measure μ on Γ. This is easily shown to be the Wiener measure μ_w. The last part of the theorem follows by interpolation [28], [39]. □

The last result suggests that some of the well-known properties of the Wiener integral E might generalise to the Feynman maps $\mathcal{F}^s$. We have already seen that the translation formula for Wiener integrals generalises to the Feynman maps. The next most obvious potential candidate for generalisation is the Cameron-Martin formula [6].

Theorem 3 (Abstract Cameron-Martin Formula)

Let $(1+K) : H \to H$ be a linear injection with K trace-class and $\det(1+K) \neq 0$, det being the Fredholm determinant. Let $g : H \to \mathbb{C}$ and define $g_{1+K} : H \to \mathbb{C}$ by $g_{1+K}[\gamma] = g[(1+K)\gamma]$.

It is convenient to denote by $(e_s^K g_{1+K})$

$$[e_s^K g_{1+K}][\gamma] = \exp\left\{\frac{i}{s}(K\gamma,\gamma) + \frac{i}{2s}(K\gamma,K\gamma)\right\} g_{1+K}[\gamma], \tag{4.3}$$

so that $(e_s^K g_{1+K}) : H \to \mathbb{C}$. Then, if $g \in \mathcal{F}^s(P_\infty H)$, $(e_s^K g_{1+K}) \in \mathcal{F}^s(P_\infty H)$ and

$$\mathcal{F}^s[e_s^K g_{1+K}] = |\det(1+K)|^{-1} \mathcal{F}^s[g] \tag{4.4}$$

Proof

By definition we have

$$\mathcal{F}_n^s[e_s^K g_{1+K}] = (2\pi i s\Delta t)^{-\frac{n}{2}} \int \exp\left\{\frac{i\|(1+K)P_n\gamma\|^2}{2s}\right\} g_{1+K}[P_n\gamma]\, d^n\gamma, \tag{4.5}$$

where the integration variables are $\Delta\gamma_j = (\gamma_{j+1}-\gamma_j)$, $j=0,1,..,n-1$ and $\gamma_j = \gamma\left(\frac{jt}{n}\right)$, $j = 0,1,2,..,n$.

We change integration variables to $\Delta\gamma'_j = (f_j, \gamma')$, $j = 0,1,2,...n-1$, where $\gamma' = (1+K)P_n\gamma$ and $f_0,...,f_{n-1}$ form an orthonormal basis for $(1+K)P_nH$. Using the reproducing kernel it is a simple matter to show that the Jacobian determinant for this change of variable is

$$|J_n| = \left|\det(1_n + P_nKP_n + P_nK^*P_n + P_nK^*KP_n)\right|^{-1/2}$$

1_n being the (nxn) identity.

Further $P_n \xrightarrow{s} 1$ and K trace-class imply that $\det(1_n+P_nKP_n) \longrightarrow \det(1+K) \neq 0$, as $n \to \infty$. Hence, $|J_n| < \infty$ for sufficiently large n. If we now let $Q_n : H \longrightarrow (1+K)P_nH$ be orthogonal projection, $Q_n \xrightarrow{s} 1$ as $n \to \infty$. Therefore, changing integration variables to $d(Q_n\gamma')$ and letting $n \to \infty$ gives, for $g \in \mathcal{F}(H)$,

$$\mathcal{F}^s\left[e_s^K g_{1+K}\right] = \left|\det(1+K)\right|^{-1} \mathcal{F}^s[g]\ . \tag{4.6}$$

Corollary 1

$$\mathcal{F}^s(P_\infty H) \supset \bigcup_K e_s^K [\mathcal{F}(H)]_{1+K} \qquad \operatorname{im} s \leqslant 0,\ s \neq 0, \tag{4.7}$$

where the union is over all trace-class K with $(1+K) : H \to H$ an injection, $\det(1+K) \neq 0$.

In most practical cases one requires a Cameron-Martin formula for a linear transformation $(1+K) : H \to H$ with

$$(K\gamma)(\tau) = \int_0^t K(\tau,\sigma)\, \gamma(\sigma)d\sigma, \tag{4.8}$$

where the kernel $K(\tau,\sigma)$ is discontinuous across the diagonal $\tau = \sigma$. It is not easy to give simple sufficient conditions on such a kernel to ensure that the corresponding linear transformation is trace-class. Nevertheless, we can still prove a Cameron-Martin formula for a linear transformation with a 'C^0 discontinuous kernel'.

Definition

$K(\cdot,\cdot)$ is said to be a C^0 discontinuous kernel if $K(\tau,\sigma) = K^{\pm}(\tau,\sigma)$, for $\tau \gtrless \sigma$, where $K^+(\tau,\sigma)$ is continuous in the variable (τ,σ) in the closed triangle $0 \leqslant \sigma \leqslant \tau \leqslant t$ and $K^-(\tau,\sigma)$ is continuous in the variable (τ,σ) in the closed triangle $0 \leqslant \tau \leqslant \sigma \leqslant t$.

We define $k(\cdot,\cdot)$ by

$$k(\tau,\sigma) = \begin{cases} K^+(\tau,\sigma), & \tau > \sigma, \\ K^-(\tau,\sigma), & \tau < \sigma, \end{cases} \qquad k(\sigma,\sigma) = \frac{1}{2}\left\{K^+(\sigma,\sigma_-) + K^-(\sigma,\sigma_+)\right\}. \tag{4.9}$$

Then from Hadamard's inequality we can deduce

$$\mathrm{Det}(1+K) = 1 + \frac{1}{1!}\int_0^t k(\sigma_1,\sigma_1)d\sigma_1 + \frac{1}{2!}\int_0^t\int_0^t \begin{vmatrix} k(\sigma_1,\sigma_1)k(\sigma_1,\sigma_2) \\ k(\sigma_2,\sigma_1)k(\sigma_2,\sigma_2)\end{vmatrix} d\sigma_1 d\sigma_2 + \ldots < \infty. \tag{4.10}$$

We point out here that $\mathrm{Det}(1+K)$ is not necessarily $\det(1+K)$, the Fredholm determinant of $(1+K)$. The following theorem is nevertheless valid.

Theorem 3′ (Cameron-Martin formula for C^0 discontinuous kernels)

Let $(1+K) : H \longrightarrow H$ be a linear injection with C^0 discontinuous kernel, $\mathrm{Det}(1+K) \neq 0$ and K Hilbert-Schmidt. Then, if $g \in \mathcal{F}(H)$, $e_s^K g_{1+K} \in \mathcal{F}^s(P_\infty H)$ and

$$\mathcal{F}^s[e_s^K g_{1+K}] = |\mathrm{Det}(1+K)|^{-1} \mathcal{F}^s[g]. \tag{4.11}$$

Proof

Although the proof is quite long it merely consists of showing that under the given hypotheses $\lim_{n\to\infty} \det(1_n + P_n K P_n) = \mathrm{Det}(1+K)$. This is done by extending part of the original Cameron-Martin formula proof [6]. The crucial step consists of showing that

$\det(1_n + P_n K P_n) = \det(\delta_{ij} + A^n_{ij})_{i,j = 0,1,..,n-1}$ where

$$A^n_{ij} = \int_{\frac{(j-1)t}{n}}^{\frac{(j+1)t}{n}} K\left(\frac{it}{n}, \sigma\right)\left[1 - \left|\frac{n\sigma}{t} - j\right|\right] d\sigma$$ and the sequence

$\left\{ \frac{n}{t} A^n_{\rho_n(\tau)\ \rho_n(\sigma)} \right\}$ is uniformly bounded, with $\lim_{n\to\infty} \frac{n}{t} A^n_{\rho_n(\tau)\rho_n(\sigma)} = k(\tau,\sigma)$, with

$\rho_n(\sigma) = \left[\frac{n\sigma}{t}\right]$, [ ] being integer part [41]. □

Lemma 3

For the C^o discontinuous $(1+K) : H \to H$ with $(K\gamma)(\tau) = \int_\tau^t K(\sigma)\gamma(\sigma)d\sigma$

$$\mathrm{Det}(1+K) = \exp\left\{ \frac{1}{2} \int_0^t K(\sigma)d\sigma \right\}. \tag{4.12}$$

Proof

The second term of Det (1+K) is clearly $\left\{\frac{1}{2} \int_0^t K(\sigma)d\sigma\right\}$. Let us consider the third term. Interchanging rows and columns $\begin{vmatrix} k(\sigma_1,\sigma_1) & k(\sigma_1,\sigma_2) \\ k(\sigma_2,\sigma_1) & k(\sigma_2,\sigma_2)\end{vmatrix}$ is seen to be a symmetrical function of (σ_1,σ_2). Hence we obtain

$$\frac{1}{2!}\int_0^t\int_0^t \begin{vmatrix} k(\sigma_1,\sigma_1) & k(\sigma_1,\sigma_2) \\ k(\sigma_2,\sigma_1) & k(\sigma_2,\sigma_2)\end{vmatrix} d\sigma_1 d\sigma_2 = \int_0^t d\sigma_2 \int_0^{\sigma_2} d\sigma_1 \begin{vmatrix} k(\sigma_1,\sigma_1) & k(\sigma_1,\sigma_2) \\ k(\sigma_2,\sigma_1) & k(\sigma_2,\sigma_2)\end{vmatrix} d\sigma_1 d\sigma_2. \tag{4.13}$$

However, for the kernel in question, $k(\sigma_2,\sigma_1) = 0$, for $\sigma_1 < \sigma_2$. Hence, we arrive at

$$\text{r.h.s.} = \int_0^t d\sigma_2 \int_0^\sigma d\sigma_1\ k(\sigma_1,\sigma_1)\ k(\sigma_2,\sigma_2) = \frac{1}{2!}\left\{ \frac{1}{2}\int_0^t K(\sigma)d\sigma\right\}^2. \tag{4.14}$$

The remaining terms can be dealt with in a like manner. □

Now let G(z) be an entire function of z, $G(z) = \sum_{n=0}^{\infty} a_n z^n$. Let $g \in \mathcal{S}(H)$ and define

$G(g) = \sum_{n=o}^{\infty} a_n g^n : H \to \mathbb{C}$, $G(g) \in \mathcal{F}(H)$. Then from the Cameron-Martin formula for a C^o discontinuous $(1+K) : H \to H$ with $\mathrm{Det}(1+K) \neq 0$ we obtain

$$\mathcal{F}^s[\, e_s^K\, G(g)_{1+K}] = |\mathrm{Det}(1+K)|^{-1} \mathcal{F}^s[G(g)] = |\mathrm{Det}(1+K)|^{-1} \sum_{n=o}^{\infty} a_n \mathcal{F}^s[g^n]. \tag{4.15}$$

r.h.s. being convergent because $|\mathcal{F}^s[g^n]| \leqslant \|g^n\|_o \leqslant \|g\|_o^n$. We shall use this result in conjunction with the above lemma in the next section.

Until now we have been content to discuss the mathematical properties of the Feynman maps $\mathcal{F}^s$. We now go on to discuss some of the applications to non-relativistic quantum mechanics of the Feynman map $\mathcal{F}^1 = \mathcal{F}$. Most of our applications will depend on the Cameron-Martin formula above and the Feynman-Itô formula of the next section. This Feynman-Itô formula expresses the wave-function solution of the time-dependent Schrödinger equation as a Feynman integral.

5. The Feynman-Itô Formula

We begin by discussing the Feynman-Dirac conjecture for a spinless non-relativistic quantum particle moving in a potential V on the real line $\mathbb{R}^1$. Initially we choose units so that the particle mass, $m = 1$, and Planck's constant divided by 2π, $\hbar = 1$.

Let $(P_n\gamma)$ denote the polygonal path defined by

$$(P_n\gamma)(\tau) = \gamma_j + \left(\tau - \frac{jt}{n}\right)\left(\gamma_{j+1} - \gamma_j\right)\frac{n}{t}, \qquad \frac{jt}{n} \leqslant \tau < \frac{(j+1)t}{n}, \tag{5.1}$$

$j=0,1,2,\dots,(n-1)$, where initially γ_j are fixed and arbitrary, save for γ_n and $\gamma_n = 0$. Denote by $(P_n\gamma + X)$ the polygonal path

$$(P_n\gamma + X)(\tau) = (P_n\gamma)(\tau) + X, \qquad \tau \in (0,t), \tag{5.2}$$

for a constant X. Also let $S_{cl}[P_n\gamma + X]$ denote the classical action of the unit mass particle in the potential V in $\mathbb{R}^1$

$$S_{cl}[P_n\gamma + X] = \sum_{j=o}^{n-1} \frac{(\gamma_{j+1} - \gamma_j)^2}{2\Delta t} - \int_0^t V[P_n\gamma + X]\, d\tau, \tag{5.3}$$

where $\Delta t = t/n$.

Now let $\psi(X,t)$ be the quantum mechanical amplitude for the particle of mass unity in the potential V to be at X at time t. Then, following earlier work of Dirac, Feynman conjectured

$$\psi(X,t) = \lim_{n\to\infty} N_n \int d^n\gamma \exp\left\{iS_{cl}[P_n\gamma + X]\right\} \psi(\gamma_o + X, 0), \tag{5.4}$$

where $N_n = (2\pi i\Delta t)^{-\frac{n}{2}}$ is a normalisation constant and $d^n\gamma = d\gamma_o \ldots d\gamma_{n-1}$, each integration being from $-\infty$ to $+\infty$ [11], [14].

If one replaces $\int_0^t V[P_n\gamma + X]\,d\tau$ by its Riemann sum approximation $\sum_{j=o}^{n-1} V[\gamma_j + X]\Delta t$, one obtains the approximate Feynman-Dirac conjecture proved by Nelson using the Lie-Kato-Trotter product formula [7], [27]. The corresponding result for the diffusion (heat) equation is the Feynman-Kac formula first proved in a seminal paper of Kac [19], [20].

If one does not make the Riemann sum approximation for the potential term, then the exact Feynman-Dirac conjecture is equivalent to

$$\psi(X,t) = ⨍ [\exp\{-i \int_0^t V[\gamma(\tau) + X]\,d\tau\}\phi[\gamma(0) + X]], \tag{5.5}$$

where $\psi(\cdot,0) = \phi(\cdot)$. The proof of this result is the content of the next theorem.

<u>Theorem 4</u>

The (weak) solution of the Schrödinger equation

$$i\frac{\partial\psi}{\partial t} = -\frac{1}{2}\frac{\partial^2\psi}{\partial X^2} + V[X]\,\psi \tag{5.6}$$

with Cauchy data $\psi(X,0) = \phi(X) = \int \exp(i\alpha X)d\nu(\alpha) \epsilon L^2(\mathbb{R}^1)$ with a real-valued (anharmonic) potential $V = AX^2 + BX + C + \int \exp(i\alpha X)d\mu(\alpha)$; μ,ν being measures of bounded variation on $\mathbb{R}^1$, $A > 0$, is

$$\psi(X,t) = ⨍ [\exp\{-i \int_0^t V[\gamma(\tau) + X]\}\,\phi[\gamma(0) + X]]. \tag{5.7}$$

<u>Proof</u>

We put $A = w^2/2$ and $B = C = 0$. The apparently more general result is no more difficult to establish. The key element in the proof is the evaluation of the Feynman integral $I(X,\alpha_o)$.

$$I(X,\alpha_o) = ⨍[\exp\{-i\int_0^t \frac{w^2}{2}(X + \gamma(\tau))^2 d\tau\}\exp\{i\sum_{j=1}^{n}\alpha_j(X + \gamma(\sigma_j))\}\exp\{i\alpha_o(X+\gamma(\sigma_o))\}]. \tag{5.8}$$

Let us show how to evaluate this Feynman integral when $t < \pi/2w$. We shall use the Cameron-Martin formula.

We define $(1+K) : H \to H$ by $(K\gamma)(\tau) = -w \int_\tau^t \tan(w\sigma)\gamma(\sigma)\,d\sigma$. Then a simple integration by parts gives

$$(K\gamma,\gamma) + \frac{1}{2}(K\gamma,K\gamma) = \frac{-w^2}{2}\int_0^t \gamma^2(\sigma)\,d\sigma. \tag{5.9}$$

Hence, using our earlier notation,

$$e_1^K[\gamma] = \exp\left\{\frac{-iw^2}{2}\int_0^t \gamma^2(\sigma)\,d\sigma\right\} \tag{5.10}$$

and from above lemma

$$\mathrm{Det}(1+K) = [\cos(wt)]^{\frac{1}{2}}. \tag{5.11}$$

Define the continuous function $\gamma^\sigma(\cdot) \in H$ by

$$\gamma^\sigma(\tau) = -\cos(w\sigma)\ \ln\left[\frac{\sec(w\tau)+\tan(w\tau)}{\sec(wt)+\tan(wt)}\right], \text{ for } \tau > \sigma. \tag{5.12}$$

Then it can be shown that, for $\sigma \in [0,t]$, $\gamma \in H$,

$$[(1+K)^{-1}\gamma](\sigma) = (\gamma^\sigma,\gamma) \tag{5.13}$$

and, for $\sigma \leqslant \tau$, $\sigma,\tau \in [0,t]$,

$$(\gamma^\sigma,\gamma^\tau) = \frac{\cos(w\sigma)[\sin w(t-\tau)]}{w\cos(wt)} = g_o(\sigma,\tau). \tag{5.14}$$

Here the significance of g_o is that $g_o(\sigma,\tau)$ is the Green's function of the Sturm-Liouville differential operator $-\left(\frac{d^2}{d\tau^2}+w^2\right)$ with boundary conditions $\frac{dg_o(\sigma,\tau=0)}{d\tau} = 0$, $g_o(\sigma,\tau=t) = 0$.

A simple calculation using the Cameron-Martin formula and the above inner products now gives, for $t < \pi/2w$,

$$I(X,\alpha_o) = [\cos(wt)]^{-\frac{1}{2}} \exp\left\{\frac{-iwX^2}{2}\tan(wt)\right\}\exp\left\{i\sum_{j=o}^{n}\frac{\alpha_j\cos(w\sigma_j)X_j}{\cos(wt)}\right\}\exp\left\{\frac{-i}{2}\sum_{o}^{n}\alpha_j\alpha_k g_o(\sigma_j,\sigma_k)\right\}. \tag{5.15}$$

An explicit calculation yields the same result for $t \neq \left(n+\frac{1}{2}\right)\pi/w$, $n = 0,1,2,\ldots$

We now follow Albeverio and Høegh-Krohn [1]. Let $\Lambda_o(X)$ denote the normalised ground state wave-function of $H_o = \left(-\frac{1}{2}\frac{\partial^2}{\partial X^2} + \frac{w^2X^2}{2}\right)$, $\Lambda_o(X) = \left(\frac{w}{\pi}\right)^{\frac{1}{4}} \exp\left(\frac{-wX^2}{2}\right)$. Define

$$\exp\{i\alpha X(\sigma)\} = \exp\{-i\sigma H_o\} \exp\{(i\alpha X)\} \exp\{i\sigma H_o\} . \tag{5.16}$$

Then, putting $\sigma_o = 0, \sigma_n = t$, an explicit determination of r.h.s. yields, for $0 \leqslant \sigma_1 \leqslant \sigma_2, \ldots \leqslant \sigma_{n-1} \leqslant t$,

$$(\Lambda_o, \exp\left\{i \sum_{j=1}^{n} \alpha_j X(\sigma_j)\right\} \exp(-itH_o)\Lambda_o)_{L^2} = \int \Lambda_o^*(X)\; I(X,\alpha_o) \left|\tilde{\Lambda}(\alpha_o)dx\, d\alpha_o\right._{\sigma_o=0,\sigma_n=t} , \tag{5.17}$$

where $\sim$ is Fourier transform and we are using the known value of l.h.s. for harmonic oscillator Hamiltonians H_o,

$$\text{l.h.s.} = \exp\left\{-\frac{itw}{2}\right\} \exp\left\{-\frac{1}{4w}\sum_1^n \alpha_j\, e^{-i|t_j-t_k|w}\, \alpha_k\right\} . \tag{5.18}$$

Let $H = H_o + V(X)$, where $V(X) = \int \exp(i\alpha X)d\mu(\alpha)$ and let $g(X) = \int \exp(i\alpha X)d\nu(\alpha)$, $f(X) = \int \exp(i\alpha X)d\nu'(\alpha)$; μ, ν and ν' being of bounded absolute variation on $\mathbb{R}^1$. Then a consequence of the last equation is that

$$(\Lambda_o, f \exp(-itH)g\Lambda_o)_{L^2} = \int f(X)\Lambda_o^*(X) \mathcal{F}\left[\exp\left\{-i\int_0^t \frac{w^2}{2}[X+\gamma(\tau)]^2 d\tau\right\}\right.$$

$$\left.\exp\left\{-i\int_0^t V[X+\gamma(\tau)]d\tau\right\} g[X+\gamma(0)]\Lambda_o[X+\gamma(0)]\right] , \tag{5.19}$$

where the Feynman integral on r.h.s. is by definition the L^2-valued function of X

$$\sum_{n=0}^{\infty} \frac{1}{n!}\mathcal{F}\left[\exp\left\{-i\int_0^t \frac{w^2}{2}[X+\gamma(\tau)]^2 d\tau\right\}\left\{-i\int_0^t V[X+\gamma(\tau)]d\tau\right\}^n g[X+\gamma(0)]\Lambda_o[X+\gamma(0)]\right]$$

$$= [\exp\{-itH\}g\Lambda_o][X] , \tag{5.20}$$

both sides being given by the norm convergent Dyson series for $[\exp\{-it(H_o+V)\}g\Lambda_o][X]$.

Since $\mathcal{S}(\mathbb{R})\Lambda_o$ is dense in $L^2(\mathbb{R})$ the result follows for $t \neq \left(n+\frac{1}{2}\right)\pi$, $n = 0,1,2\ldots$. For other values of t the result follows from continuity considerations [1][41]. □

6. The Quasiclassical Representation and Maslov's Results

In this section we show how the celebrated results of Maslov on the classical mechanical limit of quantum mechanics can be obtained from the polygonal path formulation of the Feynman integral [23]-[25]. The derivation uses the quasiclassical representation of the next theorem and the Cameron-Martin formula above [36].

Theorem 5

Let $\psi(x,t)$ be the solution of the Schrödinger equation

$$i\hbar \frac{\partial\psi}{\partial t} = \frac{-\hbar^2}{2m}\frac{\partial^2\psi}{\partial x^2} + V(x)\psi, \tag{6.1}$$

with Cauchy data $\psi(x,0) = \phi(x) \in L^2(\mathbb{R}^1)$, where V is a anharmonic potential and ϕ is the Fourier transform of a measure of bounded absolute variation. Let $X_{cl}(\tau) \in C^2(0,t)$ be the real solution of

$$m\ddot{X}_{cl}(\tau) = -\frac{\partial V}{\partial X}[X_{cl}(\tau)], \qquad \tau \in [0,t], \tag{6.2}$$

with $X_{cl}(t) = x$ and define the classical action $S_{cl} = \int_0^t \frac{m}{2}\dot{X}^2_{cl}(\tau)d\tau - \int_0^t V[X_{cl}(\tau)]d\tau$. Then

$$\exp\{-iS_{cl}/\hbar\}\psi_\hbar(x,t) = \mathcal{F}\left[\exp\left\{\frac{-i}{\hbar}\int_0^t \Delta^2 V d\tau\right\}\exp\left\{-i\left(\frac{m}{\hbar}\right)^{\frac{1}{2}}\frac{dX_{cl}}{d\tau}(0)\Gamma(0)\right\}\phi\left[X_{cl}(0)+\left(\frac{\hbar}{m}\right)^{\frac{1}{2}}\Gamma(0)\right]\right], \tag{6.3}$$

where

$$\Delta^2 V = V\left[X_{cl}(\tau) + \left(\frac{\hbar}{m}\right)^{\frac{1}{2}}\Gamma(\tau)\right] - V[X_{cl}(\tau)] - \left(\frac{\hbar}{m}\right)^{\frac{1}{2}}\Gamma(\tau)\frac{\partial V[X_{cl}(\tau)]}{\partial X}. \tag{6.4}$$

Proof

The proof utilises the translation invariance of $\mathcal{F}$ under $\gamma \to \gamma + a$, $\gamma \in H$, fixed $a \in H$, $f \in \mathcal{F}(H)$,

$$\mathcal{F}[f[\gamma + a]] = \exp\left\{\frac{i}{2}a^2\right\}\mathcal{F}\left[\exp\left\{-i(a,\gamma)\right\}f[\gamma]\right]. \tag{6.5}$$

First we observe that

$$\left.\begin{array}{l} i\hbar \dfrac{\partial\psi}{\partial t} = -\dfrac{\hbar^2}{2m}\dfrac{\partial^2\psi}{\partial x^2} + V(x)\psi \\ \\ \text{with } \psi(x,0) = \phi(x) \end{array}\right\} \Leftrightarrow \left\{\begin{array}{l} i\dfrac{\partial\Psi}{\partial t} = -\dfrac{1}{2}\dfrac{\partial^2\Psi}{\partial X^2} + \dfrac{1}{\hbar}V\left[\left(\dfrac{\hbar}{m}\right)^{\frac{1}{2}}X\right]\Psi \\ \\ \text{with } X = \left(\dfrac{m}{\hbar}\right)^{\frac{1}{2}}x,\ \Psi(X,0) = \phi\left(\left(\dfrac{\hbar}{m}\right)^{\frac{1}{2}}X\right). \end{array}\right. \tag{6.6}$$

We now express the wave-function solution of the Schrödinger equation on r.h.s. above as a Feynman integral by using the Feynman-Itô formula. Applying translation formula above with $a(\tau) = \left(\frac{m}{\hbar}\right)^{\frac{1}{2}}[X_{cl}(\tau) - x]$ the result follows after an integration by parts. □

In the quasiclassical representation we can choose $\frac{dX_{cl}}{d\tau}(0)$ for convenience depending upon the exact form of the initial wave-function ϕ. With the appropriate choice of $\frac{dX_{cl}(0)}{d\tau}$, for sufficiently regular V and ϕ, the quasiclassical representation gives a formal power series for ψ in ascending powers of $\hbar$.

Let us see how this procedure works in the particular case $\psi(x,0)=\exp\{iS_o(x)/\hbar\}\psi_o(x)$, where S_o is real and S_o,ψ_o are independent of $\hbar$. This initial data corresponds to an initial particle distribution $\rho_o(x) = |\psi_o(x)|^2$ and to a limiting value of the probability current $j_{\hbar=0} = S_o'(x)\rho_o(x)/m$, giving an initial particle flux associated with the velocity field $S_o'(x)/m$.

Now let $x[x_o,p_o,\tau]$ be the (unique) solution of the classical equations

$$m\frac{d^2x}{d\tau^2} = -\frac{\partial V[x]}{\partial x} \tag{6.7}$$

with $m\frac{dx}{d\tau}(0) = p_o$, $x(0) = x_o$. We shall assume that the classical problem satisfies:-

(1) The solution $x = x[x_o,p_o,t]$ exists and is unique for $t \in [0,T]$.

(2) The equation $x = x[x_o,S_o'(x_o),t]$ can be solved for x_o to yield n distinct solutions $x_o^i = x_o^i(x,t)$, $i = 1,2,..,n$.

(3) There are n solutions $X_i = X_i(x,\tau)$ of the equation

$$m\frac{d^2X_i(x,\tau)}{d\tau^2} = -\frac{\partial V[X_i(x,\tau)]}{\partial X}, \qquad \tau \in [0,t], \tag{6.8}$$

with $X_i(x,t) = x$, $X_i(x,0) = x_o^i(x,t)$, $m\frac{dX_i(x,0)}{d\tau} = S_o'[X_i(x,0)]$, $i=1,2,..,n$.

Specifically $X_i(x,\tau) = x[x_o^i(x,t), S_o'[x_o^i(x,t)],\tau]$, $\tau \in [0,t]$, $i=1,2,..,n$.

(4) For $i = 1,2,..,n$, $X_i(x,\tau)$, $\tau \in [0,t]$, does not pass through any focus of the classical problem so that $\frac{\partial X_i(x,\tau)}{\partial x_o^i} \neq 0$, $\tau \in [0,t]$, $i = 1,2,..,n$. (We shall relax this assumption in a moment.)

<u>Corollary 2</u>

Under the above assumptions, for $\psi_o \in C_o^\infty$, the first term ψ_Q in the quasiclassical representation for the wave-function solution $\psi_\hbar$ with initial data $\psi_\hbar(x,0) = \exp\{iS_o(x)/\hbar\}\,\psi_o(x)$ is given by

$$\psi_Q(x,t) = \sum_{j=1}^{n} \exp\left\{\frac{i\,\tilde{S}[X_j]}{\hbar}\right\} \psi_o\left[x_o^j(x,t)\right] \left|\frac{\partial x_o^j(x,t)}{\partial x}\right|^{\frac{1}{2}}, \tag{6.9}$$

where $\tilde{S}[X_j] = S_o[x_o^j(x,t)] + \int_0^t \left\{\frac{m}{2}\dot{X}_j^2(x,\tau) - V[X_j(x,\tau)]\right\}d\tau.$

Proof

Before considering the quasiclassical representation we construct a partition of unity so that $\psi_o = \sum_{i=1}^{n} \psi_o^i$, where $\psi_o^i(x') = \psi_o(x')$, $|x'-x_o^i(x,t)| < \varepsilon_i$ and $\psi_o^i(x') = 0$, $|x'-x_o^i(x,t)| > \delta_i$, $\delta_i > \varepsilon_i$ and $N(x_o^i,\varepsilon_i) \cap N(x_o^j,\delta_j) = \Phi$, $i,j = 1,2,..,n$, $i \neq j$. We now substitute $\psi_o = \sum_{j=1}^{n} \psi_o^j$ in the Feynman-Itô formula for the Schrödinger equation (6.6).

We form a quasiclassical representation for each of the terms ψ_o^j in the sum. For the jth term we choose $X_{cl}(\tau) = X_j(x,\tau)$ and proceed as above. In this way the jth term becomes an expansion about $X_j(x,\tau)$. Letting $\hbar \to 0$ in each of these terms gives, after a little manipulation,

$$\psi_Q(x,t) = \sum_{j=1}^{n} \exp\left\{\frac{i}{\hbar}\tilde{S}[X_j]\right\} \psi_o[x_o^j(x,t)]\,\mathcal{F}_j, \tag{6.10}$$

where $\mathcal{F}_j$ denotes the Feynman integral

$$\mathcal{F}_j = \mathcal{F}\left[\exp\left\{-\frac{i}{2m}\int_0^t \gamma^2(\tau)V''[X_j(x,\tau)]d\tau\right\}\exp\left\{\frac{i}{2m}\gamma^2(0)S_o''[X_j(x,0)]\right\}\right]. \tag{6.11}$$

To complete the proof we evaluate each of the Feynman integrals $\mathcal{F}_j$. Since all the integrals are evaluated in the same way we drop the index j.

Define the Jacobi field $J(\sigma) = \frac{\partial X}{\partial x_o}(x,\sigma)$. Partially differentiating w.r.t. x_o the classical equations of motion for $X(x,\sigma)$, we see that

$$\ddot{J}(\sigma) + m^{-1}V''[X(x,\sigma)]J(\sigma) = 0, \qquad \sigma \in [0,t], \tag{6.12}$$

with $J(\sigma = 0) = 1$ and $\dot{J}(\sigma = 0) = m^{-1}S_o''[X(x,0)]$.

We now consider the C^o discontinuous $(1+K) : H \to H$ with $(K\gamma)(\tau) = \int_\tau^t \dot{J}(\sigma)J^{-1}(\sigma)\gamma(\sigma)d\sigma$.

A simple integration by parts gives

$$e^{K}_{1}[\gamma] = \exp\left\{-\frac{i}{2m}\int_0^t \gamma^2(\tau)V''[X(x,\tau)]d\tau\right\}\exp\left\{\frac{i}{2m}\gamma^2(0)S_o''[X(x,0)]\right\}. \tag{6.13}$$

As in the proof of the Feynman-Itô formula define $\gamma^{\sigma}(\cdot)\;\varepsilon\; H$ by $\frac{d\gamma^{\sigma}}{d\sigma'}(\sigma') = \theta(\sigma'-\sigma)J(\sigma)J^{-1}(\sigma')$. Then we have

$$[(1+K)^{-1}\gamma](\sigma) = (\gamma^{\sigma},\gamma). \tag{6.14}$$

Moreover, we have

$$(\gamma^{\sigma},\gamma^{\tau}) = J(\sigma)J(\tau)\int_{\sigma\vee\tau}^{t} J^{-2}(\sigma')d\sigma' = g_x(\sigma,\tau), \tag{6.15}$$

where $g_x(\sigma,\tau)$ is the Green's function of the Sturm-Liouville differential operator $-\left(\frac{d^2}{d\sigma^2}+m^{-1}V''[X(x,\sigma)]\right)$, with boundary conditions $\frac{dg_x(\sigma=0,\tau)}{d\sigma} - m^{-1}S_o''[X(x,0)]g_x(\sigma=0,\tau) = 0$, $g_x(\sigma=t,\tau) = 0$.

From Lemma 3, since $\int_0^t \dot{J}(\sigma)J^{-1}(\sigma)d\sigma = \ln\left[\frac{\partial X(x,t)}{\partial x_o}\right]$, we obtain using the Cameron-Martin formula

$$\oint \left[\exp\left\{-\frac{i}{2m}\int_0^t \gamma^2(\tau)V''[X(x,\tau)]d\tau\right\}\exp\left\{\frac{i}{2m}\gamma^2(0)S_o''[X(x,0)]\right\}\right] = \left|\frac{\partial X}{\partial x_o}(x,t)\right|^{-\frac{1}{2}}. \tag{6.16}$$

The final result follows from the implicit function theorem. □

The last result is not valid if any of the trajectories X_j pass through a focus. For in this case $(1+K)$ is no longer C^o discontinuous. However, Corollary 3 is valid if we replace assumption (4) by assumption (4)$'$.

(4)$'$ For $j=1,2,..,n$, $X_j(x,\tau)$, $\tau\;\varepsilon\;[0,t]$, passes through only a finite number of foci, τ_k, $k=1,2,..,N_j$, so that $\frac{\partial X_j(x,\tau_k)}{\partial x_o^j} = 0$, $\tau_k\;\varepsilon\;(0,t)$, $k=1,2,—,N_j$, $j=1,2,..,n$.

Corollary 3

Under assumptions (1), (2), (3), (4)$'$, for $\psi_o\,\varepsilon\, C_o^{\infty}$, the first term ψ_Q in the quasi-classical representation for $\psi_{\hbar}$ with initial data $\psi_{\hbar}(x,0) = \exp\{iS_o(x)/\hbar\}\psi_o(x)$ is given by

$$\psi_Q(x,t) = \sum_{j=1}^{n}\exp\left\{\frac{i}{\hbar}\tilde{S}[X_j]\right\}\psi_o[x_o^j(x,t)]\left|\frac{\partial x_o^j(x,t)}{\partial x}\right|^{\frac{1}{2}}\exp\left\{\frac{-i\pi}{2}M(X_j)\right\}, \tag{6.17}$$

where $M(X_j)$ is the Maslov index of X_j i.e. $M(X_j)$ is the number of zeros of

$\dfrac{\partial X_j(x,\tau)}{\partial x_o^j}$ as τ varies on the interval $(0,t)$, $j = 1,2,\ldots,n$.

Proof

When X_j passes through a finite number of foci we show how the polygonal path formulation of the Feynman integral leads to

$$\mathcal{F}\left[\exp\left\{-\frac{i}{2m}\int_0^t \gamma^2(\tau)V''[X_j(x,\tau)]\,d\tau\right\}\exp\left\{\frac{i}{2m}\gamma^2(0)S_o''[X_j(x,0)]\right\}\right] = \exp\left\{\frac{-i\pi M(X_j)}{2}\right\}$$

$$\left|\frac{\partial x_o^j(x,t)}{\partial x}\right|^{\frac{1}{2}}, \tag{6.18}$$

$M(X_j)$ being the Maslov index of X_j, which in this case is just the number of foci on X_j. The evaluation method given here is partly based on a method of evaluating Wiener integrals described by Gelfand and Yaglom [16]. The method is originally due to Montroll [26]. Again we drop the index j.

It is convenient to denote by $p(\tau)$, $p(\tau) = m^{-1}V''[X(x,\tau)]$. We assume $p(\tau)$ is continuous on $[0,t]$.

Then explicit substitution for $(P_n\gamma)$ and the mean value theorem for integrals give

$$\mathcal{F}_n\left[\exp\left\{-\frac{i}{2m}\int_0^t V''[X(x,\tau)]\gamma^2(\tau)\,d\tau\right\}\exp\left\{\frac{i}{2m}S_o''[X(x,0)]\gamma^2(0)\right\}\right] = (2\pi i\Delta t)^{-\frac{n}{2}}$$

$$\int \exp\left\{\frac{i}{2\Delta t}\gamma^T D^n \gamma\right\} d^n\gamma, \tag{6.19}$$

where $\gamma^T = (\gamma_0,\gamma_1,\ldots,\gamma_{n-1})$, $\gamma_j = \gamma\left(\frac{jt}{n}\right)$, $j = 0,1,\ldots,n-1$, $\Delta t = t/n$,

$$D^n = \begin{bmatrix} 1-(\Delta t)^2q_0/3 & 1-(\Delta t)^2p_0/6 & 0 & 0 & \ldots 0 & 0 \\ -1-(\Delta t)^2p_0/6 & 2-(\Delta t)^2q_1/3 & -1-(\Delta t)^2p_1/6 & 0 & \ldots 0 & 0 \\ 0 & -1-(\Delta t)^2p_1/6 & 2-(\Delta t)^2q_2/3 & -1-(\Delta t)^2p_2/6 & \ldots 0 & 0 \\ - & - & - & - & \ldots - & - \\ 0 & 0 & \cdots\cdots & -1-(\Delta t)^2p_{n-2}/6 & 2-(\Delta t)^2q_{n-1}/3 & -1-(\Delta t)^2p_{n-1}/6 \\ 0 & 0 & \cdots\cdots & 0 & -1-(\Delta t)^2p_{n-1}/6 & 2-(\Delta t)^2q_n/3 \end{bmatrix} \tag{6.20}$$

$q_j = p_j + p_{j-1}$, $j = 0,1,2,\ldots,n$, $p_j = p(\tau'_j)$, some $\tau'_j \in \left[\frac{jt}{n}, \frac{(j+1)t}{n}\right]$, $j = 0,1,2,\ldots,$ $n-1$, $p_n = 0$ and $p-1 = 3(m\,\Delta t)^{-1} S_o''[X(x,0)]$ [36].

We now make repeated use of the result $\int_{-\infty}^{\infty} \exp\{-zt^2\}dt = \left(\frac{\pi}{z}\right)^{\frac{1}{2}}$, re $z \geqslant 0$, where the $(\ )^{\frac{1}{2}}$ refers to the branch of $(\ )^{\frac{1}{2}}$ which is positive on the positive reals. Then, if the eigenvalues of D^n are λ_j, $j = 1,2,\ldots,n$, $\lambda_j \neq 0$, from above we obtain

$$(2\pi i\Delta t)^{-\frac{n}{2}} \int \exp\left\{\frac{i}{2\Delta t} \gamma^T D^n \gamma\right\} d^n\gamma = (2\pi i\Delta t)^{-\frac{n}{2}} \prod_{j=1}^{n} \left(\frac{2\pi\Delta t}{-i\lambda_j}\right)^{\frac{1}{2}} = \prod_{j=1}^{n} \left(\frac{1}{\lambda_j}\right)^{\frac{1}{2}} = |\det D^n|^{-\frac{1}{2}}$$
$$\exp\left\{\frac{-i\pi}{2} N_-(n)\right\}, \tag{6.21}$$

$N_-(n)$ being the number of negative eigenvalues of D^n. We must now take the limit as n tends to infinity.

Firstly let D^n_k be the minor of order $(k+1)$ in the top left hand corner of D_n. Then we have

$$\frac{(D^n_{k+1} - 2D^n_k + D^n_{k-1})}{(\Delta t)^2} + \frac{(p_k + p_{k+1})}{3} D^n_k + \frac{p_k}{3} D^n_{k-1} = \frac{(\Delta t)^2}{36} p_k^{\,2} D^n_{k-1}, \tag{6.22}$$

for $2 \leqslant k \leqslant n-1$ and

$$D^n_o = 1 - \frac{(\Delta t)^2}{3} p_1 - \frac{\Delta t}{m} S_o''[X(x,0)], \quad \frac{(D^n_1 - D^n_o)}{\Delta t} = \frac{S_o''[X(x,0)]}{m} + O(\Delta t). \tag{6.23}$$

We can in principle, for fixed n, solve the above difference equation for $D^n_{n-1} = \det(D^n)$. We require $\lim_{n\to\infty} |D^n_{n-1}|$ and $\lim_{n\to\infty} N_-(n)$.

Define $D^n(\cdot) \in C^2[0,t]$ by $D^n(k\Delta t) = D^n_k$, $k = 0,1,2,\ldots,n-1$. We see that as $n \to \infty$ $D^n(\tau) \to D(\tau)$, where $D(\tau)$ is the unique solution of

$$\ddot{D}(\tau) + m^{-1} V''[X(x,\tau)]D(\tau) = 0, \qquad \tau \in [0,t], \tag{6.24}$$

with $D(0) = 1$ and $\dot{D}(0) = m^{-1} S_o''[X(x,0)]$. From the last corollary $D(\tau)$ is the Jacobi field $D(\tau) = J(\tau) = \frac{\partial X(x,\tau)}{\partial x_o}$ and $\lim_{n\to\infty} |D^n_{n-1}| = \left|\frac{\partial X(x,t)}{\partial x_o}\right|$.

We now see how to determine $N_-(n)$. We simply complete the square in the exponential integrand. Starting from the top left hand corner of D^n, we see that the coefficients of the new square variables are just D^n_o, D^n_1/D^n_o, $D^n_2/D^n_1, \ldots,$ D^n_{n-1}/D^n_{n-2} i.e. $D^n(0)$, $D^n(\Delta t)/D^n(0), \ldots,$ $D^n\left(\frac{n-1t}{n}\right)/D^n\left(\frac{n-2t}{n}\right)$. Hence, for sufficiently large n and 'suitably smooth' D^n, $N_-(n)$ is just the number of sign changes of $D^n(\cdot)$ on the interval $[0,t)$.

We have seen that $D^n(\tau) \to \frac{\partial X(x,\tau)}{\partial x_o}$ as $n \to \infty$. The function $\frac{\partial X(x,\tau)}{\partial x_o} \in C^2[0,t]$ has only a finite number of simple zeros at $\tau = t_k \in (0,t)$, $k = 1,2,\ldots,N$. (The zeros have to be simple otherwise $J(\tau) = \frac{\partial X(x,\tau)}{\partial x_o}$, being the solution of a linear second order differential equation, would be identically zero.) We now assume $D^n(\tau) \to J(\tau)$ and $D^{n'}(\tau) \to J'(\tau)$ uniformly on $[0,t]$. Then, since $J(\tau) \in C^2[0,t]$ has only a finite number of simple zeros, $D^n(\tau)$ is 'suitably smooth' for large n and $N_-(n) \to M(X)$, the number of (simple) zeros of $\frac{\partial X(x,\tau)}{\partial x_o}$ on the interval $[0,t)$. Combining these results proves the corollary. □

7. The Quasiclassical Propagator

In this final section we show how the polygonal path formulation of the Feynman integral and the quasiclassical representation lead to Maslov's results for the propagator and the flat-space semiclassical expansion of DeWitt [10], [23] - [25]. We finally use Maslov's quasiclassical propagator to give an elementary (possibly new) derivation of the Bohr-Sommerfeld quantisation for a particle in a potential well.

If we put the initial data $\phi = \delta_\xi$ in the quasiclassical representation, we obtain the representation for the Green's function

$$G(x,\xi,t) = \exp\{i\, S_{cl}/\hbar\} \oint \Bigg[\exp\left\{\frac{-i}{\hbar}\int_0^t \Delta^2 V\, d\tau\right\}\exp\left\{-i\left(\frac{m}{\hbar}\right)^{\frac{1}{2}} \frac{dX_{cl}}{d\tau}(0)\gamma(0)\right\}$$

$$\delta\left[\left(\frac{\hbar}{m}\right)^{\frac{1}{2}}\gamma(0) + X_{cl}(0) - \xi\right]\Bigg] . \tag{7.1}$$

This equation expresses the Green's function as a Feynman integral with an integrand containing the Dirac δ-function. Evidently we are required to choose $X_{cl}(0) = \xi$. Let us see how this is possible. We make similar assumptions to those made in the last section. We assume:-

(1)″ The solution $x = x[x_o,p_o,t]$ exists and is unique for $t = [0,T]$.

(2)″ The equation $x = x[x_o,p_o,t]$ can be solved for p_o to yield n distinct solutions $p_o^j = p_o^j(x_o,x,t)$, $j = 1,2,\ldots,n$.

(3)″ There are n solutions $X_j = X_j(x,\tau)$ of the equation

$$m\frac{d^2X_j(x,\tau)}{d\tau^2} = -\frac{\partial V[X_j(x,\tau)]}{\partial X}, \qquad \tau \in [0,t] , \tag{7.2}$$

with $X_j(x,t) = x, X_j(x,0) = \xi$. Specifically $X_j(x,\tau) = x[\xi,p_o^j(\xi,x,t),\tau]$, $\tau \in [0,t]$, $j = 1,2,\ldots,n$.

(4)″ For $j = 1,2,\ldots,n$, $X_j(x,\tau)$ passes through only a finite number of foci τ_k, $k = 1,2,\ldots,N_j$ as τ varies on the interval $(0,t)$ so that $\frac{\partial X_j(x,\tau_k)}{\partial p_o^j} = 0$,

$\tau_k \in (0,t)$, $k = 1,2,\ldots,N_j$, $j = 1,2,\ldots,n$.

If each of the X_j makes a contribution G_j to G as in the last section, according to the above, we have

$$G_j(x,\xi,t) = \left(\frac{m}{\hbar}\right)^{\frac{1}{2}} \exp\{i\, S[X_j]/\hbar\} \oint \left[\exp\left\{\frac{-i}{\hbar}\int_0^t \Delta^2 V_j d\tau\right\}\delta[\gamma(0)]\right], \tag{7.3}$$

where $S[X_j] = \int_0^t \left\{\frac{m}{2}\dot{X}_j^2(x,\tau) - V[X_j(x,\tau)]\right\}d\tau$ and $\Delta^2 V_j = V\left[X_j(x,\tau) + \left(\frac{\hbar}{m}\right)^{\frac{1}{2}}\gamma(\tau)\right]$ $- \left(\frac{\hbar}{m}\right)^{\frac{1}{2}}\gamma(\tau)\frac{\partial V[X_j(x,\tau)]}{\partial X} - V[X_j(x,\tau)]$. Hence, formally, the quasiclassical propagator G_Q is given by

$$G_Q(x,\xi,t) = \left(\frac{m}{\hbar}\right)^{\frac{1}{2}} \sum_{j=1}^{n} \exp\{i\, S[X_j]/\hbar\} \oint \left[\exp\left\{\frac{-i}{2m}\int_0^t \gamma^2(\tau)V''[X_j(x,\tau)]d\tau\right\}\delta[\gamma(0)]\right]. \tag{7.4}$$

We simply evaluate the Feynman integrals on r.h.s. above as in the last section. When $X_j(x,\tau)$ does not pass through a focus this can be done most effectively by using the Cameron-Martin formula. Again we shall drop the index j. First we require an elementary lemma.

<u>Lemma 4</u>

$\forall\, a,b \in H$, $a \neq 0$,

$$\oint[\exp\{i(b,\gamma)\}\delta[(a,\gamma)]] = (2\pi i\|a\|^2)^{-\frac{1}{2}} \exp\left\{\frac{-i}{2\|a\|^2}[\|a\|^2\|b\|^2 - (a,b)^2]\right\}. \tag{7.5}$$

<u>Proof</u>

Put $f[\gamma] = \exp\{i(b,\gamma)\}\delta[(a,\gamma)]$. Then an explicit calculation gives, for $a \neq 0$,

$$\oint_n[f] = [2\pi i(a,P_n a)]^{-\frac{1}{2}} \exp\left\{\frac{-i}{2(a,P_n a)}[(a,P_n a)(b,P_n b) - (a,P_n b)(a,P_n b)]\right\}. \tag{7.6}$$

The result follows from Lemma 1. □

Now let $J(\sigma) = \left.\frac{\partial x[\xi,p_o,\sigma]}{\partial \xi}\right|_{p_o=p_o(\xi,x,t)}$. Then $J(\sigma)$ satisfies the Jacobi field equation

$$\ddot{J}(\sigma) + m^{-1}V''[X(x,\sigma)]J(\sigma) = 0 , \tag{7.7}$$

with $J(\sigma = 0) = 1$, $\dot{J}(\sigma = 0) = m^{-1}\frac{\partial p_o(\xi,x,t)}{\partial \xi} = -m^{-1}\frac{\partial^2 S(x,\xi,t)}{\partial \xi^2}$, $S(x,\xi,t) = S[X]$.

We define $(1+K): H \to H$ by $(K\gamma)(\tau) = \int_\tau^t \dot{J}(\sigma)J^{-1}(\sigma)\gamma(\sigma)d\sigma$ and a simple calculation gives

$$e_1^K[\gamma]\,\delta[\gamma(0)] = \exp\left\{-\frac{i}{2m}\int_0^t \gamma^2(\sigma)V''[X(x,\sigma)]d\sigma\right\}\delta[\gamma(0)] . \tag{7.8}$$

Moreover we can show that

$$[(1+K)^{-1}\gamma](\sigma) = (\gamma^\sigma,\gamma) , \quad \sigma \in [0,t] , \tag{7.9}$$

where in this case

$$(\gamma^\sigma,\gamma^\tau) = g_X(\sigma,\tau) , \tag{7.10}$$

$g_X(\sigma,\tau)$ being the Green's function of the Sturm-Liouville differential operator $-\left(\frac{d^2}{d\tau^2} + m^{-1}V''[X(x,\sigma)]\right)$, with boundary conditions $\frac{dg_X(\sigma=0,\tau)}{d\sigma} + m^{-1}\frac{\partial^2 S(x,\xi,t)}{\partial \xi^2}g_X(\sigma=0,\tau)=0$, $g_X(\sigma=t,\tau) = 0$.

Hence the Cameron-Martin formula gives in the absence of foci

$$\mathcal{F}\left[\exp\left\{-\frac{i}{2m}\int_0^t \gamma^2(\sigma)V''[X(x,\sigma)]d\sigma\right\}\delta[\gamma(0)]\right] = |J(t)|^{-\frac{1}{2}}\,[\delta[(\gamma^0,\gamma)]] \tag{7.11}$$

and, using Lemma 4 and the above innerproducts,

$$\mathcal{F}\left[\exp\left\{-\frac{i}{2m}\int_0^t \gamma^2(\sigma)V''[X(x,\sigma)]d\sigma\right\}\delta[\gamma(0)]\right] = e^{-\frac{i\pi}{4}}|2\pi g_X(0,0)J(t)|^{-\frac{1}{2}} . \tag{7.12}$$

When this expression is substituted back into the above formula for the quasi-classical propagator, after a little further calculation, we obtain Maslov's propagator below with zero Maslov indices [13].

We can now, equally easily, see the origin in our formulation of DeWitt's flat-space semiclassical expansion. We simply evaluate the Feynman integral

$$\mathcal{F}\left[\exp\left\{i\sum_{k=1}^n \alpha_k\gamma(\sigma_k)\right\}\exp\left\{-\frac{i}{2m}\int_0^t \gamma^2(\sigma)V''[X(x,\sigma)]d\sigma\right\}\delta[\gamma(0)]\right], \text{ for } \sigma_k \in [0,t], k=1,2,\ldots,n,$$

as above. A short calculation using the Cameron-Martin formula, Lemma 4 and the above inner products gives (see over)

$$\mathcal{F}\left[\exp\left\{i\sum_{k=1}^{n}\alpha_k\gamma(\sigma_k)\right\}\exp\left\{-\frac{i}{2m}\int_0^t\gamma^2(\sigma)V''[X(x,\sigma)]\,d\sigma\right\}\delta[\gamma(0)]\right] = e^{-\frac{i\pi}{4}}\left|2\pi g_X(0,0)J(t)\right|^{-\frac{1}{2}}$$

$$\times\exp\left\{-\frac{i}{2}\sum_{1}^{n}\alpha_k\,\mathcal{G}_X(\sigma_k,\sigma_i)\alpha_i\right\}, \tag{7.13}$$

where $\mathcal{G}_X(\cdot,\cdot)$ is the Feynman-Green's function

$$\mathcal{G}_X(\sigma,\tau) = g_X(\sigma,\tau) - g_X(\sigma,0)g_X^{-1}(0,0)g_X(0,\tau)\ . \tag{7.14}$$

$\mathcal{G}_X(\sigma,\tau)$ is the Green's function of the Sturm-Liouville differential operator

$-\left(\frac{d^2}{d\sigma^2} + m^{-1}V''[X(x,\sigma)]\right)$, with boundary conditions $\mathcal{G}_X(\sigma = 0,\tau) = 0$, $\mathcal{G}_X(\sigma = t,\tau) = 0$. [13].

Formally equating the coefficients of the various α-products on r.h.s. and l.h.s. of above equation now enables us to evaluate all the remaining Feynman integrals which arise as the coefficients of the powers of $\hbar$ in the quasiclassical representation for the propagator. The resulting formal power series is just the flat-space semiclassical expansion of DeWitt [10]. Here we have derived this semiclassical expansion from the quasiclassical representation. In the original paper the inspired choice of the Feynman-Green's function as covariance for the semiclassical expansion was made by choosing (apparently from Markov considerations) a certain linear transformation. This covariance arises naturally here from the quasiclassical representation.

Let us now see how, even in the presence of foci, a direct evaluation of the Feynman integral leads to the Maslov propagator. We argue as in the last section, dropping the index j. Taking into account the presence of the $\delta[\gamma(0)]$ term gives

$$\mathcal{F}_n\left[\exp\left\{-\frac{i}{2m}\int_0^t V''[X(x,\tau)]\gamma^2(\tau)\,d\tau\right\}\delta[\gamma(0)]\right] = (2\pi i\ \Delta t)^{-\frac{1}{2}}[\det D^{n-1}]^{-\frac{1}{2}}\ , \tag{7.15}$$

where D^{n-1} is the $(n-1)\times(n-1)$ matrix obtained from the matrix D^n of the last section by deleting the first row and the first column. We put $C_k^{n-1} = \Delta t\ D_k^{n-1}$, where D_k^{n-1} is the minor of order k in the top left hand corner of D^{n-1}. Then, just as in the last section,

$$\frac{\left(C_{k+1}^{n-1} - 2C_k^{n-1} + C_{k-1}^{n-1}\right)}{(\Delta t)^2} + \frac{(p_k + p_{k+1})C_k^{n-1}}{3} + \frac{p_k C_{k-1}^{n-1}}{3} = \frac{(\Delta t)^2}{36}\,p_k{}^2 C_{k-1}^{n-1}\ , \tag{7.16}$$

for $2 \leqslant k \leqslant (n-2)$ and

$$c_1^{n-1} = \Delta t\left[2 - \frac{(\Delta t)^2}{3} q_1\right], \qquad \frac{\left(c_2^{n-1} - c_1^{n-1}\right)}{\Delta t} = 1 + O[(\Delta t)^2] . \tag{7.17}$$

As before we define $C^{n-1}(\cdot) \in C^2(0,t)$ by $C^{n-1}\left(\frac{jt}{n-1}\right) = c_j^{n-1}$, $j = 1,2,\ldots,n-1$. Then $C^{n-1}(\tau) \to C(\tau)$ as $n \to \infty$, where

$$\ddot{C}(\tau) + m^{-1}V''[X(x,\tau)]C(\tau) = 0, \qquad \tau \in [0,t], \tag{7.18}$$

$C(0) = 0$ and $\dot{C}(0) = 1$. The unique solution of this equation is

$$C(\tau) = m \frac{\partial x}{\partial p_o}[\xi, p_o(\xi,x,t),\tau] = m \frac{\partial X(x,\tau)}{\partial p_o} .$$

Hence, if $M(X)$ is the number of zeros of $\frac{\partial X(x,\tau)}{\partial p_o}$ on the interval $(0,t)$, arguing as in the last section,

$$\mathcal{F}\left[\exp\left\{-\frac{i}{2m}\int_0^t V''[X(x,\tau)]\gamma^2(\tau)d\tau\right\}\delta[\gamma(0)]\right] = (2\pi i m)^{-\frac{1}{2}}\left|\frac{\partial X(x,t)}{\partial p_o}\right|^{-\frac{1}{2}} \exp\left\{\frac{-i\pi M(X)}{2}\right\} . \tag{7.19}$$

In spite of appearances this is consistent with the above value of the Feynman integral. Combining these results gives Corollary 4 in agreement with Maslov.

<u>Corollary 4</u>

Let the quasiclassical propagator $G_Q(x,\xi,t)$ be the lowest order term in the quasi-classical representation for the Green's function of the time-dependent Schrödinger equation. Then

$$G_Q(x,\xi,t) = \sum_{j=1}^{n} (2\pi i\hbar)^{-\frac{1}{2}} \exp\left\{\frac{i}{\hbar} S[X_j] - \frac{i\pi}{2} M(X_j)\right\} \left|\frac{\partial X^j(x,t)}{\partial p_o^j}\right|^{-\frac{1}{2}}, \tag{7.20}$$

$M(X_j)$ being the Maslov index of the classical trajectory X_j, satisfying $X_j(x,t) = x$, $X_j(x,0) = \xi$, $j = 1,2,\ldots,n$.

It seems fitting to conclude by giving a simple derivation of the Bohr-Sommerfeld quantisation for the classically periodic system of a particle in a potential well V. We emphasise first that n and $M(X_j)$ in above equation are functions of time. We write $M(X_j) = M_j(t)$ and $S[X_j] = S_j(x,\xi,t)$. We tacitly assume that, for some finite α, $n = O(t^\alpha)$ as $t \to \infty$. Then, since $\frac{\partial X_j}{\partial p_o^j}$ can only have simple zeros, for classically periodic systems, $G_Q(x,\xi,\cdot)$ is a tempered distribution with support in $[0,\infty)$. Partly imitating Gutzwiller we take the Fourier (Laplace) transform with respect to t of $G_Q(x,\xi,t)$ [17],[29],[30],[42].

Let δ be an infinitesimal positive quantity. Then we consider $\tilde{G}_Q(x,\xi,E+i\hbar\delta)$.

$$\tilde{G}_Q(x,\xi,E+i\hbar\delta) = \int_0^\infty \exp\left\{\frac{iEt}{\hbar} - \delta t\right\} G_Q(x,\xi,t)\,dt. \tag{7.21}$$

Letting $\hbar \sim 0$ and using the 1-dim principle of stationary phase gives

$$\tilde{G}_Q(x,\xi,E+i\hbar\delta) \sim \sum_{t_k} \sum_{j=1}^{n} \exp\left\{\frac{i}{\hbar}(S_j^k + Et_k) - \frac{i\pi}{2} M_j(t_k) - \delta t_k\right\} \left|\frac{\partial X_j(x,t_k)}{\partial p_o^j}\right|^{-\frac{1}{2}} \left[\frac{\partial^2 S(x,\xi,t_k)}{\partial t_k^2}\right]^{-\frac{1}{2}},$$

where $S_j^k = \int_0^{t_k} \left\{\frac{m}{2}\dot{X}_j^2(x,\tau) - V[X_j(x,\tau)]\right\} d\tau$ and $t = t_k$ is determined by $\frac{\partial S_j(x,\xi,t)}{\partial t} + E = 0.$

i.e. $\frac{m}{2}\dot{X}_j^2(x,t_k) + V[X_j(x,t_k)] = E$, with $X_j(x,t_k) = x$. This last condition simply fixes the energy E of the trajectory X_j which requires $(p_o^j)^2 = 2m[E - V(\xi)]$. It is easy now to show that $(S_j^k + Et_k) = \int_0^{t_k} \sqrt{2m[E - V(X_{jk})]}\,dX_{jk}$, the line integral being along the trajectory $X_{jk} = x[\xi, p_o^j, \tau]$, $\tau \in [0, t_k]$, with $(p_o^j)^2 = 2m[E - V(\xi)]$.

We assume that the potential well V has two simple turning points x_1 and x_2, the unique solutions of the equation $V(\cdot) = E$, where $V(x) < E$, $x_1 < x < x_2$; $V(x) > E$, $x > x_2$ or $x < x_1$. For definiteness choose $x_1 < \xi < x < x_2$. There are two contributions to the j summation $p_j = \pm\sqrt{2m[E - V(\xi)]}$. For the plus sign the corresponding k summation is over the values

$$t_k = T_i, T_i + T_r, T + T_i, T + T_i + T_r, 2T + T_i, \ldots$$

$T = T(E)$ being the classical period, $T_i = T_i(E)$ being the time of travel from ξ to x, $T_r = T_r(E)$ being the time of travel from x to x, undergoing a single reflection at x_2. For the minus sign the corresponding k summation is over the values

$$t_k = T - T_i - T_r,\ T - T_i,\ 2T - T_i - T_r,\ 2T - T_i,\ 3T - T_i - T_r, \ldots\ .$$

Hence, using the periodicity and summing the above geometric series, we obtain, as $\hbar \sim 0$,

$$\tilde{G}_Q(x,\xi,E+i\hbar\delta) \sim \left\{ \sum_{\tau=T_i, T_i+T_r} \exp[-\delta\tau + i\Phi^+(E,\tau)] \left|\frac{\partial X^+(x,\tau)}{\partial p}\right|^{-\frac{1}{2}} \left[\frac{\partial^2 S^+(x,\xi,\tau)}{\partial \tau^2}\right]^{-\frac{1}{2}} \right.$$

$$\left. + \sum_{\tau=T-T_i-T_r, T-T_i} \exp[-\delta\tau + i\Phi^-(E,\tau)] \left|\frac{\partial X^-(x,\tau)}{\partial p}\right|^{-\frac{1}{2}} \left[\frac{\partial^2 S^-(x,\xi,\tau)}{\partial \tau^2}\right]^{-\frac{1}{2}} \right\} [1 - \exp[-\delta T + i\Phi(E,T)]], \tag{7.22}$$

where $\Phi^{\pm}(E,\tau) = \hbar^{-1}\int_0^{\tau}\sqrt{2m[E-V(x^{\pm})]}\,dx^{\pm} - \frac{\pi}{2}M^{\pm}(\tau)$, $x^{\pm} = x[\xi,\pm\sqrt{2m[E-V(\xi)]},\sigma]$, for

$\sigma \in [0,\tau]$, $M^{\pm}(\tau)$ being the corresponding Maslov index. Here, since $M(T) = 2$, $\frac{\partial x}{\partial p}$ vanishing at the turning points x_1 and x_2 ,

$$\Phi^{\pm}(E,T) = 2\hbar^{-1}\int_{x_1}^{x_2}\sqrt{2m[E-V(x)]}\,dx - \pi \stackrel{\text{def}}{=} \Phi(E,t). \tag{7.23}$$

We now consider $\tilde{G}_Q(x,\xi,E+i\hbar\delta)$ as a function of the real variable E. As we increase E, for infinitesimal δ, the modulus of r.h.s. of above equation exhibits a series of pronounced peaks at the values E_j determined by $\Phi(E_j,T) = 2j\pi$, $j = 0,1,2,\ldots$, the value of $\tilde{G}_Q(x,\xi,E+i\hbar\delta)$ being $O(\delta^{-1})$ at each of these peaks.

However, if we assume $V(x)$ is sufficiently smooth and $V(x) \to \infty$ as $|x| \to \infty$, $H = \left[\frac{-\hbar^2}{2m}\frac{d^2}{dx^2} + V(x)\right]$ is self-adjoint on a suitable domain in $L^2(\mathbb{R}^1)$. Also, the spectrum of H on $L^2(\mathbb{R})$, S is discrete, $S = \{E_o,E_1,\ldots\}$, where $E_j \nearrow \infty$. Assuming there is no degeneracy,

$$G(x,\xi,t) = \sum_{j=0}^{\infty} e^{\frac{-iE_jt}{\hbar}}\,\phi_j(x)\phi_j^*(\xi), \tag{7.24}$$

ϕ_j being the normalised eigenfunction of H corresponding to eigenvalue E_j, $j = 0,1,2,\ldots$. It follows that

$$\tilde{G}(x,\xi,E+i\hbar\delta) = \sum_{j=0}^{\infty}\frac{i\hbar\phi_j(x)\phi_j^*(\xi)}{[E-E_j+i\delta\hbar]}, \tag{7.25}$$

$\tilde{G}(x,\xi,E+i\hbar\delta)$ being just the kernel of the resolvent $i\hbar\,[E+i\delta\hbar-H]^{-1}$.

We, therefore, identify the peaks E_j with the eigenvalues E_j' of the Hamiltonian H. This gives an elementary derivation of the classic Bohr-Sommerfeld-Maslov quantisation condition

$$\Phi(E,T) = 2j\pi \Longleftrightarrow \int_{x_1(E)}^{x_2(E)}\sqrt{2m(E-V(x))}\,dx = \left(j+\frac{1}{2}\right)\pi\hbar, \quad j = 0,1,2,\ldots . \tag{7.26}$$

Letting $\delta \to 0$, the residues of $\tilde{G}_Q(x,\xi,E+i\hbar\delta)$ at the pole-eigenvalues above give the W.K.B. eigenfunctions of $H = \left[\frac{-\hbar^2}{2m}\frac{d^2}{dx^2} + V(x)\right]$ as expected. It is gratifying that the Feynman integral provides such a complete calculational scheme in nonrelativistic quantum mechanics.

References

1. S. Albeverio and R. Høegh-Krohn, 'Mathematical Theory of Feynman Path Integrals' (Springer Lecture Notes in Maths 523, Berlin-Heidelberg-New York, 1976).
2. S. Albeverio and R. Høegh-Krohn, Invent. Math. 40, 59-106 (1977).
3. K. Brock, 'On the Feyman integral', Aarhus University Preprint (1976).
4. R. H. Cameron, J. Mathematical Phys. 39, 126-141 (1960).
5. R. H. Cameron, J. anal. Math. 10, 287-361 (1962).
6. R. H. Cameron and W. T. Martin, Trans. Amer. Math. Soc. 58 No. 2, 184-219 (1945).
7. Ph. Combe, G. Rideau, R. Rodriguez and M. Sirugue-Collin, 'On some mathematical problems in the definition of Feynman path integral', Marseille CPT Preprint (1976).
8. C. DeWitt Morette, Comm. Math. Phys. 28, 47-67 (1972).
9. C. DeWitt Morette, Comm. Math. Phys. 37, 63-81 (1974).
10. C. DeWitt Morette, Ann. Phys. 97, 367-399 (1976).
11. P. A. M. Dirac, 'Quantum Mechanics' (Oxford U.P., London, 1930).
12. P. A. M. Dirac, Rev. Modern Phys. 17, 195-199 (1945).
13. K. D. Elworthy and A. Truman, 'Classical mechanics, the diffusion (heat) equation and the Schrödinger equation on a Riemannian manifold' in preparation.
14. R. P. Feynman and A. R. Hibbs, 'Quantum Mechanics and Path Integrals' (McGraw-Hill, New York, 1965).
15. R. P. Feynman, Rev. Modern Phys. 20, 367-387 (1948).
16. I. M. Gelfand and A. M. Yaglom, J. Mathematical Phys. 1, 48-69 (1960).
17. M. C. Gutzwiller, J. Mathematical Phys. 12, 343-358 (1971).
18. K. Itô in 'Proceedings of 5th Berkeley Symposium on mathematical statistics and probability'. (University of California Press, Berkeley, 1967) Vol. II part 1 p.145.
19. M. Kac, 'Probability in the Physical Sciences' (Interscience, New York, 1959) Chap. IV.
20. M. Kac, Trans. Amer. Math. Soc. 65, 1-13 (1949).
21. P. Kree, 'Holomorphie et theorie des distributions en dimension infinie', Conference a Campinas, Aout 1975.
22. A. Maheshwari and C. DeWitt Morette, 'Path integration in nonrelativistic quantum mechanics' to appear in Physics Reports C.
23. V. P. Maslov, Zh. Vych. Mat. 1, 113-128 (1961).
24. V. P. Maslov, Zh. Vych. Mat. 1, 638-663 (1961).
25. V. P. Maslov, 'Theorie des Perturbations et Methods Asymptotiques' (Dunod, Paris, 1972).
26. E. W. Montroll, Commun. Pure Appl. Math. 5, 415-453 (1952).
27. E. Nelson, J. Mathematical Phys. 5, 332-343 (1964).
28. M. Reed and B. Simon, 'Fourier Analysis, Self-adjointness' (Academic, New York, 1975).
29. L. Schwartz, Medd. Lunds. Mat. Sem. Supplementband p.196 (1952).
30. G. R. Screaton and A. Truman, J. Mathematical Phys. 14, 982-985 (1973).
31. E. C. Titchmarsh, 'The Theory of Functions' (Oxford U.P., London, 1952) p.419.

32. J. Tarski, Ann. Inst. H. Poincaré, A15, 107-140 (1971).

33. J. Tarski, Ann. Inst. H. Poincaré, A17, 313-324 (1972).

34. J. Tarski in 'Complex Analysis and its Applications' Vol. III (IAEA, Vienna, 1976).

35. J. Tarski, 'Definitions and selected applications of Feynman-type integrals', Hamburg University Preprint, 1974.

36. A. Truman, J. Mathematical Phys. 17, 1852-1862 (1976).

37. A. Truman, J. Mathematical Phys. 18, 1499-1509 (1977).

38. A. Truman, J. Mathematical Phys. 18, 2308-2315 (1977).

39. A. Truman, J. Mathematical Phys. 19, 1742-1750 (1978).

40. A. Truman in 'Vector Space Measures and Applications I', Proceedings Dublin 1977 (Springer Lecture Notes in Maths 644, Berlin-Heidelberg-New York, 1978).

41. A. Truman, 'The Feynman Map and the anharmonic oscillator' in preparation.

42. A. Voros, 'Semi-classical approximations' D.Ph. T/74 57 CEA Saclay Report.

- SECTION II -

WEYL QUANTIZATION OF CLASSICAL SPIN SYSTEMS

QUANTUM SPINS AND FERMI SYSTEMS

Ph. COMBE [*] , R. RODRIGUEZ [*]
M. SIRUGUE-COLLIN [**] , M. SIRUGUE

C.N.R.S. - LUMINY - Case 907
Centre de Physique Théorique
F-13288 MARSEILLE CEDEX 2 (France)

and

[*]Université d'Aix-Marseille II, UER de Luminy
[**]Université de Provence

INTRODUCTION

In this talk, we propose a somewhat new approach to the treatment of the quantization of spins and fermion systems, avoiding the usual techniques based on Grassmann algebras. Our aim, in the following, is to show that bosons and fermions can be embedded within the same mathematical framework, the commutation and anti-commutation quantum relations of the two systems being extracted from the central extensions of abelian groups.

For spins and Fermi systems, the underlying group is discrete and has all its elements of order two. The study of its central extensions has been initiated by Mackey [1] .

We shall construct all these central extensions satisfying natural physical assumptions. Moreover, the Weyl algebras built on these extensions are unique up to isomorphism.

The Spin operators and Fermi operators are then considered as Weyl quantization of functions on the dual group which is compact, the transition from spin to Fermi operators being an isomorphism of this group preserving the Haar measure.

This allows a formulation of path integral in phase space which in some sense is closer to the usual formulation for spin and Fermi systems in terms of Grassmann algebras [2] .

§1 - GROUP STRUCTURE FOR LATTICE SPIN SYSTEMS AND FERMI SYSTEMS

A natural generalization of canonical commutation relations is to consider an abelian group $\mathcal{G}$ and a central extension $\mathcal{G} \otimes_{\xi} \mathbb{T}$ of $\mathcal{G}$ by the torus $\mathbb{T}$; indeed, such an extension is associated with a multiplier viz , a function $\xi : \mathcal{G} \times \mathcal{G} \longrightarrow \mathbb{T}$ which satisfies the following relation :

(1.1) $$\xi(g_1,g_2)\,\xi(g_1g_2,g_3) = \xi(g_1,g_2g_3)\,\xi(g_2,g_3)\ , \quad \forall g_1,g_2,g_3 \in G\ .$$

Now, if U is a unitary representation of $G \underset{\xi}{\otimes} T$ which satisfies

(1.2) $$U(g,\alpha) = \alpha\, U(g,1)\ , \quad \forall g \in G\ , \alpha \in T\ .$$

Then, one has the following relation :

(1.3) $$U(g)\,U(g') = \xi(g,g')\,\xi(g',g)^{-1}\, U(g')U(g)\ ,$$

where $U(g) = U(g,1)$.

The $U(g)$'s are the generalizations of the usual Weyl operators

(1.4) $$U(p,q) = \exp i(pQ - qP)\ ,$$

$(p,q) \in R^{2n}$ and (1.3) is a generalization of

(1.5) $$U(p,q)\,U(p',q') = \exp(i(pq' - qp'))\,U(p',q')\,U(p,q)\ .$$

Now, if ξ and ζ are two multipliers on G such that there exists an automorphism α of G and a function λ from G to T such that

(1.6) $$\xi(g,g') = \lambda(g)\,\lambda(g')\,\lambda(gg')^{-1}\,\zeta(\alpha(g),\alpha(g'))\ ,$$

then there exists a $*$-isomorphism β from $G \underset{\xi}{\otimes} T$ onto $G \underset{\zeta}{\otimes} T$ such that

(1.7) $$\beta(g,\alpha)_\xi = (\alpha(g), \lambda(g)\alpha)_\zeta$$

In particular, if α is the identity, ξ and ζ are called similar.

Actually, the unitary representations of $G \underset{\xi}{\otimes} T$ satisfying (1.2) are in one to one correspondence with the $*$ representations of a canonical C^*-algebra $\overline{\Delta(G,\xi)}$, [3] , which is generated by the elements δ_g , $g \in G$ which satisfy

(1.9) $$\delta_g\, \delta_{g'} = \xi(g,g')\, \delta_{gg'}\ ,$$

(1.10) $$\delta_g^* = \xi(e,e)^{-1}\,\xi(g,g^{-1})^{-1}\,\delta_{g^{-1}}\ .$$

Let us remark here that one can always choose a multiplier ξ' similar to ξ such that

$$\xi'(e,e) = \xi'(g,g^{-1}) = 1 \quad .$$

As we said, the case of the ordinary Quantum Mechanics corresponds to $\mathcal{G} = R^{2n}$ and $\xi(p,q\,;p',q') = \exp\left(\frac{i}{2}(pq'-p'q)\right)$ with the convention $\hbar = 1$ ($\hbar$ Planck constant).

In the following we shall consider the group $\mathcal{F}_\Lambda \times \mathcal{F}_\Lambda$ where Λ is an at most countable set and $\mathcal{F}_\Lambda$ is the abelian group of the finite subsets of Λ equipped with the symmetric difference (denoted Δ). If Λ is infinite countable, this group is discrete locally compact. We shall denote by $\mathcal{P}_\Lambda \times \mathcal{P}_\Lambda$ its dual which is compact, $\mathcal{P}_\Lambda$ being isomorphic to the group of all subsets of Λ . Let us remark that for the symmetric difference, all elements of $\mathcal{F}_\Lambda$ or $\mathcal{P}_\Lambda$ are of order two.

The central extensions of $\mathcal{F}_\Lambda \times \mathcal{F}_\Lambda$ provide the canonical anticommutation relations (C.A.R) as well as commutation relations for spin systems on a lattice. Moreover, all physical interesting central extensions of $\mathcal{F}_\Lambda \times \mathcal{F}_\Lambda$ are isomorphic.

Actually, if ξ is a multiplier on the above group, the bicharacter defined as $b_\xi(g,g') = \xi(g,g')\,\xi(g',g)^{-1}$ for $g,g' \in \mathcal{F}_\Lambda \times \mathcal{F}_\Lambda$, completely characterizes the central extension of $\mathcal{F}_\Lambda \times \mathcal{F}_\Lambda$ up to isomorphism. In other words, two bicharacters b_ξ and b_ζ associated with two multipliers ξ and ζ are equal if and only if ξ and ζ are similar. Moreover, if b_ξ and b_ζ are two bicharacters on $\mathcal{F}_\Lambda \times \mathcal{F}_\Lambda$ such that there exists an automorphism γ on $\mathcal{F}_\Lambda \times \mathcal{F}_\Lambda$ which satisfies

$$b_\xi(\gamma(g),\gamma(g')) = b_\zeta(g,g') \ , \ \forall g,g' \in \mathcal{F}_\Lambda \times \mathcal{F}_\Lambda \ ,$$

then $\overline{\Delta(\mathcal{F}_\Lambda \times \mathcal{F}_\Lambda, \xi)}$ and $\overline{\Delta(\mathcal{F}_\Lambda \times \mathcal{F}_\Lambda, \zeta)}$ are $\ast$-isomorphic.

Now, we have the following :

<u>Proposition (1.11)</u>

Let Λ a countable set, with even or infinite cardinality $|\Lambda|$. Then, there are only four bicharacters on $\mathcal{F}_\Lambda \times \mathcal{F}_\Lambda$ which are

i) symmetric

ii) such that $b((X,Y);(X,Y)) = 1 \ , \ \forall (X,Y) \in \mathcal{F}_\Lambda \times \mathcal{F}_\Lambda$,

iii) non degenerate

iv) invariant under the group of finite permutations of points in Λ

v) invariant under the permutations of X and Y ,

namely

$$b^S((X,Y);(X',Y')) = (-1)^{|X\cap Y'|+|Y\cap X'|} ,$$
$$b^F((X,Y);(X',Y')) = (-1)^{|X\cap X'|+|Y\cap Y'|+(|X|+|Y|)(|X'|+|Y'|)}$$
$$b^\varphi((X,Y);(X',Y')) = (-1)^{|X\cap X'|+|Y\cap Y'|+|X||X'|+|Y||Y'|}$$
$$b^\sigma((X,Y);(X',Y')) = (-1)^{|X\cap Y'|+|X'\cap Y|+|X||Y'|+|X'||Y|}$$

If $|\Lambda|$ is odd, b^φ and b^σ are degenerate.

The two first conditions are necessary conditions for b to be of the form

$$b_\xi(g,g') = \xi(g,g')\,\xi(g',g)^{-1} \ , \ \forall g,g' \in \mathfrak{F}_\Lambda \times \mathfrak{F}_\Lambda \ ,$$

with ξ a multiplier. Note that if b is symmetric, it is also skew symmetric.

Now, let us remark that every bicharacter χ on $\mathfrak{F}_\Lambda$, invariant by the group of finite permutations of points in Λ is of the form :

(1.12) $$\chi(X,Y) = (-1)^{\alpha|X\cap Y|+\beta|X||Y|} ,$$

where α and β are 0 or 1 . Then, inspection of conditions i) ii) iii) and v) , among all possibilities for the parameters α and β retains the ones which correspond to the bicharacters given in Proposition (1.11).

Now, if one restricts to the subsets of Λ reduced to one point, one can realize that b^S and b^σ , b^F and b^φ correspond respectively to the commutation relations of quantum spin systems on a lattice and to canonical anti-commutation relations.

Let us call $\delta^i_{X,Y}$ for $i = S, \sigma, F$ and φ the elements of the Weyl algebra $\overline{\Delta(\mathfrak{F}_\Lambda\times\mathfrak{F}_\Lambda, \xi^i)}$, b^i corresponding to ξ^i . Then one has, for X and Y restricted to one point, say $\{i\}$, or ϕ , the empty set :

(1.13)
$$\delta^S_{i\phi}\,\delta^S_{j\phi} = \delta^S_{j\phi}\,\delta^S_{i\phi} \ ,$$
$$\delta^S_{\phi i}\,\delta^S_{\phi j} = \delta^S_{\phi j}\,\delta^S_{\phi i} \ ,$$

$$\delta^s_{i\phi}\,\delta^s_{\phi j} = (1-2\delta_{ij})\,\delta^s_{\phi j}\,\delta^s_{i\phi}\,, \tag{1.13}$$

$$\delta^s_{\phi i}\,\delta^s_{j\phi} = (1-2\delta_{ij})\,\delta^s_{j\phi}\,\delta^s_{\phi i}\,.$$

Let us write also the anticommutation relation for $\delta^F_{X,Y}$ corresponding to a Fermi system

$$\delta^F_{i\phi}\,\delta^F_{\phi j} = -\,\delta^F_{\phi j}\,\delta^F_{i\phi}\,,$$

$$\delta^F_{\phi i}\,\delta^F_{j\phi} = -\,\delta^F_{j\phi}\,\delta^F_{\phi l}\,, \tag{1.14}$$

$$\delta^F_{i\phi}\,\delta^F_{j\phi} = (2\delta_{ij}-1)\,\delta^F_{j\phi}\,\delta^F_{i\phi}\,,$$

$$\delta^F_{\phi i}\,\delta^F_{\phi j} = (2\delta_{ij}-1)\,\delta^F_{\phi j}\,\delta^F_{\phi i}\,.$$

The four different b 's of the last proposition are associated with the following multipliers which are such that

$$\xi((X,Y);(X,Y)) = 1 \tag{1.15}$$

$$\xi^s((X,Y);(X',Y')) = i^{-|X\cap Y|-|X'\cap Y'|+|X\Delta X'\cap Y\Delta Y'|}\,(-1)^{|Y\cap X'|}\,, \tag{1.16}$$

$$\xi^F((X,Y);(X',Y')) = i^{-|X\Delta\theta_1(X\Delta Y)\cap Y\Delta\theta_1(X\Delta Y)|}\,. \tag{1.17}$$

$$\cdot\, i^{-|X'\Delta\theta_1(X'\Delta Y')\cap Y'\Delta\theta_1(X'\Delta Y')+|X\Delta X'\Delta\theta_1(X\Delta X'\Delta Y\Delta Y')\cap Y\Delta Y'\Delta\theta_1(X\Delta X'\Delta Y\Delta Y')|}$$

$$\cdot\,(-1)^{|Y\Delta\theta_1(X\Delta Y)\cap X'\Delta\theta_1(X'\Delta Y')|}\,.$$

$$\xi^\varphi((X,Y);(X',Y')) = i^{|X\cap\theta_2(X)|+|Y\cap\theta_2(Y)|} \tag{1.18}$$

$$i^{|X'\cap\theta_2(X')|+|Y'\cap\theta_2(Y')|-|X\Delta X'\cap\theta_2(X\Delta X')|-|Y\Delta Y'\cap\theta_2(Y\Delta Y')|}$$

$$(-1)^{|X'\cap\theta_2(X)|+|Y'\cap\theta_2(Y)|}\,.$$

$$\xi^\sigma((X,Y);(X',Y')) = i^{-|X||Y|-|X\cap Y|-|X'||Y'|-|X'\cap Y'|} \tag{1.19}$$

$$i^{|X\Delta X'||Y\Delta Y'|+|X\Delta X'\cap Y\Delta Y'|}\,(-1)^{|X'||Y|+|X'\cap Y|}\,.$$

where θ_1 and θ_2 are the following homomorphisms of $\mathcal{P}_\Lambda$

$$\theta_1(\{x_1\}) = \phi \quad ,$$
$$\theta_1(\{x_i\}) = \{x_j \in \Lambda \, ; \, j < i \} \, , \quad \forall i > 1 \, ,$$

(1.20)
$$\theta_2(\{x_1\}) = \phi \, ,$$
$$\theta_2(\{x_{2k+1}\}) = \{x_j \in \Lambda \; ; \; j < 2k+1\} \, , \quad k > 0 \, ,$$
$$\theta_2(\{x_{2k}\}) = \{x_j \in \Lambda \; ; \; j \leq 2k \} \, , \quad k \geq 1 \, .$$

for an arbitrary given order on the points in Λ .

Actually, in the case where $|\Lambda|$ is even or infinite, one can prove that all these extensions are isomorphic. More precisely

<u>Theorem</u> (1.21)

If $|\Lambda|$ is even or infinite, then if ξ_i $i = 1,2$ are any two multipliers previously described, there exists an automorphism γ_{12} of $\mathcal{P}_\Lambda \times \mathcal{P}_\Lambda$ and a function f_{12} from $\mathcal{P}_\Lambda \times \mathcal{P}_\Lambda$ in the torus such that

$$\alpha_{12}(\delta^1_{X,Y}) = f_{12}(X,Y) \, \delta^2_{\gamma_{12}(X,Y)} \quad ,$$

defines a $*$-isomorphism of $\overline{\Delta(\mathcal{P}_\Lambda \times \mathcal{P}_\Lambda , \xi_1)}$ onto $\overline{\Delta(\mathcal{P}_\Lambda \times \mathcal{P}_\Lambda \cdot \xi_2)}$.

The proof rests on the fact that, if b^1 and b^2 are any two bicharacters on $\mathcal{P}_\Lambda \times \mathcal{P}_\Lambda$ defined in (1.11), then there exists an automorphism γ_{12} of $\mathcal{P}_\Lambda \times \mathcal{P}_\Lambda$ such that

$$b^2((X,Y);(X',Y')) = b^1(\gamma_{12}(X,Y) \, ; \, \gamma_{12}(X',Y')) \tag{1.22}$$

The structure of γ_{SF} and $\gamma_{\varphi\sigma}$ is especially simple, indeed they can be written in terms of the homomorphisms θ_1 and θ_2 of $\mathcal{P}_\Lambda$ defined in (1.20), viz

(1.23)
$$\gamma_{SF}(X,Y) = (X \Delta \theta_1(X \Delta Y) \, , \, Y \Delta \theta_1(X \Delta Y)) \quad ,$$

$$\gamma_{\varphi\sigma}(X,Y) = (X \Delta \theta_2(X \Delta Y) \, , \, Y \Delta \theta_2(X \Delta Y)) \quad .$$

One can then deduce (1.22) by direct calculation. The structure of γ_{ij} (and γ_{ij}^{-1}) for the other choices of pairs, namely $(S\varphi)$, $(F\varphi)$, $(S\sigma)$ and $(F\sigma)$ is not of the form (1.23). We can give explicitly the action of $\tau_{F\varphi}$ (and $\tau_{F\varphi}^{-1}$) on the generators, the action of the other τ 's being given by suitable products of τ_{SF} , $\tau_{\varphi\sigma}$ and $\tau_{F\varphi}$. Let us remark that $\tau_{FS}^2 = \tau_{\varphi\sigma}^2 = i$ (the identity).

Thus, we have, if $E_k = \{x_{2i} \in \Lambda \,;\, i \leq k\}$, $k > 0$,

$O_k = \{x_{2i-1} \in \Lambda \,;\, i \leq k\}$, $k > 0$ and

$N_k = \{x_i \in \Lambda \,;\, i \leq k\}$, $k > 0$,

$$\tau_{F\varphi}(\{x_{2i+1}\}, \phi) = (E_i \Delta \{x_{2i+1}\}, E_i),$$
$$\tau_{F\varphi}(\{x_{2i+2}\}, \phi) = (E_i, E_i \Delta \{x_{2i+1}\}),$$
$$\tau_{F\varphi}(\phi, \{x_{2i+1}\}) = (O_{i+1} \Delta \{x_{2i+2}\}, O_{i+1}),$$
$$\tau_{F\varphi}(\phi, \{x_{2i+2}\}) = (O_{i+1}, O_{i+1} \Delta \{x_{2i+2}\}).$$

and

$$\tau_{F\varphi}^{-1}(\{x_{2i+1}\}, \phi) = (\{x_{2i+1}\}, N_{2i}),$$
$$\tau_{F\varphi}^{-1}(\{x_{2i+2}\}, \phi) = (N_{2i+2}, \{x_{2i+1}\}),$$
$$\tau_{F\varphi}^{-1}(\phi, \{x_{2i+1}\}) = (\{x_{2i+2}\}, N_{2i}),$$
$$\tau_{F\varphi}^{-1}(\phi, \{x_{2i+2}\}) = (N_{2i+2}, \{x_{2i+2}\}).$$

One can see that $\overline{\Delta(H_{F_\Lambda} \times H_{F_\Lambda}, \xi)}$ for any ξ in (1.16).(1.19) is isomorphic to the U.H.F. algebra.

§2. FIELD OPERATORS

Using previous results, we shall describe the observables of a quantum spin system or a Fermi system as the Weyl quantized of functions on a "classical phase space" according to the Weyl procedure.

Let us remark that for $|\Lambda|$ infinite, we loose the Mackey Von Neumann uniqueness theorem, namely there exists inequivalent representations of $\overline{\Delta(H_{F_\Lambda} \times H_{F_\Lambda}, \xi)}$. Hence it is necessary to be careful about the class of functions we quantize within a given representation.

Weyl operators in $\overline{\Delta(\mathcal{P}_\Lambda \times \mathcal{P}_\Lambda, \xi^S)}$ correspond to the following operators :

(2.1) $$\delta^S_{X,Y} = i^{-|X \cap Y|} \bigotimes_{i \in X} \sigma^i_x \bigotimes_{j \in Y} \sigma^j_y$$

σ^i_x, σ^j_y are the 2x2 spin matrices.

For $\overline{\Delta(\mathcal{P}_\Lambda \times \mathcal{P}_\Lambda, \xi^F)}$, they are :

(2.2) $$\delta^F_{X,Y} = i^{|\theta_1(X \Delta Y)| - |X \Delta \theta_1(X \Delta Y) \cap Y \Delta \theta_1(X \Delta Y)|} \cdot (-1)^{|Y \cap \theta_1(Y)|} \prod^{<}_{i \in X} b(e_i) \prod^{<}_{j \in Y} b(f_j) .$$

where the products are in an increasing order and

$$b(e_i) = \frac{1}{2}(b^+_i + b^-_i)$$
$$b(f_i) = \frac{1}{2i}(b^+_i - b^-_i)$$

$b^\pm_i$ are the usual creation and annihilation operators of a Fermi system.

Now, let us consider a function f on $\mathcal{P}_\Lambda \times \mathcal{P}_\Lambda$ with Fourier transform

(2.3) $$\mathcal{F}f(X,Y) = \int_{\mathcal{P}_\Lambda \times \mathcal{P}_\Lambda} d\hat{X}\, d\hat{Y}\, (-1)^{|X \cap \hat{Y}| + |\hat{X} \cap Y|} f(\hat{X}, \hat{Y})$$

$d\hat{X}\, d\hat{Y}$ is the normalized Haar measure on the compact group $\mathcal{P}_\Lambda \times \mathcal{P}_\Lambda$; then one is tempted to define the quantized $Q(f)$ of f according to the Weyl procedure

(2.4) $$Q(f) = \int_{\mathcal{P}_\Lambda \times \mathcal{P}_\Lambda} dX\, dY\, \mathcal{F}f(X,Y)\, \delta_{XY}$$

where $dX\, dY$ is the (discrete) measure on $\mathcal{P}_\Lambda \times \mathcal{P}_\Lambda$, and the δ_{XY} 's are the generators of $\overline{\Delta(\mathcal{P}_\Lambda \times \mathcal{P}_\Lambda, \xi)}$. The definition (2.4) is quite formal ; so, we give in the following a class of functions especially interesting since it can be quantized in any representation.

Theorem (2.5)

Let $\mathcal{B}_\Lambda$ be the algebra of functions on $\mathcal{P}_\Lambda \times \mathcal{P}_\Lambda$ which are Fourier transforms of a ℓ_1 function on $\sharp_\Lambda \times \sharp_\Lambda$, then

$$(2.6) \qquad Q^\xi(f) = \int_{\sharp_\Lambda \times \sharp_\Lambda} dX\, dY \;\mathcal{F}f(X,Y)\, \delta^\xi_{XY}$$

defines an element of $\overline{\Delta(\sharp_\Lambda \times \sharp_\Lambda, \xi)}$.

Moreover

$$Q^\xi(f)^* = Q^\xi(f^*) \quad \text{where} \quad f^*(X,Y) = \overline{f(X,Y)}$$

$$(2.7) \qquad Q^\xi(f_1)\, Q^\xi(f_2) = Q^\xi(f_1 \circ f_2)$$

where $f_1 \circ f_2 = \mathcal{F}^{-1}(\mathcal{F}f_1 \times \mathcal{F}f_2)$

The twisted convolution product $\times$ of two functions f and g being defined as

$$f \times g\,(X,Y) = \int_{\sharp_\Lambda \times \sharp_\Lambda} dX'dY'\, \xi((X,Y);(X',Y'))\, f(X',Y')\, g(X\Delta X', Y\Delta Y')$$

Furthermore

$$Q^\xi(1) = 1$$

$$(2.8) \qquad Q^\xi(\chi_{XY}) = \delta^\xi_{XY}$$

where

$$\chi_{XY}(\hat{X}', \hat{Y}') = (-1)^{|X \cap \hat{Y}'| + |Y \cap \hat{X}'|} ,$$

$$X, Y \in \sharp_\Lambda \;,\quad \hat{X}', \hat{Y}' \in \mathcal{P}_\Lambda .$$

The definition (2.6) makes sense since, for the considered class of function $\mathcal{F}f$ belongs to $\ell_1(\sharp_\Lambda \times \sharp_\Lambda)$ and $Q^\xi(f)$ belongs to $\overline{\Delta(\sharp_\Lambda \times \sharp_\Lambda, \xi)}^{\|\cdot\|_1}$ ([3]) which is a subalgebra of $\overline{\Delta(\sharp_\Lambda \times \sharp_\Lambda, \xi)}$.

Moreover, in (2.7), $\mathcal{F}(f_1 \circ f_2)$ belongs to $\ell_1(\sharp_\Lambda \times \sharp_\Lambda)$ since $\ell_1(\sharp_\Lambda \times \sharp_\Lambda)$ is a Banach space for the $\times$ product. (2.8) are straightforward.

Let us remark here that $\mathcal{P}_\Lambda \times \mathcal{P}_\Lambda$ being compact, every continuous function on $\mathcal{P}_\Lambda \times \mathcal{P}_\Lambda$ can be uniformly approximated by functions in $\mathcal{B}_\Lambda$.

Now, according to theorem (1.20), there exists an automorphism γ of $\sharp_\Lambda \times \sharp_\Lambda$ which connects δ^F_{XY} and δ^S_{XY} . It is of the form

$$\alpha\,(\delta^{S}_{XY}) = \delta^{F}_{\gamma(X,Y)}$$

where

$$\gamma(X,Y) = (X \Delta \theta_1 (X \Delta Y),\ Y \Delta \theta_1 (X \Delta Y))\ ,\quad X, Y \in \not{\mathcal{P}}_\Lambda\ .$$

Let us define $^{t}\gamma$ the transpose of γ, which is an automorphism of $\mathcal{P}_\Lambda \times \mathcal{P}_\Lambda$ preserving the Haar measure ; $^{t}\gamma$ links the two quantizations according to the formula

(2.9) $$\alpha(Q^{F}(f)) = Q^{S}(f \circ {}^{t}\gamma)\ ,\qquad \forall f \in \mathcal{B}_\Lambda\ .$$

§3 - QUASI-CLASSICAL STATES ON $\overline{\Delta(\not{\mathcal{P}}_\Lambda \times \not{\mathcal{P}}_\Lambda, \xi^{S})}$

Some states of $\overline{\Delta(\not{\mathcal{P}}_\Lambda \times \not{\mathcal{P}}_\Lambda, \xi^{S})}$ are of interest in what follows. It is the convex, and weakly closed set of "quasi-classical states" ω which are defined by

(3.1) $$\omega\,(\delta^{S}_{XY}) = 0 \qquad \text{if} \quad X \neq Y\,.$$

If we remark that the multiplier ξ^{S} is such that $\xi^{S}((X.X);(Y.Y)) = 1$, then, for each quasi-classical state, there exists a probability measure μ on $\mathcal{P}_\Lambda \times \mathcal{P}_\Lambda$ such that

(3.2) $$\omega(\delta^{S}_{XY}) = \mathcal{F}\mu\,(X,Y)\quad ,\ X, Y \in \not{\mathcal{P}}_\Lambda$$

This leads immediately to the formula :

$$\omega_\mu(Q(f_1) \ldots Q(f_n)) = \int_{\mathcal{P}_\Lambda \times \mathcal{P}_\Lambda} d\mu\,(\hat{x}\,\hat{y})\,(f_1 \circ \ldots \circ f_n)(\hat{x},\hat{y})$$

One can construct explicitly the representation associated with this state, the representation space being $\ell_2(\not{\mathcal{P}}_\Lambda) \otimes L_2(\mathcal{P}_\Lambda, d\nu)$ where the measure ν is given in terms of μ by

$$\mathcal{F}\mu\,(X,Y) = \delta(X \Delta Y)\ \hat{\mathcal{F}}\nu(X)$$

where $\hat{\mathcal{F}}$ is the Fourier transform on $\mathcal{P}_\Lambda$, and δ is the Kronecker symbol in X and Y .

Among the quasi-classical states one finds the following well known product states :

$$\omega_{\{a_i\}}(\delta^S_{XY}) = \prod_{x_i \in \Lambda} \{(1-|\{x_i\} \cap X|)(1-|\{x_i\} \cap Y|) + |\{x_i\} \cap X||\{x_i\} \cap Y| a_i\}$$

where a_i is any function from $x_i \in \Lambda$ into $[-1, +1]$ and especially

- the Fock state where $a_i = -1$, $\forall i$.
- the anti-Fock state where $a_i = 1$, $\forall i$.
- the central state where $a_i = 0$, $\forall i$.

For this last state, the measure μ is the Haar measure on $\mathcal{P}_\Lambda \times \mathcal{P}_\Lambda$.

§4 - FEYNMAN PATH INTEGRAL

We just indicate how it is possible to use this formalism to write a path integral for Fermi systems which avoids the Grassmann algebra formalism. For the sake of simplicity, we consider explicitly spin systems. Other cases can be treated using the isomorphism between the different algebras we have considered.

Moreover, we restrict ourselves to $|\Lambda| < \infty$ in order to avoid to be too precise on the class of functions we consider. In this case $\not{P}_\Lambda$ and $\mathcal{P}_\Lambda$ are isomorphic. Let us nevertheless make the distinction between the two. Furthermore, in this case, there is by the Mackey-von Neumann theorem and up to a quasi-equivalence, only one representation of $\overline{\Delta(\not{P}_\Lambda \times \not{P}_\Lambda, \xi^S)}$. We choose for instance the following one. The representation space being the space of functions from $\not{P}_\Lambda$ into $\mathbb{C}$ endowed with the scalar product

$$(4.1)\qquad (f \mid g) = \sum_{X \in \not{P}_\Lambda} \overline{f(X)}\, g(X)$$

The π representation is given by

$$(4.2)\qquad \left(\pi(\delta^S_{XY}) f\right)(Z) = i^{-|X \cap Y|} (-1)^{|X \cap Z|} f(Z \Delta Y)$$

Then, there exists two orthonormal sets of eigenvectors $|\tilde{X})$ and $|Y\rangle$, $X, Y \in \not{P}_\Lambda$ which are in the π representation eigenvectors of $\delta^S_{X\phi}$ and $\delta^S_{\phi Y}$ respectively and verify

(4.3)

i) $\langle \tilde{X} | Y) = 2^{-|\Lambda|/2} (-1)^{|X \cap Y|}$

ii) $\mathbb{1} = 2^{-|\Lambda|/2} \sum_{X,Y \in \mathcal{P}_\Lambda} (-1)^{|X \cap Y|} |\tilde{X})(Y|$

They are given by $Y(Z) = \delta(Y \Delta Z) = \begin{cases} 1 & \text{if } Z = Y \\ 0 & \text{otherwise} \end{cases}$

and $\tilde{X}(Z) = 2^{-|\Lambda|/2} (-1)^{|Z \cap X|}$

Let us give the matrix elements of $Q^s(h)$ given by

(4.4) $$Q^s(h) = 2^{-|\Lambda|} \sum_{S,T \in \mathcal{P}_\Lambda} \mathcal{F}' h(S,T)\, \delta^s_{ST}$$

where $\mathcal{F}'$ is the normalised Fourier transform on $\mathcal{P}_\Lambda \times \mathcal{P}_\Lambda$

(4.5) $$\mathcal{F}' h(S,T) = 2^{-|\Lambda|} \sum_{\hat{U},\hat{V} \in \mathcal{P}_\Lambda} (-1)^{|\hat{U} \cap T| + |\hat{V} \cap S|} h(\hat{U},\hat{V})$$

One has $Q^s(1) = \mathbb{1}$; in a first step, we have

(4.6) $$(Y | \pi(\delta^s_{X'Y'}) X) = 2^{-|\Lambda|/2}\, i^{-|X' \cap Y'|} (-1)^{|X' \cap Y| + |Y' \cap X| + |X \cap Y|}$$

This last relation allows to write, in the π representation for $Q^s(h)$ given in (4.4), the following relation

(4.7) $$(Y | (1 + i\epsilon Q^s(h)) \tilde{X}) = 2^{-\frac{5}{2}|\Lambda|} (-1)^{|X \cap Y|}$$

$$\sum_{\substack{S,T \in \mathcal{P}_\Lambda \\ \hat{U},\hat{V} \in \mathcal{P}_\Lambda}} i^{-|S \cap T|} (-1)^{|S \cap \hat{V} \Delta Y| + |T \cap \hat{U} \Delta X|} (1 + i\epsilon\, h(\hat{U},\hat{V}))$$

We shall now derive a path integral formula for the quantum spin system in a way very similar to the one used in ordinary Quantum Mechanics [4].

Indeed, as usual, let us introduce the approximation of the exponential

$$\exp(itQ^s(h)) = \lim_{N \to \infty} \left(1 + \frac{it}{N} Q^s(h)\right)^N$$

which corresponds to a regular time slicing. For given initial and final states $|Y_i)$ and $|Y_f)$, one has using (4.7) :

(4.8) $$(Y_i \mid \exp(it\, Q^s(h))\, Y_f)$$

$$= \lim_{N\to\infty} 2^{-\frac{N|\Lambda|}{2}} \sum_{\substack{X_0,\dots X_{N-1} \in \mathcal{P}_\Lambda \\ Y_0,\dots Y_N \in \mathcal{P}_\Lambda}} (Y_i \mid (1 + \tfrac{it}{N} Q^s(h))\, \tilde{X}_0)$$

$$(Y_1 \mid (1 + \tfrac{it}{N} Q^s(h))\, \tilde{X}_1) \dots\dots (Y_{N-1} \mid (1 + \tfrac{it}{N} Q^s(h))\, \tilde{X}_{N-1})$$

$$(-1)^{|X_0 \cap Y_1| + \dots + |X_{N-1} \cap Y_N|} \; \delta(Y_N \Delta Y_F)\, \delta(Y_0 \Delta Y_i)$$

(4.9) $$= \lim_{N\to\infty} 2^{-3N|\Lambda|} \sum_{\substack{X_0 \dots X_{N-1} \in \mathcal{P}_\Lambda \\ Y_0 \dots Y_N \in \mathcal{P}_\Lambda}} (-1)^{\sum_{k=1}^{N} |X_{k-1} \cap Y_{k-1} \Delta Y_k|}$$

$$\sum_{\substack{S_1 \dots S_N \in \mathcal{P}_\Lambda \\ T_1 \dots T_N \in \mathcal{P}_\Lambda}} \sum_{\substack{\hat{U}_1 \dots U_N \in \mathcal{P}_\Lambda \\ \hat{V}_1 \dots V_N \in \mathcal{P}_\Lambda}} \prod_{i=1}^{N} i^{-|S_i \cap T_i|} (-1)^{|S_i \cap \hat{V}_i \Delta Y_{i-1}| + |T_i \cap \hat{U}_i \Delta X_{i-1}|}$$

$$\cdot \exp\Big(\frac{it}{N} \sum_{i=1}^{N} h(\hat{u}_i, \hat{v}_i)\Big)\; \delta(Y_0 \Delta Y_i)\, \delta(Y_N \Delta Y_f)$$

where one takes moreover the usual approximation of the exponential for the classical function $h(\hat{U},\hat{V})$.

Thus, let us define the function μ_N on $(\mathcal{P}_\Lambda \times \mathcal{P}_\Lambda)^N$ by

(4.10) $$\mu_N(\hat{U}_1, \hat{V}_1, \dots, \hat{u}_N, \hat{v}_N, Y_i, Y_F)$$

$$= 2^{-2N|\Lambda|} \sum_{\substack{X_0 \dots X_{N-1} \in \mathcal{P}_\Lambda \\ Y_0 \dots Y_N \in \mathcal{P}_\Lambda}} \sum_{\substack{S_1, \dots, S_N \in \mathcal{P}_\Lambda \\ T_1, \dots, T_N \in \mathcal{P}_\Lambda}} (-1)^{\sum_{k=1}^{N} |X_{k-1} \cap Y_{k-1} \Delta Y_k|}$$

$$\prod_{i=1}^{N} i^{-|S_i \cap T_i|} (-1)^{|S_i \cap \hat{V}_i \Delta Y_{i-1}| + |T_i \cap \hat{U}_i \Delta X_{i-1}|} \delta(Y_0 \Delta Y_i)\, \delta(Y_N \Delta Y_f)$$

The summation on the X_i 's may be performed to give

$$2^{N|\Lambda|} \prod_{k=1}^{N} \delta(Y_{k-1} \Delta Y_k \Delta T_k)$$

and the summation on the T_i 's gives

(4.11)
$$\mu_N(\hat{U}_1,\hat{V}_1,\dots,\hat{U}_N,\hat{V}_N;Y_I,Y_f)$$

$$= 2^{-N|\Lambda|}\sum_{\substack{Y_0\cdots Y_N\in \mathcal{P}_\Lambda\\ S_1\cdots S_N\in\mathcal{P}_\Lambda}}\prod_{i=1}^{N} i^{-|S_i\cap Y_i\Delta Y_{i-1}|}(-1)^{|S_i\cap\hat{V}_i\Delta Y_{i-1}|}$$

$$\cdot(-1)^{|Y_{i-1}\Delta Y_i\cap\hat{U}_i|}\quad\delta(Y_0\Delta Y_I)\,\delta(Y_N\Delta Y_f)$$

which may be written, according to the formula

(4.12)
$$\sum_{S\in\mathcal{P}_\Lambda} z^{|S\cap A|}(-1)^{|S\cap B|}$$

$$= 2^{|\Lambda|}\left(\frac{1+z}{2}\right)^{|A|}\left(\frac{1-z}{1+z}\right)^{|B|}\Theta(B;A)$$

where $\Theta(B;A)=\begin{cases}1 & \text{if}\\ 0 & \text{otherwise}\end{cases}$ and $z\in\mathbb{C}$,

(4.13)
$$\mu_N(\hat{U}_1,\hat{V}_1,\dots\dots,\hat{U}_N,\hat{V}_N;Y_I,Y_f)$$

$$=\sum_{Y_0,\dots,Y_N\in\mathcal{P}_\Lambda}\prod_{i=1}^{N}\left(\frac{1-i}{2}\right)^{|Y_{i-1}\Delta Y_i|}\left(\frac{1+i}{1-i}\right)^{|\hat{V}_i\Delta Y_{i-1}|}$$

$$\Theta(\hat{V}_i\Delta Y_{i-1};Y_{i-1}\Delta Y_i)\,(-1)^{|Y_{i-1}\Delta Y_i\cap\hat{U}_i|}\delta(Y_0\Delta Y_I)\,\delta(Y_N\Delta Y_f)$$

With μ_N given in (4.13), we have the following formula

$$(4.13) \qquad (\Psi_I \mid \exp(it\, Q^s(h))\, \Psi_f)$$

$$= \lim_{N\to\infty} 2^{-N|\Lambda|} \sum_{\substack{\vartheta_1,\ldots,\vartheta_N \in \mathcal{P}_\Lambda \\ \hat{\vartheta}_1,\ldots,\hat{\vartheta}_N \in \mathcal{P}_\Lambda}} \mu_N(\hat{\vartheta}_1,\vartheta_1,\ldots,\hat{\vartheta}_N,\vartheta_N;\Psi_I,\Psi_f)\, e^{\frac{it}{N}\sum_{i=1}^{N} h(\hat{\vartheta}_i,\vartheta_i)}$$

which is of the type of a discrete approximation of a Feynman path integral on the phase space $\mathcal{P}_\Lambda \times \mathcal{P}_\Lambda$ and allows to say that one can write path integrals for different statistics in a unified way.

References

[1] G.W. Mackey, Acta Math. 99, 265 (1958)

[2] F.A. Berezin, The Method of Second Quantization. Academic Press (1966)

[3] J. Manuceau, M. Sirugue, D. Testard, A. Verbeure, Commun.math.Phys. 32 231 (1973)

[4] M. Mizrahi, Journ. Math. Phys. 16, 2201-6 (1975)

FEYNMAN PATH INTEGRAL AND THEORY OF FORMS

by Paul KREE

Definitions [1] [3] of Feynman path integral $\int f \, dm$ use Parseval relation : if the function f is Fourier transform of a measure μ, then $\int f \, dm = \int \hat{\mu} dm = \int \hat{m} \, d\mu$. The technique presented bellow use also some kind of Parseval relation, but Fourier transform of measures is replaced by **Laplace** transform of analytical functionals and corresponding anticommutative "measures". More precisely $\varepsilon \in \{-1, +1\}$ is fixed, and functions are replaced by ε-symmetrics forms, i.e. symmetric forms for bosons ($\varepsilon = -1$), antisymmmetric forms for fermions ($\varepsilon = +1$) ; and measures are replaced by coforms. In our talk we have presented the duality theory [4] [5] concerning some space $\Gamma_\varepsilon(Y)$ of forms $f(y)$ on a complete nuclear space Y, and some space $F_\varepsilon(Y')$ of forms $\varphi(\eta)$ on strong dual Y' ; duality is denoted $\langle f,\varphi\rangle$ or $\langle f(y), \varphi(\eta)\rangle$. A coform m on Y is a linear continuous form on the space $\Gamma_\varepsilon(Y)$ of test forms on Y, and Laplace transform $m \to \hat{m}$ realize a bijection $\Gamma'_\varepsilon(Y) \to F_\varepsilon(Y')$. Therefore, Feynman integral $\int f \, dm$ can be defined by $\langle f,\hat{m}\rangle$. This technique of duality brackets permits to developp a symbolic calculus for quantum operators, $\varepsilon = \pm 1$. Scattering operators can be computed explicitely in some simple cases : see § 3. In order to avoid all topological difficulties in the present text, Y is an algebraic dual equipped with the weak topology : contrepart of this simplification is that only cylindrical forms on Y can be integrated. In the present text we introduce also partial contraction
$\mathrm{II} = \langle f(x,y), \hat{m}(\eta)\rangle = \int f(x,y) \, m(y)$ where f and $\hat{m}$ are forms defined on $X \times Y$ and Y' resp. Result II of this operation is a form defined on X : see § 1.
As a direct application, we developp in § 2 an ε-canonical formalism in classical mechanic and classical field theory, giving a method of quantization. For $\varepsilon = -1$, this corresponds to the usual Hamiltonian formalism, but we add an hilbertian structure, an we work with corresponding conjugate variables z and $\bar{z} = w \, q \pm ip$. For $\varepsilon = +1$, formulas are exactly the same, but meaning are different because, usual differential calculus is replaced by anticommutative differential calculus. In finite dimension, this gives a new spin theory. In infinite dimension this gives a positive classical hamiltonian to Dirac equation and new connection between spin and statistic. § 4 shows how symbolic calculus written in § 3 using duality brackets, can be written using symbols of integration, i.e. in terms of Feynman path integrals. Present work is dedicated to Jean Leray.

§ 1. ALGEBRAIC STUDY OF ε-SYMMETRIC FORMS.

(1.1) ε-symmetric forms tensors.

Let E be a complex vector space over a field $\mathbb{K}$ of characteristic zero. A tensor $t = \Sigma\, t_k \in \oplus\, ({}_k\otimes E) = \otimes E$ is ε-symmetric if t is symmetric for $\varepsilon = -1$ (resp. antisymmetric if $\varepsilon = +1$). Space of ε-symmetric tensors on E is denoted $T_\varepsilon(E)$ or $\odot E$; subspace of homogeneous tensors of degree k is denoted $T^k_\varepsilon(E)$ or ${}_k\odot E$. By convention ${}_o\odot E = \mathbb{K}$. For any permutation $\alpha \in \mathfrak{S}_k$, $sg_\varepsilon\, \alpha$ denotes the signature of α if $\varepsilon = +1$ (resp the scalar + 1 if $\varepsilon = -1$). Operator Sym_ε of ε-symmetrization is a projector $\otimes E \to \odot E$; for example, operator Sym_ε

maps $x_1 \otimes \ldots \otimes x_k \in {}_k\otimes E$ on

(1.2) $$x_1\, x_2 \ldots x_k = \frac{1}{k!} \sum_{\alpha \in \mathfrak{S}_k} (sg_\varepsilon\, \alpha) x_{\alpha(1)} \otimes \ldots \otimes x_{\alpha(k)}$$

Sometimes, for $\varepsilon = +1$, this tensor is denoted $x_1 \wedge \ldots \wedge x_k$, and ${}_k\odot E$ is denoted ${}_k\wedge E$. Space $T_\varepsilon(X)$ is an associative algebra for operation of ε-symmetric tensorial product $x : u \to tu = Sym_\varepsilon(t \otimes u)$. The twist $t \to t^\vee$ in $T_\sigma(E)$ in the linear involution such that $(x_1\, x_2 \ldots x_k)^\vee = x_k\, x_{k-1} \ldots x_1$. Let X and X' be two vector spaces in duality.Natural and twisted dualities between ${}_k\otimes X$ and ${}_k\otimes X'$ are defined by :

(1.3) $$\langle x_1 \otimes \ldots \otimes x_k,\ \xi^1 \otimes \ldots \otimes \xi^k\rangle_{k,nat} = \langle x_1, \xi^1\rangle \ldots \langle x_k, \xi^k\rangle$$

and $\langle t_k, \theta_k\rangle_k = \langle t_k, \overset{\vee}{\theta}_k\rangle_{k,nat}$. Same notations are used for induced duality between ${}_k\odot X$ and ${}_k\odot X'$. Natural and twisted dualities between $T_\varepsilon(X)$ and $T_\varepsilon(X')$ are defined by

(1.4) $$\langle t,u\rangle_{nat} = \sum_{k=o}^{\infty} k!\, \langle t_k,u_k\rangle_{k,nat}$$

(1.5) and $$\langle t,u\rangle = \sum_{k=o}^{\infty} k!\, \langle t_k,u_k\rangle_k = \langle t,u^\vee\rangle_{nat}$$

(1.6) ε-symmetric form.

Let X and E be two vector spaces. The space $\mathcal{F}^k_\varepsilon(X\ ;\ E)$ of k-homogeneous ε-symmetric forms on X valueted in E in the space of ε-symmetric k-linear maps $f_k : X^k \to E$. The space of ε-symmetric vectorial forms on X valueted in E is $\mathcal{F}_\varepsilon(X,E) = \prod_{k=o}^{\infty} \mathcal{F}^k_\varepsilon(X,E)$ If $Y = \mathbb{K}$, we write simply, $\mathcal{F}_\varepsilon(X) = \Pi\, \mathcal{F}^k_\varepsilon(X)$. The twist $f \to f^\vee$ is naturally defined in $\mathcal{F}_\varepsilon(X,E)$. An element (f_k) of $\mathcal{F}_\varepsilon(X,Y)$ is written $f = \Sigma\, f_k$. The dual of $T_\varepsilon(X)$ is $\mathcal{F}_\varepsilon(X)$. Natural and twisted dualities between

$T_\varepsilon(X)$ and $\mathcal{F}_\varepsilon(X)$ are defined in the following way.
For $t = \Sigma\, t_k$ and $t_k = x_{k,1}\, x_{k,2} \dots x_{k,k}$:

(1.7) $$\langle f,t\rangle_{nat} = \sum_{k=o}^{\infty} k!\; f_k(x_{k,1},\dots, x_{k,k})$$

(1.8) $$\langle f,t\rangle \;\; = \Sigma\, k!\; f_k(x_{k,k},\dots, x_{k,1}) = \langle f,\check{t}\rangle_{nat} = \langle \check{f},t\rangle_{nat}$$

Space $\mathcal{F}_\varepsilon(X)$ is an algebra for the ε-symmetric product, and $\mathcal{F}_\varepsilon(X,E)$ is a module on $\mathcal{F}_\varepsilon(X)$. Forms f on X, g on Y... are sometimes denoted $f(x)$, $g(y)$,..., even if the value of a form in a point is not defined in general. Inverse image of g by a linear map $\ell : X \to Y$ is denoted $g(\ell x)$ or simply $\underline{g}$ or g is ℓ is defined by the context. If $\beta : E \times F \to G$ is a bilinear map, contracted ε-symmetric product $\langle\langle f(x)\; g(x)\rangle\rangle$ of $f \in \mathcal{F}_\varepsilon(X,E)$ and $g \in \mathcal{F}_\varepsilon(X,F)$ is a vectorial form $\in \mathcal{F}_\varepsilon(X,G)$. For example, β can be a bilinear form of duality $E \times E' \to \mathbb{K}$.

(1.9) Derivation.

Let X and X' be two vector spaces in duality. Symbol $X'_\alpha \subset\subset X'$ means that X'_α is a finite dimensional subspace of X'. Let i_α(resp s_α) be the canononical injection $X'_\alpha \subset X'$ (resp projection of X on $X_\alpha = X/(X'^{\perp}_\alpha)$. A duality is defined between X_α and X'_α by $\langle s_\alpha x,\xi\rangle = \langle x, i_\alpha \xi\rangle$. Image of the canonical imbedding $T_\varepsilon(X') \to \mathcal{F}_\varepsilon(X)$ is denoted $\mathcal{F}_{\varepsilon\text{-cyl}}(X)$ because any φ in this image is cylindrical, i.e. of type $\varphi(x) = \varphi_\alpha(s_\alpha x)$ with $\varphi_\alpha \in \mathcal{F}_\varepsilon(X_\alpha)$. Use of twisted duality $\langle f,\varphi\rangle = \langle f(x),\, \varphi(\xi)\rangle$ is motivated by (1.17)... For ψ fixed in $T_\varepsilon(X) \simeq \mathcal{F}_{\varepsilon\text{-cyl}}(X')$, operator of left derivation $f \to \psi(D)\, f$ in $\mathcal{F}_\varepsilon(X)$ is defined as the transpose of operator of right product by $\psi : \varphi \to \varphi\psi$ in $T_\varepsilon(X)$. In order to define global derivative $D^k f$ of $g \in \mathcal{F}_\varepsilon(X,F)$ for $\varepsilon = \pm 1$, we suppose first that F is the algebraic dual E' of a space E. We recall that for any space G, the vector space $\mathcal{L}(G,E')$ of linear maps $G \to E'$ is isomorphic to $(E \otimes G)'$. We denote $\vec{\xi}^k \in \mathcal{F}^k(X', {}_k\bigcirc X')$ the vectorial form $(\xi^1;\dots;\xi^k) \to \xi^1\,\xi^2 \dots \xi^k$. The dual of space $H = T_\varepsilon(X) \otimes ({}_k\bigcirc X) \otimes E$ is the space $H' = \mathcal{F}_\varepsilon(X, \mathcal{F}^k_\varepsilon(X,E))$. We use as bilinear form of duality between H and H', the product of following duality forms : the bilinear form $\langle\; ,\;\rangle$ between $T_\varepsilon(X)$ and $\mathcal{F}_\varepsilon(X)$, form $\langle\; ,\;\rangle_k$ between $\bigcirc X$ and $\mathcal{F}^k_\varepsilon(X)$, and natural duality form on $E \times E'$. Linear operator $\mathcal{F}_\varepsilon(X,E') \to H'$ of left derivation $g \to D^k g$ is defined as the transpose of operator $H \to T_\varepsilon(X) \otimes E$ of contracted right multiplication by $\vec{\xi}^k$, with respect to the natural bilinear map $\beta : (({}_k\bigcirc X) \otimes E) \times ({}_k\bigcirc X') \to E$. This implies that the derivative $D^k\, g_\ell$ of an homogeneous form g_ℓ of degree ℓ is an homogeneous form of degree ℓ-k, (and vanish if $k > \ell$) and

(1.10) $(D^k g_\ell(x_{\ell-k};\ldots;x_1))(y_1,\ldots,y_k) = \frac{\ell!}{\ell-k!} g_\ell(y_1;\ldots;y_k;x_{\ell-k};\ldots;x_1)$

(1.11) This explicit expression of $D^k g_\ell$ shows that the restriction of $D^k g_\ell$ to any $X_\alpha \subset\subset X$, depends only from the restriction of g_ℓ to X_α.

(1.12) Examples.

a/ Let X and X' be two vector spaces in duality. The canonical two form $x.\xi$ on $X \times X'$ maps $((x;\xi);(x';\xi'))$ on the scalar $\frac{1}{2}(<x,\xi'> - \varepsilon<x',\xi>)$ Exponential $e^{x.\xi}$ is the form $\sum_{n=o}^{\infty} (x.\xi)^n n!^{-1}$ on $X \times X'$.

b/ If X as finite dimension d, we use dual basis (e_j) and (ε^j) and corresponding coordinates (x^j) and (ξ_j)in X and X'.The set of multi-indices $j = (j_1,\ldots,j_d)$ used in distribution theory on $\mathbb{R}^d$ is denoted bellow $J(-1, d)$. Exterior algebra use the set $J(+1, d)$ of subsets j of $\{1,\ldots, d\}$; such subset is characterized by a strictly increasing sequence $j = (j_1,\ldots, j_k)$; length of j is $|j| = k$; x^j means 1 if $j = \phi$ and $x^{j_1}\Lambda\ldots\Lambda x^{j_k}$ if not. For $j \in J(\varepsilon,d)$, left differentiation operator $f \to \xi^j(D) f$ is denoted ∂^j : this agree with distribution theory and [2]. Symbol $(j)_\varepsilon$ means 1 if $\varepsilon = -1$ and $(-1)^{|j|(|j|-1)/2}$ if $\varepsilon = +1$. Symbol $j!$ denotes $j_1!\ldots j_d!$ if $\varepsilon = -1$ and 1 if $\varepsilon = +1$. We have for example :

(1.13) $x.\xi = \sum_1^d x^j \xi_j$; $<x^j, (\xi^{j'})^\vee> = \delta_{j,j'}\, j!$; $e^{x.\xi} = \sum_{j \in J(\varepsilon,d)} j!^{-1} (x^j)^\vee \xi^j$

(1.14) Contraction on dual letters.

Let E,G,G' be vector spaces, G and G' are in duality. For $R(x,y) \in \mathcal{F}_\varepsilon(E \times G)$ and $t = t(\eta) \in T_\varepsilon(G)$ we define now contraction $II = <R(x,y), t(\eta)> \in \mathcal{F}_\varepsilon(E)$ on dual letters y and η. We note that $\mathcal{F}^n_\varepsilon(E \times G) = (T^n_\varepsilon(E \times G))' = \bigoplus_{i=o}^{n} (T^{n-i}(E) \otimes T^i_\varepsilon(G))'$. Therefore, we introduce the space $\mathcal{F}^{n-i,i}_\varepsilon(E \times G)$ of n-linear forms on $(E)^{n-i} \times G^i$, separately ε-symmetric with respect to $(n-i)$ first arguments $\in E$, and last i arguments in G. Element $R_{n-i,i}$ of this space, associated to R, is defined by :

(1.15) $R_{n-i,i}(x_1,\ldots x_{n-i}, y_{n-i+1}\ldots y_n) = \binom{n}{i} R(\underline{x}_1,\ldots, \underline{x}_{n-i}, \underline{y}_{n-i+1}\ldots \underline{y}_n)$

with $\underline{x}_m = (x_m,0)$ for $m \leq n-i$ and $\underline{y}_m = (0,y_m)$ for $m > n-i$.

. Then $R = \sum_i \mathrm{Sym}_\varepsilon\, \underline{R}_{n-i,i}$. For any $i \in \{0,\ldots n\}$, canonical multilinear map $\mathcal{F}^{n-i,i}_\varepsilon(E \times G) \times (G)^i \to \mathcal{F}^{n-i}_\varepsilon(E)$ define, by universal property of tensor product, a bilinear map $R_{n-i,i}$; $t \to <R_{n-i,i}(x,y), t(\eta)>_i$ from $\mathcal{F}^{n-i,i}_\varepsilon(E \times G)\times(_i OG)$ to $\mathcal{F}^{n-i}_\varepsilon(E)$. If $t = \Sigma\, t_i$, ε-symmetric form, $II = \Sigma\, II_k$ is defined by :

$$(1.16) \qquad \forall k \quad II_k(x) = \sum_{i=o}^{\infty} i!\ \langle R_{k,i}(x,y),\ t_i(\eta)\rangle_i$$

For some ε-symmetric form $\varphi \in \mathcal{F}_\varepsilon(G')$, $\langle R(x,y), \varphi(\eta)\rangle \in \mathcal{F}_\varepsilon(E)$ is also defined by same formula. For example, for any vector space X, $\forall f \in \mathcal{F}_\varepsilon(X)$:

$$(1.17) \qquad \langle e^{x.\eta}\ ,\ f(y)\rangle = f(x)$$

If X,X' ; Y,Y' are two paars of vector spaces in duality, for $f \in \mathcal{F}_\varepsilon(X)$, $g \in \mathcal{F}_\varepsilon(Y)$, $\varphi \in T_\varepsilon(X)$, $\psi \in T_\varepsilon(Y)$, we have :

$$(1.18) \qquad \langle f(x)\ g(y),\ \psi(\eta)\ \varphi(\xi)\rangle = \langle f,\varphi\rangle\ \langle g,\psi\rangle\ .$$

§ 2. CANONICAL FORMALISMS IS MECHANICS AND FIELD THEORY.

We use notations of § 1, but now, $\mathbb{K} = \mathbb{R}$ or $\mathbb{C}$, E is a complete nuclear space and X is the strong dual of a complete nuclear space Y. If $(p_u, u \in U)$ is a filtering increasing family of semi-norms defining the topology of X, u is identified with the absolute polar of semi ball $\{p_u < 1\}$ and X_u denotes the subspace of X generated by X ; X_u is a Banach space with unit ball u ; Y is semi reflexive and X is the inductive limit of Banach spaces X_u [8]. We work with the space $F_\varepsilon(X,E)$ of hypocontinuous ε-symmetric forms $f = \Sigma f_k$ defined on X with values in E. This means that for any $(k,u) \in \mathbb{N} \times U$, restriction of f_k to $(X_u)^k$ is continuous. We suppose that X is complex and equipped with a conjugation $z \to \bar{z}$. Generic point of antivector space $\bar{X}$ is denoted $\bar{z}$: additive group of vector space $\bar{X}$ coïncide with X but multiplication of $\bar{z}$ by a scalar λ is $\lambda\bar{z} = \overline{\bar{\lambda}z} \neq \bar{\lambda}\ \bar{z}$; $\bar{X}$ is not isomorphic with X. Coming back to X, the underlying real vector space is denoted X_r. In the theory of several complex variables $z_j = x_j + i\ y_j$, it is many times very convenient to write $f(\bar{z},z)$ a complex valued differentiable function on $(\mathbb{C}^d)_r$; and if f is real analytic, f can be considered as the restriction of an holomorphic function $f(\bar{z},z')$ defined in some open subset of $\bar{\mathbb{C}}^n \times \mathbb{C}^n$. We extend this in the infinite dimensional case for $\varepsilon = \pm 1$. Triplets are used, in order to avoid difficulties of domains concerning observables.

(2.1) Phase triplet.

a/ Let $\mathcal{Z} = (S \subset Z \overset{j}{\subset} {}'S)$ be à triplet of complex spaces :injection j of separable Hilbert space Z in nuclear complete space $'S$ is continuous with dense range ; and adjoint of j identified the strong antidual S of $'S$ to a dense subpace of Z. Spaces S and Z are equipped with compatible conjugations $z \to \bar{z}$. Phase triplet defined by $\mathcal{T}$ is $\mathcal{T}_r = (S_r \subset Z_r \subset (S_r)')$. Note that dim $Z < \infty$ implies $S = Z = {}'S$. Scalar product (z,z_1) in Z is sesquilinear i.e. linear with respect to z_1 and antilinear with respect to z ; we write simply $\bar{z}\, z_1$. The sesquilinear form defining antiduality **between** S and 'S extend this scalar product.

b/ Complexified phase triplet is deduced from $\mathcal{T}_r$ by complexification : $\mathcal{T}_r^c = (V \subset W \subset {}'V)$ with $V = \bar{S} \times S$, $W = \bar{Z} \times Z$, $'V = S' \times {}'S$.

c/ Canonical form φ_ε on S_r or Z_r is following ε-symmetric homogeneous 2-form

(2.2) $$\varphi_\varepsilon \begin{pmatrix} \bar{z} & \bar{z}_1 \\ z & z_1 \end{pmatrix} = \begin{cases} 2 \operatorname{Im} \bar{z}\, z_1 & \text{if } \varepsilon = -1 \\ 2i \operatorname{Re} \bar{z}\, z_1 & \text{if } \varepsilon = +1 \end{cases}$$

Canonical form φ_ε admits following extension
$((\bar{z}\,;\, z')\,;\,(\bar{z}_1, z')) \longrightarrow i(z'\,\bar{z}_1 + \varepsilon\, \bar{z}\, z_1')$ on spaces V and W

(2.3) Canonical isomorphism $\lambda_\varepsilon : (\bar{z},z') \to (iz', \varepsilon i\bar{z})$ of V on $\bar{V} = \bar{S} \times S$, is such that $\varphi_\varepsilon(v\,;\,v_1) = \langle\lambda_\varepsilon v, v_1\rangle$ for arbitrary $v = (\bar{z},z')$ and $v_1 = (\bar{z}_1, z_1') \in V$. Therefore, if $\varepsilon = -1$, λ_ε coïncide with well known symplectic isomorphism ; λ_ε defines by linear continuous extensions, isomorphisms $W \to \bar{W}$ and $'V \to {}'\bar{V} = V'$.

(2.4) Observables.

Let E be an l.c.s. with conjugation ; an observable on phase triplet $\mathcal{T}_r$ with values in E is an ε-symmetric form $f = f(\bar{z},z') = \sum_0^N f_k \in F_\varepsilon(V,E)$; f is characterized by its restriction $f(\bar{z},z)$ to S_r. If $E = \mathbb{C}$, f is called a scalar observable. In general, left derivative of f is a vectorial form $Df \in F_\varepsilon(V,L(V,E))$. Observable f as two partial derivatives :

(2.5) $$Df = (D_{\bar{z}} f, D_z f) \in F_\varepsilon(V,L(\bar{S},E)) \times F_\varepsilon(V,L(S,E))$$

(2.6) Canonical gradient of observable f is $^{\#}Df = \lambda_\varepsilon^{-1}\, Df \in F_\varepsilon(V,L(\bar{V},E))$.
For $\varepsilon = -1$, (2.3) implies that $^{\#}Df$ coïncide with the familiar symplectic gradient. Canonical gradient as two components :

(2.7) $^{\#}Df = (-i\varepsilon\, D_z f, -i\, D_{\bar{z}} f) \in F_\varepsilon(V,L(S,E)) \times F_\varepsilon(V,L(\bar{S},E))$ An observable f is called regular if $Df \in \mathcal{F}_\varepsilon(V,L('V,E))$. Let f and g be two observables such that, one at least is regular, one at least is scalar. Then, Poisson bracket is

$\{f,g\} = \langle\langle Df\ {}^{\#}Dg\rangle\rangle$; contraction is defined with respect to the natural bilinear map $X \times L(X,E) \to E$ with $X = V$ or $'V$. Using (2.5) and (2.7)

(2.8) $$\{f,g\} = -\, i\, \varepsilon \langle\langle D_{\bar z} f \quad D_z g\rangle\rangle - i \langle\langle D_z f \quad D_{\bar z} g\rangle\rangle$$

A conjugation $f = \sum f_j \to f^\star = \sum f_j^\star$ is defined in $F_\varepsilon(V,E)$ by

$$f_j^\star(\bar v_1, v_1' ; \ldots ; \bar v_j, v_j') = \overline{f_j(\bar v_j', v_j ;\ldots; \bar v_1', v_1)}$$

A scalar observable f, homogeneous of degree two is called positive if f is **self** conjugate and satisfies :

(2.9) $$\forall\ \phi \in T_\varepsilon(S)\ ;\ \langle e^{\bar z . z'}\ f(\bar z, z'),\ \phi(\bar z')\ \phi^\star(z)\rangle \geqslant 0$$

A mechanical system of type ε is a paar $(\mathcal{C},H)$ where energy H(or hamiltonian) is a regular self conjugate and scalar observable $H(v,t)$ depending eventually on time. A motion $v(t) \in \mathcal{C}^1(\mathbb{R},V)$ of the system satisfied Hamilton equation

(2.10) $$v^\cdot(t) = {}^{\#}DH(v(t),t)$$

If H does not depend on t, system is called isolated and we suppose then that H is positive. If $\varepsilon = +1$, $H = \sum_1^N H_j$, ${}^{\#}DH_j \in F_1^{j-1}(V,V)$ and R H S of (2.10) means $\sum_1^N {}^{\#}DH(v(t)^{j-1},t)$. Therefore, in this theory, for $\varepsilon = +1$, Hamiltonians are of type $H_1 + H_2$. We suppose also that motion exists and is unique for any initial data in V. Using (2.5) and (2.7), (2.10) can be written :

(2.11) $$\bar v^\cdot(t) = -i\ \varepsilon\ D_z\ \overline{H(v(t)}, v(t)\ ;\ t)\ ;\ v^\cdot(t) = -i\ D_{\bar z}\ H(\overline{v(t)}, v(t), t)$$

(2.12) <u>Proposition.</u>

Let α be a canonical transform in phase triplet $\mathcal{C}$, i.e. α is a bijective map, α and $\alpha^{-1} = \beta$ are C^1, and preserve canonical two form φ_ε. Let $H'(v,t) = H(\beta\ v,t)$. Then $t \to v(t)$ is a motion of system $(\mathcal{C},H)$ if and only if $t \to w(t) = (\alpha \circ v)(t)$ is a motion of system $(\mathcal{C},H')$.

In fact for any $\delta u \in S_r$, and any $v \in S_r$

$$\varphi_\varepsilon(\alpha'(v)\ {}^{\#}DH(v,t),\ \alpha'(v)\delta u) = \varphi_\varepsilon({}^{\#}DH(v,t),\delta u)$$

$$= \langle DH(v,t),\delta u\rangle = \langle DH'(\alpha\ v,t),\alpha'(v)\delta u\rangle = \varphi_\varepsilon({}^{\#}DH'(\alpha\ v,t),\ \alpha'(v)\delta u)$$

Therefore ${}^{\#}DH'(\alpha\ v,t) = \alpha'(v)\ {}^{\#}DH(v,t)$. If $t \to v(t)$ is a motion of $(\mathcal{C},H)$, $w^\cdot(t) = \alpha'(v(t))\ v^\cdot(t) = \alpha'(w(t))\ {}^{\#}DH(v(t),t) = {}^{\#}DH'(w(t),t)$ and $t \to w(t)$ is a motion of $(\mathcal{C},H')$.

(2.13) For ε and $\varepsilon' \in \{-1, +1\}$, product of two mechanical systems of type ε and ε' resp. can be defined naturally. In physic, fields and particles are governed by equations of motion EM. In order to find a mechanical system corresponding to a given EM, following method can be used a) write canonical form of EM b) Internal space S of triplet $\mathcal{T}$ is a space of smooth solutions of free equation of motion FEM; note that S is not unique if dim $S = \infty$. c) Canonical two form φ_ε is given by FEM. d) Central Hilbert space Z muss be stable under action of natural invariance groups.

(2.14) Examples.

a) Rotator in oriented euclidean space E_3 ; EM is $\xi\,(t) = C \times \xi(t)$ where $\times$ denotes vectorial product and C is fixed in E_3. Introducing $t \to \zeta(t) \in Z = E_3^c$, and conjugation $\zeta \to \bar{\zeta}$ canonical form of E M is :

$$\zeta^\cdot(t) = C \times \zeta(t),\ \bar{\zeta}^\cdot(t) = -\,C \times \bar{\zeta}(t).$$

b) Let M be an affine Minkovski space with (s+1) dimensions, s = 0, 1,2 or 3. Choosing an origin in M and coordinates, generical point $x \in M$ as coordinates x^j and $x = (x^o, x')$ with $x^o = t$. Klein Gordon real classical field $\varphi = \varphi(x) = \varphi(t,x')$ on M with mass $m > 0$, source $\gamma = \gamma(t,x')$, and spring force $q = q(t,x') \in \mathcal{S}(M)$, satisfies following canonical EM :

(2.15) $\varphi^\cdot(t,x') = \pi(t,x')$; $\pi^\cdot(t,x') = (\Delta - m^2 + q(t,x'))\,\varphi(t,x') + \gamma(t,x')$.

(2.16) Therefore $H(\varphi,\pi,t) = \frac{1}{2} <\pi,\pi>_o + \frac{1}{2} <(-\Delta + m - q(t,.))\,\varphi,\varphi>_o - <\gamma(t,.),\varphi>_o$.

c) Let $\omega > 0$ and $B = \begin{bmatrix} -i\omega & 0 \\ 0 & i\omega \end{bmatrix}$. Following differential system is a finite dimensional model for Dirac system

$$\psi^\cdot(t) = B\,\psi(t)\ ;\quad \overline{\psi(t)}^\cdot = -\,B\,\overline{\psi(t)}\ .$$

(217) Proposition.

a) Mechanical system corresponding to (2.14-a) is (Z_r, H) where hermitian space $Z \simeq \mathbb{C}^3$ is equipped with scalar product $(\zeta,\zeta^1) = \frac{1}{2}\sum \bar{\zeta}_j\,\zeta_j^1$ and

$$(\zeta\ ;\ \zeta^1) \xrightarrow{H} \frac{i}{2}[(\zeta,\ C \times \zeta^1) - (\zeta^1,\ C \times \zeta)]\ \text{or}$$

$$H(\bar{\zeta},\zeta) = i\,\bar{\zeta}_1 \wedge (C_2\,\zeta_3 - C_3\,\zeta_1) + i\,\bar{\zeta}_2 \wedge (C_3\,\zeta_1 - C\,\zeta_3) + i\,\bar{\zeta}_3 | C_1\zeta_2 - C_2\zeta_1)$$

For φ and $\psi \in F_1(\mathbb{C}^3)$, $2\xi = \zeta + \bar{\zeta}$, $\{\varphi(\xi), \psi(\xi)\}_+ = -\,2i \sum_{k=1}^{3} \partial_k\,\varphi \wedge \partial_k\,\psi$

In particular $\{\xi_j, \xi_k\}_+ = -\,2i\,\delta_{j,k}$ for $j\ ;\ k = 1,2,3$

b) In the dual of vector space M, positive hyperboloïd of mass m is $\{k = (k^o, k'),\ k^o = \omega(k') = (k'^2 + m^2)^{1/2}\}$; image of invariant measure μ on $H\uparrow$ by chart $k \to k'$ is $\mu(k') = (2\,\omega(k'))^{-1}\, dk'$; and image of Schwartz space S of $H\uparrow$ is $\mathcal{S}(\mathbb{R}^s)$. Fourier transform (F.T) and partial FT on M are defined by $\varphi(x) \to \varphi(k) = (2\pi)^{-(s+1)/2} \int e^{ikx}\, \varphi(x)\, dx$ and $\varphi(t,x') \to \varphi(t,k')$ with $kx = k^o x^o - k'x'$. Then, mechanical system of type -1 corresponding to (2.14-b) is

$$(\mathcal{T}_r, H) \text{ with } \mathcal{T} = (S\mu \subset L^2(H\uparrow, \mu)\ \mu \subset {}'S),$$

$H(t) = H_o + H_\gamma(t) + H_q(t)$ with $H_o(\bar z, z) = \bar z\, \omega\, z$, $H_\gamma(\bar z, z, t) = -\overline{\gamma(t,.)}z - \gamma(t,.)\bar z$ and

$$H_q(\bar z, z, t) = -2^{-1}(2\pi)^{-s/2} \int_{\mathbb{R}^{2s}} \frac{q(t,k'+k'')\overline{z(k'')}\,\overline{z(k')} + q(t,k'-k'')z(k'')\overline{z(k')}\, dk'dk''}{4\omega(k')\,\omega(k'')} + \text{im. conj.}$$

For $f \in {}'S$, define $a(\bar f) : z \to \bar f z$ and $a^\star(f) : z \to f\bar z$.
Then observables $a(\bar f)$ and $a^\star(f)$ are regular for $f \in S$; and for $g \in S$

$$\{a(\bar f), a(\bar g)\} = \{a^\star(f), a^\star(g)\} = 0 \ ;\ \{a(\bar f), a^\star(g)\} = -i\, \bar f\, g.$$

c) If $s = 0$, positive hyperboloïd of mass $m = \omega$, is point $k = \omega$ of $\mathbb{R}$, and μ is Dirac measure of mass $(2\omega)^{-1}$. Mechanical system of type $\varepsilon = +1$ corresponding to (2.14.c) is (Z_r, H) with $Z = L^2(H\uparrow, \mathbb{C}^2) = \{\zeta = (\zeta^+, \zeta^-) \in \mathbb{C}^2$; $2\omega \|\zeta\|^2 = |\zeta^+|^2 + |\zeta^-|^2$, $H = \omega(\bar\zeta^+ \wedge \zeta^+ + \bar\zeta^- \wedge \zeta^-)$ satisfies (2.9). Moreover $\{\zeta^+, \bar\zeta^+\}_+ = \{\zeta^-, \bar\zeta^-\}_+ = -i$; $\{\zeta^+, \zeta^+\}_+ = \{\zeta^+, \zeta^-\}_+ = \dots = 0$.

. For this last example, and for Dirac equation, classical Hamiltonian H is an exterior form ; and H is positive !! By (2.9), (3.6) (3.7) and (3.13) quantum hamiltonian $\hat H$ is a positive self adjoint operator. (2.17-a) shows how anticommutative Poisson brackets defined in [2] for $\varepsilon = +1$ in dimension 3, can be derived from definition (2.8) valid for arbitrary dimension and $\varepsilon = \pm 1$.

3. SYMBOLIC CALCULUS.

We start with a phase triplet $\mathcal{T}_r$ with $\mathcal{T} = (S \subset Z \subset S)$. The topology of S is locally convex inductive limit of topologies of all finite dimensional subspaces of S ; and 'S is the algebraic antidual of S, equipped with the weak topology. Conjugate triplet is $\bar{\mathcal{T}} = (\bar S \subset \bar Z \subset S')$. Fock space is $\bigoplus_{k=o}^{\infty} Z_\varepsilon^k$, scalar product in Z_ε^k is denoted $(.,.)_{nat}$; and any $\phi_k \in Z_\varepsilon^k$ can be identified with following

ε-symmetric form, homogeneous of degree k on $\bar{Z}$:

$(\bar{z}_1 ; \dots ; \bar{z}_k) \to (z_1 z_2 \dots z_k, \phi_k)_{k,nat}$. Bellow, Z_ε^k is identified with this space of forms on $\bar{Z}$. Triplet centered on holomorphic representation of Fock space:

(3.1) $FH_\varepsilon(\bar{Z}) = \{\Phi = \sum_{k=o}^{\infty} \Phi_k \; ; \Phi_k \in Z_\varepsilon^k \; ; \|\Phi\|_z = \sum \|\Phi_k\|^2 k! < \infty\}$.

(3.2) is $\mathcal{F}_\varepsilon \mathcal{T} = (T_\varepsilon(S) \simeq \mathcal{F}_{\varepsilon\text{-cyl}}(S') \subset FH_\varepsilon(\bar{Z}) \subset \mathcal{F}_\varepsilon(\bar{S}))$

(3.3) Antiduality between internal and external spaces of this triplet, extend the scalar product of $FH_\varepsilon(\bar{Z})$. We define antilinear isomorphisms $t \to t^\star$: $T_\varepsilon(S) \to T_\varepsilon(\bar{S})$ and $f \to f^\star$: $\mathcal{F}_\sigma(S) \to \mathcal{F}_\sigma(\bar{S})$, combining twist and conjugation in $\mathcal{T}$:

$$(x_1 x_2 \dots x_k)^\star = \bar{x}_k \bar{x}_{k-1} \dots \bar{x}_1 \quad \text{and} \quad f_k^\star(\bar{x}_1, \dots \bar{x}_k) = \overline{f(\bar{x}_k, \dots x_1)}.$$

(3.4) <u>Proposition and definition of kernel.</u>

Vector space Op of linear operators $\hat{Q}:\mathcal{F}_{\varepsilon\text{-cyl}}(S) \to \mathcal{F}_\varepsilon(\bar{S})$ is isomorphic to $\mathcal{F}_\varepsilon(\bar{S} \times S)$. This isomorphism maps any $\hat{Q} \in$ Op on a form $\tilde{Q}(\bar{z},z')$ such that $\forall\, \Phi$ and $\psi \in T_\varepsilon(S)$

(3.5) $$(\hat{Q}\,\Phi)\;(\bar{z}) = \langle \tilde{Q}(\bar{z},z'),\; \Phi(\bar{z}')\rangle$$

(3.6) $$(\psi,\hat{Q}\,\Phi) = \langle \psi^\star,\; \hat{Q}\,\Phi\rangle = \langle \Phi(\bar{u}')\; \psi^\star(u),\; \tilde{Q}(\bar{u},u')\rangle$$

Form $\tilde{Q}$ is called the kernel of $\hat{Q}$.

Princip of the proof. In fact, vector space Op is isomorphic to the space of sesquilinear forms on $T_\varepsilon(S) \times T_\varepsilon(S)$. Using (3.3), Op is isomorphic to the space of bilinear forms on $T_\varepsilon(\bar{S}) \times T_\varepsilon(S)$. By universal property of tensor product, Op is isomorphic to the algebraic dual of $T_\varepsilon(\bar{S}) \otimes T_\varepsilon(S) \simeq T_\varepsilon(\bar{S} \times S)$, and therefore to $\mathcal{F}_\varepsilon(\bar{S} \times S)$.

Let Opi (resp Opf) be the space of linear operators $T_\varepsilon(S) \to T_\varepsilon(S)$ (resp $T_\varepsilon(S) \to FH_\varepsilon(\bar{Z})$). Kernel theorem [4] characterize kernels of operators $\hat{Q} \in$ Op belonging to Opi and Opf. The <u>symbol</u> (or normal form) is following ε-symmetric form on $\bar{S} \times S$

(3.7) $$Q(\bar{z},z') = \tilde{Q}(\bar{z},z')\, e^{-\bar{z}.z'} = e^{-\bar{z}.z'}\, \tilde{Q}(\bar{z},z')$$

Because $(\exp(\bar{z}.z'))\, \exp(-\bar{z}.z') = 1$, symbol map $\hat{Q} \to Q(\bar{z},z')$, is an isomorphism of Op on $\mathcal{F}_\varepsilon(\bar{S} \times S)$.

(3.8) Proposition.

Let $\hat{Q} \in Op$. If Z_α denotes any complex finite dimensional subspace of S, let $\hat{Q}_\alpha = r_\alpha \hat{Q} i_\alpha$, where i_α(resp r_α) denotes the canonical injection of $T_\varepsilon(Z_\alpha)$ in $T_\varepsilon(S)$ (resp surjection of $\mathcal{F}_\varepsilon(\bar{S})$ on $\mathcal{F}_\varepsilon(\bar{Z}_\alpha)$. Then the kernel (resp symbol)of $\hat{Q}_\alpha$ is the restriction to $\bar{Z}_\alpha \times Z_\alpha$ of the kernel (resp symbol) of $\hat{Q}$.

Princip of the proof.

Because restriction of form $\exp - \bar{z}.z'$ on $\bar{S} \times S$ is the form $\exp - \bar{z}.z'$ on $\bar{Z}_\alpha \times Z_\alpha$ and because restrictions of the product of two forms on $\bar{S} \times S$ to $\bar{Z}_\alpha \times Z_\alpha$, coïncide with the product of restrictions, it is sufficient to prove assertion concerning kernels. Let Op_α be the space of linear operators $T_\varepsilon(Z_\alpha) \to \mathcal{F}_\varepsilon(\bar{Z}_\alpha)$. Then proposition (3.8) can be proven applying (3.4) to Op and to Op_α.

(3.9) Proposition.

With notations of (3.8), let $z^1, z^2 \ldots z^d$ be orthonormal coordinates in Z_α ; $\dim Z_\alpha = d$. Suppose that

$$Q_\alpha(\bar{z},z') = \sum_{k,\ell \in J\varepsilon(d)} Q_{k,\ell}\, \bar{z}^k\, z'^{\ell}$$

with some constants $Q_{k,\ell}$. Then for any $i \in J\,\varepsilon(d)$

$$Q_\alpha(\bar{z}^i) = \sum_{j \in J\varepsilon(d)} Q_{j,i}\; i!\; (i)_\varepsilon\; \bar{z}^j$$

Princip of the proof. A priori $\hat{Q}_\alpha(\bar{z}^i) = \sum a_j\, \bar{z}^j$.

Constant a_j is computed applying (3.6) with $\phi = \bar{z}^i$ and $\psi = \bar{z}^j$ and using (1.13).

For $\varepsilon = -1$, using topology, domain $T_\varepsilon(S)$ can be replaced by a bigger space containing coherent states $e^{z'}: \bar{z} \to \exp \bar{z}\, z'$; and formal serie giving kernel $\tilde{Q}$ converge. Therefore, kernel is given by formula $\tilde{Q}(\bar{z},z') = (\hat{Q}(e^{z'}))\,(\bar{z})$. In general, for $\varepsilon = \pm 1$, this equality of scalars is replaced by an equality of forms :

(3.10) Theorem.

For $\varepsilon = \pm 1$ and any $\hat{Q} \in Op$, we have

$$Q(\bar{z},z') = (Q(e^{\bullet . z'}))\,(\bar{z})$$

where for any $Z_\alpha \subset\subset Z$, restriction R_α of right hand side R to $\bar{Z}_\alpha \times Z_\alpha$

is

$$R_\alpha(\bar{z},z') = \lim_{n\to\infty} \left[(\hat{Q}_\alpha \otimes I_\alpha) \left(\sum_{k=o}^{n} k!^{-1} (\bar{\bullet}.z')^k \right) \right] (\bar{z})$$

where $\hat{Q}_\alpha \otimes I_\alpha$ is the tensorial product of $\hat{Q}_\alpha$ and identity operator in space $T_\varepsilon(\bar{Z}_\alpha)$. Moreover :

(3.11) $$\forall \phi \in T_\varepsilon(S) \; ; \; (\hat{Q}\,\Phi)(\bar{z}) = \langle \phi(\bar{u}') \quad e^{\bar{z}.u}, \; \tilde{Q}(\bar{u},u')$$

(3.12) $$(\hat{Q}\,\phi)\;(\bar{z}) = \langle \phi(\bar{u}' + \bar{z})\; e^{\bar{z}.u}, \; Q(\bar{u},u')\rangle$$

Proof of (3.10) use (3.4), 3.8), (3.9), and anticommutative Taylor formula [4] [5]. For $f \in S$, creation and annihilation operators are defined by $a^\star(f) : \phi \to f\;\phi$ and $a(\bar{f}) : \phi \to \bar{f}(D)\;\phi = \partial_{\bar{f}}\,\phi$. Symbols of these operators are resp. $f\bar{z}$ and $\bar{f}z$. For example $\varepsilon = +1$, (3.10) shows that restriction of the kernel of $\hat{Q} = a^\star(f)$ to $\bar{Z}_\alpha \times Z_\alpha$ is for any $Z_\alpha \subset\subset Z$ containing f : $\tilde{Q}_\alpha(\bar{z},z') = (f\bar{z}) \wedge (\exp \bar{z}.z')$. Therefore $Q(\bar{z},z') = f\bar{z}$.

(3.13) . For any linear map $\ell : S \to S$, collection of linear maps ${}_k{}^0\,\ell$ in spaces ${}_k{}^0\,S$ define an operator $\Gamma\ell$ in $T_\varepsilon(S)$ and application of (3.10) shows that symbol of $\Gamma\ell$ is $(\Gamma\ell)\;(\bar{z},z') = \bar{z}.\;\ell z'$. Free hamiltonians are of this type. In view of (3.10), we say for $\varepsilon = +1$ that $\exp \bar{z}.z'$ is a coherent superstate. Even if $\dim\; Z < \infty$, this extension of coherent state is different from Pelemedov extension [7] because $\exp \bar{z}.z'$ in not in Fock space. For $\varepsilon = \pm 1$, α and $\beta \in S$, operators $\exp\; a^\star(\beta)$ and $\exp\; a(\bar{\alpha}) \in$ Opi are defined by their symbols $\exp \beta\,\bar{z}$ and $\exp \bar{\alpha}\,z'$. For $\varepsilon = -1$, we obtain using (3.12) :

$$((\exp\; a(\bar{\alpha}))\phi)\;(\bar{z}) = \langle \phi(\bar{u}' + \bar{z}),\; e^{\bar{\alpha}u'}\rangle = \phi(\bar{\alpha} + \bar{z})$$

(3.14) If $\varepsilon = +1$, form $\phi(\bar{\alpha} + \bar{z})$ is defined by last relation. Therefore :

$$\phi(\bar{\alpha} + \bar{z}) = \phi + \alpha(D)\;\phi$$

We use now the topological theory [4], [5] of symbolic calculus. We consider particular case where $\varepsilon = -1$, and where familly P of weights defined in [4], [5] is $P = P_H$. Triplet (3.2) is now following triplet of symmetric forms

$$F_{-1}(\mathcal{Z}) = (\text{Exp } S' \subset FH_{-1}(\bar{Z}) \subset H(\bar{S} \times S))$$

The space Op of linear continuous operators $\hat{Q} : \text{Exp } S' \longrightarrow H(\bar{S})$ is equipped with the topology of uniform convergence on equicontinuous subsets of Exp S' ; and $\text{Op} \simeq H(\bar{S} \times S)$. We denote Ope the space of linear continuous maps of external space ; space of adjoints maps is a space Opi of linear operators in internal space Exp S'. Kernel theorem [4] characterized kernels of operators belonging to Opi, to Ope...

(3.15) For $\hat{L} \in$ Opi (resp Op) and $\hat{Q} \in$ Op (resp Ope) product $\hat{A} = \hat{Q}\,\hat{L}$ is defined in Op, and we prove in [4] that symbol of $\hat{A}$ is

(3.16) $$A(\bar{z},z') = \sum_{k=o}^{\infty} k!^{-1} \langle\langle D^k_{z'},\ Q(\bar{z},z')\ D^k_{\bar{z}}\ L(\bar{z},z')\rangle\rangle$$

These results are used in the proof of following theorem.

(3.17) <u>Theorem.</u>

Let $S = \mathcal{S}(H\uparrow)$. Consider a scalar quantum boson field on $M(\dim s = 0,1,2$ or $3)$ with Hamiltonian $H(t) = H_o + H_\gamma(t)$ defined by (2.17.b) Then, kernel of scattering operator $\hat{S}$ is

(3.18) $$\tilde{S}(\bar{z},z') = (e^{z'},\hat{S}\,e^{z'}) = \exp\Big[i\int_{-\infty}^{+\infty} d\theta \int \overline{\gamma(\theta,k')}\ e^{-i\theta\omega(k')} z'(k')\ \frac{dk'}{2\omega(k')} + i\int_{-\infty}^{+\infty} d\theta \int \gamma(\theta,k') e^{i\theta\omega(k')}\overline{z(k')}\frac{dk'}{2\omega(k')} - \int_{-\infty}^{+\infty} d\omega \int_{-\infty}^{u} d\theta \int \overline{\gamma(u,k')}\gamma(\theta,k') e^{i\omega(k')(\theta-u)} \frac{dk'}{2\omega(k')}\Big]$$

Princip of the proof. Let $t' \leqslant t''$ and let $U(t'',t')$ be evolution operator of Schrödinger equation $: i\psi'(t) = H(t)\,\psi(t)$. Proposition 3.3 of [4] exposé 5, shows that a familly of operators converge in nuclear complete space Op, if and only if their kernels converge in some space of holomorphic functions on $\bar{S} \times S$. Therefore, the problem is reduced to verify the convergence for $t' \to -\infty$ and $t'' \to +\infty$, of kernels (given page 5.25 of [4]) of operators $(\exp it\, H_o)\ U(t'',t')\ (\exp - it\, H_o)$.

Using formulas (3.16) and (2.8), we can compute the distorsion between commutator $[\hat{Q},\hat{L}]$ of two quantum operators $\hat{Q}$ and $\hat{L}$ and Poisson bracket of classical observables $Q(\bar{z},z')$ and $L(\bar{z},z')$ defined by symbols of $\hat{Q}$ and $\hat{L}$:

(3.19) <u>Proposition.</u>

$$[\hat{Q},\hat{L}]\ (\bar{z},z') - i\ \{Q,L\}\ (\bar{z},z') =$$

$$= \sum_{k=2}^{\infty} k!^{-1}(\langle\langle D^k_{z'},Q\ \ D^k_{\bar{z}}\ L\rangle\rangle - \langle\langle D^k_{\bar{z}}\ Q\ D^k_{z'},\ L\rangle\rangle)$$

In particular if d^oQ or $d^oL < 2$:

(3.20) Symbol of commutator $[\hat{Q},L] = i\ \{Q,L\}$.

This proposition permits to pass rigourously from classical field equations to quantum fields equations. For example :

(3.21) Corollary

Let $\Phi(x')$ be the free boson field at time $t = 0$. With notations of proposition (3.17), quantum field defined for $t \geqslant 0$ by

(3.22) $$\Phi(x',t) = U(t,0)^{*}\ \Phi(x')\ U(t,0)$$

satisfies following evolution equation

(3.23) $$\partial_{tt}\ \Phi = \Delta\ \Phi - m^2\ \Phi + \gamma(t)$$

Proof.

Note that $\phi(x')$ is defined as an element $\in$ Op, even if $\phi(x')$ is not an unbounded operator in Fock space !! Note that right hand side of (3.22) is defined as an element in Op, because $U(t,0) \in$ Opi $\cap$ Ope ; and Φ is a C^{∞} function with values in complete nuclear space Op, even if $\Phi(t,x')$ is not an unbounded operator in the Fock space !! Now :

$$i^2\ \partial_{tt}\ \Phi(x',t) = U(t,0)^{*}\ [H(t),\ [H(t),\ \Phi(x')]]\ U(t,0)$$

and application of (3.19) gives (3.23). Same kind of computations can be done if $H(t) = H_o + H_\gamma(t)$ is replaced by $H(t) + H_q(t)$.

§ 4. FERMI INTEGRALS AND FEYNMAN INTEGRALS.

(4.1) Definition and proposition.

a) Let X be a vector space ; an ε-symmetric coform M on algebraic dual X' of X is a linear form on $T_\varepsilon(X) \simeq \mathcal{F}_{\varepsilon\text{-cyl}}(X')$. For any $\varphi \in T_\varepsilon(X)$ we write $< M, \varphi > = \int \varphi(\xi)\ M(\xi)$. Space of coforms on X' is denoted $\mathcal{F}'_{\varepsilon\text{-cyl}}(X')$.

b) For any coform M on X', there exists one form $\hat{M} = \sum \hat{M}_k$ on X such that for any $\varphi = \Sigma\, \varphi_k$:

$$< M, \varphi > = \sum k!\ < \hat{M}_k, \varphi_k >_k .$$

$\hat{M}$ is called the Laplace transform of M. Inverse Laplace transform is denoted $f = f(x) \mapsto \delta f$.

If Y is a secund vector space and if g is an ε-symmetric form on $X' \times Y$ we set

$$\int f(\xi,y)\ M(\xi) = \langle f(\xi,y),\ \hat{M}(x)\rangle$$

if RHS is defined as a contraction. In particular using (1.17) :

$$\int e^{\xi.x}\ M(\xi) = \hat{M}(x)$$

This shows that $\hat{M}$ coïncide for $\varepsilon = -1$ with usual Laplace transform. Operations on coforms are studied in [4]. For any $\alpha \in X^{\star}$, translated coform $t_{-\alpha} M$ is defined by

$$\forall \varphi \in T_{\varepsilon}(X)\ ;\ \langle t_{-\alpha}M,\varphi\rangle = \int \varphi(\xi+\alpha)\ M(\xi)$$

A Fermi coform on $X^{\star}$ is a non vanishing antisymmmetric coform on $X^{\star}$ invariant by all translations.

(4.2) Theorem.

Let X be a vector space over a field of characteristic zéro. If $\dim X < \infty$, all Fermi coform M on the algebraic dual X' of X are proportional. If $\dim X = +\infty$ there exists no Fermi coforms, and therefore no Fermi integral on X'.

Proof.

a) For any $\alpha \in X'$, for any coform M on X', Laplace transform of $t_{-\alpha}M$ is $\hat{M}(x) \wedge e^{ax}$. In fact, (4.1) shows that space of coforms on X' can be identified with the space of forms on X. Definition of translations of coforms means that linear map $M \to t_{-\alpha} M$ is the transpose of $\varphi \to \varphi + \partial_{\alpha} \varphi = \varphi + \alpha(D)\ \varphi$. But this last map is the transposed map of $f(x) \to f(x) + f(x) \wedge ax = f \wedge e^{a}$ and assertion is proven.

b) Using a), a coform M on X' is Fermi if and only if $\hat{M} \neq 0$ and $\hat{M} = \hat{M} \wedge \alpha$ for any $\alpha \in X'$.

c) If $\dim X = n < \infty$, coordinates can be used in X and using b), we see that Laplace transform of a Fermi coform is an homogeneous exterior form of degree n on X. Let $\psi \in F^{n}_{+1}(X')$ be such that $\langle\hat{M},\psi\rangle = 1$. Any exterior form φ on X' can be written $\varphi = \lambda\ \psi + \varphi'$ with $d^{o}\ \varphi' < n$ and

$$\langle M, \varphi\rangle = \langle \hat{M}, \varphi\rangle = \langle \hat{M}, \lambda\ \psi\rangle + 0 = \lambda .$$

This means that $\int \varphi(\xi)\ M(\xi)$ is an usual Fermi integral [2].

d) Suppose that M is a Fermi coform on X' and $\dim X' = +\infty$. Then $\hat{M} \neq 0$, and restriction of $\hat{M}$ to any $X_\alpha \subset\subset X$ muss be homogeneous of degree $\dim X_\alpha$; this is impossible.

(4.3) Interpretation of some Feynman integrals in terms of coforms.

a) Let Z a separable complex Hilbert space with conjugation ; $\varepsilon = \pm 1$. Canonical normal coform ν' on algebraic dual of $\bar{Z} \times Z$ is defined by Laplace transform $e^{\bar{z}.z'}$ defined on $\bar{Z} \times Z$; ν' is represented by a gaussian measure on the dual of Z_r for $\varepsilon = -1$. b) With notations of § 3, the cokernel $\delta \tilde{Q}$ (resp cosymbol δQ) of $\hat{Q} \in Op$ is inverse Laplace transform of $\hat{Q}$(resp Q). Therefore, formulas (3.5), (3.11)... can be written, substituing brackets, by symbols of integration c). For example evolution operator of free boson field as cokernel $M = \delta(\exp(\bar{z}\, e^{it\omega} z'))$. This shows that M is represented by a gaussian measure rotating in a complex space [6]. More precisely, M is represented by the image of ν' by the transpose of linear map $(\bar{z},z') \to (\bar{z}\ ;\ e^{it\omega} z')$ in S. Formula (3.16) shows that cokernel $\delta \tilde{S}$ of $\hat{S}$ can be represented by a Dirac measure.

=-=-=-=-=-=-=-=-=-=-=-=-=

REFERENCES

[1] S. Albeverio and R.J. Hoegh Krohn.
Mathematical theory of Feynman Path Integral. Lecture Notes in mathematics n° 523. Springer Verlag (1976).

[2] F.A. Berezin and M.S. Marinov.
Particle spin dynamics as the Grassmann variant of classical mechanics. Annals of Physics 104 p. 336-362 (1977).

[3] P. Krée.
Théorie des distributions et calculs différentiels sur un espace de Banach. Séminaire Lelong. Analyse 1974-75 p. 163-192. Lecture Notes in Mathematics n° 524 (1976).

[4] P. Krée.
Séminaire sur les équations aux dérivées partielles en dimension infinie 3ème année 1976-1977. Publié par le secrétariat mathématique de l'Institut H. Poincaré.

[5] P. Krée.
Trois notes (à paraître) aux Comptes Rendus. Série A. juin 1978.

[6] P. Krée.
Théorie des distributions et holomorphie en dimension infinie Mathematics studies n° 12 - North Holland 1977.

[7] A.M. Pelemedov.
Generalized coherent states and some of their applications. Sov. Phys. Usp. 20 (9) september 1977 p. 703-720.

[8] L. Schwartz.
Théorie des distributions à valeurs vectorielles. Ann Inst. Fourier, Grenoble, t. 7, 1957, p. 16-141.

- SECTION III -

CARACTERISATION DE PROCESSUS PAR LA METHODE DES SPECIFICATIONS LOCALES

Ph. Blanchard
Universität Bielefeld
D-4800 BIELEFELD

§0. Introduction

La méthode des spécifications locales de Dobrushin-Föllmer permet de construire les processus stochastiques associés à un système de probabilités conditionelles. Ces processus s'obtiennent en construisant la frontiere de Dynkin-Martin relative à la specification locale envisagée. Nous esquisserons brièvement cette construction dans le cas de processus gaussiens, de processus correspondant au mouvement brownien d'une particule soumise à l'action d'une force dérivant d'un potentiel et enfin de processus de Poisson généralisés.

§1. Spécifications locales et frontière de Dynkin-Martin

Nous allons rappeler la définition d'une spécification locale et décrire quelques resultats de Föllmer renvoyant à son seminaire [1] et à [2] pour la plupart des détails techniques et démonstrations ainsi que pour les références. Soient donnés un espace mesurable $(\Omega, \hat{F})$, un ensemble I d'indices filtrant à droite pour une relation transitive $<<$, une famille décroissante de sous-tribus de $\hat{F}$ notée $(\hat{F}_\alpha)_{\alpha\in I}$ c'est à dire telle que l'on ait $\hat{F}_\alpha \supset \hat{F}_\beta$ si $\alpha << \beta$.

Une spécification locale se définit alors comme une famille $\Pi = (\Pi_\alpha)_{\alpha\in}$ de noyaux de probabilité sur $\Omega \times \hat{F}$ (c · à · d telle que $\Pi_\alpha(\omega, \cdot)$ soit une loi de probabilité sur $(\Omega \times \hat{F})$) jouissant des propriétés suivantes:

1) $\Pi_\alpha (\cdot, A)$ est $\hat{F}_\alpha$-mesurable $\forall A \in \hat{F}$

2) $\Pi_\alpha (\cdot, A) = \mathbb{1}_A(\cdot) \quad \forall A \in \hat{F}_\alpha$

3) $\Pi_\alpha (\omega, A) = \int \Pi_\alpha (\omega, d\eta)\, \Pi_\beta(\eta, A)$

$\forall \alpha >> \beta \quad \forall \omega \in \hat{F}$

On dira qu'un état macroscopique P (c'est à dire une loi de probabilité sur $(\Omega,\hat{F})$) est un état de Gibbs pour la spécification locale $\Pi = (\Pi_\alpha)_{\alpha\in I}$ si

$$E_P [A \mid \hat{F}_\alpha] (\omega) = \Pi_\alpha(\omega, A) \quad P \ \ p\cdot s$$

et cela $\forall \alpha \in I$ et $\forall A \in \hat{F}$.

Autrement dit si P est un état de Gibbs associé à la spécification locale $\Pi = (\Pi_\alpha)_{\alpha \in I}$ alors $\Pi_\alpha(\omega, A)$ n'est rien d'autre que l'espérance conditionelle de P etant donnée la sous σ-algèbre $\hat{F}_\alpha$.

Par $G(\Pi)$ on désigne l'ensemble des états de Gibbs associés à la spécification locale Π. Le problème est donc etant donnée une spécification locale $(\Omega, \hat{F}, (\hat{F}_\alpha)_{\alpha \in I})$ d'essayer de déterminer les états de Gibbs associés. La notion de frontière de Dynkin-Martin permet de résoudre dans certain cas ce problème comme nous le verrons à l'aide des trois exemples annoncés.

Par $E(\Pi)$ on désigne l'ensemble des limites faibles

$$\rho_\omega(\cdot) = \underset{\alpha}{\text{w-lim}}\ \Pi_\alpha(\omega, \cdot\,)$$

c'est à dire telles que

$$\rho_\omega(f) = \lim_\alpha \Pi_\alpha(\omega, f)$$

pour toute fonction f continue bornée sur Ω, ce que l'on notera $f \in C_b(\Omega)$. $E(\Pi)$ sera appellée la <u>frontière de Dynkin-Martin</u> associée à Π. Un résultat de Föllmer [1] affirme de plus que si $\forall f \in C_b(\Omega)$ alors $\Pi_\alpha(\cdot, f) \in C_b(\Omega)$ on a l'inclusion suivante: $E(\Pi) \subset G(\Pi)$. Autrement dit dans cette situation si $E(\Pi)$ n'est pas vide $G(\Pi)$ ne l'est aussi pas.

<u>Remarque</u>: Supposons qu'il existe au moins un état de Gibbs pour la famille $(\Pi_\alpha)_{\alpha \in I}$. Alors pour $\alpha << \beta$ on a $\hat{F}_\beta \subset \hat{F}_\alpha$ et par conséquent $\forall f \in L(P)$

$$E_p(f \mid \hat{F}_\beta) = E_p(E_p(f \mid \hat{F}_\alpha) \mid \hat{F}_\beta)$$

ce qui équivant à $\alpha << \beta \Rightarrow \Pi_\beta = \Pi_\beta \circ \Pi_\alpha$. La condition 3) est donc nécessaire pour l'existence d'un état de Gibbs.

<u>Remarque</u>: La construction qui vient d'être décrite ne fait que formaliser une méthode fréquemment employée en mécanique statistique, où l'on considère souvent une situation du type suivant:

V_n est un domaine fini de $\mathbb{R}^d$. On fixe la configuration ω du système à l'exterieur de V_n (autrement dit ω joue le rôle d'une condition aux limites). Soit alors ω' une configuration à l'intérieur de V_n dans de nombreux modèles on est alors en mesure de calculer l'énergie $E_n(\omega', \omega)$ de la configuration ω', ω étant fixé à l'extérieur et l'hypothèse de Gibbs permet alors d'écrire la mesure d'équilibre $\Pi_n(\omega, \cdot)$. Après quoi on essaie de déterminer l'ensemble de toutes les limites possibles quand $V_n \to \mathbb{R}^d$ et quand on fait varier la condition aux limites ω. Autrement dit on essaie de déterminer la frontière de Martin-Dynkin $E(\Pi)$ associée à $\Pi = (\Pi_n)_{n \in \mathbb{N}}$.

§2 Processus sur $\Omega = \{ \varphi \mid \varphi \in C([0,1], \mathbb{R}) \quad \varphi(o) = o \}$ [2]

Dans ce paragraphe Ω désignera l'espace de Banach des fonctions réelles continues sur [o,] et nulles à l'origine muni de la norme de la convergence uniforme

$$\| \varphi \| = \sup_{t \in [o,1]} | \varphi(t) |$$

Nous allons maintenant esquisser la construction de deux modèles de spécifications locales $(\Omega, \hat{F}, (\hat{F}^i_n)_{n \in \mathbb{N}} \quad \Pi_i = (\Pi^i_n) \quad i = 1,2)$ pour lesquelles il sera possible de déterminer les frontières de Dynkin-Martin associées $E(\Pi^i) \quad i = 1,2$. Nous renvoyons à [2] pour la plupart des démonstrations.

Soit $\Pi_t : \Omega \to \mathbb{R}$ avec $\Pi_t(\varphi) = \varphi(t)$. Nous notons $\hat{F}$ la tribu borélienne de Ω, cette tribu est aussi engendrée par les $\Pi_t(\varphi)$ où t parcourt une suite dense dans [0,1].

Par μ nous désignons la mesure de Lebesgue sur [0,1] et par $W(\mu)$ la mesure de Wiener sur $(\Omega,\hat{F})$ c'est à dire la mesure associée au processus gaussien normal de variance $\mu([s,t]) = t - s$.

On introduit alors les quantités suivantes:

$$I_n = \{ 0, \frac{1}{2^n}, \frac{1}{2^{n-1}}, \dots\dots, 1 \}$$

$$[0,1] = \bigcup_n I_n$$

$$S_n(\varphi) = 2^n \sum_{k=1}^{2^n} \left[\Pi_{\frac{k}{2^n}}(\varphi) - \Pi_{\frac{k-1}{2^n}}(\varphi) \right]^2$$

On peut remarquer que $2^n S_n(\varphi)$ n'est rien d'autre que l'action associée au chemin polygonal $\tilde{\varphi}$ tel que $\tilde{\varphi}(\tau) = \varphi(t_{n,k}) \quad \tau = t_{n,k} = \frac{k}{2^n} \quad k = 1,2,\dots 2^n$. On a donc

$$2^n S_n(\varphi) = \int_0^1 \tilde{\varphi}^2(\tau)\, d\tau$$

Afin de définir les deux spécifications locales annoncées $\Pi^i \quad i = 1,2$ on commence par introduire un espace (Ω',B) isomorphe à $(\Omega,\hat{F})$. L'espace Ω' est le sous-espace de $\mathbb{R}^{\mathbb{N}}$ constitué par les suites réelles $C = (C_m)_{m \in \mathbb{N}}$ qui vérifient les deux propriétés suivantes

(A) $\lim_{m \to o} c_m \, p(m) = 0$

(B) $||| C ||| = \sup_m |p(m) \, C_m| < +\infty$

Dans ces deux relations p(m) est defini par

$$p(m) = (2)^{\frac{-(n-1)}{2}} \quad \text{si} \quad 2^{n-1} < m \leq 2^n \; n \leq 1 \quad p(1) = 1$$

On démontre alors ([2] Proposition 1) qu'il existe une application lineaire bijective Φ de Ω sur Ω' telle que Φ et Φ^{-1} soient continues et mesurables. Pour tout $\varphi \in \Omega$ $\Phi(\varphi)$ est une suite réelle dans Ω'. On écrit ensuite

$$\Omega' = \Omega'_n \times \hat{\Omega}'_n$$

où Ω'_n désigne l'ensemble des suites de Ω' dont seuls les 2^n élements ne sont pas identiquement nuls. Nous notons maintenant B_n (respectivement $\hat{B}_n$) les tribus engendrées par les ensembles cylindriques à base dans Ω'_n(respectivement $\hat{\Omega}'_n$). Les images de ces tribus par Φ^{-1} sont F_n et $\hat{F}_n$. Schématiquement on a la situation suivante:

$$\begin{array}{ccccc} \Omega' & = & \Omega'_n & \times & \hat{\Omega}'_n \\ & & \downarrow & & \downarrow \\ B(\mathbb{R}^{2^n}) & \approx & B_n & & \hat{B}_n \\ & & \downarrow & & \downarrow \\ \Phi^{-1}(B_n) & = & F_n & & \hat{F}_n = \Phi^{-1}(\hat{B}_n) \end{array}$$

Autrement dit on a $\Phi(\varphi) = (\Phi_n(\varphi), \hat{\Phi}_n(\varphi))$ avec $\Phi_n(\varphi) \in \Omega'_n$ et $\hat{\Phi}_n(\varphi) \in \hat{\Omega}'_n$. D'autre part F_n est la σ-algèbre engendrée par les Π_t pour lesquels $t \in I_n$.

Nous sommes maintenant en mesure de définir deux systèmes de spécifications locales. Soient $\varphi \in \Omega$, $A_1 \in F_n$, $A_2 \in \hat{F}_n$ quelconques ; posons alors

$$\Pi_n^1 (\varphi, A_1 \cap A_2) = \varepsilon_{\hat{\Phi}_n(\varphi)} (\varphi(A_2)) \otimes \lambda^n_{S_n(\varphi)} (\Phi(A_1))$$

où $\lambda^n_R(\cdot)$ est la mesure de Haar sur la sphère $S^{2^n-1}(o,\sqrt{R})$ et $\mathcal{E}_a(\cdot)$ la mesure de Dirac en a sur $(\hat{\Omega}'_n, \hat{B}_n)$.

Quant à la seconde specification locale on la définit par

$$\Pi_n^2 (\varphi, A_1 \cap A_2) = \mathcal{E}_{\hat{\Phi}_n(\varphi)} (\Phi(A_2)) \otimes C_n \int_{\Phi(A_1)} dx \, e^{-\frac{1}{2} \|x\|^2}$$

où $C_n e^{-\frac{1}{2}\|x\|^2} dx$ est une mesure de probabilité sur $(\Omega'_n, B_n) \approx (\mathbb{R}^{2^n}, B(\mathbb{R}^{2^n}))$.
Avant d'énoncer les résultats il reste encore à definir une famille décroissante de sous-tribus de $\hat{F}$. On définit

$$\hat{F}_n^1 = \sigma\left\{ \hat{F}_n , S_n(\cdot) \right\}$$

autrement dit $\hat{F}_n^1$ est la σ-algèbre emgendrée par les éléments de $\hat{F}_n$ et par les ensembles de la forme $\{ \varphi \mid \sigma_n(\varphi) \in B \;\; B \text{ ouvert de } \mathbb{R} \}$. En utilisant la forme relativement explicite des Π^i il est facile d'établir le résultat suivant:

Theorème 1 [2]

$(\Omega, F, (\Pi_n^1)_{n\in \mathbb{N}}, (\hat{F}_n^1)_{n\in \mathbb{N}})$ et $(\Omega, F, (\Pi_n^2)_{n\in \mathbb{N}}, (\hat{F}_n)_{n\in \mathbb{N}})$ sont des spécifications locales au sens de Dobrushin-Föllmer.

D'autre part il est facile de se convaincre que si $f \in C_b(\Omega)$ alors $\Pi_n^i(\cdot, f) \in C_b(\Omega)$ ce qui entraine alors

$$E(\Pi_i) \subset G(\Pi_i)$$

Plus précisement on peut démontrer

Theorème 2 [2]

On a $E(\Pi^2) = G(\Pi^2) = G_{ex}(\Pi^2) = \{W(\mu)\}$ où $G_{ex}(\Pi^2)$ désigne l'ensemble des états de Gibbs extremaux.

Theorème 3 [2]

On a les propriétés suivantes:

a) $\exists \text{ w-}\lim\limits_{n\to\infty} \Pi_n^1(\cdot, \varphi) \Leftrightarrow \exists \lim\limits_{n\to\infty} \frac{1}{2^n} S_n(\varphi) \equiv \sigma^2(\varphi)$

b) Au cas où la limite existe on a alors

$$\text{w-}\lim_{n\to\infty} \Pi_n^1(\cdot, \varphi) = W(\sigma^2(\varphi)\cdot\mu)$$

c) $G_{ex}(\Pi^1) = \{ W(\sigma^2\mu) \mid 0 \leq \sigma^2 < +\infty \}$

Pour démontrer les theorème 2 et 3 on utilise le résultat classique de Prohorov [5].

Theorème 4 (Prohorov)

Soit Ω un espace metrique et $M_1(\Omega)$ l'ensemble des mesures de probabilité sur Ω. Soit $\{\mu_n\}_{n\in \mathbb{N}} \subset M_1(\Omega)$ telle que $\forall \varepsilon > 0 \ \exists K_\varepsilon$ compact dans Ω telle que

$$\mu_n(\Omega \setminus K_\varepsilon) < \varepsilon \quad \forall n$$

Alors la suite $\{\mu_n\}_{n\in \mathbb{N}}$ possède une sous-suite convergente

Posons $\mu_n = \Pi_n^i$ $i = 1,2$. A l'aide de la forme explicite des Π_n^i on peut démontrer l'estimation suivante

$$E_{\mu_n}[\,|\varphi(t)-\varphi(s)| \leq C\,|S-t|^{\frac{1}{2}-\varepsilon}] \geq 1 - \eta \quad \forall \eta > 0$$

En désignant par $K \subset \Omega$ l'ensemble des fonctions φ pour lesquelles cette estimation est valable on note que $\varphi \in K$ est Hölder-continue et donc que K est equicontinu et borné. Par le theorème d'Ascoli-Arzela il en résulte que K est compact et nous sommes en mesure d'appliquer le théorême de Prohorov.

Interprétation physique.

Le theorême 3 affirme que $W(\sigma^2(\varphi)\ \mu)(\cdot) = \lim \Pi_n^1(\varphi, \cdot)$ et il résulte de la définition de Π_n^1 que la mesure de Wiener $W(\sigma^2(\varphi)\mu)(\cdot)$ est essentiellement obtenue comme limite d'une suite de mesures de Haar concentrées sur des sphères dans des espaces dont la dimension croît. Or puisque $\sigma_n^2(\varphi) = 2^{-n}\ S_n(\varphi)$ nous pouvons envisager $W(\sigma^2(\varphi)\ \mu)(\cdot)$ comme limite d'une suite de mesures qui sont essentiellement de la forme

$$\delta\left[\ \frac{1}{2}\int_0^1 \dot{\tilde{\varphi}}\ (\tau)\ d\tau - E_n\ \right]$$

avec $E_n = \frac{1}{2}\ 2^n\ \sigma_n^2(\varphi)$

Formellement nous voyons ainsi qu'à la limite, la mesure de Wiener $W(\sigma^2(\varphi)\ \mu)(\cdot)$ est de la forme

$$W(\sigma^2(\varphi)\cdot\mu)(\cdot) \sim \delta\left[\ \frac{1}{2}\int_0^1 \dot{\varphi}^2(\tau)\ d\tau\ - E\ \right]$$

où E est une quantité infinie (précisément comme E_n quand $n \to \infty$). Dans cette interprétation, la spécification locale Π^1 est l'analogue de la distribution microcanonique de la mécanique statistique classique. Mais la specification locale $\Pi^2(\varphi,\cdot)$ converge elle aussi vers la mesure de Wiener $W(\mu)$, ce qui ne fait que traduire que formellement la mesure de Wiener n'est rien d'autre que $e^{-\frac{1}{2}\int_0^1 \dot{\varphi}(\tau)\ d\tau}\ d\varphi$ ce que nous érirons

$$W(\mu) \sim e^{-\frac{1}{2}\int_0^1 \dot{\varphi}\ (\tau)\ d\tau}\ d\varphi$$

Dans la language de la mécanique statistique on dira donc que Π^2 est l'analogue de la distribution canonique et les theorêmes 2 et 3 s'interprêtent comme exprimant l'équivalence des ensembles de la mécanique statistique.

Généralisation

Nous allons maintenant considérer la situation où la particule est soumise à un champ de forces dérivant d'un potentiel V. Sous des conditions assez générales sur V [3] le processus associé que nous noterons η satisfait à l'équation differentielle stochastique

$$d\eta(t) = \beta(\eta(t)) + dw(t)$$

dans laquelle $w(t)$ désigne le mouvement brownien ordinaire et β est une fonction connue du potentiel. Supposant $t > s$, il résulte de l'équation différentielle stochastique que

$$\eta(t) - \eta(s) - \int_s^t \beta\,(\eta(\tau))\,d\tau = w(t) - w(s)$$

le membre de droite de cette équation se laisse construire à l'aide des spécifications locales Π^i $i = 1,2$ que nous avons précédement considerées. Soit $T : \Omega \to \Omega$ telle que

$$(T\varphi)(t) = \varphi(t) + \int_0^t \beta\,(\varphi(\tau))\,d\tau$$

cette transformation induit une transformation $dW \to dW_T$ de la mesure de Wiener et dw_T n'est rien d'autre sous les hypothèses faites que la mesure associée au processus η. Dans ce cas comme dans le "cas libre" on obtient deux représentations pour W_T

$$W_T = \lim_n \Pi_n^i(T) \qquad i = 1,2$$

et on obtient ainsi formellement l'équivalence entre les mesures suivantes

$$\delta\,[\ \frac{1}{2}\int_0^1 \dot{\varphi}^2(\tau)\,d\tau + \int_0^1 V(\varphi(\tau))\,d\tau - \text{const}]\ d\varphi$$

$$\text{const.}\quad e^{-\int_0^1 V(w(\tau))\,d\tau}\ dw(\omega)$$

L'argument de la fonction δ " est l'action classique totale le long du chemin φ . Donc en présence d'une perturbation V les résultats obtenus par la méthode des spécifications locales s'interprètent encore comme exprimant l'equivalence des ensembles microcanonique et canonique de la mécanique statistique classique.

§3 Processus de Poisson généralises [4]

Nous allons maintenant considérer le cas d'un processus stochastique qui s'avérera être de la classe des processus de Poisson generalisés. On peut dire heuristiquement que ce processus est localement un processus de Poisson défini par le nombre de "particules" de la réalisation considerée. Plus precisément un tel processus est défini de la manière suivante: on désigne par X un espace topologique de Hausdorff localement compact et par B la σ-algèbre des sous-espaces boréliens de X. Un sous-ensemble de X sera dit borné si son adhérence est compacte. Soit alors B_0 l'ensemble de tous les boréliens bornés de X. Par $M(X)$ on désignera l'ensemble de toutes les mesures de Radon non négatives sur X et par F la plus petite σ-algèbre de sous-ensembles de M tells que les fonctions

$$N(\cdot,B) : M \to \mathbb{R}$$
$$N(\mu,B) = \mu(B) \qquad B \in B_0$$

soient mesurables. Pour $G \in B_o$ on notera F_o la sous-tribu engendrée par les $N(\cdot,G')$ avec $G' \subseteq G$. Autrement dit F_G s'interprètera comme la sous-tribu associée aux événements localisés dans G. Si ρ est une mesure de Radon sur (X,B) alors on entend par processus de Poisson d'intensité ρ la mesure de probabilité P_ρ sur (M,F) telle que si $G_1 \ldots\ldots G_n$ sont des ensembles boréliens deux à deux disjoints les variables aléatoires $N(\cdot, G_1), \ldots\ldots N(\cdot, G_n)$ soient indépendantes et distribuées suivant une loi de Poisson d'intensité $\rho(G_i)$

Pout tout $G \in B_o$ introduisons la projection

$$T_G\, p = \text{Res}_G\, \mu$$

et

$$P_{\rho,G}(\varphi) = P_\rho(\varphi_o T_G) \quad \forall\, \varphi \in L^1(P_\rho)$$

Soit alors $M^{\cdot\cdot}$ l'ensemble de touts les mesures de Radon ponctuelles c'est à dire somme de mesures de la forme ε_x (mesure de Dirac en x) et $F^{\cdot\cdot} = M^{\cdot\cdot} \cap F$. On sait que le processus P_ρ est concentré sur $M^{\cdot\cdot}$. $P_{\rho,G}$ n'est rien d'autre que le processus de Poissson d'intensité ρ restreint à G. On a alors

$$P_{\rho,G}(\varphi) = e^{-\rho(G)} \sum_{n \geq o} \frac{1}{n!} \int_G \cdots \int_G \varphi\,(\varepsilon_{x_1} + \ldots + \varepsilon_{x_n})\, \rho(dx_1) \ldots \rho(dx_n)$$

pour toute fonction φ bornée mesurable sur M. Rappelons que la transformée de Laplace $\hat{P}$ de P déterminé P et que la transformée de Laplace d'un processus de Poisson P_ρ est donnée par

$$\hat{P}_\rho(if) = \exp\,[\,\rho(\exp(-f) - 1]$$

pour toute fonction f continue à support compact sur X. Pour toute mesure ρ on définit alors

$$\Pi_G(\mu, A_1 \cap A_2) = \mathbb{1}\, A_2\,(\mu) \quad P_{\rho,G}(A_1 \mid N(\cdot,G) = \mu(G))$$

avec $\mu \in M^{\cdot\cdot}$, $G \in B_o$, $A_1 \in F^{\cdot\cdot}_G$, $A_2 \in F^{\cdot\cdot}_{X \setminus G}$ et on entend par $\hat{F}^{\cdot\cdot}_{X\setminus G}$ la tribu engendrée par tous les ensembles de la forme

$$\{\, \mu \in M^{\cdot\cdot} \mid N(\mu,G) = n \,\} \cap A \quad A \in F^{\cdot\cdot}_{X\setminus G} \quad n \in \mathbb{N}$$

Il n'est alors pas difficile de se convaincre que $\Pi = (\Pi_G)_{G \in B_o}$ est une spécification locale au sens de Dobrushin-Föllmer. Autrement dit on a:

- $\Pi_G(., A)$ est $\hat{F}^{\cdot\cdot}_{X\setminus G}$ mesurable $\forall\, A \in F^{\cdot\cdot}$
- $\Pi_G(\cdot, A) = \mathbb{1}_A \quad \forall\, A \in \hat{F}_{X\setminus G}$
- $\Pi_{G'}\, \Pi_G = \Pi_{G'} \qquad \forall\, G, G' \in B_o \quad G \leq G'$

On envisage alors une suite croissante $\{G_n\}_{n \in \mathbb{N}}$ de sous ensembles tels que

$$G_1 \subseteq G_1^o \subseteq G_2 \subseteq \ldots\ldots$$

et on définit

$$\lim_{n \to \infty} \frac{N(\mu, G_n)}{\rho(G_n)} = Y(\mu)$$

$$\sigma\,\{\mu)(\cdot) = \lim_{n \to \infty} \Pi_{G_n}(\mu, \cdot)$$

En condisérant la forme explicite de la transformée de Laplace $\Pi_{Gn}(\mu, \cdot)$ (if) N.X. Xanh et H. Zessin [4] ont démontré le théorème suivant qui permet de caractériser la frontière de Dynkin-Martin associée et pour la démonstration duquel on renvoie au travail original.

Theorème 5 [4]

Tout etat de Gibbs canonique associé à la spécification locale Π satisfait

$$\lim_{n \to \infty} \Pi_{G_n}(\mu, \cdot) = P_{Y(\mu)\cdot\rho} \qquad P \ \ p\cdot s$$

et on a

$$G_{ex}(\Pi) = \{ P_{z\cdot\rho} \quad z \in [o,+\infty] \} \text{ si } \rho(X) = +\infty$$

$$G_{ex}(\Pi) = \{ X_n \left(\frac{\rho^n}{\rho(X)^n} \right) \quad n = o,1,2,\ldots\} \text{ si } \rho(X) < +\infty$$

où $X_n(\tau)$ désigne l'image de τ par rapport à X_n

$$X_n : (x_1 \ldots\ldots x_n) \rightarrow \varepsilon_{x_1} + \ldots\ldots + \varepsilon_{x_n} \qquad x_j \in X$$

Remarque

La quantité $\dfrac{N(\mu,G_n)}{\rho(G_n)}$ s'interprète comme la densité de particules en G_n et

$$\Pi_G(\mu, A) = P_{\rho,G}\,(A \mid \{ \mu' \mid \mu'(G) = \mu(G)\} \ )$$

n'est rien d'autre que l'espérance conditionelle de l'événement A étant donné l'ensemble des μ' qui ont le même nombre de points (c · a · d de particules) que μ dans G.

References

[1] Föllmer H. Phase transitions and Martin boundary. In: Sem. Probab. IX Lecture Notes in Math. 465 Berlin-Heidelberg-New York, Springer (1975)

[2] Ph. Blanchard, C. Pfister "Processus gaussiens, equivalences d'ensembles et spécification locale",Z. Wahrscheinlichkeitstheorie verw. Gebiete 44 177-19o (1978)

[3] S. Albeverio, R. Høegh-Krohn "Dirichlet forms and diffusion processes on rigged Hilbert spaces,"Z. Wahrscheinlichkeitstheorie verw. Gebiete 4o 1-57 (1977)

[4] N.X. Xanh, H. Zessin "Martin Dynkin boundary of mixed Poisson Processes,"Z. Wahrscheinlichkeitstheorie verw. Gebiete 37 191-2oo (1977)

[5] P. Billingsley "Convergence pf Probability measures",John Wiley, New York (1968)

RENORMALIZATION GROUP APPROACH TO THE HIERARCHICAL MODEL

Pierre Collet *

Harvard University

Cambridge , Massachusetts 02138/USA

ABSTRACT

The Hierarchical Model is a one dimensional ferromagnetic spin model with long range forces . Using the invariance of this model under a suitable renormalization group , we analyse the temperature behavior of the system . In particular , we prove the existence of a non-trivial fixed point and compute its critical indices using a local linearization of the renormalization transformation around the fixed point . Some global results (cross-over) are also described , and they provide a description of the phases near the critical point . This talk summarizes some of the results of joint work with J-P.Eckmann [C.E.] .

* Work supported by the University of Geneva .

I) Introduction

The Hierarchical Model is a one dimensional lattice system of classical spins (real identically distributed random variables) coupled by long-range forces . The Hamiltonian of a system of 2^n spins is given by

$$H_{2^n}(S_1 , S_2 , \dots , S_{2^n}) = H^{(h)}_{2^n} (S_1, S_2 , \dots , S_{2^n}) + \sum_{i=1}^{2^n} \text{Log} (f(S_i)) \quad ,$$

where f is the probability density of each individual spin , and $H^{(h)}_{2^n}$ is the Hierarchical Hamiltonian . The detailed structure of $H^{(h)}_{2^n}$ is quite involved [Dy , C.E] , this is a non translation invariant quadratic Hamiltonian which depends on a coupling constant c . The critical dimension (defined according to Ginzburg and Landau [G.L.]) is given by

$$d_c = 2 - \log_2 c \quad .$$

The partition function $Z_{\beta,c,2^n,f}$ and the thermal expectation $< \quad >_{\beta,c,2^n,f}$ for a system of 2^nspins , with individual probability density f , coupling constant c , and inverse temperature β are defined by

$$Z_{\beta,c,2^n,f} = \int\cdots\int e^{-\beta H_{2n} (S_1 , \dots , S_{2n})} \prod_{i=1}^{2^n} dS_i$$

and

$$< g(.,\ldots,.) >_{\beta,c,2^n,f} = \frac{1}{Z_{\beta,c,2^n,f}} \int\cdots\int g(S_1,\ldots,S_{2^n})$$

$$\cdot\ e^{-\beta H_{2^n}(S_1,\ldots,S_{2^n})} \prod_{i=1}^{2^n} dS_i \ .$$

The free energy $F_{\beta,c,2^n,f}$ is defined by

$$F_{\beta,c,2^n,f} = \frac{1}{2^n} \operatorname{Log} Z_{\beta,c,2^n,f} \ .$$

The following relations [Ba] are consequences of the particular spin-spin interaction of our model

$$Z_{\beta,c,2^n,f} = Z_{\beta,c,2^{n-1},\mathbf{N}_\beta(f)} \ ,$$

$$F_{\beta,c,2^n,f} = \frac{1}{2} F_{\beta,c,2^{n-1},\mathbf{N}_\beta(f)} \ , \quad (1)$$

$$< \left(\frac{S_1+\ldots+S_{2^n}}{(2/\sqrt{c})^n} \right)^q >_{\beta,c,2^n,f} = < \left(\frac{S_1+\ldots+S_{2^{n-1}}}{(2/\sqrt{c})^{n-1}} \right)^q >_{\beta,c,2^{n-1},\mathbf{N}_\beta(f)}$$

$$\forall\ q \in \mathbb{N} \quad (2)$$

where $\mathbf{N}_\beta(f)(x) = \left[\frac{2}{c}\ e^{\beta x^2/2} \int f^\beta(x/\sqrt{c} - y)\ f^\beta(x/\sqrt{c} + y)\ dy \right]^{1/\beta}$

Formulas (1) and (2) are the basic ingredients for the renormalization group analysis . They relate expectations of some mean spin observables (here a qth moment) in a model with 2^n spins to the expectation of the same observable in a model with half the number of spins, same temperature , same coupling constant , but the new one spin distribution $N_\beta(f)$. The well known program of Kadanoff and Wilson [K.W] is realized here in a particularly simple , explicit , and exact way :

two systems with different number of degrees of freedom are related by a change of some parameter .

Here we change f , and the transformation

$$f \longrightarrow N_\beta(f)$$

will of course be called the renormalization transformation . Iterating (2) for example , we obtain

$$< \left(\frac{S_1+\dots+S_{2n}}{(2/\ c)^n}\right)^q >_{\beta,c,2^n,f} \quad = \quad < \left(\frac{\sqrt{c}\cdot S}{2}\right)^q >_{\beta,c,1,N_\beta(f)} \quad .$$

In other words , if we know $N_\beta(f)$, we are able to compute the expectation of a mean spin observable in a system with 2^nspins . We are particularly interested in the thermodynamic limit $n \to +\infty$, and we therefore need the behavior of N_β for large n . In the simplest case , this behavior is governed by the fixed points of the transformation N_β . In Section II) , we state a theorem on the existence of a non-trivial fixed point of N_β . The local behavior of N_β around this fixed point is then studied . We are then able to give a precise definition of a phase transition in the Hierarchical Model , and to compute the associated critical indices . Section III) is devoted to the behavior of the model when the temperature is not the critical temperature . From the solution of a global problem , known as a cross-over in the physical literature , we are able to give a detailed description of the phases of the system.

II) Fixed point and local behavior of the renormalization transformation .

Instead of N_β , we shall study the somewhat simpler transformation T given by

$$T = S_\beta N_\beta S_\beta^{-1}$$

where

$$(S_\beta g)(x) = \left(\left\{\frac{2-c}{\beta\ c}\right\} \frac{4\pi}{c} \right)^{\frac{1}{2}} e^{x^2/2} \quad g\left(x\left\{\frac{2-c}{\beta\ c}\right\}^{\frac{1}{2}}\right)$$

and

$$T(\phi)(x) = \frac{1}{\sqrt{\pi}} \int_{-\infty}^{\infty} e^{-u^2} \quad \phi(zc^{-\frac{1}{2}}-u) \quad \phi(zc^{-\frac{1}{2}}+u)\ du \quad .$$

T has three obvious fixed points

- $\phi=0$ which is not interesting since this function cannot be normalized ,
- $\phi=$ cte , which is related as we shall see to the high temperature behavior of the model ,
- $\phi=1$, which through S_β gives $g(x) = S_\beta(g)(x)$ = a Gaussian , the celebrated " trivial " gaussian fixed point .

This last fixed point does not depend on c while T obviously does . We can therefore look for new fixed points using bifurcation theory . A simple analysis shows that for $c = \sqrt{2}$ a new branch of fixed points is very likely to cross the known branch $\phi\equiv 1$. This is easily proved in perturbation theory (with $\varepsilon = \sqrt{2} - c$ as perturbation parameter) , but an analytic proof seems quite difficult since the most powerful implicit function theorem (Nash-Moser type theorem [L.Z]) do not seem to work in this case . However , using a partial resummation of the perturbation expansion and the hypercontractivity property of the tangent map DT(1,.) , we were able to prove the following theorem

THEOREM 1 . For $\varepsilon=\sqrt{2}-c$ sufficiently small , T has a fixed point ϕ_ε in L_∞ given by

$$\phi_\varepsilon = 1 - \varepsilon\theta H_4 + \mathcal{O}_{L_\infty}(\varepsilon^2)$$

where $\theta=\frac{\text{Log}2}{144(\sqrt{2}-1)^2}$, and H_4 is the fourth hermite polynomial . ϕ_ε is a positive function bounded by cte.$e^{-\varepsilon x^4/2}$. Moreover , $S_\beta(\phi_\varepsilon)$ is the probability density of a real random variable which is not infinitly divisible .

We now state a proposition which explains why we are interested in the new fixed point instead of the old one $\phi \equiv 1$.

PROPOSITION 2 . Let $\varepsilon=\sqrt{2}-c$ be sufficiently small . The following properties hold in the Hilbert space $L_2(e^{-(1-c^{-1})x^2}dx)$

i) For $\varepsilon<0$, DT(1,.) has two eigenvalues greater than one .

ii) For $\varepsilon>0$, DT(1,.) has three eigenvalues greater than one .

iii) For $\varepsilon>0$, $DT(\phi_\varepsilon,.)$ has two eigenvalues greater than one .

In other words , if c is less than $\sqrt{2}$ the new fixed point ϕ_ε is more stable than the old one .

We now investigate the behavior of the transformation T in a neighborhood of ϕ_ε . This behavior is given by the two following theorems

THEOREM 3 . In a sufficiently small neighborhood of ϕ_ε in L_∞ , T has a stable manifold W_s and an unstable manifold W_u (see fig. 1) . Moreover

i) Codim $W_s=2$, and the tangent space to W_s at ϕ_ε is the stable spectral subspace of $DT(\phi_\varepsilon,.)$.

ii) dim $W_u = 2$ and the tangent space to W_u at ϕ_ε is the unstable spectral subspace E_u of $DT(\phi_\varepsilon,.)$.

THEOREM 4 . There is a C^1 diffeomorphism U of a neighborhood V of ϕ_ε in L_∞ such that

$$UTU^{-1}(\cdot) = DT(\phi_\varepsilon,.) + R(\cdot)$$

where R is a contraction from V to $E_s \cap V$ where E_s is the stable spectral subspace of $DT(\phi_\varepsilon,.)$.

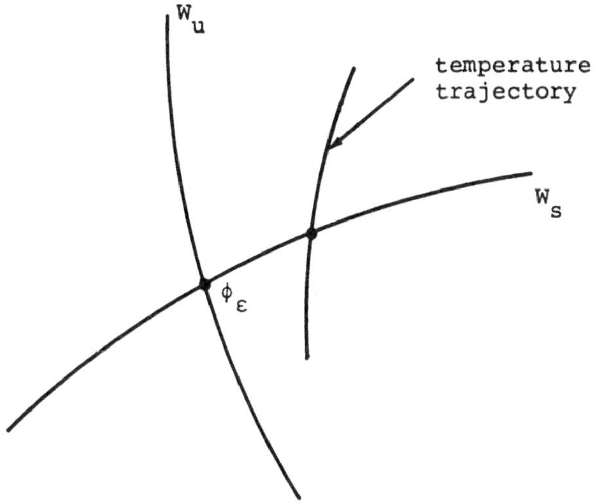

Fig. 1 . Stable and unstable manifold .

We are now able to give a precise geometrical definition of a phase transition (in the Hierarchical Model) , and of the associated critical indices .

Assuming a model is given , that is to say f and c are fixed , the temperature trajectory is the curve in L_∞ defined by

$$\beta \longrightarrow \phi_\beta(x) = (S_\beta f^\beta)(x) \quad .$$

The number $1/\beta_c$ is a critical temperature if for $\beta = \beta_c$ the temperature trajectory intersects transversally the stable manifold W_s (notice that for technical reasons , we have to assume that ϕ_{β_c} is in V) . We now define the critical index of a thermodynamical quantity at the critical temperature $T_c = 1/\beta_c$.

<u>Definition</u> . Let G be a thermodynamical quantity such that $G(T_c) = 0$ or $G(T_c) = \infty$ .The critical index Ind_G of G at the critical temperature T_c is defined by the following limit (if this limit exists)

$$\mathrm{Ind}_G = \lim_{T \to T_c} \frac{\mathrm{Log}|G(T)|}{\mathrm{Log}|T-T_c|}$$

<u>Remark</u> . This definition reads formally $G(T) \sim |T-T_c|^{\mathrm{Ind}_G}$ if $T \to T_c$.

We are now going to compute the critical index associated with the free energy $F_{\beta,c,f}$ defined by

$$F_{\beta,c,f} = \lim_{n \to +\infty} F_{\beta,c,2^n,f} \quad .$$

This limit exists as a consequence of the G.H.S. inequality [Ru] and moreover satisfies

$$F_{\beta,c,f} = \tfrac{1}{2} F_{\beta,c,N_\beta(f)}$$

as is easily seen from eq. (1) .

For the following purposes , it will be easier to work in the space where T is nearly a linear operator . We therefore define a new function $\hat{F}_{\beta,c,g}$ by

$$\hat{F}_{\beta,c,US_\beta f^\beta - \phi_\varepsilon} = F_{\beta,c,f} \quad .$$

This new function $\hat{F}_{\beta,c,g}$ satisfies

$$\hat{F}_{\beta,c,g} = \tfrac{1}{2}\,\hat{F}_{\beta,c,DT(\phi_\varepsilon)(g)} + R(g) \quad .$$

If e_0 and e_1 are the two unstable eigenvectors of $DT(\phi_\varepsilon,.)$ (with eigenvalues $\lambda_0 > \lambda_1$ respectively) , we have

$$US_\beta f^\beta = \phi_\varepsilon + g_o(\beta-\beta_c)\, e_0 + g_1(\beta-\beta_c) + r_\beta$$

where g_0 and g_1 are C^1 , and r_β is a vector in E_s .

Therefore the nth iterate of $US_\beta f^\beta$ is given by

$$(\, DT(\phi_\varepsilon)\,)^n(US_\beta f^\beta-\phi_\varepsilon) = g_0(\beta-\beta_c)\lambda_0^n\, e_0 + g_1(\beta-\beta_c)\lambda_1^n\, e_1 + r_{\beta,n} \;,$$

where $r_{\beta,n} \to 0$ if $n \to +\infty$.

For every sequence (β_n) such that $g_0(\beta_n-\beta_c)\lambda_0^n \xrightarrow[n\to+\infty]{} a\neq 0$ we have

$$\hat{F}_{\beta_n,c,US_\beta f^\beta-\phi_\varepsilon} = 2^{-n}\;\hat{F}_{\beta_n,c,X_n}$$

where $X_n = g_0(\beta_n-\beta_c)\lambda_0^n\, e_0 + g_1(\beta_n-\beta_c)\lambda_1^n\, e_1 + r_{\beta,n}$.

One can show that $\hat{F}_{\beta_c,c,ae_0} \neq 0$ if $a \neq 0$, therefore

$$\frac{\text{Log}\, F_{\beta_n,c,f}}{\text{Log}|\beta_n^{-1}-\beta_c^{-1}|} = \frac{\text{Log}\,\hat{F}_{\beta_n,c,US_{\beta_n}f^\beta-\phi_\varepsilon}}{\text{Log}|\beta_n^{-1}-\beta_c^{-1}|} = \frac{-n\text{Log}2 + \text{Log}\,\hat{F}_{\beta_n,c,X_n}}{\text{Log}|\beta_n-\beta_c| - \text{Log}|\beta_n\beta_c|}$$

$$\xrightarrow[n\to+\infty]{} \text{Log}\;2\;/\text{Log}\;\lambda_0 \quad .$$

Other critical indices can be computed by this method . If $M_{\beta,h}$ is the magnetization at temperature β^{-1} and in the magnetic field h , and if χ_β is the magnetic susceptibility , we obtain

$$\text{"}\beta\text{"} = \lim_{\substack{\beta\to\beta_c\\ h\to 0}} \frac{\text{Log}\, M_{\beta,h}}{\text{Log}|\beta-\beta_c|} = \tfrac{1}{2}\,\text{Log}\; c\;/\;\text{Log}\;\lambda_1$$

$$\text{"}1/\delta\text{"} = \lim_{h\to 0} \frac{\text{Log}\, M_{\beta_c,h}}{\text{Log}|h|} = \tfrac{1}{2}\,\text{Log}\; c\;/\;\text{Log}\;(\;2/\sqrt{c}\;)$$

$$"-\gamma" = \lim_{\beta \to \beta_c} \frac{\text{Log } \chi_\beta}{\text{Log}|\beta - \beta_c|} = \tfrac{1}{2} \text{ Log } (c/2) \ / \text{ Log } \lambda_1 \ .$$

The eigenvalue λ_1 has a perturbation expansion with respect to the parameter $\varepsilon = \sqrt{2} - c$, from which we deduce analogous expansions for the critical indices . These expansions are known in the physical literature as the ε expansions . In the particular case of the Hierarchical Model , a numerical analysis strongly indicates that these expansions are Borel summable [C.E.H.] .

III) The case $\beta \neq \beta_c$.

The preceding section was entirely devoted to an analysis of the model at the critical temperature . We now investigate what happens when $\beta \neq \beta_c$ (but for technical reasons , $\beta \sim \beta_c$) . This is a very difficult problem which has only been solved for the Ising Model using a duality argument between high and low temperature [M.M.S.] . In the Hierarchical Model , this problem can be solved if we have enough information on

$$T^n(S_\beta f^\beta) \quad \text{for} \quad n \to +\infty \quad \text{and} \quad \beta \neq \beta_c \ .$$

In other words , we have to follow the unstable manifold W_u to infinity . Intuitively , one can expect the following behavior

$$T^n(S_\beta f^\beta) \to \text{another fixed point if } \beta \neq \beta_c \ .$$

This result is made precise in the following Theorems .

THEOREM 5 . If $\beta < \beta_c$ ($T > T_c$) , but β is near β_c

$$T^n(S_\beta f^\beta)\left((c/2)^{n/2} . \right) \xrightarrow[n \to +\infty]{} \text{a Gaussian}$$

in the following sense:

the moments of $\dfrac{S_1 + \ldots + S_{2^n}}{2^{n/2}} \longrightarrow$ the moments of a gaussian random variable .

Remarks . 1) This theorem is similar to the central limit theorem which is always true in free systems .

2) Due to a different scaling , this Gaussian measure is the δ fixed point we already found .

Proof of Theorem 5 . The proof is quite intricate , and we shall only indicate the most important ideas . First of all , the starting point is the function $S_\beta f^\beta$ which has the following form

$$(S_\beta f^\beta) \sim e^{-\alpha x^2} e^{-\varepsilon x^4} \qquad \alpha << \varepsilon \quad .$$

The action of T on a Gaussian is explicitly known and given by

$$T(e^{-\alpha(.)^2})(x) = \text{cte } e^{-2\alpha x^2/c}$$

Therefore

$$T^n (e^{-\alpha(.)^2})(x) \sim e^{-(2/c)^n \alpha x^2} \quad ,$$

and δ is the limit of the corresponding sequence of normalized functions . In order to build a complete proof , one has to carefully control the behavior of some remainder terms . Instead of doing this , we present in Fig 2.a a numerical analysis of the iterations .

THEOREM 6 . If $\beta>\beta_c$ ($T<T_c$) but β is near β_c , there is a unique positive real number $\mu(\beta)$ such that

$$T^n(S_\beta f^\beta)((c/2)^n (\pm\mu(\beta) 2^{n/2})) \xrightarrow[n\to+\infty]{} \text{a gaussian}$$

in the following sense:

the moments of $\dfrac{S_1 +\ldots+ S_{2n}}{2^{n/2}} \pm \mu(\beta)\, 2^{n/2} \longrightarrow$ the moments of a Gaussian random variable .

Remarks . 1) $\dfrac{|S_1|+\ldots+|S_{2n}|}{2^n} \longrightarrow \mu(\beta)$ which is the magnetization .

2) The variance of the Gaussian random variable is the magnetic susceptibility χ_β .

3) $\mu(\beta) \to 0$ and $\chi_\beta \to +\infty$ if $\beta \to \beta_c^+$ (i.e. $T \to T_c^-$) .

<u>Proof of Theorem</u> 6 . Again , we only indicate the most important ideas . The starting point is now

$$(S_\beta f^\beta)(x) \quad \underset{-}{\sim} \quad e^{\alpha x^2} \quad e^{-\varepsilon x^4} \qquad \alpha > 0 \ , \quad \alpha << \varepsilon \quad .$$

The action of T on this function is approximately given by

$$T \ (S_\beta f^\beta)(x) \quad \underset{-}{\sim} \quad \text{cte } e^{2\alpha x^2/c} \quad e^{-\varepsilon x^4} \qquad .$$

Therefore ,

$$T^n \ (S_\beta f^\beta)(x) \quad \underset{-}{\sim} \quad \text{cte} \quad e^{(2/c)^n \alpha \, x^2} \qquad e^{-\varepsilon x^4} \qquad .$$

Let n_0 be defined by

$$\alpha^2 \ (2/c)^{2n_0} \quad = \quad 4\varepsilon \qquad .$$

It is easy to check that if n is less than n_0 , $e^{(2/c)^n \alpha x^2} \quad e^{-\varepsilon x^4}$ has only one maximum , while $e^{(2/c)^{n_0} \alpha x^2} \quad e^{-\varepsilon x^4}$ is flat on a distance of order $\varepsilon^{-\frac{1}{4}}$. Moreover , $e^{(2/c)^n \alpha x^2} \quad e^{-\varepsilon x^4}$ has two (symmetric) maxima if n is greater than n_0 . The transition from one maximum to two maxima is in fact very sudden as we shall see from the numerical analysis .

We now take $n = n_0 + 1$ and apply the transformation T once more . The new function has now three maxima , and each time we apply T the number of maxima increases . The graph of a particular iterate typically consists of two big symmetric peaks , with some small peaks in between . However , these small peaks are so small that they can be controlled and in fact disappear in the limit . This is quite different from the free case where such little peaks , once appeared , sum up to a Gaussian . We present in Fig 2.c a numerical analysis of the iterations . The very flat curve obtained when n is equal to n_0 is particularly well visible . We also notice that for $n = n_0 + 1$ the curve has two well-separated peaks .

Some successive iterations for $\beta = \beta_c$ are presented on Figure 2.b . We have also done the numerical analysis for a model where the magnetic field h is not equal to zero . The results are shown in Fig 3.a,b,c . The new phenomenon (if we neglect the distorsions) is that every curve is approximately centered around the magnetization $2^n \mu(\beta,h)$.

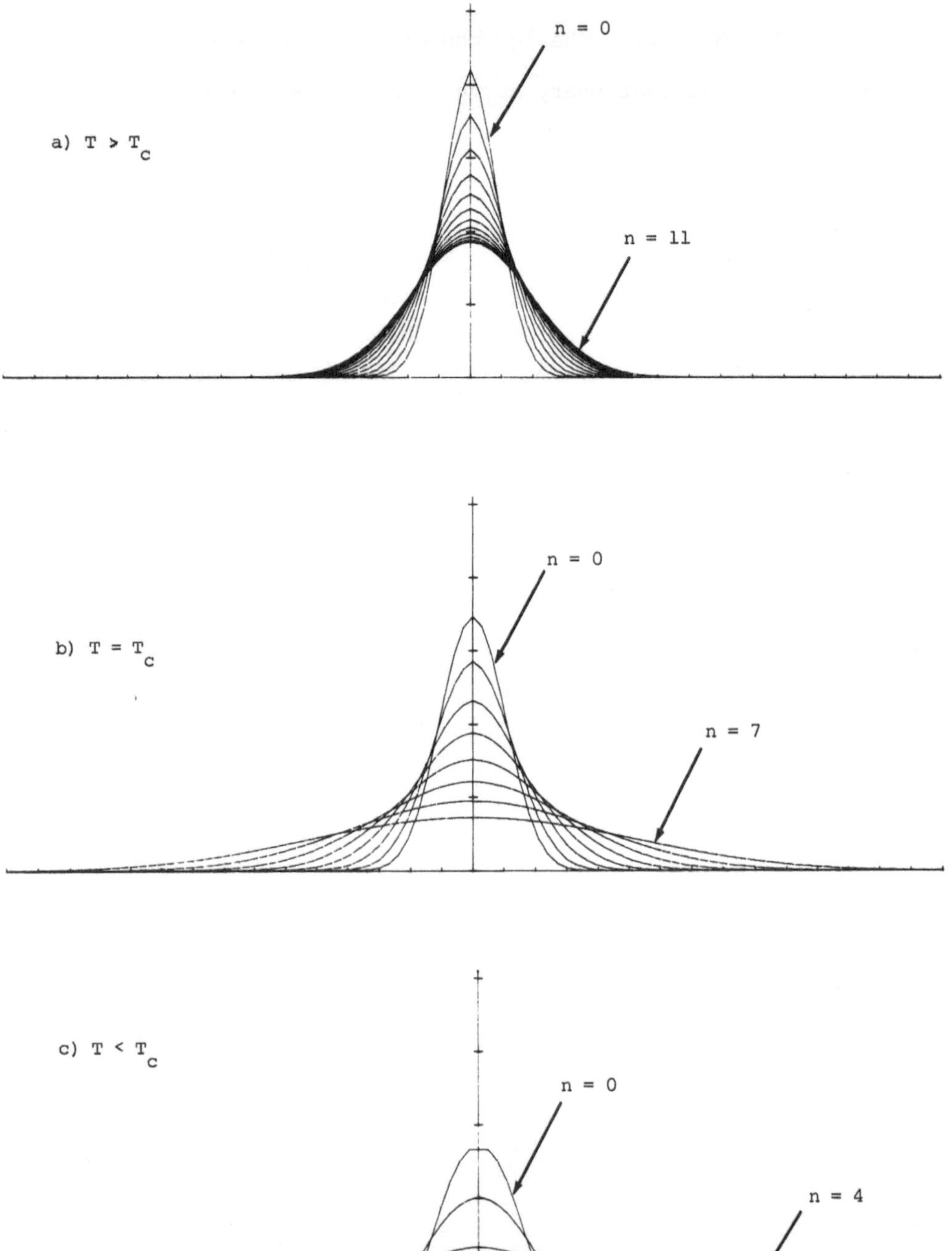

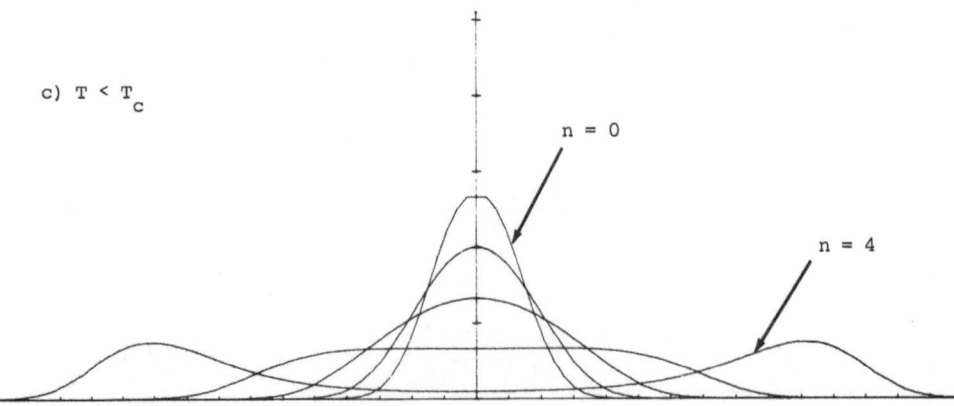

Fig. 2 .The spin density for h=0

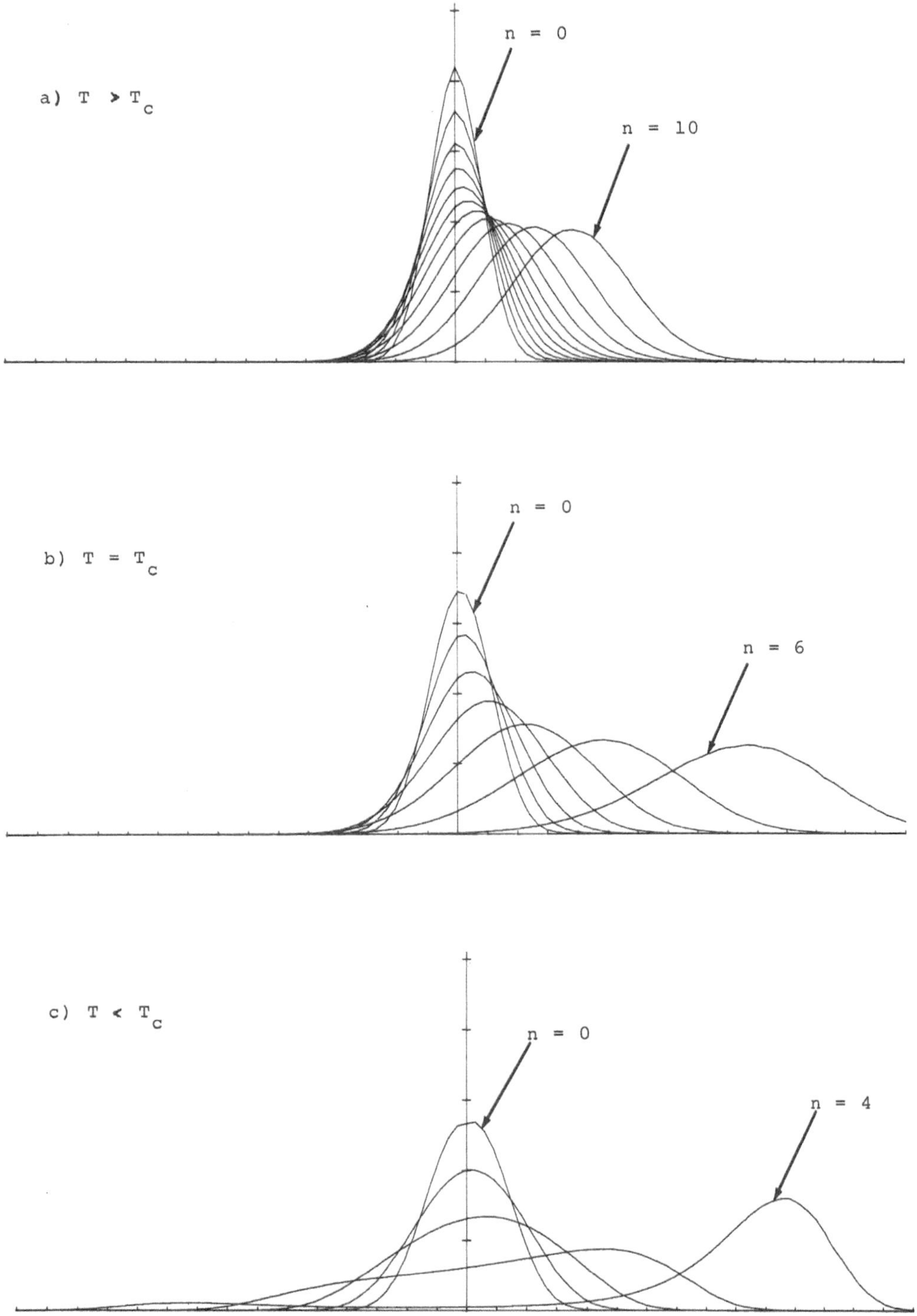

Fig. 3 . The spin density for h>0

REFERENCES .

[Ba] . G.A. BAKER : Ising model with a scaling interaction . Phys. Rev. B 5 , 2622 (1972) .

[C.E.] . P.COLLET,J-P.ECKMANN : A renormalization group analysis of the Hierarchical Model in statistical mechanics . Lecture Notes in Physics N° 74 , Springer , Berlin ,Heidelberg , New York , 1978.

[C.E.H.] . P.COLLET,J-P.ECKMANN,B.HIRSBRUNNER : A numerical test of Borel summability in the ε-expansion of the Hierarchical Model . Physics Letters 71B 385 (1977) .

[Dy] . F.J. DYSON : An Ising ferromagnet with discontinuous long-range order . Commun.Math.Phys. 21 269 (1971) .

[G.L.] .L.D. LANDAU : Collected papers by L.D. Landau , ed. D. ter Haar , Gordon and Breach New York (1975) .

[K.W.] . J.KOGUT,K.G.WILSON : Phys. Rep. 12C (1974) .

[L.Z.] . S.LOJASIEWICZ,E.ZEHNDER : An inverse function theorem in Frechet-spaces preprint Bochum (1977) .

[M.M.S.] . A.MESSAGER,S.MIRACLE-SOLE : Equilibrium states of two-dimensional Ising model in the two-phase region Commun.Math.Phys. 40 187 (1975) .

[Ru] . D.RUELLE : Statistical Mechanics . W.A.Benjamin inc , New York (1969) .

SPINNING PARTICLES AND RELATIVISTIC PARTICLES IN THE FRAMEWORK OF NELSON'S STOCHASTIC MECHANICS*

by

DANIELA DOHRN

FRANCESCO GUERRA

PATRIZIA RUGGIERO

Institute of Physics**, University of Salerno, Salerno

October 1978

* Expanded version of a talk given by F.G. at the Conference "Mathematical Problems in Feynman Path Integral",

Marseille, May 22-26, 1978.

Work supported by CNR.

** Postal address: Istituto di Fisica dell'Università, Via Vernieri 42,

84100 Salerno

1. INTRODUCTION.

In previous papers (see [1,2] and references quoted there) we have extended the formulation of Nelson's stochastic mechanics [3] to general Lagrangian systems on Riemannian manifolds. From a general point of view stochastic mechanics provides an alternative approach to quantization of classical dynamical systems, based on purely classical concepts of probability and stochastic differential equations. This alternative approach has been particularly useful in the study of relativistic quantum field theory, where the Euclidean formulation of Schwinger [4], Symanzik [5] and Nelson [6] has made possible relevant progress in the last years (see for example [7], [8] and references quoted there).

Partly motivated by the problems of incorporation of spin into the stochastic mechanics framework and its relativistic extension, there have been attempts to extend stochastic mechanics to general Riemannian manifolds (with definite or indefinite metric) [9], [10]. In [2] we have given the general formulation of stochastic mechanics on Riemannian manifolds with positive definite metric. A crucial rôle is played by the concept of geodesic correction to stochastic parallel displacement [11], which is essential for a natural geometric definition of mean forward and backward derivative of a tensor field.

In this note we show how to apply the methods of [2] in order to have the stochastic mechanics of a charged spinning particle interacting with a given electromagnetic field. We describe the inner degrees of freedom of the particle through the SU(2) group (compare with [9] and [10] where the rotation group is employed), this gives immediately a decomposition of the Schrödinger equation, arising in this particular case from the methods of [2], into a family of Pauli-type equations for each spin. The interaction with the electromagnetic field is unable to change the spin of the particle, according with usual physical interpretation. Our formalism can represent the basis for a natural and rigorous introduction of Feynman-Kac path integrals for spinning particles (even for half-integer spins!). In the second part of the paper we consider the problem of extending the methods of [2] to the case of a non positive definite metric. This situation will cover for example the case of a relativistic particle in a given gravitational and electromagnetic field. For the sake of simplicity we consider only the case of a scalar relativistic particle in a given electromagnetic field. By following the ideas contained in [12] we will show how to bild up a stochastic description of the motion of a relativistic particle leading to the usual Klein-Gordon equation.

The paper is organized as follows. In Section 2 we recall some basic elementary notions of differential geometry on SU(2), which will be employed in the following. Section 3 is devoted to stochastic mechanics on SU(2). Finally in Section 4 we consider the general case of stochastic mechanics of spinning particles, by taking SU(2) as inner space describing the rotational degrees of freedom, according to ideas of Bopp and Haag [13] and Wheeler [14] . Turning to the problem of relativistic extension, in Section 5 we discuss Markov property in the relativistic framework. We rely on ideas contained in Feynman picture of relativistic quantum mechanics [15] where the virtual trajectories, which contribute to transition probability amplitudes, can move backward in time and are interpreted in such case as trajectories of antiparticles moving forward.

In Section 6 we present a general formulation [12] of relativistic stochastic mechanics based on a generalization of the theory of stochastic differential equations, which is made necessary by the fact that infinitesimal displacements along the trajectories can have space-like or time-like character. This formulation should provide the basis for a rigorous introduction of Feynman-Kac type path-integrals in the relativistic case.

In conclusion we whould like to thank Profs. E. Onofri, L. de la Peña-Auerbach, T. Toyoda and J.P. Vigier for useful discussions on the subject.

The second named author (F.G.) would like to thank the Organizing Committee for the kind hospitality extended to him in Marseille.

2. DIFFERENTIAL GEOMETRY ON SU(2).

The SU(2) group is made of all two by two unitary matrices with determinant equal to one. We find convenient to exploit the well known isomorphism of SU(2) with S_3 (the three-dimensional spherical surface in $\mathbb{R}^4$). Therefore we write $SU(2) \ni g \equiv \{g^i\}$, $i = 0,1,2,3, g^i \in \mathbb{R}$, $g^i g^i = 1$. Any element of SU(2) can be written in the form

$$g = g^0 \mathbb{1}_2 + i\,\vec{\sigma}\cdot\vec{g} \equiv g^0 \mathbb{1}_2 + i\,\sigma^\alpha g^\alpha \quad , \quad \alpha = 1,2,3 \,, \tag{1}$$

where $\mathbb{1}_2$ is the unit two by two matrix and $\{\sigma^\alpha\}$ are the Pauli matrices

$$\sigma^1 = \begin{pmatrix} 0 & 1 \\ 1 & 0 \end{pmatrix} \quad , \quad \sigma^2 = \begin{pmatrix} 0 & -i \\ i & 0 \end{pmatrix} \quad , \quad \sigma^3 = \begin{pmatrix} 1 & 0 \\ 0 & -1 \end{pmatrix} . \tag{2}$$

Notice

(3) $\sigma^{\alpha} = \sigma^{\alpha +}$, $\sigma^{\alpha}\sigma^{\beta} + \sigma^{\beta}\sigma^{\alpha} = 2\delta^{\alpha\beta}$, $\sigma^{\alpha}\sigma^{\beta} = -i\,\varepsilon^{\alpha\beta\gamma}\sigma^{\gamma} + \delta^{\alpha\beta}$,

where $\varepsilon^{\alpha\beta\gamma}$ is the completely antisymmetric Ricci tensor, i.e. $\varepsilon^{123} = 1$, etc, so that the vector product in $\mathbb{R}^3$ is given by

(4) $$(\vec{u} \times \vec{v})^{\alpha} = \varepsilon^{\alpha\beta\gamma} u^{\beta} v^{\gamma} \quad , \quad \alpha, \beta, \gamma = 1, 2, 3 .$$

For g,h $\in$ SU(2) we clearly have

(5) $$\begin{cases} g^{+} = g^{0} - i\,\vec{\sigma}\cdot\vec{g} , \\ \det g = g^{i} g^{i} = 1 , \\ gh = (g^{0}h^{0} - \vec{g}\cdot\vec{h}) + i\,\vec{\sigma}\cdot(g^{0}\vec{h} + h^{0}\vec{g} + \vec{g}\times\vec{h}) , \\ g^{-1} = g^{+} \quad , \; g g^{+} = 1 \end{cases}$$

The tangent space $TSU(2)_g$ at g is made of vectors $t \equiv \{t^i\}$ such that

(6) $$t^{i} g^{i} = 0 \qquad \text{for} \quad t \in TSU(2)_g .$$

The Lie algebra su(2) of SU(2) is the tangent space at the origin, su(2) = $TSU(2)_e$, e= 1_2.

Since $e^0 = 1$, $e^{\alpha} = 0$, then $t \in$ su(2) means $t^0 = 0$, $t = i\vec{\sigma}.\vec{t}$. As it is well known the Lie brackett is related to the vector product

(7) $$[t, t'] = 2i\,\vec{\sigma}\cdot(\vec{t} \times \vec{t}') \quad , \qquad t, t' \in su(2).$$

The differential geometry on SU(2) is simply given by the elementary differential geometry on S_3. We could introduce local charts such that g^{α} are the independent parameters varying in the spherical region $g^{\alpha} g^{\alpha} \leq 1$, and $g^0 = \sqrt{1-g^{\alpha} g^{\alpha}}$ in the region where $g^0 \geq 0$ and $g^0 = -\sqrt{1-g^{\alpha} g^{\alpha}}$ where $g^0 \leq 0$. In general

(8) $$\frac{\partial g^0}{\partial g^{\alpha}} = -\frac{g^{\alpha}}{g^0} .$$

But we find convenient, in order to have explicit covariance, to retain all g^i, i=0,1,2,3, restricting them by the condition

(9) $$g^{i} g^{i} = 1.$$

Consequently we define the differential operators in the following way. If ϕ is a scalar field on SU(2) then

(10) $$\nabla^i\phi \equiv \left(\delta^{ij} - g^i g^j\right)\frac{\partial\phi}{\partial g^j} ,$$

where the subtracted term assures that only variations of ϕ along S_3 enter in the definition of $\nabla^i\phi$ and moreover $\nabla^i\phi$ is tangent to S_3 in g, i.e.

(11) $$g^i \nabla^i \phi = 0 .$$

For a vector field $f \equiv f(g)$ on SU(2) we have

(12) $$f^i(g)\, g^i = 0 .$$

Therefore if we move along a curve $g(\tau)$ with tangent $t(\tau)=dg(\tau)/d\tau$, we have

(13) $$g^i df^i = - f^i dg^i$$

We define the substantial variation

(14) $$\delta f^i \equiv df^i - g^i g^j df^j = df^i + g^i f^j dg^j ,$$

and the covariant derivative ∇ such that

(15) $$\frac{\delta f^i}{\delta\tau} = t^j \nabla^j f^i .$$

A simple calculation shows

(16) $$\nabla^j f^i = \left(\delta^{jk} - g^j g^k\right)\frac{\partial f^i}{\partial g^k} + g^i f^j .$$

Analogously for tensor fields

(17) $$\nabla^j F^{il} = \left(\delta^{jk} - g^j g^k\right)\frac{\partial F^{il}}{\partial g^k} + g^i F^{jl} + g^l F^{ij} \qquad , \text{ etc.}$$

In general

(18) $$g^j \nabla^j = 0 .$$

We define also the Laplace-Beltrami operator

(19) $$\Delta \equiv \nabla^j \nabla^j$$

and the Laplace-de Rham Δ_D such that

(20) $$\Delta_D \nabla^i = \nabla^i \Delta_D .$$

Δ_D and Δ coincide on scalar functions but on vector functions one has for example

(21) $$(\Delta_D f)^i = \Delta f^i + R^{ij} f^j ,$$

where R^{ij} is the curvature tensor

(22) $$R^{ij} = 2(g^i g^j - \delta^{ij}) .$$

On SU(2) the geodesic lines correspond to circles of radius one on S_3. Their explicit equations are easily found. In fact for the general geodesic $g=g(\tau)$ starting at $g(0)$ for $\tau=0$ with tangent $t(0)$ one has the equation

(23) $$\delta t = 0 \quad i.e. \quad dt^i/d\tau + g^i t^j t^j = 0 .$$

Since $t^j t^j \equiv \omega^2, \omega > 0$, stays constant, we have

$$(24)\qquad \frac{d^2 g^i}{d\tau^2} + \omega^2 g^i = 0$$

and the solution

$$(25)\qquad \begin{cases} g^i(\tau) = g^i(0)\cos\omega\tau + t^i(0)\,\mathrm{sen}\,\omega\tau/\omega \\ t^i(\tau) = -\omega g^i(0)\,\mathrm{sen}\,\omega\tau + t^i(0)\cos\omega\tau. \end{cases}$$

Notice

$$(26)\qquad g^i(0)\,g^i(\tau) = t^i(0)\,t^i(\tau)/\omega^2 = \cos\omega\tau .$$

The explicit expression of the parallel displacement of a vector along the geodesic is also easily found in the form

$$(27)\qquad f^i(\tau) = f^i(0) - \eta\left[g^i(0)\,\mathrm{sen}\,\omega\tau/\omega + t^i(0)(1-\cos\omega\tau)/\omega^2\right],$$

where $\eta = f^j t^j$

Notice

$$(28)\qquad f^i(\tau) - f^i(0) = \eta\,\omega^{-2}\left[t^i(\tau) - t^i(0)\right].$$

Clearly under parallel transport the component of f orthogonal to the plane of the geodesic stays unchanged, while the component parallel to the tangent t rotates whit t.

In [11] we have introduced the notion of geodesic correction to parallel displacement and in [2] we have exploited this notion in the frame of stochastic mechanics on Riemannian manifolds.

It is very simple to find the explicit form of the geodesic displacement $\bar{f}(\tau)$ by exploiting the definition given in [11] and the parallel displacement (27). A simple calculation shows

$$(29)\qquad \bar{f}^i(\tau) = f^i\cos\omega\tau - f^j t^j g^i(0)\,\mathrm{sen}\,\omega\tau/\omega .$$

Therefore the geodesic correction is given by

$$(30)\qquad \delta_g f^i(\tau) = \bar{f}^i(\tau) - f^i(\tau) = \left[f^j(0)\,t^j(0)\,t^i(0)/\omega^2 - f^i(0)\right](1-\cos\omega\tau).$$

3. STOCHASTIC MECHANICS ON SU(2).

We follow the general scheme given in [2]. We consider a dynamical system with configuration variables given by $g \equiv \{g^i\}$, $i=0,1,2,3$, $g^i g^i=1$, and Lagrangian

(31) $$\mathcal{L}(g,\dot{g},t) = \frac{1}{2}\dot{g}^i\dot{g}^i - \frac{\lambda}{2} g^i g^i + B^i \dot{g}^i ,$$

where $\dot{g}^i = dg^i/dt$, λ is a Lagrangian multiplier which enforces $g^i g^i=1$ and B^i is a vector field on SU(2) such that

(32) $$g^i B^i(g) = 0 \quad , \quad \nabla^i B^i = 0 .$$

In the following section B^i will be related to the magnetic field.

Let us introduce the covariant acceleration a^i

(33) $$a^i = \ddot{g}^i + g^i \dot{g}^j \dot{g}^j .$$

Then the Euler equations of motion are given by

(34) $$a^i + (\nabla^j B^i - \nabla^i B^j)\dot{g}^j + \partial B^i/\partial t = 0 .$$

In order to introduce stochastic mechanics, we promote g(t) to a stochastic process taking values on SU(2) and satisfying the system of stochastic differential equations

(35) $$dg^i(t) = b^i_{(+)}(g(t),t)\,dt + dw^i(t) .$$

Here $b^i_{(+)}$ is a nonrandom vector field to be specified below and $dw^i(t)$ are the increments in the time interval dt of the process of Brownian motion on SU(2). The complete specification of $dw^i(t)$ is given by

(36) $$\begin{cases} E_t(dw^i) = -\frac{3}{2}\beta g^i\,dt , \\ E_t(dw^i dw^j) = \beta(\delta^{ij} - g^i g^j)\,dt \end{cases}$$

where E_t is the conditional expectation at time t and β is a constant.

Mean forward and backward derivatives are introduced as in [2]. In particular we find

(37) $$(D_{(\pm)} g)^i = b^i_{(\pm)}$$

with

(38) $$b^i_{(+)} - b^i_{(-)} = 2\beta \nabla^i \log\sqrt{\rho} ,$$

where ρ is the invariant density of the process such that

(39) $$\langle F(g(t),t) \rangle = \int F(g,\tau)\,\rho(g,\tau)\,d\mu(g) ,$$

(40) $$d\mu(g) = 2\,\delta(1 - g^i g^i)\,dg$$

with dg as "Lebesgue" measure on $\mathbb{R}^4$.

We have also for the vector field f

(41) $$(D_{(\pm)} f)^i = \frac{\partial f^i}{\partial t} + b^j_{(\pm)} \nabla^j f^i \pm \frac{\beta}{2}(\Delta_3 f)^i$$

Then, following the general scheme of [2], we end up with the Schrödinger equation

(42) $$i\beta \frac{\partial \psi}{\partial t} = -\frac{\beta^2}{2}\left(\nabla^j - \frac{i}{\beta} B^j\right)\left(\nabla^j - \frac{i}{\beta} B^j\right)\psi ,$$

where the wave function $\psi = \psi(g,t)$ is given by

(43) $$\psi(g,t) = \sqrt{\rho(g,t)}\, \exp\left[\frac{i}{\beta} S(g,t)\right],$$

(44) $$f^i + B^i = \nabla^i S .$$

4. STOCHASTIC MECHANICS OF NONRELATIVISTIC SPINNING PARTICLES.

Our basic assumption is that the configuration space of a nonrelativistic spinning particle is given by $R^3 \times SU(2)$, where R^3 describes the translational degrees of freedom and SU(2) the rotational ones. In order to justify the choice of SU(2) instead of the traditional O(3), let us make the following elementary considerations. Consider a rigid body with a fixed point which is connected through elastic strings to some distant fixed points (three or more). It is immediately seen that the positions of the body are given by the universal covering group SU(2), if we take into account not only the "geometric" position of the body, given by Euler angles, but also the topological state of entanglement of the strings. For example a rotation of 2π around some axis will produce superpositions of the strings which cannot be topologically disentangled (if we assume that the strings cannot cross each other). On the other hand a further rotation of 2π (therefore 4π in total) produces further superpositions, so that the strings reach a state which can be topologically disentangled and reduced to the initial one. Therefore we assume that the position of a particle must take into account its essential entanglement with the surroundings [14]. Such assumption traduces the idea of a particle as state of excitation of a field whose"tails" provide the entanglement with the surroundings. This gives support to the view that the rotational degrees of freedom are described by SU(2). Similar ideas appear also in [13].

Therefore we assume that $R^3 \times SU(2)$ is the configuration space of a particle of mass m, charge e, interacting with a magnetic field with vector potential $\vec{A}$ and a scalar field V. Stochastic mechanics of such a system can be immediately written down by following Nelson's prescription for the translational degrees of freedom and the method given in previous Section for the rotational degrees of freedom.

Then for the wave function $\psi(\vec{x},g,t)$, $x \in R^3$, $g \in SU(2)$, $t \in R$ we have the general Schrödinger equation

$$(45) \quad i\hbar \frac{\partial \psi}{\partial t} = -\frac{\hbar^2}{2m}\left(\partial^\alpha - i\frac{e}{\hbar}A^\alpha\right)\left(\partial^\alpha - i\frac{e}{\hbar}A^\alpha\right)\psi + V(x)\psi - \frac{\hbar\beta}{2}\left(\nabla^j - \frac{i}{\beta}B^j\right)\left(\nabla^j - \frac{i}{\beta}B^j\right)\psi ,$$

∂^α are the derivatives in R^3 and ∇ the covariant derivatives on SU(2) introduced in Sections 2 and 3.

The parameter β is defined by

$$(46) \quad \beta = \hbar / 4I$$

where I is the inertial moment of the body.

The field B on SU(2) is related to the magnetic field $\vec{H}$ through

(47) $\hbar B^0 = -\mu\, \vec{g}\cdot\vec{H}\ , \quad \hbar \vec{B} = \mu\, g^0 \vec{H} + \mu\, \vec{g}\times\vec{H}\ ,$

where μ is the magnetic moment of the particle.

Notice

(48) $\nabla^j B^j = 0 \quad \text{and} \quad \hbar^2 B^i B^i = \mu^2 \vec{H}^2 .$

The most interesting aspect of (45) is the following factorization property.

Let $R^{(j)}_{mm'}$, $j=0,\frac{1}{2},1,\ldots,$ $m,m'=-j,-j+1,\ldots,j$, be the matrices of the j representation of SU(2) and consider the Peter-Weyl decomposition

(49) $\psi(\vec{x},g,t) = \sum_{j,m,m'} R^{(j)}_{mm'}(g)\, \psi_{mm'}(\vec{x},t)\ .$

Through straigthforward calculations we can prove

(50) $-\nabla^k\nabla^k R^{(j)}_{mm'}(g) = 4j(j+1)\, R^{(j)}_{mm'}(g)\ ,$

(51) $i\hbar B^k \nabla^k R^{(j)}_{mm'}(g) = -\mu\, R^{(j)}_{mm''}(g)\, \vec{\Sigma}^{(j)}_{m''m'}\cdot\vec{H}\ ,$

Where $\vec{\Sigma}$ are the appropriate rotation matrices, for example

(52) $\vec{\Sigma}^{(0)} = 0\ , \quad \vec{\Sigma}^{(1)} = \vec{\sigma}$, etc.

Putting (49) into (45) and exploiting (50) and (51), we see immediately that the Schrödinger equation (45) factorizes into indipendent Pauli like equations for each spin

(53) $i\hbar\, \partial\psi^{(j)}_{mm'}/\partial t = C^{(j)}\psi^{(j)}_{mm'} - \mu\, \vec{H}\cdot\vec{\Sigma}^{(j)}_{m'm''}\, \psi^{(j)}_{mm''}\ ,$

(54) $C^{(j)} = -\frac{\hbar^2}{2m}\left(\partial^\alpha - i\frac{e}{\hbar}A^\alpha\right)\left(\partial^\alpha - i\frac{e}{\hbar}A^\alpha\right) +$
$+ \hbar^2 j(j+1)/2I + 2I\mu^2\hbar^{-2}\vec{H}^2 + V\ .$

Therefore we have seen that it is possible to give a formulation of quantum mechanics of spinning particles based on the general theory of stochastic mechanics. The theory holds also for particles with half integer spin and may provide the basis for a rigorous formulation of Feynman-Kac path integrals in this case.We plan to report about work in progress on this subject in a future occasion.

5. THE PROBLEM OF RELATIVISTIC MARKOV PROCESSES

The main problem arising in the construction of Markov diffusion processes satisfying relativistic covariance is to overcome the consequences of general theorems [16], [17] which forbid the existence of non trivial diffusion processes in configuration space with relativistic covariance and with the usual Markov property.

In order to fix ideas let us consider the simple case of a single scalar charged particle performing a diffusion in space-time.We describe the position $x(\tau)$ in Minkowski space-time as a function of an invariant parameter τ,which in the classical case can be taken as proper time.

The usual definition of Markov property would be the following.For each value of τ let us consider, in the probability space of $x(\tau)$, the σ-algebras $\Sigma_{\leq\tau}, \Sigma_\tau, \Sigma_{\geq\tau}$ generated by events $x(\tau)$,with $\tau' \leq \tau$, $\tau'=\tau$, $\tau'\geq\tau$ respectively.Let $E_{\leq\tau}$, E_τ, $E_{\geq\tau}$ be the associated conditional expectations.Then Markov property tells us, roughly speaking, that the past ($\tau'\leq\tau$) and the future ($\tau'\geq\tau$) are conditionally indipendent if the present ($\tau'=\tau$) is known.The precise mathematical definition is

$$(55) \qquad E_{\leq\tau} E_{\geq\tau} = E_{\geq\tau} E_{\leq\tau} = E_\tau \quad , \forall\tau .$$

However it can be easily shown (see for example [16] [17]that as a consequence of (55) the process $x(\tau)$ must be trivial, the only possibility is given by straigth line trajectories for $x(\tau)$ without any kind of effective diffusion.

But we must notice that (55) is only a direct extension to proper time τ of the form of Markov property valid for processes labelled by the physical time t. Therefore there are serious doubts that the relevant Markov property must be expressed in the form given by (55).In fact in any case the process is observed in the physical Minkowski space-time and τ is only an auxiliary invariant parameter introduced in order to have a more explicit formulation of relativistic covariance. On the other hand if we consider the Feynman picture of relativistic quantum mechanics [15] we must take into account also the fact that the virtual trajectories, contributing to transition amplitudes, can move backward in time, beeing interpreted in this case as trajectories of antiparticles moving forward.It seems reasonable to assume the possibility of relativistic diffusion processes having analogous properties.But if trajectories can move backward in the physical time then any consideration based on concepts of past, present and future related to τ is completely devoid of observational meaning.

We have considered the problem of relativistic Markov processes [12] inside the framework of the general program [1,2] of extending Nelson stochastic mechanics [3] to general dynamical systems.Therefore the processes we have considered have connections with relativistic quantum mechanics, in particular with the Feynman path-integral formulation [15], in the same way as the generalized random fields of Euclidean field theory are connected with usual quantum field theory (see [6], [7], [1]).The main lines of our approach will be described in the next Section.For related work based on different points of view see [10] and [18].

Let us end this Section with a sketch of a possible formulation of the relevant Markov property in the relativistic case as a meaningful replacement of (55).

For any space-like three-dimensional "surface" σ in the Minkowski space-time let $A_\sigma^\pm$ be respectively the regions of the future and the past with respect to σ. Given a diffusion process $x(\tau)$ let us consider for any region A of space-time the σ-algebra Σ_A in the probability space of $x(\tau)$ generated by events $x(\tau) \in A$ for any τ. Roughly speaking in order to be sure whether an event of Σ_A occured or not it is enough to look only at the portions of trajectories intersecting A, irrespective of the proper time τ. This definition of the relevant σ-algebras makes clear that τ is a redundant parameter and that all observations refer to regions of the physical space-time.

Introduce the conditional expectations E_σ, $E_\sigma^\pm$ with respect to the σ-algebras associated to σ and $A_\sigma^\pm$ respectively. Then the proposed physical Markov property is the following

$$(56) \qquad E_\sigma^+ E_\sigma^- = E_\sigma^- E_\sigma^+ = E_\sigma \quad , \quad \forall \sigma \text{ space-like.}$$

It would be interesting to see whether the process introduced in the next Section satisfy this property.

Let us now turn to the relativistic generalization of Nelson stochastic mechanics.

6. RELATIVISTIC GENERALIZATION OF NELSON STOCHASTIC MECHANICS.

For the sake of simplicity we consider only the case [12] of a single scalar charged particle of mass m and charge e moving in a given external electromagnetic potential $\phi^\mu(x)$. The general case of particles with spin is treated in a forthcoming paper.

In the classical case let $x(\tau)$ be the position of the particle in Minkowski space-time M^4 as a function of proper time τ. We have $x \equiv \{x^\mu\} \equiv \{x^0 = ct, \vec{x}\} \in M^4$, $\tau \in \mathbb{R}$ and our metric is such that $x^2 = x^\mu x_\mu = c^2t^2 - \vec{x}^2$. The equation of motion

$$(57) \qquad m \frac{d^2 x^\mu}{d\tau^2} = \frac{e}{c} F^\mu{}_\nu \frac{dx^\nu}{d\tau}$$

can be derived through a variational principle from the invariant Lagrangian

$$(58) \qquad \mathcal{L}(x,\dot{x}) = \frac{1}{2} m g_{\mu\nu} \dot{x}^\mu \dot{x}^\nu + \frac{e}{c} g_{\mu\nu} \dot{x}^\mu \phi^\nu ,$$

where $g_{\mu\nu} = \mathrm{diag}\,(1,-1,-1,-1)$, $\mu,\nu = 0,1,2,3$, $\dot{x}^\mu = dx^\mu/d\tau$ and $F_{\mu\nu} = \partial_\mu\phi_\nu - \partial_\nu\phi_\mu$.

We have considered in [2] the stochastic mechanics arising from general Lagrangian systems. However the methods of [2] cannot be exploited directly in this case because the quadratic term in (58) is not positive definite. On the other hand we do not assume Markov property in the form (55), therefore no trouble arises if we take also into account that small displacements Δx along the trajectories can be characterized through their space-like or time-like nature, in the cases $(\Delta x)^2 \leq 0$ or ≥ 0. The whole procedure of stochastic quantization can be described as follows (see also [12]).

First of all, following Nelson's basic idea, we promote the classical configuration variables $x^\mu(\tau)$ to a random process $x^\mu(\tau)$, taking values in the Minkowski space-time $\mathcal{M}^4$, labelled by an invariant parameter τ. We assume also that $x^\mu(\tau)$ transforms like a vector under Lorentz transformations (relativistic covariance). The next assumptions, following [3] and [2], would be to write stochastic differential equations for the increments dx^μ in the interval $d\tau$, but this must not be done since we have left out the formulation (55) of Markov property, which would lead us only to trivial results. We replace stochastic differential equations by the following assumptions concerning the conditional expectations of the increments dx^μ. It is immediately seen that in the case of positive definite metric our assumptions are equivalent to the usual stochastic differential equations of [3] and [2].

Let $\Delta\tau > 0$ and define $\Delta^\pm x^\mu = \pm\left[x^\mu(\tau \pm \Delta\tau) - x^\mu(\tau)\right]$. We denote by $E(.\mid u)$ the conditional expectation under the condition that the event u happens.

We assume that there are velocity vector fields $b^\mu_{(\pm)}(x,\tau)$, $x \in \mathcal{M}^4$, $\tau \in \mathbf{R}$, for the given process $x(\tau)$, and a constant $\hbar$ (for all processes), such that in the limit $\Delta\tau \to 0^+$ we have the following four covariant conditions (for each choice of upper and lower indices $\pm$)

$$(59)\quad \begin{cases} E\left(\dfrac{\Delta^\pm x^\mu}{\Delta\tau}\,\Big|\, u_t^\pm\right) + E\left(\dfrac{\Delta^\mp x^\mu}{\Delta\tau}\,\Big|\, u_s^\mp\right) \to b^\mu_{(\pm)}(x,\tau), \\[2ex] E\left(\dfrac{\Delta^\pm x^\mu\,\Delta^\pm x^\nu}{\Delta\tau}\,\Big|\, u_t^\pm\right) - E\left(\dfrac{\Delta^\mp x^\mu\,\Delta^\mp x^\nu}{\Delta\tau}\,\Big|\, u_s^\mp\right) \to \dfrac{\hbar}{m}\, g^{\mu\nu}, \end{cases}$$

where $u^\pm_{ts}$ are respectively the events

$$(60)\quad \begin{cases} u_t^\pm \equiv \{\, x(\tau) = x\,,\ (\Delta^\pm x)^2 \geq 0 \,\}, \\ u_s^\pm \equiv \{\, x(\tau) = x\,,\ (\Delta^\pm x)^2 \leq 0 \,\}. \end{cases}$$

Conditions (59)(60) replace the stochastic differential equations employed in [3] and [2] .

In order to introduce the third dynamical assumption (Nelson's form of the second principle of dynamics) we must give the definition of mean forward and backward derivatives $D_{(\pm)}$.

Our definition is the following. For a smooth function $F=F(x,\tau)$ we define

$$(61)\quad \left(D_{(\pm)}F\right)(x,\tau) = \lim_{\Delta\tau\to 0^+} \left[E\left(\frac{\Delta^{\pm}F}{\Delta\tau}\,\middle|\, u_t^{\pm}\right) + E\left(\frac{\Delta^{\mp}F}{\Delta\tau}\,\middle|\, u_s^{\mp}\right)\right],$$

where

$$(62)\quad \Delta^{\pm}F = \pm\left[F\left(x(\tau\pm\Delta\tau),\tau\pm\Delta\tau\right) - F\left(x(\tau),\tau\right)\right]$$

and $u^{\pm}_{t,s}$ are as in (60).

It is immediately seen, by exploiting (61) and (59), that

$$(63)\quad \left(D_{(\pm)}F\right)(x,\tau) = \frac{\partial F}{\partial\tau} + b^{\mu}_{(\pm)}\partial_{\mu}F \pm \frac{\hbar}{2m}\partial_{\mu}\partial^{\mu}F\,,$$

in particular

$$(64)\quad D_{(\pm)}x^{\mu} = b^{\mu}_{(\pm)}\,.$$

Let us now introduce the invariant density $\rho=\rho(x,\tau)$ of the process, such that

$$(65)\quad E(F) = \int F(x,\tau)\,\rho(x,\tau)\,dx\,.$$

It is very simple to prove, starting from the definition (63), that for smooth functions F and G one has

$$(66)\quad \frac{d}{d\tau}E(FG) = E\left(F D_{(+)}G\right) + E\left(D_{(-)}F\,G\right).$$

From (66), (65) and (63) one can immediately derive the following relation between $b_{(+)}$ and $b_{(-)}$,

$$(67)\quad \delta b^{\mu} \equiv \frac{1}{2}\left(b^{\mu}_{(+)} - b^{\mu}_{(-)}\right) = \frac{\hbar}{m}\partial^{\mu}\log\sqrt{\rho}$$

and the following equation for ρ

$$\frac{\partial \rho}{\partial \tau} = - \partial_\mu \left(\rho b^\mu \right) , \tag{68}$$

where

$$b^\mu = \frac{1}{2} \left(b^\mu_{(+)} + b^\mu_{(-)} \right) . \tag{69}$$

Now the basic dynamical assumption, suggested by Nelson's analogous proposal in the non relativistic case [3] , is that the equation of motion (57) must hold in the smoothed form

$$m \frac{1}{2} \left(D_{(+)} D_{(-)} + D_{(-)} D_{(+)} \right) x^\mu(\tau) = \frac{e}{c} F^\mu{}_\nu \frac{1}{2} \left(D_{(+)} + D_{(-)} \right) x^\nu(\tau) . \tag{70}$$

Conditions (59) and (70) are the basic assumptions made on the stochastic process $x^\mu(\tau)$.

We assume that there is a scalar function S such that

$$b^\mu + \frac{e}{m} \phi^\mu = \frac{1}{m} \partial^\mu S \tag{71}$$

and define

$$\psi(x,\tau) = \sqrt{\rho(x,\tau)} \, \exp \frac{i}{\hbar} S(x,\tau) . \tag{72}$$

Then, if we exploit (64) and (67), we see immediately, following the same method as in [3] and [2] , that equations (68) and (70) are equivalent to the following equation of Schrödinger type for ψ

$$i\hbar \frac{\partial \psi}{\partial \tau} = \frac{\hbar^2}{2m} \left(\partial_\mu - i \frac{e}{\hbar} \phi_\mu \right) \left(\partial^\mu - i \frac{e}{\hbar} \phi^\mu \right) \psi . \tag{73}$$

It would be easy, following the methods of [2] based on geodesic correction to stochastic parallel displacement, to consider also the case where in (58) the metric $g_{\mu\nu}$ is space-time dependent (i.e. for a given external gravitational field). We would end up with an equation of the type (73) where ∂_μ would be interpreted as covariant derivatives with respect the given metric $g_{\mu\nu}$.

The connection of (73) with the Klein-Gordon equation can be easily obtained through the standard position (see [15])

$$\psi(x,\tau) = \exp \left[\frac{i}{2} \frac{mc^2}{\hbar} \tau \right] \varphi(x) . \tag{74}$$

Then (73) reduces to

(75) $$-\left(\partial_\mu - i\frac{e}{\hbar}\phi_\mu\right)\left(\partial^\mu - i\frac{e}{\hbar}\phi^\mu\right)\varphi = \frac{mc^2}{\hbar^2}\varphi ,$$

and ρ and $b_{(\pm)}$ do not depend on τ, therefore (68) reduces to the relativistic continuity equation

(76) $$\partial_\mu \left(\rho b^\mu\right) = 0 .$$

Notice that ρb^0 is not positive definite. In fact the physical interpretation of ρb^μ is as a four-current density and the trajectories which move backward in time are interpreted as antiparticles and give negative contributions to the density ρb^0 .

Thus we have seen that by abandoning Markov property in the unphysical form (55) and by replacing stochastic differential equations with the conditions (59) it is possible to give a stochastic foundation, following Nelson's original ideas, also to relativistic quantum mechanics.

The extension to spinning particles of our stochastic scheme poses some challenging problems which we plan to discuss in a future note.

REFERENCES.

[1] F. Guerra and P. Ruggiero: Phys. Rev. Lett., 31, 1022 (1973).
[2] D. Dohrn and F. Guerra: Lettere al Nuovo Cimento 22, 121 (1978).
[3] E. Nelson: Phys. Rev. 150, 1079 (1966).
[4] J. Schwinger: Phys. Rev., 115, 721 (1959).
[5] K. Symanzik: J. Math. Phys., 7, 510 (1966).
[6] E. Nelson: J. Funct. Anal., 12, 97 (1973).
[7] B. Simon: The $P(\phi)_2$ Euclidean (Quantum) Field Theory, Princeton N.J. 1974.
[8] G.F. Dell'Antonio, S.Doplicher and G. Jona-Lasinio (eds.), Mathematical Problems in Theoretical Physics, Springer Verlag, Berlin, 1978.
[9] T.G. Dankel jr.: Arch. Rational Mech. Anal., 37, 192 (1971).
[10] J.P. Caubet: Le Mouvement Brownien Relativiste, Springer Verlag, Berlin, 1976.
[11] D. Dohrn and F. Guerra: Geodesic Correction to Stochastic Parallel Displacement of Tensors, to appear in the Proceedings of the Conference on Stochastic Behaviour in Classical and Quantum Hamiltonian Systems, Como, June 20-24, 1977 edited by G. Casati and J. Ford.
[12] F. Guerra and P. Ruggiero: A Note on Relativistic Markov Processes, Lettere al Nuovo Cimento, to appear.
[13] F. Bopp and R. Haag: Z. Naturforschg, 5a, 644 (1950).
[14] C.W. Milner,K.S. Thorne and J.A. Wheeler: Gravitation, W.H. Freeman and Co. San Francisco, 1970.
[15] R.P. Feynman: Phys. Rev., 80, 440 (1950).
[16] R.M.Dudley: Arkiv för Math., 6, 241 (1965).
[17] R. Hakim: J. Math. Phys, 9, 1805(1968).
[18] J.P. Vigier: Model of Quantum Statistics in Terms of a Fluid with Irregular Stochastic Fluctuations Propagating at the Velocity of Light: A Derivation of Nelson's Equations, Preprint of Institut Henri Poincaré, Paris, 1978.

CONSTRUCTION OF A CLASS OF CHARACTERISTIC FUNCTIONALS

by

R. Gielerak, W. Karwowski

Institute of Theoretical Physics, Wroclaw, University

Wroclaw, Poland

and

L. Streit

Fakultät für Physik, Universität Bielefeld, Bielefeld

Germany

This note presents a construction of generalized random fields (prop. 1 and 2). In particular if we base our construction on a given Euclidean invariant field, the new one will have the same invariance. Similarly the properties of T-positivity and clustering carry over to the new models. A Gaussian input will give rise to non-Gaussian fields such as e.g. the "ultralocal" ones.

<u>Proposition 1:</u> Let $\rho_o \geq 0$ and for $\nu \geq 1$ ρ_ν be positive distributions on $D(\mathbb{R}^{\nu\cdot d})$ such that

$$\rho_o + \sum_{\nu=1}^{\infty} \frac{1}{\nu!} < \rho_\nu, f^{\otimes\nu} > \equiv < e^f >_\rho$$

is absolutely convergent whenever $f \in D(\mathbb{R}^d)$.

For $x \in \mathbb{R}^d$ let μ_x be positive measures on $\mathbb{R}$ such that

$$\psi_f(x) \equiv \int_{-\infty}^{\infty} e^{if(x)\lambda}\, d\mu_x(\lambda) \quad \text{is in} \quad C_o(\mathbb{R}^d) \quad \text{for any}$$

real $f \in D(\mathbb{R}^d)$.

Then $<e^{\psi_f}>_\rho$ is a positive definite functional an $C(\mathbb{R}^d)$.

<u>Proof:</u> To see that $<e^{\psi_f}>_\rho$ is well-defined one uses the fact that the ρ_ν are positive distributions, hence measures with a unique positive extension to $C_o(\mathbb{R}^{\nu\cdot d})$ [1] . This makes $<\rho_\nu, \psi_f^{\otimes\nu}>$ well-defined since $\psi_f^{\otimes\nu} \in C_o(\mathbb{R}^{\nu\cdot d})$.

The absolute series convergence follows if we majorize ψ_0 by $g \in D(\mathbb{R}^d)$:

$$\sum_\nu \frac{1}{\nu!} \mid <\rho_\nu, \psi_f^{\otimes\nu}> \mid \leq \sum_\nu \frac{1}{\nu!} <\rho_\nu, \psi_0^{\otimes\nu}> \leq <e^g>_\rho < \infty.$$

$<e^{\psi_f}>_\rho$ will be positive definite if $\psi^{\otimes\nu}$ is so for all $x \in \mathbb{R}^{\nu\cdot d}$. But for arbitrary finite sequences of $a_\alpha \in \mathbb{C}$ and real $f_\alpha \in C(\mathbb{R}^d)$

$$\sum_{\alpha,\beta} a_\alpha^* a_\beta \psi^{\otimes\nu}_{f_\alpha - f_\beta}(x_1,\ldots,x_\nu) =$$

$$= \sum_{\alpha,\beta} a_\alpha^* a_\beta \prod_{j=1}^{\nu} \int d\mu_{x_j}(\lambda_j)\, e^{i\lambda_j(f_\alpha(x_j) - f_\beta(x_j))} =$$

$$= \prod_{j=1}^{\nu} (\int d\mu_{x_j}(\lambda_j))\, \Big| \sum_\alpha a_\alpha\, e^{i\sum_{j=1}^{\nu}\lambda_j f_\alpha(x_j)} \Big|^2 \geq 0 \ .$$

The following specifications of ρ resp ψ may serve to illustrate the conditions imposed in prop. 1

Example 1: Let φ be any generalized random field obeying the 1st GKS inequality [2] , i.e.

$$< \prod_{j=1}^{n} \varphi(f_i) > \geq 0 \quad \text{if} \quad 0 \leq f_i \in D(\mathbb{R}^d)$$

and the exponential bound $<e^{\varphi(f)}> < \infty \quad \forall f \in D$

and set $\rho_\nu(x_1,\ldots,x_\nu) = <\varphi(x_1)\cdots\varphi(x_\nu)>$.

These conditions are in particular true for even $P(\varphi)_2$ Euclidean quantum field theories [2] [3] . However, prop. 1 does not require the ρ_ν to be the moments of a generalized random field.

Example 2: Let $g(x)$ be a "cutoff function" :

$$0 \leq g \in C_0(\mathbb{R}^d)$$

and $\mu_x(A) = g(x)\mu(A)$ for some probability measure μ . As consequence then

$$\psi_f(x) = g(x) \int e^{i\lambda f(x)}\, d\mu(\lambda) \ .$$

For these ψ_f it is easy to show that $<e^{\psi_f}>_\rho$ is the Fourier transform of a measure.

Proposition 2: Let ρ be as in prop. 1 and ψ_f as in ex. 2. Then

$$C_{g\mu}(f) = <e^{\psi_f}>_\rho \,/\, <e^g>_\rho$$

is a normalized continuous positive definite functional on $C(\mathbb{R}^d)$, and on $S(\mathbb{R}^d)$ the Fourier transform of a finite positive measure on the space of tempered distributions:

$$C_{g\mu}(f) = \int_{S'} e^{i<\chi,f>}\, d\,\nu\,(\chi)$$

Proof: Let $f_n \to f$ in $C(\mathbb{R}^d)$, hence in particular $f_n(x) \to f(x)$ uniformly on supp g .

Then also $\psi_{f_n}(x) = g(x) \int d\mu(\lambda)\, e^{i\lambda f_n(x)} \to g(x) \int d\mu(\lambda)\, e^{i\lambda f(x)}$ uniformly in x , since g is bounded and characteristic functions are uniformly continuous on the whole real line. Hence $\psi_{f_n}^{\otimes\nu} \to \psi_f^{\otimes\nu}$ in $C_o(\mathbb{R}^{\nu\cdot d})$ and

$$<\rho_\nu, \psi_{f_n}^{\otimes\nu}> \to <\rho_\nu, \psi_f^{\otimes\nu}>$$

since positive distributions extend to C_o as continuous functionals.

Now use the fact that

$$| \psi_f(x) | \leq g(x) .$$

As a result the series convergence of $<e^{\psi_f}>_\rho$ is uniform in f so that we have continuity:

$$<e^{\psi_{f_n}}>_\rho = \sum_\nu \frac{1}{\nu!} <\rho_\nu, \psi_{f_n}^{\otimes\nu}> \to \sum_\nu \frac{1}{\nu!} <\rho_\nu, \psi_f^{\otimes\nu}> = <e^{\psi_f}>_\rho .$$

Positive definiteness has been established in prop. 1, and since $f_n \to f$ in S implies that $f_n \to f$ in C , the conditions of the Bochner-Minlos theorem are fullfilled qed.

It is desirable to study the limit as the cutoff function g approaches a constant, e.g. with a view toward the construction of stationary processes. As in many examples we allow for the ρ_ν to be approximated by a sequence $\rho_\nu^{(n)}$.

As cutoff functions we consider positive $g_n \in C_o(\mathbb{R}^d)$ such that for any $r > 0$ there is an $n_o(r)$ such that $g_n(x) = g_o > 0$ for all $|x| \leq r$ and all $n \geq n_o(r)$.

Proposition 3: Assume $\rho_\nu^{(n)}$ as in prop.1 and such that for some sequence of cutoff functions g_n as above there exists the limit

$$\lim_{n\to\infty} \frac{<e^{f+g_n}>_{\rho^{(n)}}}{<e^{g_n}>_{\rho^{(n)}}} \equiv E_{g_o}(f)$$

as a continuous linear functional on $C_o(\mathbb{R}^d)$. Then $\lim_{n\to\infty} C_{g_n\mu}(f) \equiv C_{g_o\mu}$ is the characteristic functional of a measure (on $D'(\mathbb{R}^d)$ for $f \in D(\mathbb{R}^d)$) .

Proof: For any $f \in C_o(\mathbb{R}^d)$ consider $n \geq n_o(r)$ where $r \geq |x|$ for all $x \in$ supp f.

Set

$$\psi_{n,f}(x) \equiv g_n(x) \int d\mu(\lambda)\, e^{i\lambda f(x)}$$

$$= g_n(x) \left(\int d\mu(\lambda)\, e^{i\lambda f(x)} - 1\right) + g_n(x)$$

$$= g_o \int d\mu(\lambda)\, e^{i\lambda f(x)} - g_o + g_n(x)$$

$$= \Phi_f(x) + g_n(x)$$

with $\Phi_f(x) = g_o \int d\mu(\lambda)\,(e^{i\lambda f} - 1)$ in $C_o(\mathbb{R}^d)$.

Hence

$$\lim_{n\to\infty} C_{g_n\mu}(f) = \lim_{n\to\infty} \frac{\langle e^{\psi_{n,f}}\rangle_{\rho^{(n)}}}{\langle e^{g_n}\rangle_{\rho^{(n)}}}$$

$$= \lim_{n\to\infty} \frac{\langle e^{\Phi_f + g_n}\rangle_{\rho^{(n)}}}{\langle e^{g_n}\rangle_{\rho^{(n)}}}$$

$$= E_{g_o}(\Phi_f)\ .$$

This is continuous as f varies in $D(\mathbb{R}^d)$ since

$$f \to \Phi_f$$

is continuous on $C_o(\mathbb{R}^d)$. Normalization and positive definiteness carry over from $C_{g_n\mu}$ to the limit, hence the proposition follows by the Bochner-Minlos theorem.

<u>Example 3:</u> Let ρ_ν be given by a Gaussian process:

$$\langle e^f\rangle_\rho = e^{\frac{1}{2}(f,Kf)}$$

with the Kernel $K(x,y) = \rho_2(x,y)$ of K a symmetric distribution.

Then (for $g_o = 1$)

$$E(f) = \lim_{n\to\infty} \frac{e^{\frac{1}{2}(f+g_n,\, K(f+g_n))}}{e^{\frac{1}{2}(g_n,\, Kg_n)}}$$

$$= e^{\frac{1}{2}(f,Kf)} \lim_{n\to\infty} e^{(g_n,Kf)}$$

exists if K is such that $\int dx\, dy\, K(x,y)\, f(y)$ is finite.

In this case then

$$C_\mu(f) = E\,(\int(e^{i\lambda f} - 1)\, d\mu(\lambda))$$

$$= \exp\{\tfrac{1}{2}\int dx\, dy\, K(x,y)[\iint(e^{i\lambda_1 f(x)} - 1)(e^{i\lambda f_2(y)} - 1)\, d\mu(\lambda_1)\, d\mu(\lambda_2)$$

$$+ 2\int d\mu(\lambda)\,(e^{i\lambda f(y)} - 1)\}$$

$$= \exp\{\tfrac{1}{2}\int dx\, dy\, K(x,y)\iint(e^{i\lambda_1 f(x) + i\lambda_2 f(y)} - 1)\, d\mu(\lambda_1)\, d\mu(\lambda_2)\}$$

If we specialize the original process to be white noise, i.e.

$$K(x,y) = \delta(x-y)$$

we obtain Levy-Khintchin type expressions for "ultra local" fields [4]

$$C_\mu(f) = \exp\{\tfrac{1}{2}\int dx \int (e^{i(\lambda_1+\lambda_2)f(x)} - 1)\, d\mu(\lambda_1)\, d\mu(\lambda_2)\}\ .$$

These also arise for $\rho_\nu \equiv \rho_0{}^\nu$ =const, since then

$$<e^{\psi_f}>_\rho = \sum_{\nu=0}^{\infty} \frac{\rho_0{}^\nu}{\nu!} \int dx\, g(x) \int d\mu(\lambda)\, e^{i\lambda f(x)}$$

$$= \exp\{\int dx\, g(x) \int d\mu(\lambda)\, e^{i\lambda f(x)}\}$$

so that

$$C_{g\mu}(f) = e^{\int dx\, g(x) \int d\mu(\lambda)(e^{i\lambda f(x)} - 1)}.$$

<u>Proposition 4:</u> Further to prop. 3 assume that $E_{g_0}(f)$ arises from an Euclidean invariant process χ

$$E_{g_0}(f) = \int_{D'} e^{<\chi,f>}\, d\,\nu_E(\chi)\ .$$

Then $C_{g_0\mu}$ defines a Euclidean invariant process φ. For $f \in D$

$$C_{g_0\mu}(f) = \int_{D'} e^{i<\varphi,f>}\, d\,\nu_c(\varphi)$$

If χ clusters, so does φ.
If χ obeys T-positivity, so does φ.

<u>Remarks:</u> As e.g. Ex. 3 shows the new process φ differs from the original one (χ) in a non trivial fashion. However one should not expect "new" Minkowski space quantum field theories from it since - at least for local transformations $f \to \psi_f$ -

correlations are altered only at coincident points and these do not enter into the Osterwalder-Schrader construction.

Proof: By definition

$$C_{g_0\mu}(f) = E_{g_0}(\Phi_f)$$

where $\Phi_f(x) = g_0 \int d\mu(\lambda)\ (e^{i\lambda f(x)} - 1)$.

Euclidean transformations (Λ,a) act on f as

$$f(x) \to f^{(\Lambda,a)}(x) = f(\Lambda x+a) .$$

This changes Φ_f to $\Phi_{f^{(\Lambda,a)}}(x) = \Phi_f(\Lambda x+a) = \Phi_f^{(\Lambda,a)}(x)$.

So that $C_{g_0\mu}$ is invariant if E_{g_0} is.

To investigate clustering let

$$\text{supp } f_1 \cap \text{supp } f_2 = \emptyset$$

In this case $\Phi_{f_1+f_2} = \Phi_{f_1} + \Phi_{f_2}$.

Hence

$$C_{g_0\mu}(f_1+f_2^{(1,a)}) = E_{g_0}(\Phi_{f_1} + \Phi_{f_2}^{(1,a)}) \to$$

$$\xrightarrow[|a|\to\infty]{} E_{g_0}(\Phi_{f_1})\ E_{g_0}(\Phi_{f_2}) = C_{g_0\mu}(f_1)\ C_{g_0\mu}(f_2)$$

if E_{g_0} clusters.

T-positivity for the process χ can be expressed as

$$0 \le \sum_{\alpha\beta} a_\alpha^* a_\beta < (e^{<\chi,f_\alpha>})^* e^{<\chi,\theta f_\beta>}>$$

for functions f_α with $\text{supp } f_\alpha \subset \mathbb{R}_+^d = \{x:x_0 > 0\}$ and with $(\theta f)(x_0,\ldots,x_{d-1}) \equiv f(-x_0,x_1,\ldots x_{d-1})$; ie,

$$0 \le \sum_{\alpha\beta} a_\alpha^* a_\beta E_{g_0}(f_\alpha^* + \theta f_\beta) .$$

Now replace f_α by Φ_{f_α} (with f_α real) and use $\Phi_f^* = \Phi_{-f}$

and $$(\theta\Phi_f)(x) = \Phi_{\theta f}(x)$$

to obtain

$$0 \leq \sum_{\alpha,\beta} a_\alpha^* a_\beta E_{g_0} (\Phi_{-f_\alpha} + \Phi_{\theta f_\beta})$$

and since supp (-f) ∩ supp (θf) = ∅

$$0 \leq \sum_{\alpha,\beta} a_\alpha^* a_\beta E_{g_0} (\Phi_{-f_\alpha + \theta f_\beta})$$

$$= \sum_{\alpha,\beta} a_\alpha^* a_\beta C_{g_0\mu} (-f_\alpha + \theta f_\beta)$$

$$= \sum_{\alpha,\beta} a_\alpha^* a_\beta < (e^{i<\varphi,f_\alpha>})^* e^{i<\varphi,\theta f_\beta>} >$$

qed.

References

[1] I.M. Gelfand, N.Ya. Vilenkin: Generalized Functions, vol. 4, Applications of Harmonic Analysis. Acad.Press, New York (1964), Ch.II § 2.1.

[2] B. Simon: The $P(\varphi)_2$ Euclidean (Quantum) Field Theory. Princeton University Press (1976) § VIII.3

[3] J. Fröhlich: Helv. Phys. Acta 47, 265 (1974)

[4] J. Klauder: Functional Techniques and their Application in Quantum Field Theory. In Lectures in Theoretical Physics vol. XIV (Boulder 1973)

TOPICS ON EUCLIDEAN CLASSICAL FIELD EQUATIONS WITH UNIQUE VACUUA

Jeffrey Rauch*
Department of Mathematics

and

David N. Williams
The Harrison M. Randall Laboratory of Physics

The University of Michigan
Ann Arbor, Michigan 48109

ABSTRACT: We discuss the real classical field equation

$$(-\Delta+\mu^2)\varphi + \lambda F(\varphi) = f ,$$

where $\mu^2 > 0$, $\lambda \geq 0$, $F \in C^\infty(\mathbb{R})$, and $aF(a) \geq 0$ for all $a \in \mathbb{R}$, and where the source function f belongs to various function spaces contained in the Sobolev space $H_{-1}(\mathbb{R}^d)$. We review a number of results whose proofs are to appear elsewhere, on existence of a solution $\varphi \in H_1(\mathbb{R}^d)$, the correspondences between the function spaces of f and φ (regularity properties), contractivity properties and uniqueness of φ, and analytic dependence on λ and functional differentiability in f. We mention some of the ideas from the proofs, including a few alternate methods we have not discussed elsewhere. We prove a new result on positivity preservingness: if f is a nonnegative measure in $H_{-1}(\mathbb{R}^d)$, then $0 \leq \varphi \in H_1(\mathbb{R}^d)$, a result that is also valid for $\mu^2 = 0$. Finally, we give an intuitive interpretation of some of the results for the case of $d = 1$ dimension.

* Work supported in part by the NSF under Grant No. MCS 7701748

I. INTRODUCTION

The Feynman path integral approach has achieved some stunning conceptual and technical successes in its Euclidean incarnation, where the mathematical problem of its definition is not such an interesting issue (neglecting the always serious question of renormalization). Almost from the outset, "Q-space" methods were important in the Glimm-Jaffe constructive quantum field theory program. Then came the "Markov revolution," as Simon calls it [1], which was preceded by the work of Schwinger and Symanzik, and which erupted as a powerful, path space theory following the work of Nelson. The history and references may be found in Simon's book [1]: which also documents a good sampling of the contribution of Euclidean path integral techniques to the maturation of constructive field theory in two space-time dimensions.

Currently, the Euclidean path integral is an essential conceptual tool in instanton physics; we look upon the fact that its use there is still far from realizing its technical potential as an opportunity for the future.

That path integrals are naturally dominated by the solutions of classical field equations, in the regime of semi-classical approximations, is familiar; and it is these solutions we want to discuss. Let us study the real, scalar, Euclidean classical field equation

$$(-\Delta+\mu^2)\varphi + \lambda F(\varphi) = f, \qquad \mu^2 > 0, \qquad \lambda \geq 0, \tag{1}$$

where the external source $f:\mathbb{R}^d \to \mathbb{R}$ is prescribed, and belongs generically to the real Sobolev space $H_{-1}(\mathbb{R}^d)$ for reasons that emerge in Sec. II. The real Sobolev spaces are defined in terms of the Fourier transform by

$$H_s(\mathbb{R}^d) = \left\{ \psi \in \mathrm{Re}\,\mathcal{S}'(\mathbb{R}^d) : (\mu^2+k^2)^{s/2}\hat{\psi}(k) \in L^2(\mathbb{R}^d) \right\},$$

normed by

$$\|\psi\|_{H_s} = \langle \psi, (\mu^2-\Delta)^s \psi \rangle^{1/2}.$$

For nonnegative integers s, H_s is the space of distributions that belong, together with all derivatives through order s, to L^2.

The interaction function $F:\mathbb{R} \to \mathbb{R}$ is assumed to obey $F \in C^\infty$ (sometimes we need only $F \in C$) and $aF(a) \geq 0$ for all $a \in \mathbb{R}$. The effect of the latter condition is that the interaction energy U, where $U'(\varphi) = \mu^2\varphi + \lambda F(\varphi)$, has a unique minimum at $\varphi = 0$. It turns out that this enforces a unique vacuum (the solution for $f = 0$), $\varphi = 0$; and our discussion does not handle the very interesting case of field equations with instanton or soliton solutions.

On the other hand, our discussion does admit strongly nonlinear interaction functions F, with arbitrary growth at large φ; and we are able to make a rather detailed correspondence between the regularity and fall off properties of f and those of φ. Among these, if f belongs to the Schwartz test function space $\mathcal{S}(\mathbb{R}^d)$, then so does φ. That is one of the natural choices for f when the solution of the classical field equation arises as the classical limit of the functional derivative of the connected generating functional of the Euclidean quantum field theory; here the unconnected generating functional may be regarded as the Laplace transform of the normalized, interacting Euclidean path space measure, which is defined on the dual space $\mathrm{Re}\,\mathcal{S}'$. An application of this result in [2] showed that the renormalized, one loop quantum correction to φ remains in $\mathcal{S}$, in case $d = 4$ and $F = \varphi^3$.

With the further hypothesis that F be nondecreasing, we find that the solution is unique, not only at $f = 0$, but at any fixed f in H_{-1}. Furthermore, the solution becomes analytic in λ for $0 \leq \lambda < \infty$ and has functional derivatives of all orders in f. The neighborhood of analyticity at $\lambda = 0$ includes negative values, with a radius which generally shrinks as the size of f increases. At any value of $\lambda \in [0,\infty)$, the radius of analyticity depends on the particular function space norm with respect to which the analyticity of φ is defined. We have not been able to show, for example, that a fixed $f \in \mathrm{Re}\,\mathcal{S}(\mathbb{R}^d)$ gives rise to a uniform analyticity neighborhood of $\{\lambda \in [0,\infty)\}$ for each seminorm in the family that defines the topology of $\mathcal{S}$: it is an open question whether one exists.

The functional differentiability in f for $F = \varphi^3$ and $d = 4$ was applied in [2], where it was found that functional derivatives of all orders of the connected generating functional exist and are analytic at $\lambda \in [0,\infty)$ through the renormalized, one loop correction.

The plan of the rest of our discussion is the following. In Sec. II we list our results as five theorems, divided according to the topics of existence of solutions, regularity of solutions, contraction estimates and uniqueness, smooth dependence on (λ, f), and positivity preservingness. We include there a number of remarks, with sketches of selected ideas from the proofs of Theorems 1-4. The details of the proofs may be found in [3] , except that we mention a few alternative methods in Sec. II that we have not described elsewhere. Remarks (1.i) and (2.iv) mention open questions.

Theorem 5 on positivity preservingness is new, and we are pleased to acknowledge that it arose from a question posed by R. Høegh-Krohn at this conference. That result is proved in Sec. III. Let us also note that Theorem 2.i already contains a sharp version of positivity preservingness.

Section IV gives an intuitive interpretation of uniqueness at zero f , existence at slightly negative λ, and positivity preservingness, in the case of $d = 1$ dimension, where the theory comes under the Newtonian mechanics of a one-dimensional, point particle. The interpretation at slightly negative λ is based on a remark of C. Itzykson during the conference, which we are happy to acknowledge.

Since the work in [3] has gone to press, we have learned of a paper by I. V. Volovich [4] , which studies a class of Euclidean classical field equations similar to that considered here, with regularity and falloff assumptions not identical to ours.

II. STATEMENT OF RESULTS

Theorem 1 (Existence of Solutions). If $F \in C(\mathbb{R})$ and $aF(a) \geq 0$ for all $a \in \mathbb{R}$, then for any $f \in H_{-1}(\mathbb{R}^d)$ there is at least one $\varphi \in H_1(\mathbb{R}^d)$ which is a weak solution of the classical field equation in the sense that $F(\varphi) \in L^1_{loc}$ and Eq. (1) is satisfied in the sense of distributions. In addition, $\varphi F(\varphi) \in L^1(\mathbb{R}^d)$ and $F(\varphi) \in H_{-1}(\mathbb{R}^d)$.

Remarks. (1.i) Formally, $\varphi F(\varphi) \geq 0 \Rightarrow \langle \varphi, F(\varphi) \rangle \geq 0$. Thus, multiplying the classical field equation by φ and integrating gives

$$\|\varphi\|^2_{H_1} \leq \langle \varphi, f \rangle \leq \|\varphi\|_{H_1} \|f\|_{H_{-1}}, \tag{2}$$

so $f = 0 \Rightarrow \varphi = 0$, which means that φ is unique in H_1 for zero source. It remains an open question whether φ is unique in H_1 for nonzero $f \in H_{-1}$ under the above hypotheses on F. A rigorous non-uniqueness result for $-\Delta\varphi - \mu^2\varphi + \lambda\varphi^3 = f$, with the "wrong" sign of the mass term, in case φ is uniformly bounded but not zero at infinity, can be found in the work of Volovich [4].

(1.ii) The proof in [3] justifies the a priori (i.e., independent of λ and F) bound $\|\varphi\|_{H_1} \leq \|f\|_{H_{-1}}$ derived formally above. This bound is one of the strong reasons that $f \in H_{-1}$ and $\varphi \in H_1$ is natural. In an earlier (unpublished) approach to the existence theory, we used the stronger hypotheses $d \leq 4$ and F polynomial bounded to prove an a priori bound $\|\varphi\|_{H_2} \leq c\,\|f\|_{L^2}$. The Schauder continuity trick to prove existence can then be realized as follows: Let $\Lambda = \{\lambda \leq 0:$ a solution $\varphi \in H_2$ exists for fixed $f \in L^2\}$. If furthermore $F(0) = 0$ and $F \in C^\infty$, Sobolev estimates and the contraction mapping principle can be used to show that, given a solution $\varphi(\lambda_0, f) \in H_2$ for $\lambda_0 \in \Lambda$, there are solutions $\varphi(\lambda_0+\epsilon, f) \in H_2$ for $0 < |\epsilon| \leq$ some $R(\lambda_0, f)$. This shows that Λ is open in $[0,\infty)$. The contraction mapping principle argument also yields analyticity in some complex neighborhood of λ_0, if F is sufficiently analytic.

Next, one takes a sequence of solutions $\varphi(\lambda_n, f)$ with $\lambda_n \rightarrow \lambda \in [0,\infty)$. Because of the a priori bound, one can extract a weakly convergent subsequence $\varphi(\lambda_n, f) \rightharpoonup \varphi \in H_2$, then use Sobolev embedding theorems to show that a subsequence $\lambda_n F(\varphi_n) \rightarrow \lambda F(\varphi)$ in the sense of distributions. That shows that Λ is closed in $[0,\infty)$.

Since $\lambda = 0 \in \Lambda \Rightarrow \Lambda \neq \phi$, and Λ is open and closed in $[0,\infty)$, one finds the existence of a solution in H_2 for each $f \in L^2$ and $\lambda \in [0,\infty)$.

(<u>1.iii</u>) The proof in [3] uses Galerkin's method and degree theory (see [3] for references). One approximates H_1 by a sequence $V_k \subset V_{k+1} \subset C_0^\infty(\mathbb{R}^d)$, $\dim V_k < \infty$, $V_k \to H_1$. Then one defines $\Pi_k : L^2 \to V_k$ as L^2 orthogonal projection, and one defines

$$C_0^\infty \ni \psi \longmapsto T(\psi) = (-\Delta+\mu^2)\psi + \lambda F(\psi) \in C_0$$

together with

$$V_k \ni \psi \longmapsto \Pi_k T(\psi) - \Pi_k f \in V_k .$$

Then the fact that $aF(a) \geq 0$, combined with degree theory for finite dimensions, implies the existence of a zero, $\varphi_k \in V_k$, $\Pi_k T(\varphi_k) - \Pi_k(f) = 0$, obeying a uniform bound $\|\varphi_k\|_{H_1} \leq c\|f\|_{H_{-1}}$. A convergent subsequence may then be extracted by the compactness of local Sobolev embeddings, much as before. The advantages of this method are the weaker assumptions on F, the directness of the existence proof for any $\lambda \geq 0$, and that it works for $f \in H_{-1}(\mathbb{R}^d)$ without restriction on d.

<u>Theorem 2 (Regularity of Solutions).</u> In addition to the hypotheses of Theorem 1 let $F \in C^\infty(\mathbb{R})$. Suppose $\varphi \in H_1(\mathbb{R}^d)$, $F(\varphi) \in H_{-1}(\mathbb{R}^d) \cap L^1_{loc}(\mathbb{R}^d)$, and φ satisfies the classical field equation (1).

(i) (<u>L^p regularity</u>) If $1 < p \leq \infty$ and $f \in L^p(\mathbb{R}^d)$ then $\varphi \in L^p(\mathbb{R}^d)$ and $\mu^2\|\varphi\|_{L^p} \leq \|f\|_{L^p}$. If $f \in L^\infty$ then

$$\varphi \in \bigcap_p W^{2,p}_{loc}(\mathbb{R}^d) \subset \bigcap_{0<\alpha<1} C^{1+\alpha}(\mathbb{R}^d), \qquad \lim_{|x|\to\infty} \varphi(x) = 0, \text{ and}$$

$$\min\{0, \operatorname{ess\,inf} f\} \leq \mu^2\varphi \leq \max\{0, \operatorname{ess\,sup} f\}.$$

(ii) (<u>Smooth regularity</u>) If $f \in C^{k+\alpha}(\mathbb{R}^d) \cap L^\infty(\mathbb{R}^d)$ then $\varphi \in C^{k+2+\alpha}(\mathbb{R}^d)$ for any integer $k \geq 0$ and $\alpha \in (0,1)$. If $f \in \mathcal{S}(\mathbb{R}^d)$ then $\varphi \in \mathcal{S}(\mathbb{R}^d)$.

(iii) (<u>H_s regularity, small s</u>) If $s \in [-1,1]$ and $f \in H_s(\mathbb{R}^d) \cap L^\infty$ then $\varphi \in H_{s+2}(\mathbb{R}^d)$.

(iv) (<u>Weighted spaces, small s</u>) For any integer k and $s \in \mathbb{R}$ let $\mathcal{S}_{k,s} = \{\eta \in H_s(\mathbb{R}^d) : x^\alpha \eta \in H_s(\mathbb{R}^d) \text{ if } |\alpha| \leq k\}$. If $s \in [-1,1]$ and $f \in \mathcal{S}_{k,s} \cap L^\infty$ then $\varphi \in \mathcal{S}_{k,s+2}$.

In addition to the above hypotheses assume that the number of dimensions $d \leq 5$. Then

(v) (<u>H_s regularity</u>) If $s \geq -1$ and $f \in H_s(\mathbb{R}^d)$ then $\varphi \in H_{s+2}(\mathbb{R}^d)$.

(vi) (<u>Weighted spaces</u>) If $k \geq 0$ is an integer, $s \in [-1,\infty)$, and $f \in \mathcal{S}_{k,s} \cap L^\infty$, then $\varphi \in \mathcal{S}_{k,s+2}$.

<u>Remarks.</u> (<u>2.i</u>) The most important estimate is

$$\mu^2 \|\varphi\|_{L^\infty} \leq \|f\|_{L^\infty}.$$

It makes the growth of F at infinity irrelevant if $f \in L^\infty$.

(<u>2.ii</u>) The result $f \in H_s \Rightarrow \varphi \in H_{s+2}$ goes by Sobolev cranking and bootstrapping.

(<u>2.iii</u>) The proof that $f \in \mathcal{S} \Rightarrow \varphi \in \mathcal{S}$ in [3] uses a maximum principle argument and an induction on "elliptic estimates." Another straightforward method (unpublished) is an induction based on multiplying the field equation by x^α and pushing it through the Laplacian.

(<u>2.iv</u>) It is an open question whether the restriction $d \leq 5$ in Theorem 2.v and 2.vi is an artifact of the proof.

<u>Theorem 3 (Contraction Estimates).</u> Suppose that $F \in C(\mathbb{R})$, $F(0) = 0$, and F is nondecreasing. For $j = 1,2$ let φ_j be solutions, in the sense of Theorem 1, of the field equation (1) with sources $f_j \in H_{-1}(\mathbb{R}^d)$. The

(i) $\mu^2 \|\varphi_1 - \varphi_2\|_{H_1} \leq \|f_1 - f_2\|_{H_{-1}}$.

(ii) If for some $p \in (1,\infty]$ f_1 and f_2 are in $L^p(\mathbb{R}^d)$, then

$$\mu^2 \|\varphi_1 - \varphi_2\|_p \leq \|f_1 - f_2\|_p.$$

<u>Remarks.</u> (<u>3.i</u>) The monotonicity condition means that not only does the interaction energy U such that $U'(\varphi) = \mu^2\varphi + \lambda F(\varphi)$ have a unique minimum at $\varphi = 0$, it also has only positive curvature.

(<u>3.ii</u>) The first inequality results formally from the identity

$$(\varphi_1-\varphi_2)\,[(-\Delta+\mu^2)(\varphi_1-\varphi_2)+\lambda F(\varphi_1)-\lambda F(\varphi_2)] \;=\; (\varphi_1-\varphi_2)(f_1-f_2)$$

together with monotonicity,

$$(\varphi_1-\varphi_2)[F(\varphi_1)-F(\varphi_2)] \geq 0 ,$$

which formally yields upon integration

$$\|\varphi_1-\varphi_2\|_{H_1}^2 \leq \langle \varphi_1-\varphi_2, f_1-f_2 \rangle \leq \|\varphi_1-\varphi_2\|_{H_1} \|f_1-f_2\|_{H_{-1}} .$$

This bound enforces uniqueness in H_1 for any f in H_{-1}; it is a second reason that $(\varphi, f) \in H_1 \times H_{-1}$ is natural (see Remark (1.ii)).

Theorem 4 (Smooth Dependence on λ, f). In addition to the hypotheses of Theorem 3 assume that $F \in C^{\infty}(\mathbb{R})$ and that $s+2 > d/2$.

(i) (Differentiable H_s dependence) For $f \in H_s(\mathbb{R}^d)$ and $\lambda \geq 0$ let $\varphi(\lambda, f) \in H_{s+2}(\mathbb{R}^d)$ be the unique solution of the field equation (1). The map $(\lambda, f) \mapsto \varphi(\lambda, f)$ is infinitely differentiable on $[0,\infty) \times H_s(\mathbb{R}^d)$ with values in $H_{s+2}(\mathbb{R}^d)$.

(ii) (Analytic dependence on λ) Suppose in addition to the above hypotheses that F is the restriction to $\mathbb{R}$ of a function analytic on a complex neighborhood of $\mathbb{R}$. Then for fixed f the map $\lambda \mapsto \varphi(\lambda, f)$ is real analytic on $[0,\infty)$ with values in $H_{s+2}(\mathbb{R}^d)$, that is, for each $\lambda_0 \in [0,\infty)$ there exist $\varphi_n \in H_{s+2}(\mathbb{R}^d)$, $n = 0, 1, \ldots,$ and $r > 0$ such that $\varphi(\lambda, f) = \sum(\lambda-\lambda_0)^n \varphi_n$ is convergent for $|\lambda-\lambda_0| < r$.

(iii) (Weighted spaces $\mathcal{S}_{k,s}$) Suppose k is a nonnegative integer and that the hypotheses of (i) are in force. Then the map $(\lambda, f) \mapsto \varphi(\lambda, f)$ is an infinitely differentiable function on $[0,\infty) \times \mathcal{S}_{k,s}$ with values in $\mathcal{S}_{k,s+2}$. If in addition F is real analytic on $\mathbb{R}$, then the map $\lambda \mapsto \varphi(\lambda, f)$ is real analytic on $[0,\infty)$ with values in $\mathcal{S}_{k,s+2}$.

Remarks. (4.i) Consider the map

$$[0,\infty) \times H_{s+2} \times H_s \ni (\lambda, \varphi, f) \mapsto G(\lambda, \varphi, f) = (-\Delta+\mu^2)\varphi + \lambda F(\varphi) - f \in H_s .$$

The proof in [3] is based on the continuity and functional differentiability, for $F \in C^{\infty}(\mathbb{R}^n)$ and $s > d/2$, of the map

$$H_s \times \dots \times H_s \ni (\varphi_1, \dots, \varphi_n) \longmapsto F(\varphi_1, \dots, \varphi_n) \in H_s ,$$

due to Schauder's lemma, combined with the infinite dimensional implicit function theorem applied to $G(\lambda, \varphi, f) = 0$, which yields that $\varphi(\lambda, f)$ is C^∞ in λ and continuous and functionally differentiable in f. For analyticity at $\lambda \in [0, \infty)$ when F is real analytic, one applies the implicit function theorem to $\Gamma(\lambda, \varphi) \equiv G(\lambda, \varphi, f_0)$.

(4.ii) An earlier (unpublished) proof essentially amounted to proving the implicit function theorem by applying the Sobolev cranking machinery (for polynomial bounded F) and the contraction mapping principle. The latter principle gives directly the strong convergence of the expansion of $\varphi(\lambda, f)$ in series of powers of $(\lambda - \lambda_0)$ near λ_0.

Theorem 5 (Positivity Preservingness). Let $f \in H_{-1}(\mathbb{R}^d)$ be a nonnegative measure. Let $\varphi \in H_1(\mathbb{R}^d)$ be a weak solution of the classical field equation (1) in the sense of Theorem 1, with $F \in C(\mathbb{R})$, $aF(a) \geq 0$ for all $a \in \mathbb{R}$, $F(\varphi) \in L^1_{loc}$, and $\mu^2 \geq 0$. Then $\varphi \geq 0$.

Remarks. (5.i) An example of an f that obeys Theorem 5 is

$$f = \delta(x_1)\, g(x_2, \dots, x_d), \qquad 0 \leq g \in L^2(\mathbb{R}^{d-1}).$$

(5.ii) Theorem 5 admits $\mu^2 = 0$.

(5.iii) Besides the result in Theorem 2.i, there is an elementary proof in case $\varphi \in C^2(\mathbb{R}^d)$ and $\varphi(x) \to 0$ as $|x| \to \infty$. Suppose that at some point x, $\varphi(x) < 0$. Then

$$\Delta \varphi = \mu^2 \varphi + \lambda F(\varphi) - f \leq 0,$$

because $F(\varphi) \leq 0$ and $-f \leq 0$. Since $\varphi(\infty) = 0$, φ has a negative minimum at some point x_0; and at all such points $\Delta\varphi(x_0) \leq 0$. This is impossible. The argument admits $\mu^2 = 0$.

III. PROOF THAT POSITIVITY IS PRESERVED

The method is based on a variant of Lemma 1 in [3], which we adapt in a more or less self-contained way as follows. Let $k:\mathbb{R}\to\mathbb{R}$ obey

$$\begin{aligned} k(a) &= 0, \qquad a \geq 0, \\ &= -a, \qquad a \leq -\epsilon < 0, \\ k \in C^{\infty}(\mathbb{R}), &\qquad k'(a) \leq 0, \qquad k''(a) \geq 0. \end{aligned}$$

Since k and all of its derivatives are uniformly Lipschitz continuous and vanish at zero, their compositions with H_1 belong to H_1; i.e., $k^{(n)}(\varphi) \in H_1(\mathbb{R}^d)$. Moreover, $k^{(n)}(\varphi) \in L^{\infty}$ for $n \geq 1$. Furthermore, $fk'(\varphi) \leq 0$ is well-defined and true in the sense of distributions, because for any $0 \leq g \in \mathcal{D}(\mathbb{R}^d)$ we have $0 \geq gk'(\varphi) \in H_1$; and the middle expression of

$$0 \geq \langle gk'(\varphi), f\rangle = \langle g, k'(\varphi)f\rangle$$

is a valid pairing for $f \in H_{-1}$, which makes $k'(\varphi)f$ a nonpositive measure.

The main step in the proof is to show that

$$-\Delta k(\varphi) \leq fk'(\varphi) \leq 0. \tag{3}$$

The interpretation is well-defined in the sense of distributions, because $k(\varphi) \in H_1 \Rightarrow k(\varphi) \in L_1^{loc} \subset \mathcal{D}'$. Note that all terms in the formal equations

$$\begin{aligned} \Delta k(\varphi) &= k''(\varphi)\nabla\varphi\cdot\nabla\varphi + k'(\varphi)\Delta\varphi \\ &= k''(\varphi)\nabla\varphi\cdot\nabla\varphi + k'(\varphi)[\mu^2\varphi + \lambda F(\varphi) - f] \end{aligned} \tag{4}$$

are well-defined in the sense of distributions, because $k''(\varphi) \in L^{\infty}$, $\nabla\varphi\cdot\nabla\varphi \in L^1$, $k'(\varphi) \in H_1$, $\Delta\varphi \in H_{-1}$, and $k'(\varphi)[\mu^2\varphi + \lambda F(\varphi)] \in L_1^{loc}$. The second equality is innocuous, while the first can be proved after a short argument based on the distribution identity $\nabla k(\varphi) = k'(\varphi)\nabla\varphi$ plus the fact that $\mathcal{D}$ is strongly dense in H_1. We get (3) from (4) by dropping the nonnegative terms $k''(\varphi)\nabla\varphi\cdot\nabla\varphi$, $\mu^2 k'(\varphi)\varphi$, and $k'(\varphi)F(\varphi)$.

We now prove that the properties

$$\Delta k(\varphi) \geq 0, \qquad 0 \leq k(\varphi) \in L^2,$$

imply that $k(\varphi) = 0$, from which it is immediate that $\varphi \geq 0$. To see that $k(\varphi) = 0$, it suffices to see that $\chi \equiv \eta * k(\varphi) = 0$ for all $0 \leq \eta \in \mathcal{D}(\mathbb{R}^d)$. Now

$$\Delta\chi = \eta * \Delta k(\varphi) \geq 0,$$

so $\chi \in C^\infty$ is subharmonic; $\lim_{|x|\to\infty} \chi = 0$, since $\hat{\eta}\widehat{k(\varphi)} \in L^1$; and clearly $\chi \geq 0$. For any $x_0 \in \mathbb{R}^d$ and $R > 0$, let $A_R = \{x: R < |x-x_0| < R+\delta\}$, where δ is chosen so that meas $A_R = 1$. By subharmonicity

$$\chi(x_0) \leq \int_{A_R} \chi \leq \sup_{A_R} \chi \xrightarrow[R\to\infty]{} 0,$$

so $\chi(x_0) \leq 0 \Rightarrow \chi(x_0) = 0$. □

IV. INTERPRETATION FOR ONE DIMENSION

It has become customary to use the motion of a point particle under Newtonian classical mechanics as an analogy to gain intuitive insight into the nature of solutions to classical field equations in various instanton problems, where the dimension d is reduced to one, either for model purposes, or as a result of symmetry. That kind of interpretation may be applied to the unique vacuum situation as well.

Thus, let $d = 1$; and make the identification that $t = x \in \mathbb{R}$ is time, and $y = \varphi \in \mathbb{R}$ is the position of a point particle of unit mass in one space dimension. The Euclidean classical field equation becomes

$$\ddot{y}(t) = -U'[y(t)] - f(t) ,$$
$$U' \equiv -\lambda F(y) - \mu^2 y ,$$

where we have changed the sign in the definition of the potential energy U compared to that of the former field interaction energy U mentioned in the Introduction. The condition $a\lambda F(a) \geq 0$ makes U into a potential hill with at most one peak at the origin; for simplicity, we shall also impose the monotonicity condition that F be nondecreasing, whereupon the generic potential hill U has a peak at the origin, no shoulders (negative curvature), and falls away to minus infinity on both sides.

We consider smooth, rapidly decreasing external forces f and solutions $y(t)$ for the orbit of the point particle, with the aim of illustrating uniqueness at $f = 0$, existence at small negative λ, and positivity preservingness.

The falloff of the solution and its derivatives at large t means that the particle starts at the peak with zero velocity asymptotically at $t = -\infty$ and returns to the peak with zero velocity asymptotically at $t = +\infty$.

Uniqueness at $f = 0$ is a simple consequence of conservation of mechanical energy (kinetic plus potential). A particle which begins on the peak at $t = -\infty$ with zero kinetic energy and does not stay there forever must at some time be to the right of the peak with a positive velocity or to the left of the peak with negative velocity, whereupon it continues to fall forever down the hill and never returns to the peak. The only solution is therefore $y(t) = 0$.

To understand that the existence of solutions for nonzero f and

$\lambda \geq 0$ entails existence for slightly negative λ, note that for λ small and negative, the curvature and slope of the hill near $y = 0$ are dominated by the $-\mu^2 y^2/2$ term in U. Thus, near $y = 0$, the potential energy has the same structure as before, a peak at $y = 0$. The F term typically dominates the $\mu^2 y$ term at large y, even for small λ, so at large y the potential energy eventually becomes positive and climbs towards plus infinity. But the onset of this behavior can be delayed by making $|\lambda|$ smaller. No matter how large the fixed external force may be, as long as it is uniformly bounded, we know from Theorem 2.i, in case $\lambda \geq 0$, that a solution exists with $\mu^2 |y| \leq |f|$, independently of λ and F. For the same f, if λ is sufficiently small and negative, the potential energy hill looks the same as for another λF with $a\lambda F(a) \gtrsim 0$, and a slightly smaller value of $\mu^2 > 0$, out to $|y|$ less than a prescribed value. Hence a solution exists for a given f and sufficiently small, negative λ.

Finally, to understand that nonnegative external forces f entail nonnegative y, note that in such a case the external force is only leftward. Then the particle can never have been to the left of the peak; for if it starts towards the left from $t = -\infty$ with zero velocity, its kinetic energy continually increases from the combined effects of the hill and the external force; and it runs away to $y = -\infty$; while if it starts to the right and at some finite later time appears on the left, the leftward external force must have turned it around and supplied enough kinetic energy to get it back over the peak with a nonzero, leftward velocity, whence it again runs off to $-\infty$. Thus, any solution that arrives back at the peak at $t = +\infty$ can at no time have $y < 0$.

REFERENCES

1. B. Simon, The $P(\phi)_2$ Euclidean (Quantum) Field Theory, Princeton University Press, Princeton, 1974.

2. D. N. Williams, The Euclidean loop expansion for massive $\lambda:\Phi_4^4:$ Through one loop. Commun. math. Phys. 54, 193-218 (1977).

3. J. Rauch and D. N. Williams, Euclidean Nonlinear Classical Field Equations with Unique Vacuum. Commun. math Phys., to appear (1978).

4. I. V. Volovich, Classical Equations of Euclidean Field Theory. Theoretical and Mathematical Physics (translated from Russian) 34, 9-14 (1978).

NULL PLANE FIELDS AND AUTOMODEL RANDOM PROCESSES

by

L. Streit

Fakultät für Physik
Universität Bielefeld
D-4800 Bielefeld

Generalized random fields play a central role in "Euclidean" quantum field theory and a classical one in the description of (canonical) quantum fields on spacelike hyperplanes of Minkowski space. Their role and properties in null plane quantum field theory [1-12] are less well known. Two such properties are worth noting in the present context. We consider vacuum expectation values

$$E_0(f) = (\Omega, e^{iA(f)}\,\Omega)$$

and assume that the field A may be smeared with test functions concentrated on a lightlike hyperplane, so that we may set

$$\text{supp } f \subset \{\, x : x^{\perp} = o \,\},$$

using the lightlike coordinates $x^{\perp} \equiv \frac{1}{\sqrt{2}}(x^0+x^1)$, $x^{"} \equiv \frac{1}{\sqrt{2}}(x^0-x^1)$ and the two spacelike transverse ones, x^2 and x^3.

For such fields on a null plane

1. Correlations break down in the transverse directions x^2,x^3. This follows by standard arguments [5,10] from the positivity of the generator $p_{"} = \frac{1}{\sqrt{2}}(p_0-p_1)$ for lightlike translations within the hyperplane. Thus, for test functions $f(x) = h(x")\, g(x^2,x^3)$

$$E_0(f) \equiv C_h(g)$$

turns out to be the characteristic functional of a stationary process with independent values at each point [13,14].

Regrettably a complete classification of these is not yet known [14], but under the extra assumptions of transverse scaling and finiteness of n-point functions, with the scaling dimension dictated by the Lehmann representation of the two-point function, one finds that the process must be gaussian i.e. the null plane field is in the (unique) Fock representation [10].

2. This universality results from the fact that $E_o(f)$ can be viewed as the mono-directional (x^1-direction) short distance scaling fixed point of the field on the spacelike hyperplane $x^0 = o$. Indeed this latter plane is rotated to become tangent to the light cone by (first) a large boost and (second) a short distance scaling to compensate for the hyperbolic outward motion of all points. Since the boost leaves E invariant, all that remains is short distance scaling. In the limit one obtains the dilatation invariant null plane functional E_o with the boost operator in x^1-direction furnishing the unitary representation of one-dimensional dilatations. This is best illustrated by the case of massive free fields, where the spacelike functional E transforms as follows to the null plane one:

$$E(f) = e^{-\frac{1}{4}\int \frac{d^3p}{\sqrt{p^2+m^2}} |\tilde{f}|^2} \rightarrow E_o(f)$$

$$= \lim_{l\to o} e^{-\frac{1}{4}\int \frac{d^3p}{\sqrt{p_1^2+l^2(p_2^2+p_3^2+m^2)}} |\tilde{f}(p)|^2}$$

$$= e^{-\frac{1}{2}\int_o^\infty \frac{dp_1}{p_1} \int dp_2 dp_3 |\tilde{f}(p)|^2}$$

$E_o(f)$ has the characteristic properties of

- fixed point universality (no mass dependence)
- scale invariance
- breakdown of correlations in the transverse directions x^2, x^3
- long range correlations and an infrared problem in the lightlike direction which necessitates a test function space such as

$$S^{(1)} = \{ f = \frac{\partial}{\partial x^{\shortparallel}} g : g \in S \}$$

(This feature has caused considerable confusion in the discussion of nullplane quantum field theory [15, 16] but is well known in the theory of automodel processes [17, 18]).

Finally it may be worth mentioning that consideration of certain fields which are finite on spacelike hyperplanes but not so on lightlike ones still allow for the construction of a scale invariant ("automodel") fixed point if one supplements the above short distance scaling by an (infinite, as $l \rightarrow o$) multiplicative renormalization [7,19].

Non-Gaussian and non-power-of-Gaussian fixed points result. In particular the normal ordered square of a free field in two dimensional space-time leads to the infinitely divisible fixed point [7]

$$E_0(f) = \left(1 + \int_0^\infty \frac{dp}{p} \, |\tilde{f}(p)|^2\right)^{-1/2}$$

References

[1] G.R. Bart, S. Fenster: "Free Field Theories of Spin-Mass Trajectories and Quantum Electrodynamics on the Null Plane", ANL-preprint, June 1976.

[2] J.S. Bell, H. Puegg: "Null Plane Field Theory and Composite Models", CERN preprint TH 21o1, 1975.

[3] P.A.M. Dirac: "Forms of Relativistic Dynamics", Rev. Mod. Phys. 21, 392 (1949).

[4] G. Domokos: "Introduction to the Characteristic Initial Value Problem in Quantum Field Theory", Lectures at 14th Summer Inst. for Physics, Boulder, Colo., 1971.

[5] W. Driessler: "On the Structure of Fields and Algebras on Null Planes I, II", Acta Phys. Austr. 46, 63, 163 (1977).

[6] F. Jegerlehner, Helv. Phys. Acta 46, 824 (1974).

[7] W. Karwowski, L. Streit: "A Renormalization Group Model with Non-Gaussian Fixed Point", Rep. Math. Phys. 13, 1 (1978).

[8] H. Leutwyler, J.R. Klauder, L. Streit: "Quantum Field Theory on Lightlike Slabs", Il Nuovo Cim. 66A, 536 (197o)

[9] S. Schlieder, E. Seiler, Comm. math. Phys. 25, 62 (1972)

[10] L. Streit: "Lightlike Data for Quantum Field Theory", in Proc. XIIIth Winter School for Theoretical Physics, Karpacz 1976, p. 122 ff.

[11] J.H. Ten Eyck, F. Rohrlich, Phys. Rev. D9, 2237 (1974).

[12] H. Yabuki: "Null Plane Quantization in the Hamiltonian Form of the Feynman Path Integral", Kyoto University preprint RIMS-183, 1975.

[13] I.M. Gelfand, N.Y. Vilenkin: "Generalized Functions" vol 4, Academic Press, New York 1964.

[14] M.M. Rao: "Local Functionals and Generalized Random Fields with Independent Values", Theory of Probability and its Appl.. XVI, 466, (1971).

[15] T. Suzuki, S. Tameike, E. Yamada: "Some Undesirable Features of Field Theory on a Null Plane", Prog. Theor. Phys. 55, 922 (1976).

[16] N. Nakanishi, K. Yamawaki: "A Consistent Formulation of the Null Plane Field Theory", Nucl. Phys. B122, 15 (1977).

[17] R.L. Dobrushin: "Avtomodelnost i renorm-gruppa obobshčennych poley", "Gaussovskie i podčinennye Gaussovskim avtomodelnye obobshčennye slučajnye polya", to appear in Ann. Probability.

[18] G. Jona-Lasinio, Cargèse Lectures 1976.

[19] V. Enss: "Renormalization Group Limits for Wick Polynomials of Gaussian Processes", Rep. Math. Phys. 13, 87 (1978).

- SECTION IV -

DEFORMATIONS ET QUANTIFICATION

A. Lichnerowicz
Collège de France
11 Place Marcelin Berthelot
75231 Paris Cedex 05 - FRANCE

On sait qu'il est possible de donner une description complète de la mécanique classique en termes de géométrie symplectique et de crochet de Poisson. Ceci est l'essentiel du formalisme hamiltonien. Dans un programme commun avec M. Flato et J. Vey, nous avons étudié les principales propriétés et les applications des déformations de l'algèbre de Lie de Poisson. De telles déformations peuvent fournir un nouveau point de vue concernant la mécanique quantique. Je me bornerai ici, ce qui suffit à notre objet, à des systèmes dynamiques à un nombre fini de degrés de liberté; l'approche et une partie des résultats peuvent être étendues à des champs physiques.
Dérivations et déformations d'une algèbre de Lie procèdent d'une même cohomologie de l'algèbre de Lie, la cohomologie à valeurs dans l'algèbre de Lie elle-même et correspondant à la représentation adjointe. J'appellerai ici cette cohomologie la cohomologie de Chevalley de l'algèbre de Lie.

1 - Algèbres de Lie associées à une variété symplectique.

a) Soit W une variété différentiable, connexe, paracompacte de dimension paire $2n$ et classe C^∞. Tous les éléments introduits sont supposés C^∞ et nous posons $N = C^\infty(W;R)$. Une structure symplectique est définie sur W par une 2-forme F fermée et de rang $2n$. Nous notons $\mu : TW \to T^*W$ l'isomorphisme de fibrés vectoriels défini par $\mu(X) = -i(X)F$ (où $i(.)$ est le produit intérieur); cet isomorphisme s'étend naturellement aux tenseurs. Soit Λ le 2-tenseur contravariant antisymétrique $\mu^{-1}(F)$.
Une transformation infinitésimale (t.i.) symplectique est un champ de vecteurs X tel que $\mathcal{L}(X)F = 0$ (où $\mathcal{L}$ est la dérivée de Lie); c'est un automorphisme infinitésimal de la structure. Soit L l'algèbre de Lie (de dimension infinie) des t.i. symplectiques. Pour que $X \in L$, il faut et il suffit que la 1-forme $\mu(X)$ soit fermée; X est aussi dit un champ localement hamiltonien. Si $X,Y \in L$

$$(1-1) \qquad \mu([X,Y]) = d\, i(\Lambda)\, (\mu(X) \wedge \mu(Y))$$

Soit L^* le sous-espace de L défini par les images inverses des 1-formes exactes ($X_u = \mu^{-1}(du)$; $u \in N$). Un élément de L^* est un champ de vecteurs (globalement) hamiltonien. On sait (Arnold, Calabi, moi-même) que L^* est exactement l'idéal dérivé $[L,L]$ de L (chaque élément de $[L,L]$ est, par définition, une somme finie de crochets d'éléments de L); on a dim. $L/L^* = b_1(W)$, où $b_1(W)$ est le premier nombre de Betti pour l'homologie de W à supports compacts.

b) Un vecteur X définit une t.i. conforme symplectique si $\mathcal{L}(X)F = aF$, où $a \in N$. Pour $n > 1$, a est nécessairement une constante. On note L^c (algèbre de Lie des t.i. conformes symplectiques) l'algèbre de Lie des champs de vecteurs X tels qu'il existe une constante k(X) pour laquelle :

$$\mathcal{L}(X)\,F + k(X)\,F = 0 \qquad \text{ou} \qquad \mathcal{L}(X)\,\Lambda = k(X)\,\Lambda$$

L et L^* sont des idéaux de L^c. Si $X \in L^c$, on a $k(X)F = d\,\mu(X)$.
Si F est non exacte (en particulier si W est compacte), $k(X) = 0$ pour tout $X \in L^c$ et $L^c = L$.
Si F est exacte, $F = d\mu(Z)$ où $Z \in L^c$ avec $k(Z) = 1$; on a $L^c = L + C_Z$, où C_Z est le sous-espace de dimension 1 engendré par Z; on a $[L^c,L^c] = L$ et dim $L^c/L = 1$.

c) Soit $\bar{N}$ l'espace des classes d'éléments de N, modulo les constantes additives ; $\Pi : u \in N \rightarrow \bar{u} \in \bar{N}$ est la projection de N sur $\bar{N}$. L'isomorphisme naturel entre les espaces L^* et $\bar{N}$ induit sur $\bar{N}$ une structure d'algèbre de Lie définie de la manière suivante : si $\bar{u}, \bar{v} \in \bar{N}$, il résulte de (1-1) que la fonction

$$w = i(\Lambda)(d\bar{u} \wedge d\bar{v}) \tag{1-2}$$

définit une classe $\bar{w}$ qui est le crochet de $\bar{u}$ et $\bar{v}$. La fonction (1-2) est le crochet de Poisson de $\bar{u}$, $\bar{v}$ ou de deux représentants u, v dans N; nous posons $w = \{u,v\}$. Le crochet de Poisson $\{\ ,\ \}$ définit sur N une structure d'algèbre de Lie; $(N,\{\ ,\ \})$ est l'algèbre de Lie de Poisson et on a un homomorphisme de cette algèbre de Lie sur l'algèbre de Lie L^* des champs de vecteurs hamiltoniens.

2 - Dynamique classique et variétés symplectiques.

a) Considèrons un système dynamique à liaisons indépendantes du temps et n degrés de liberté. L'espace de configuration correspondant est une variété différentiable arbitraire M de dimension n. On sait que T^*M admet une structure symplectique naturelle définie par la 2-forme de Liouville-Poincaré. Pour le formalisme hamiltonien, un état dynamique du système est un point de $W = T^*M$ qui est l'espace usuel des physiciens. L'analyse des équations de la mécanique a montré depuis longtemps qu'il est essentiel de pouvoir introduire des changements de variables classiques (q^α, p_α) ne respectant pas la structure cotangente de W.
Nous sommes conduits à introduire comme espace de phase une variété symplectique (W,F) de dimension 2n. La dynamique correspondante est déterminée par une fonction $H \in N$, l'hamiltonien du système qui définit un champ de vecteurs hamiltonien X_H. Un mouvement

du système dynamique est donné , par définition , par une courbe intégrale c(t) du champ de vecteurs hamiltonien X_H , le paramètre t étant le temps. Telle est la signification géométrique des équations de Hamilton.

b) Nous pouvons adopter un autre point de vue. L'espace N admet les deux structures algèbriques suivantes :

1º une structure d'algèbre associative définie par le produit usuel des fonctions (qui est ici commutatif)

2º une structure d'algèbre de Lie définie par le crochet de Poisson.

Les dérivations du produit sont données par les champs de vecteurs; en particulier il résulte de $\{u,v\} = \mathcal{L}(X_u)v$

$$(2\text{-}1) \qquad \{ w,u,v \} = \{w,u \}.\ v + u.\ \{w,v \}$$

Considèrons une famille u_t d'éléments de N satisfaisant l'équation différentielle :

$$(2\text{-}2) \qquad du_t/dt = \{H,u_t\}$$

et prenant les valeurs initiales $u_o \in N$ pour t = 0. Il résulte de (2-1) que l'évolution de u_t dans le temps procède des trajectoires qui apparaissent dans le premier point de vue; (2-2) peut être considérée comme l'équation intrinsèque de la dynamique classique.

c) Nous avons ainsi complètement décrit la mécanique classique en termes des deux lois de composition introduites reliées par (2-1). Il est naturel d'étudier s'il est possible de déformer d'une manière convenable ces deux lois algèbriques de façon à obtenir un modèle isomorphe à la mécanique quantique usuelle. Les premiers résultats obtenus par Flato, Fronsdal, J. Vey et moi-même sont positifs.

3 - Cohomologie de Chevalley et dérivations.

a) La cohomologie de Chevalley de l'algèbre de Lie de Poisson (N,{ ,}) est définie de la manière suivante : une p-cochaîne C de N est une application p-linéaire alternée de N^p dans N, les 0-cochaînes étant identifiées aux éléments de N. Le cobord de la p-cochaîne C est la (p+1)-cochaîne ∂C définie par :

$$(3\text{-}1) \qquad \partial C(u_o,\dots,u_p) = \varepsilon^{\lambda_0\dots\lambda_p}_{0\dots p}\Big(\frac{1}{p!}\ \{u_{\lambda_o}, C(u_{\lambda_1},\dots, u_{\lambda_p})\} -$$

$$\frac{1}{2(p-1)!}\ C(\{u_{\lambda_o},u_{\lambda_1}\}, u_{\lambda_2},\dots , u_{\lambda_p})\Big)$$

où $u_\lambda \in N$ et où ε est l'indicateur antisymétrique de Kronecker. L'espace des 1-cocycles de N est l'espace des dérivations de N, l'espace des 1-cocycles exacts celui des dérivations intérieures.

Une p-cochaîne C est dite locale si, pour tout $u_1 \in N$ telle que $u_1|_U = 0$ sur un domaine U, on a $C(u_1,\dots u_p)|_U = 0$. Si C est locale, ∂C est locale.
Une p-cochaîne C est dite d-différentielle ($d \geq 1$) si elle est définie par un opérateur multidifférentiel d'ordre maximum d en chaque argument. Si C est d-différentielle, ∂C est aussi d-différentielle.

b) Nous avons établi les théorèmes non triviaux suivants :

Théorème 1 - Si T est une 1-cochaîne locale de N telle que ∂T soit d-différentielle ($d \geq 1$), T est d-différentielle. Si W est non compacte, l'hypothèse de localité peut être supprimée [3].

Théorème 2 - Toute dérivation de L^c, L, L^* est donnée par $Y \to [X,Y]$, où $X \in L^c$. Par suite les premiers espaces de cohomologie de Chevalley sont $H^1(L^c) = \{0\}$, $H^1(L) \simeq L^c/L$, $H(L^*) = L/L^*$.
Si F est non exacte, $\dim H^1(L) = 0$, $\dim H^1(L^*) = b_1(W)$
Si F est exacte, $\dim H^1(L) = 1$, $\dim H^1(L^*) = 1 + b_1(W)$.

Théorème 3 - 1° Si W est non compacte, toute dérivation $\mathcal{D}$ de N est donnée par $\mathcal{D} = \mathcal{L}(X) + k(X)$ où $X \in L^c$.
2° Si W est compacte, toute dérivation $\mathcal{D}$ de N est donnée par :

$$\mathcal{D}u = \mathcal{L}(X)u + \lambda \int_W u\,\eta$$

où $X \in L$, $\lambda \in \mathbb{R}$ et η est l'élément de volume symplectique. Il y a des dérivations non locales [1].

4 - Déformations formelles.

Je vais d'abord rappeler et étendre les éléments principaux de la théorie de Gerstenhaber [2] concernant les déformations des structures algèbriques. Je raisonnerai sur l'algèbre de Lie (N, {,}).

a) Soit $E(N;\lambda)$ l'espace des fonctions formelles de $\lambda \in C$ à coefficients dans N. Considèrons une application bilinéaire alternée $N \times N \to E(N;\lambda)$ qui donne une série formelle en λ :

$$(4\text{-}1) \qquad [u,v]_\lambda = \sum_{r=0}^{\infty} \lambda^r C_r(u,v) = \{u,v\} + \sum_{r=1}^{\infty} \lambda^r C_r(u,v)$$

où les C_r ($r \geq 1$) sont des 2-cochaînes différentielles de (N,{ , }). Ces cochaînes s'étendent à $E(N;\lambda)$ de manière naturelle. Si $u,v,w \in N$, on a

$$S\left[[u,v]_\lambda , w\right]_\lambda = \sum_{t=1}^{\infty} \lambda^t D_t(u,v,w)$$

où S est la sommation après permutation circulaire et où D_t est la 3-cochaîne

$$D_t(u,v,w) = S \sum_{r+s=t} C_r(C_s(u,v),w) \qquad (r,s \geq 0)$$

Nous dirons que (4-1) définit une déformation formelle de l'algèbre de Lie de Poisson si l'identité de Jacobi est satisfaite formellement, c'est-à-dire si $D_t = 0$ $(t = 1, 2, \ldots)$; (4-1) définit alors par extension une structure d'algèbre de Lie sur $E(N;\lambda)$. Si l'on pose :

$$E_t(u,v,w) = S \sum_{r+s=t} C_r(C_s(u,v),w) \qquad (r,s \geqslant 1)$$

On a $D_t \equiv E_t - \partial C_t$. Si (4-1) est limité à l'ordre q, <u>on a une déformation d'ordre q</u> si l'identité de Jacobi est satisfaite à l'ordre (q+1). S'il en est ainsi, E_{q+1} est mécaniquement un 3-cocycle de N. Pour qu'on puisse trouver une 2-cochaîne C_{q+1} telle que $D_{q+1} \equiv E_{q+1} - \partial C_{q+1} = 0$ il faut et il suffit que E_{q+1} soit exact; E_{q+1} définit une classe qui est <u>l'obstruction à l'ordre q+1</u> à la construction d'une déformation. Une déformation d'ordre 1 est dite une <u>déformation infinitésimale</u>. On a $E_1 = 0$ et par suite seulement $\partial C_1 = 0$, c'est-à-dire C_1 est un 2-cocycle de N.

b) Considèrons une série formelle en λ :

(4-2) $$T_\lambda = \sum_{s=0}^{\infty} \lambda^s T_s = \mathrm{Id.} + \sum_{s=1}^{\infty} \lambda^s T_s$$

où les T_s sont des opérateurs différentiels sur N; T_λ opère naturellement sur $E(N;\lambda)$. Considèrons une autre application bilinéaire alternée $N \times N \to E(N;\lambda)$ correspondant à la série formelle :

(4-3) $$[u,v]'_\lambda = \{u,v\} + \sum_{r=1}^{\infty} \lambda^r C'_r(u,v)$$

où les C'_r sont encore des 2-cochaînes différentielles. Supposons que (4-2), (4-3) soient telles qu'on ait formellement l'identité :

(4-4) $$T_\lambda [u,v]'_\lambda = [T_\lambda u , T_\lambda v]$$

A partir de formules universelles, on établit par récurrence la proposition suivante <u>Proposition - La déformation formelle (4-1) de l'algèbre de Lie de Poisson étant donnée, chaque série formelle (4-2) engendre une application bilinéaire (4-3) unique vérifiant (4-4). Cette application est une nouvelle déformation formelle qui est dite équivalente à (4-1). En particulier une déformation est dite triviale si elle est équivalente à la déformation identité ($C_r = 0$ pour tout $r \geqslant 1$)</u>

Si deux déformations sont équivalentes à l'ordre q, il apparait un 2-cocycle $\in H^2(N)$ qui est l'obstruction <u>à l'équivalence à l'ordre (q+1)</u>. En particulier deux déformations infinitésimales définies par les 2-cocycles C_1 et C'_1 sont équivalentes si $(C'_1 - C_1)$ est <u>exact</u>.

c) En ce qui concerne les déformations associatives du produit usuel de N on peut donner des définitions et résultats semblables. Une déformation associative de (N,.) est définie par une application bilinéaire $N \times N \to E(N;\nu)$ (où $\nu \in C$) donnée par :

(4-5) $$u \star_\nu v = u.v + \sum_{r=1}^{\infty} \nu^r \Gamma_r(u,v)$$

où les Γ_r sont des applications bilinéaires (ou 2-cochaînes de (N,.)) différentielles $N \times N \to N$ telles qu'on ait l'identité d'associativité

(4-6) $$(u \star_\nu v) \star_\nu w = u \star_\nu (v \star_\nu w)$$

S'il en est ainsi $\star_\nu$ définit par extension une structure d'algèbre associative sur $E(N;\nu)$. Il existe une cohomologie - dite la cohomologie de Hochschild - qui joue pour les déformations associatives le même rôle que la cohomologie de Chevalley pour les déformations d'algèbres de Lie. Une p-cochaîne de (N,.) est une application p-linéaire de N^p dans N. Le cobord de la p-cochaîne Γ est la (p+1)-cochaîne $\tilde{\partial}\Gamma$ définie par

(4-7) $$\tilde{\partial}\Gamma(u_o,..,u_p) = u_o\Gamma(u_1,..u_p)-\Gamma(u_ou_1,u_2,..u_p)+\Gamma(u_o,u_1u_2,..u_p)+\ldots+ (-1)^p \Gamma(u_o,..,u_{p-1}u_p)+(-1)^{p+1}\Gamma(u_o,..u_{p-1})u_p$$

Un 1-cocycle de (N,.) est une dérivation de cette algèbre. On montre que <u>si T est une 1-cochaîne de (N,.) telle que $\tilde{\partial}T$ soit d-différentielle ($d \geqslant 0$), T est elle-même (d+1)-différentielle</u>.

d) Nous avons montré que, pour les déformations différentielles de l'algèbre de Lie de Poisson, deux cas seulement sont possibles : ou bien toutes les 2-cochaînes C_r sont 1-différentielles ou l'ordre de différentiabilité de C_r augmente indéfiniment avec r. Je vais considérer un très intéressant exemple de déformation du second type décrit récemment par Jacques Vey. Mon point de vue diffère de celui de Vey.

5 - Le cas plat.

a) Soit (W,F) une variété symplectique. Une telle variété admet des atlas de cartes pour lesquelles F (ou Λ) a des composantes constantes (<u>cartes naturelles</u> $\{x^i\}$, (i,j,.. = 1,..,2n)).

Une <u>connexion symplectique</u> Γ est une connexion linéaire sans torsion telle que $\nabla F=0$, où ∇ est l'opérateur de dérivation covariante défini par Γ. Si $\{\Gamma^i_{jk}\}$ sont les coefficients usuels d'une connexion dans une carte naturelle $\{x^i\}$, introduisons les coefficients $\Gamma_{ijk} = F_{il}\,\Gamma^l_{jk}$. De tels coefficients $\{\Gamma_{ijk}\}$ définissent une connexion symplectique si et seulement s'ils sont complètement symétriques dans toute carte naturelle. Une variété symplectique admet une infinité de connexions symplectiques; deux telles connexions diffèrent par un 3-tenseur covariant symétrique.

b) Supposons que (W,Λ) admette une connexion symplectique <u>sans courbure</u>; s'il en est ainsi la variété (W,Λ,Γ) est dite une <u>variété symplectique plate</u>. L'exemple le plus simple est donné par le fibré cotangent de R^n, c'est-à-dire $\mathbb{R}^n \times \mathbb{R}^n$. Introduisons sur une variété symplectique plate les opérateurs bidifférentiels P^r d'ordre maximum r en chaque argument, définis de la manière suivante sur chaque domaine U d'une carte arbitraire :

$$(5\text{-}1)\qquad P^r(u,v)_{|U} = \Lambda^{i_1 j_1}\ldots\Lambda^{i_r j_r}\,\nabla_{i_1\ldots i_r}u\;\nabla_{j_1\ldots j_r}v \qquad (u,v\in N)$$

Nous posons $P^o(u,v) = u.v$. Pour $r = 1$, on obtient l'opérateur de Poisson P avec $P(u,v) = \{u,v\}$; $P^r(u,v)$ est symétrique en u,v si r est pair, antisymétrique si r est impair.

Etant donnée une fonction formelle f(z) à coefficients constants telle que $f(0) = 1$, substituons P^r à z^r dans le développement de $f(\nu z)$; on obtient une application bilinéaire $(u,v)\in N\times N \to u \star_\nu v = f(\nu P)(u,v) \in E(N;\nu)$. Cherchons à choisir f de manière à définir ainsi une déformation de l'algèbre associative (N,.). La réponse est donnée par la proposition suivante :

Proposition - Si (W,Λ,Γ) est une variété symplectique plate, il existe une seule fonction formelle du crochet de Poisson P (à un facteur constant près et à un changement linéaire près du paramètre de déformation ν) qui engendre une déformation formelle de l'algèbre associative (N,.) : c'est la fonction exponentielle.

On a ainsi :

$$(5\text{-}2)\qquad u \star_\nu v = \exp(\nu P)(u,v) = \sum_{r=0}^{\infty} (\nu^r/r!)\, P^r(u,v)$$

Il résulte de l'identité d'associativité vérifiée par (5-2) que :

$$(5\text{-}3)\qquad [u,v]_\lambda = (2\nu)^{-1}(u \star_\nu v - v \star_\nu u) = \nu^{-1}\mathrm{sh}(\nu P)(u,v) = \sum_{r=0}^{\infty} (\nu^{2r}/(2r+1)!)P^{2r+1}(u,v)$$

définit, avec $\lambda = \nu^2$, une déformation formelle de l'algèbre de Lie de Poisson. Il est remarquable que, pour $\nu = i\hbar/2$ on déduit de (5-3) un crochet $\frac{2}{\hbar}\sin(\frac{\hbar}{2}P)$ donné en 1949 par Moyal dans le contexte de la quantification de Hermann Weyl - Wigner.

La déformation (5-2) n'est jamais triviale; sinon le 2-cocycle P de Hochschild 1-différentiable serait exact et par suite le cobord d'un opérateur différentiel d'ordre 2; un tel cobord ne peut être égal à P car il est symétrique en u,v. Considérons de même le terme P^3 de (5-3). Si ce cocycle de Chevalley était exact dans la cohomologie locale, il serait le cobord d'une 1-cochaîne locale qui serait nécessairement un opérateur différentiel d'ordre 3 (théorème 1). Mais on voit qu'un tel cobord n'a pas de terme de type bidifférentiel (3,3). On peut montrer que, pour une variété symplectique plate, le second espace $H^2(N)$ de cohomologie locale de Chevalley est de dimension 1; P^3 définit une 2-classe β de cohomologie qui est un générateur pour cet espace.

On voit que les déformations (5-2) et (5-3) sont non triviales même à l'ordre 1.

6 - Généralisations.

Ces déformations peuvent-elles être généralisées à des variétés symplectiques non plates ? On voit aisément qu'on obtient pas de telles généralisations en étendant

(5-1) au cas où ∇ correspond à une connexion symplectique arbitraire Γ .

a) Si $u \in N$, notons $\mathcal{L}(X_u)\Gamma$ le 3-tenseur symétrique covariant défini à partir de la dérivée de Lie de la connexion symplectique Γ par le champ hamiltonien Γ . On a pour une carte naturelle de domaine U :

$$(6\text{-}1)\qquad (\mathcal{L}(X_u)\Gamma)_{i_1i_2i_3} = \partial_{i_1i_2i_3}u - S\,\Lambda^{kl}\,\Gamma_{ki_1i_2}\,\partial_{li_3}u - \Lambda^{kl}\,\partial_k\,\Gamma_{i_1i_2i_3}\,\partial_l u$$

La 2-cochaîne S^3_Γ définie par :

$$(6\text{-}2)\qquad S^3_\Gamma(u,v)_{|U} = \Lambda^{i_1j_1}\,\Lambda^{i_2j_2}\,\Lambda^{i_3j_3}(\mathcal{L}(X_u)\Gamma)_{i_1i_2i_3}(\mathcal{L}(X_v)\Gamma)_{j_1j_2j_3}$$

admet le même symbole principal que P^3. On a $\partial S^3_\Gamma = 0$, d'après les propriétés des dérivées de Lie. Le même raisonnement que pour le cas plat montre que le 2-cocycle S^3_Γ est toujours non exact. La 2-classe de cohomologie β définie par ce cocycle ne dépend que de la structure symplectique de la variété.

b) Introduisons les notations suivantes : Q^r est un opérateur bidifférentiel d'ordre maximum r en chaque argument, nul sur les constantes et dont le symbole principal coïncide avec celui de P^r; Q^r est supposé symétrique en u,v si r est pair, antisymétrique si r est impair. En particulier, nous prenons $Q^o(u,v) = u.v$, $Q^1(u,v) = P(u,v)$ et $Q^3 \in \beta$; J. Vey a récemment prouvé par une longue et fine étude cohomologique utilisant les résultats de Gelfand-Fuks, le théorème suivant :

Théorème 4 (Vey). Soit (W,F) une variété symplectique à troisième nombre de Betti $b_3(W)$ nul. Il existe des déformations formelles de l'algèbre de Lie de Poisson attachée à la variété de la forme :

$$(6\text{-}3)\qquad [u,v]_\lambda = \sum_{r=0}^{\infty} (\lambda^r/(2r+1)!)\,Q^{2r+1}(u,v)$$

On ne pense pas que la condition technique $b_3(W) = 0$ est nécessaire. Des formes explicites générales pour Q^{2r+1} ne sont pas connues. Cependant pour Q^3, j'ai établi le résultat suivant : il existe une connexion symplectique unique Γ telle que :

$$(6\text{-}4)\qquad Q^3 = S^3_\Gamma + \partial K$$

où K est un opérateur différentiel d'ordre $\leqslant 2$ tel que $K(1) = \text{const.}$

c) Nous dirons qu'on a sur (W,F) un $*_\nu$-produit de Vey s'il existe des Q^r tels que

$$(6\text{-}5)\qquad u *_\nu v = \sum_{r=0}^{\infty} (\nu^r/r!)\,Q^r(u,v)$$

soit associatif . Pour Q^2, j'ai montré qu'il existe une connexion symplectique unique Γ telle que

$$(6\text{-}6)\qquad Q^2 = P^2_\Gamma + \tilde{\partial} H$$

où H est un opérateur différentiel d'ordre $\leqslant 2$; la connexion est la même que celle qui figure dans (6-4) et H et K ont même partie différentielle d'ordre 2. On déduit de (6-5) par antisymétrisation une déformation (6-3) de l'algèbre de Lie de Poisson, avec $\lambda = \nu^2$. On a entre les deux lois de composition une relation semblable à (2-1)

$$[w, u *_\nu v]_{\nu^2} = [w,u]_{\nu^2} *_\nu v + u *_\nu [w,v]_{\nu^2} \qquad (6\text{-}7)$$

Ainsi le nouveau crochet définit une dérivation du $*_\nu$-produit. On peut déterminer toutes les dérivations de (6-3) et (6-5); on en déduit que pour $b_1(W) = 0$, toutes les dérivations du $*_\nu$-produit sont intérieures, c'est-à-dire données par (6-7). Des résultats analogues sont valables pour les automorphismes.

d) Le problème général de l'existence d'un $*_\nu$-produit de Vey sur (W,F) est beaucoup plus difficile que celui résolu par le théorème 4 de Vey et la réponse est inconnue. J'ai obtenu cependant des procédés de construction de tels $*_\nu$-produits pour de larges classes de fibrés cotangents de groupes et d'espaces homogènes.
Je me limiterai à l'exemple le plus simple. Considèrons la variété symplectique plate définie par le fibré cotangent de $R^n-\{0\}$, c'est-à-dire la variété $E = (R^n-\{0\})\times R^n$.
Le groupe résoluble G_2 de dimension 2 opère sur E de la manière suivante :

$$(x,y) \in E = (R^n - \{0\}) \times R^n \to (x' = e^\rho x,\ y' = e^{-\rho}(y + \sigma x)) \qquad (\rho,\sigma \in \mathbb{R})$$

Le groupe G_2 préserve la structure symplectique naturelle de E et la connexion plate. Il préserve par suite les P^r définis par (5-1). L'espace des orbites de E pour ce groupe est isomorphe à $T^* S^{n-1}$, où $S^{n-1} = SO(n)/SO(n-1)$ est la sphère de dimension (n-1). On déduit du $*_\nu$-produit invariant par G_2 défini sur E par les P^r un $*_\nu$-produit de Vey naturel sur $T^* S^{n-1}$, qui est invariant par SO(n). On peut déduire de cette méthode de quotient et de considérations de fibrations l'existence de $*_\nu$-produits naturels par exemple sur les fibrés cotangents des variétés de Stiefel et des grassmanniennes.

7 - Introduction à une théorie spectrale.

a) Considèrons une variété symplectique (W,F) admettant un $*_\nu$-produit. Nous posons $N^c = C^\infty(W;C)$ et désignons par $F(N^c;\nu)$ l'espace des séries formelles en les puissances positives, nulles, ou négatives de ν à coefficients dans N^c. Soit H l'hamiltonien classique de notre problème. Nous sommes conduits à traduire l'équation dynamique de Schrödinger de la manière suivante, où nous adoptons d'abord un point de vue formel

$$\frac{du_t}{dt} = [H,u_t]_{\nu^2} = \frac{1}{2\nu}(H *_\nu u_t - u_t *_\nu H) \qquad (u_t \in F(N^c;\nu) \times I) \qquad (7\text{-}1)$$

pour $\nu = i\hbar/2$; initialement $(u_t)_{t=0} = u_o \in F(N^c;\nu)$. Introduisons les $*$-puissances de H ($H^{(*)p} = H^{(*)p-1} * H$) et la $*$-exponentielle de H. Nous posons :

(7-2) $$\mathrm{Exp}_{\star}(Ht) = \sum_{p=0}^{\infty} (t/2\nu)^p (1/p!)\, H^{(\star)p} \in F(N^c;\nu) \times I$$

Si $u_o \in F(N^c;\nu)$, considèrons u_t définie formellement par :

(7-3) $$u_t = \mathrm{Exp}_{\star}(Ht) \star_\nu u_o \star_\nu \mathrm{Exp}_{\star}(-Ht)$$

(7-3) donne la solution formelle de (7-1) prenant la valeur u_o pour $t = 0$.

b) Considèrons maintenant le point de vue de l'analyse mathématique et donnons à ν la valeur $i\hbar/2$. Supposons que H soit tel que (7-2) converge comme série en t dans un voisinage complexe de l'origine et définisse pour un tel t une distribution $\mathrm{Exp}_{\star}(Ht)$ sur W.

Supposons que $\mathrm{Exp}_{\star}(Ht)$ admette un développement de Fourier-Dirichlet unique

(7-4) $$\mathrm{Exp}_{\star}(Ht) = \sum_{I} \Pi_\lambda\, e^{\lambda t/i\hbar}$$

où I est une suite dans $\mathbb{C}$; I donne le spectre de H et une formule de trace la multiplicité. On voit que

(7-5) $$\Pi_\lambda \star \Pi_{\lambda'} = \delta_{\lambda\lambda'}\, \Pi_\lambda \;,\quad \Sigma\, \Pi_\lambda = 1 \;,\quad H \star \Pi_\lambda = \Pi_\lambda \star H = \lambda \Pi_\lambda \;,\quad H = \Sigma \lambda\, \Pi_\lambda$$

Plus généralement, on peut chercher s'il existe un développement au sens des distributions

(7-6) $$\mathrm{Exp}_{\star}(Ht) = \int e^{t/i\hbar}\, d\mu(\lambda)$$

c'est-à-dire chercher pour $d\mu(\lambda)$ la transformée de Fourier de la distribution (en t) $\mathrm{Exp}_{\star}(Ht)$. Le spectre de H sera défini comme le spectre de cette distribution au sens de Schwartz, c'est-à-dire comme le support de la transformée de Fourier $d\mu(\lambda)$. Dans le cas discret $d\mu(\lambda) = \Sigma \Pi_{\lambda'} \delta(\lambda - \lambda')$. Le développement écrit joue le rôle de la décomposition spectrale d'un opérateur.

Un $\star$-produit est dit <u>positif</u> si pour tout $u \in N$, $\bar{u} \star u \geqslant 0$. Les $\star$-produits mis en évidence sont tous positifs. S'il en est ainsi <u>le spectre de toute fonction à valeurs réelles admettant un développement spectral au sens de (7-6) est réel</u>.

c) L'algorithme précédent, appliqué directement au cas plat, donne par exemple les niveaux d'énergie $E_m = \hbar(m + \frac{n}{2})$ et multiplicités pour <u>l'oscillateur harmonique</u> de dimension n. Pour le cas discret, la multiplicité est donnée par :

$$N_\lambda = \frac{1}{(2\Pi\hbar)^n} \int_W \Pi_\lambda\, \eta \qquad (\eta \text{ volume symplectique})$$

Pour <u>l'atome d'hydrogène</u>, on considèrera $T^{\star}S^3$ comme espace de phase et on introduira le $\star$-produit correspondant invariant par SO(4). On obtient alors le spectre complet, c'est-à-dire le spectre discret négatif ($E_m = -\frac{1}{2}(\hbar m)^{-2}$ avec multiplicité m^2) et le spectre continu positif.

Reférences

[1] A. Avez et A. Lichnerowicz C.R. Acad. Sci. Paris, t.275, A, (1972) 113-118.

[2] M. Gerstenhaber, Ann. of Math. 79, (1964) 59-103.

[3] M. Flato, A. Lichnerowicz, D. Sternheimer , Compos. Mathem. 31, (1975) 48-82; C.R. Acad. Sci. Paris, t.283, A, (1976), 19-24.

[4] J. Vey, Comm. Math. Helv. 50, (1975), 421-454.

[5] J.E. Moyal , Proc. Cambridge Phil. Soc. 45, (1949), 99-124.

[6] A. Lichnerowicz, Journ. Geom. Diff. Liège déc 1976.

[7] F. Bayen, M. Flato, C. Fronsdal, A. Lichnerowicz, D. Sternheimer, Lett. in Math. Phys. 1 (1977), 523-530; Deformation Theory and Quantization, Ann. of Phys. 111 (1978), 61-152.

[8] A. Lichnerowicz, C.R. Acad. Sci. Paris t.286, A, (1978) 49-53.

[9] A. Lichnerowicz, Lett. In Math. Phys. 1 (1977).

Geometric quantisation and the Feynman integral.

D.J. Simms
School of Mathematics
Trinity College, Dublin

The geometry of classical phase space

In this paper we show how the Feynman path integral formulation for a general finite dimensional time-dependent non-relativistic Hamiltonian system can be derived using the geometric quantisation techniques of Kostant and Souriau. A key role is played by the integral transform introduced by Blattner, Kostant and Sternberg in [1] . Indeed the basic idea of the present paper is already essentially to be found in [1], and our purpose is to present it in a form which emphasises the relationship with the classical Hamilton-Jacobi theory on the one hand, and the Feynman path integral on the other hand.

We consider a system with configuration space X, which we suppose to be a manifold of finite dimension n. We further suppose that X has a metalinear structure, see [2], so that square roots of volume elements (half-forms) are defined on X. The half-forms on X of compact support form a complex pre-Hilbert space, and we denote its completion by $L^2(X)$. We call $L^2(X)$ the space of wave-functions of the system.

The space of events is represented by the product $X \times R$ of the configuration space with the time axis. Let β be the canonical 1-form on the cotangent bundle $T^*(X \times R)$, let $\omega = d\beta$ be the symplectic 2-form, and let π be the projection map onto the base $X \times R$. The classical dynamics of the system is determined by a submanifold M of codimension 1 in $T^*(X \times R)$. Synge in [7] calls M the <u>energy surface</u> and Souriau in [6] calls it the espace d'evolution. In terms of the usual canonical coordinates q^i, p_i, t, -H on the cotangent bundle, β is $\Sigma p_i dq^i - Hdt$, ω is $\Sigma dp_i \wedge dq^i - dH \wedge dt$, and the energy surface is given by $H = \mathcal{H}(q,p,t)$ for a system with time dependent Hamiltonian function $\mathcal{H}$.

The restriction of ω to M has a 1-dimensional space of singular directions at each point of M. This gives a 1-dimensional foliation of M whose leaves are the classical paths in M. The projection of these leaves under π are the classical paths in $X \times R$. The symplectic form ω, restricted to M, is invariant along the leaves of the foliation, and the n-fold exterior product $\Omega = \omega\wedge\ldots\wedge\omega$ is a 2n-form on M which provides a volume element transverse to the classical paths in M.

For each $(x,t) \in X \times R$ we denote by F_x^t the union of all classical paths in M whose projections in $X \times R$ pass through (x,t). We suppose that for each fixed t, $F^t = \{F_x^t \mid x \in X\}$ is a foliation of M with (n+1)-dimensional leaves. The restriction of ω to each leaf of F^t vanishes. Thus F^t is a Lagrangian foliation (real polarisation) of M. The leaves of F^t are the level sets of a map ρ^t from M to X, with $\rho^t(F^t_x) = \{x\}$, and the derivative of ρ^t induces an isomorphism of the normal bundle of F^t in T*M with the pull-back under ρ^t of the cotangent bundle of X. Choose a square root $\Lambda^{\frac{1}{2}}X$ for the line-bundle of n-forms on X. Then the pull-back of $\Lambda^{\frac{1}{2}}X$ under ρ^t is a square root for the nth exterior power of the normal bundle of F^t. We denote this square root by N^t.

We denote by W^t the space of all smooth sections of N^t over M which are annihilated by the operators

$L_\zeta - \frac{i}{\hbar} < \beta, \zeta >$ for all vector fields ζ on M tangential to the leaves of F^t. Here L_ζ denotes the Lie derivative along ζ and $2\pi\hbar$ is Planck's constant. Thus W^t is the space of all sections of N^t which are covariant constant along the leaves of F^t with respect to a connection whose curvature form is $\frac{-i}{\hbar}\omega$. This is the basic construction of geometric quantisation [2], [3], [5] and [6]. We have a natural bijection

$$T^t: \Gamma(\Lambda^{\frac{1}{2}}X) \to W^t$$

where $\Gamma(\Lambda^{\frac{1}{2}}X)$ is the space of sections of $\Lambda^{\frac{1}{2}}X$, such that $(T^t\psi)(m)$ corresponds to $\psi(x)$ under the natural isomorphism of the fibre of N^t at m with the fibre of $\Lambda^{\frac{1}{2}}X$ at x, where $\pi(m) = (x,t)$.

To write down a coordinate expression for T^t, we denote by $\sigma^t: M \to M$ the map such that $\sigma^t(m)$ is the point at time t on the classical path through m. We denote by S^t the function on M whose value at m is the integral of β over the classical path in M from $\sigma^t(m)$ to m. If $\psi \in \Gamma(\Lambda^{\frac{1}{2}}X)$ has compact support in the domain of coordinates $y = (y^1,\dots,y^n)$ and $\psi = f(y)dy^{\frac{1}{2}}$ where $dy^{\frac{1}{2}}$ is the choice of a square root of $dy^1 \wedge \dots \wedge dy^n$, then

$$T^t\psi = \exp[\tfrac{i}{\hbar} S^t] \, f(Q^t)(dQ^t)^{\frac{1}{2}}$$

where Q^t denotes $y \circ \rho^t$.

Let t_1 and t_2 be two times such that the leaves of F^{t_1} and F^{t_2} intersect in the classical paths. We then have, formally, the sesquilinear pairing

$$(\cdot|\cdot): W^{t_1} \times W^{t_2} \to C^\infty(M)$$

due to Blattner, Kostant and Sternberg. For $\nu_i \in W^{t_i}$, $(\nu_1|\nu_2)$ is characterised up to sign by

$$(\nu_1 \otimes \nu_1) \wedge (\nu_2 \otimes \nu_2) = (\nu_1/\nu_2)^2\Omega.$$

We define the operator

$$T^{t_2,t_1} : \Gamma(\Lambda^{\frac{1}{2}}X) \to \Gamma(\Lambda^{\frac{1}{2}}X)$$

formally by

$$(T^{t_2,t_1}\psi|\phi) = (2\pi i\hbar)^{-n/2} \int (T^{t_1}\psi|T^{t_2}\phi)\Omega$$

where on the left we have the L_2 inner product of half-forms on X. See [4] for a discussion of the pairing.

The Feynman integral

The operator $\psi \to T^{t_2,t_1}\psi$ gives what we shall call a quasi-classical evolution of a wave function ψ from time t_1 to time t_2. The Schrödinger equation for the quantum-mechanical time evolution $t \to \psi_t$ in $L^2(X)$ is given formally by

$$\frac{\partial}{\partial t}\psi_t = \lim_{r\to o} \frac{1}{r}\left[T^{t+r,t}\psi_t - \psi_t\right].$$

Thus the quantum mechanical time evolution $t \to \psi_t$ is formally tangent at $t = t_1$ to the quasi-classical evolution $t \to T^{t,t_1}\psi_{t_1}$. It follows that the quantum evolution can, conceptually, be written as a limit of iterated transforms:

$$\psi_t = \lim_{N\to\infty} T^{t,t_{N-1}} \ldots\ldots T^{t_1,t_o}\psi_{t_o}$$

where $t_r = N^{-1}\left[rt + (N-r)t_o\right]$. To see that this is formally equivalent to the Feynman approach, we derive an explicit formula for the operator T^{t_2,t_1}.

If $\psi,\phi \in \Gamma(\Lambda^{\frac{1}{2}}X)$ and have compact support in the domain of coordinates $y = (y^1,\ldots,y^n)$ and $z = (z^1,\ldots,z^n)$ respectively:

$$\psi = f(y)dy^{\frac{1}{2}}\ , \quad \phi = h(z)dz^{\frac{1}{2}}$$

then

$$T^{t_1}\psi = \exp\left[\frac{i}{\hbar} S^{t_1}\right] f(U)(dU)^{\frac{1}{2}}$$

$$T^{t_2}\phi = \exp\left[\frac{i}{\hbar} S^{t_2}\right] h(Q)(dQ)^{\frac{1}{2}}$$

where $U = y \circ \rho^{t_1}$ and $Q = z \circ \rho^{t_2}$. Therefore

$$(T^{t_2,t_1}\psi|\phi) = (2\pi i h)^{-n/2}\int \exp\left[\frac{i}{\hbar}(S^{t_1} - S^{t_1})\right] f(U)\overline{h(Q)}(dU^{\frac{1}{2}}|dQ^{\frac{1}{2}})\Omega.$$

Let $S = S^{t_1} - S^{t_2}$, then $S = \mathcal{S}(U,Q)$ where $\mathcal{S}$ is a function of 2n variables. Also $S = S^{t_1} \circ \sigma$ where $\sigma = \sigma^{t_2}$. The canonical 1-form β is of the form $\Sigma p_j dq^j$ - Hdt where $q = z \circ \pi$, and therefore $\sigma^*\beta$ is of the form $\Sigma P_j dQ^j$. Since ω is invariant along the classical paths in M, we have

$$\omega = \sigma^*\omega = \sigma^*d\beta = d\sigma^*\beta = \Sigma dP_j \wedge dQ^j.$$

Also, on any leaf of F^{t_1}, we have

$$d\mathcal{S}(U,Q) = dS = d(S^{t_1} \circ \sigma) = \sigma^*dS^{t_1} = \sigma^*\beta = \Sigma P_j dQ^j.$$

Hence

$$\mathcal{S}_{,n+j}(U,Q) = P_j$$

and thus

$$\Omega = \omega\wedge\ldots\wedge\omega = dP \wedge dQ = \det\left[\mathcal{S}_{,n+j,k}(U,Q)\right]dU \wedge dQ.$$

Therefore

$$(dU^{\frac{1}{2}}|dQ^{\frac{1}{2}})\Omega = \{\det[S,_{n+j,k}]\}^{\frac{1}{2}} dU \wedge dQ.$$

and hence $(T^{t2,t1}\ \psi|\phi)$ is equal to

$$(2\pi i\hbar)^{-n/2}\int \exp[\tfrac{i}{\hbar}S(U,Q)]\, f(U)\overline{h(Q)}\ \{\det[S,_{n+j,k}]\}^{\frac{1}{2}} dU \wedge dQ.$$

Thus on the z coordinate domain we have

$$T^{t_2,t_1}\psi = g(z)dz^{\frac{1}{2}}$$

where

$$g(z) = (2\pi i\hbar)^{-n/2}\int \exp[\tfrac{i}{\hbar}\, S(y,z)]\, f(y)\{\det \frac{\partial^2}{\partial y^j \partial z^k}\, S(y,z)\}^{\frac{1}{2}} dy$$

Thus geometric quantisation, when applied in the way described in this paper, leads to a Feynman integral and also determines a normalising factor for the integrand.

References

1. R.J. Blattner. Quantization and representation theory, Proc. Sympos. Pure Math., vol 26, Amer. Math. Soc., Providence R.I. 1973, pp 147-165

2. V. Guillemin and S. Sternberg. Geometric asymptotics. Math surveys No 14, Amer. Math. Soc., Providence R.I. 1977.

3. B. Kostant. Quantization and unitary representations. 1. Prequantisation, Lecture Notes in Mathematics 170, Springer, Berlin 1970.

4. J. Rawnsley. On the pairing of polarizations, Comm. Math. Physics 58 (1978), 1-8.

5. D.J. Simms and N. Woodhouse. Lectures on geometric quantization. Lecture Notes in Physics no. 53, Springer, Berlin 1976.

6. J.M. Souriau. Structure des systemes dynamiques. Dunod, Paris 1970.

7. J.L. Synge. Classical dynamics. Handbuch der Physik, Vol III /1, Principles of classical mechanics and field theory, (ed.) S. Flugge, Springer, Berlin 1960.

- ALGEBRES TIERCES -

Jean-Marie SOURIAU
Université de Provence et Centre de Physique Théorique, CNRS Marseille

Nous appelons ainsi tout espace vectoriel muni d'une forme trilinéaire interne, (, ,), vérifiant les axiomes suivants :

$$\begin{cases} (x,x,x) = 0 \\ (x,y,z) - (y,x,z) = 0 \\ (x,x,(y,y,z)) - (y,y,(x,x,z))) = 2((x,x,y),y,z)) \end{cases}$$

Cette structure a certaines analogies avec la structure d'algèbre de Lie - notamment l'existence de dérivations non nulles. Elle est d'ailleurs étroitement liée à la notion d'algèbre de Lie graduée : la classification de Kac des algèbres de Lie graduées simples donne les algèbres tierces simples.

De même l'extension par Corwin, Neeman et Sternberg du théorème de Poincaré-Birkhoff-Witt aux algèbres de Lie graduées permet de donner une solution universelle au problème des <u>représentations d'algèbres tierces</u>, par injection dans une algèbre associative "enveloppante".

Nous donnons divers exemples d'algèbres tierces, notamment celles que l'on peut associer à une variété symplectique ; les problèmes de quantification formelle - qui s'interprètent en terme de représentations "cordiales" d'algèbres tierces - sont abordés.

- SECTION V -

A REASONABLE METHOD FOR COMPUTING PATH INTEGRALS ON CURVED SPACES

Cécile DeWitt-Morette
Department of Astronomy and Center for Relativity
The University of Texas
Austin, Texas 78712

A REASONABLE METHOD FOR COMPUTING PATH INTEGRALS ON CURVED SPACES

Beside their intrinsic interest, Riemannian spaces are good proving grounds for testing formalisms. Delicate ambiguities may appear on curved spaces which are wiped out on flat spaces. For instance, in 1952, Pauli observed that the WKB approximation of the propagator does not always satisfy the Schrödinger equation. On curved spaces, B. S. DeWitt showed that it "misses" it by a factor proportional to the curvature scalar R.

Pauli's remark has triggered a number of investigations: "Should one add to the Schrödinger equation a term proportional to $\hbar^2 R$, and if so how should one proceed to determine it?" To simplify the discussion we shall consider a specific case, namely a particle of mass m in a potential V. Let M be the configuration space of the system and assume that it is a Riemannian manifold. The action S of the system is then

$$S(f) = \int_T \left(\frac{m}{2}\,||\dot{f}(t)||^2 - V(f(t))\right) dt, \qquad f: T \to M.$$

Let $K(b,t_b;a,t_a)$ be the probability amplitude that the system, known to be at $a \in M$ at time t_a, be found at b at t_b. As a function of b and t_b, K is an elementary solution of the Schrödinger equation. Thus if we have an independent way of computing K on curved spaces, we can determine what the Schrödinger equation on curved spaces should be. Let

$$K = K_{WKB}\left(1 + \sum_{p=1}^{\infty} \hbar^p A_p\right)$$

be the semiclassical expansion of the propagator and let $K_n = K_{WKB}(1 + \sum_{p=1}^{n} \hbar^p A_p)$ be its nth-order approximation. The terms of order $\hbar^2 R$ in the Schrödinger equation are contributed by K_1. Feynman has shown that $K(b,t_b;a,t_a)$ can be obtained by "summing" $\exp(iS(f)/\hbar)$ over all paths $f:[t_a,t_b] \to M$ such that $f(t_a) = a$, $f(t_b) = b$. It remains only to define path integration on curved spaces... Various attempts have been made but all have encountered ambiguities, and it seems desirable to approach the problem from a different angle. Why tamper with the Schrödinger equation without being compelled to do so either for mathematical or physical reasons?

1. Assume the Schrödinger equation to be

$$i\hbar\,\partial\psi/\partial t = (-\hbar^2\Delta/2m + V)\psi \tag{1}$$

where Δ is the laplacian on curved spaces. If ψ is a function,

$$\Delta\psi = -\delta d\psi = g^{\alpha\beta}\nabla_\alpha\nabla_\beta\psi = |g|^{-1/2}\,\partial_\alpha(g^{\alpha\beta}\,|g|^{1/2}\,\partial_\beta\psi).$$

2. Find a path integral representation of the solution of (1) with Cauchy data $\psi(t_a,x) = \phi(x)$.
3. Develop methods for computing such path integrals.

The solution of this problem is based on the work of Elworthy [2] on the path integral representation of the diffusion equation on curved spaces, the work of Truman [3] on the quasiclassical representation, and the theory of prodistributions [4].

The theory of prodistributions, which I shall develop later, make it possible to extend to the Schrödinger equation the path integral representation of solutions of the diffusion equation constructed by Elworthy. It gives

$$\psi(t_b,b) = \int_{Y_+} dw_+^W(y)\ \exp(-\frac{i}{\mu^2 m}\int_{t_a}^{t_b} V(\mathrm{Dev}_b(\mu y,t))dt)\phi(\mathrm{Dev}_b(\mu y,t_a)) \tag{2}$$

This is the end product of a careful chain of arguments [5] which lack of time prevents me from describing. The explanation of all the terms which make up (2) will however indicate how it comes to be.

1. If the configuration space M is R^n, equation (2) reduces to

$$\psi(t_b,b) = \int_{Y_+} dw_+^W(y)\ \exp(-\frac{i}{\mu^2 m}\int_{t_a}^{t_b} V(b+\mu y(t))dt)\phi(b+\mu y(t_a)) \tag{3}$$

This is a Feynman-Kac type formula. It has been obtained very economically by Maheshwari from (1) by the theory of product integrals [6] combined with the theory of prodistributions [5].

Y_+ is the space of continuous paths $y: T \to R^n$ such that $y(t_b) = 0$. This is a minor change from the Kac formula where $y(t_a) = 0$, but a great practical convenience. w_+^W is the gaussian Wiener prodistribution on Y_+ defined by its Fourier transform on the dual Y_+' of Y_+.

$$y \in Y_+\ ,\ \mu \in Y_+' \iff \langle\mu,y\rangle = \int_T d\mu_\alpha(t)y^\alpha(t) < \infty$$

$$\left.\begin{aligned} \mathcal{F}w(\mu) &= \exp(-\frac{i}{2}W(\mu,\mu)) \\ W(\mu,\nu) &= \int_T d\mu_\alpha(t)\int_T d\nu_\beta(s)\ G^{\alpha\beta}(t,s) \end{aligned}\right\} \quad \text{gaussian prodistribution}$$

$$G^{\alpha\beta}(t,s) = g^{\alpha\beta}\ \inf(t_b - t,\ t_b - s) \qquad \text{Wiener prodistribution}$$

Although (2) and (3) uses the symbol dw(y) for convenience in comparing with familiar expressions, w is defined only by its Fourier transform $\mathfrak{F}w$ and integrals are computed only in terms of $\mathfrak{F}w$. More later on this point.

$\mu = \sqrt{h/m}$. Note that μ rather than $\hbar$ is the term which scales the paths y. This follows unambiguously from the use of product integrals and prodistributions which determine all factors uniquely.

The semiclassical expansion of ψ can be obtained by expanding V and ϕ in powers of μ using the method developed by Truman for the quasiclassical representation; the integrals thus obtained can be explicitly computed by the theory of prodistributions, i.e. reduced to simple integrals over finite dimensional spaces.

2. The configuration space M is an arbitrary Riemannian manifold. The development mapping introduced in the early days of Riemannian geometry has been used by Elworthy for computing path integrals on curved spaces. It is defined as follows: Let T_bM be the tangent space to the configuration space M at b. *The development mapping* is a bijection between the space of $L^{2,1}$ paths [7] on T_bM which vanish at t_b and the space of $L^{2,1}$ paths on M which are at b at t_b. It is defined as follows: Consider [8] a path z on T_bM. A path on M is said to be the development of z, if its derivative at t parallel transported to b along Dev z is equal to $\dot{z}(t)$, when it exists. When $\dot{z}(t^+) \neq \dot{z}(t^-)$, then $(\text{Dev } z)'(t^+)$ and $(\text{Dev } z)'(t^-)$ parallel transported to b are equal respectively to $\dot{z}(t^+)$ and $\dot{z}(t^-)$. The development mapping conserves angles. It follows immediately from the definition that

i. The development mapping maps a straight line on T_bM into a geodesic on M such that $(\text{Dev } z)'(t_b) = \dot{z}(t_b)$.

ii. Closed loops on T_bM are not developed into closed loops on M: a mapping from the space of paths on T_bM into the space of paths on M cannot both map closed loops into closed loops and conserve angles. A closed loop is a particular case of paths with both ends fixed. Hence a family of paths with both ends fixed is not developed into a family with both ends fixed.

The space of $L^{2,1}$ paths on T_bM vanishing at the origin is dense in the space Y_+ of continuous paths on T_bM vanishing at the origin; the space of $L^{2,1}$ paths on M going through b at t_b is dense in the space $_b(M)$ of continuous paths on M going through b at t_b. The development mapping determines a measurable map between the space of continuous paths on T_bM which vanish at t_b and the space of continuous paths on M which are at b at t_b.

Since paths are the variables of integration, we shall consider a path X on M as a mapping from $Y_+ \times T$ into M

$$\text{Dev}: (z,t) \mapsto \text{Dev}(z,t) = (\text{Dev } z)(t) .$$

All the terms in equation (2) have now been explained: the potential and the initial wave function are defined on a Riemannian manifold M but the domain of integration is the space of paths on the tangent space at b to M. Note that all the terms in equation (2) are coordinate free.

It only remains to compute (2). A difficult task but made possible by the theory of prodistributions which I shall present at long last.

A good introduction to prodistributions is the marvelous chapter of Bourbaki on "integration on topological vector spaces." Indeed this is where they were conceived. Recall the definition of a promeasure or cylindrical measure. Let X be a topological vector space, Hausdorff and locally convex. One can associate with X a family of finite dimensional spaces called a projective system. A promeasure on X is a family of bounded measures defined on its projective system. Unfortunately if one tries this scheme to define Feynman path integrals one needs to introduce families of unbounded measures on the projective system and all hell breaks loose. That is to say if we want to work within the set theory definition of measure. If on the other hand we work with measures defined by their Fourier transforms we need not be restricted to families of bounded measures. The Fourier transform of a measure, bounded or not, is a function, i.e. a distribution equivalent to a function, i.e. a mapping defined pointwise and not setwise. Thus one can define an object, which Dieudonné suggested that it be called "prodistribution" as follows. A *prodistribution* on X is a family of measures defined by their Fourier transforms on the dual spaces of the projective system; alternatively, a family of distributions defined on the projective system, hence the word prodistributions.

All right, but how does one construct a theory of integration with an object defined by its Fourier transform? Much can be done and a whole versatile path integration technology can be developed. I shall give only an example--a simple one but as useful as integration by parts in ordinary integration theory. Let $F: Y \to R$ be of the form $F = P \cdot f$ where the mapping $P: Y \to X$ is linear and continuous, then

$$\int_Y F(y)dw(y) = \int_X f(x)dw_P(x)$$

where $w_P = w \cdot \tilde{P}$.

Let us see for instance how this technique can be used to compute equation (2). To be specific we shall compute the propagator, i.e. the wave function when the initial wave function is

$$\phi(\mathrm{Dev}_b(\mu y,t_a)) = \delta(\mathrm{Dev}_b(\mu y,t_a)-a)$$

Steps to compute (2):

1. Linear mapping $b + \mu y \to q + \mu x$ where q is the path such that Dev q is the classical path Z from (a,t_a) to (b,t_b).
2. Expand in powers of μ. Set $\mathrm{Dev}(q + \mu x,t) = Y(t,x,\mu)$. Thus, for instance

$$V(Y(t,x,\mu)) = V(Z(t)) + \mu\nabla_\alpha V(Z(t))\delta Y^\alpha(t,x) + \frac{1}{2}\mu^2\nabla_\alpha\nabla_\beta V(Z(t))\delta Y^\alpha(t,x)\delta Y^\beta(t,x) + \frac{1}{2}\mu^2\nabla_\alpha V(Z(t))\delta^2 Y^\alpha(t,x) + \dots \quad (3)$$

where

$$\delta Y(t,x) = \partial_\mu Y(t,x,\mu)\Big|_{\mu=0}$$

and

$$\delta^2 Y(t,x) = \nabla_\mu \partial_\mu Y(t,x,\mu)\Big|_{\mu=0} \quad .$$

$\delta Y(\cdot,x)$ is a vector field [9] along Z generated by varying μ in $\mathrm{Dev}\,(q + \mu x)$.

$$\delta Y(\cdot,x) = \partial_\mu \mathrm{Dev}(q + \mu x)\Big|_{\mu=0} = \mathrm{Dev}'(q)x$$

3. Linear mapping $x \to \delta Y(\cdot,x)$. The integral becomes an integral on the space of vector fields along Z. Some beautiful theorems relating $\delta Y(\cdot,x)$, the parallel transports of $x(t)$ along Z and $\delta^2 Y$ due to Elworthy makes it possible to complete the calculation and the WKB approximation of the propagator on curved spaces reads

$$K_{WKB}(B;A) = \exp(i\bar{S}(B;A)/\hbar)\ (2\pi i\hbar)^{-n/2}\ |\det \partial^2\bar{S}(B;A)/\partial b^\alpha \partial a^\beta|^{1/2}\ (\mathrm{Det\ Dev}'(q))^{-1}$$

where $B = (b,t_b)$ and $A = (a,t_a)$. $\bar{S}(B;A) = S(Z)$, $\mathrm{Det\ Dev}'(q) = 1$ and the familiar result is obtained.

What have we learned? A method for computing path integrals on curved spaces. What else have we learned?

1. Expand in powers of μ , not in powers of $\hbar$.
2. Do not expand the action but the potential. The action includes a stochastic term (the kinetic energy) and a driving force (the potential energy). The kinetic energy $g_{\alpha\beta}(Q(t))\dot{Q}^\alpha(t)\dot{Q}^\beta(t)$ must be treated as a whole, as a stochastic function on curved spaces. The so-called factor ordering ambiguities and related time slicing ambiguities were only symptoms that we were using our formalisms beyond their range of definition.
3. Attempts to map families of paths with both ends fixed from tangent space to curved spaces are doomed to failure. So are formal generalizations of flat space expressions. None pick up the full complexity of $\delta^2 Y$.
4. The theory of prodistributions work well on curved spaces because it is a coordinate free formulation and because it is not constructed from short time propagators.

References

[1] In diagrammar dialect, A_1 is called the two-loop contribution.

[2] K. D. Elworthy, "Stochastic dynamical systems and their flows", to appear in Proceedings of the Conference on Stochastic Analysis, Northwestern University 1978. It is hoped that the long awaited Eells and Elworthy monograph will appear soon so that their friends do not feel guilty for using the results they freely share before publication. On the other hand they should be congratulated for not publishing their work until they have made sure that their tools have no hidden defects and until they have turned it in their minds till "it has made all smooth."

[3] A. Truman, J. Math. Phys. 17 1852 (1976) and 18 1499 (1977).

[4] C. DeWitt-Morette, A. Maheshwari and B. Nelson. Path Integration in Non-Relativistic Quantum Mechanics. Physics Reports 1979.

[5] For a detailed discussion see [DeWitt-Morette, Maheshwari, Nelson].

[6] V. Volterra and B. Hostinsky. Operations Infinitesimals Lineaires. Gauthier-Villars 1938; and J. Dollard and C. N. Friedman. Product Integration. Addison-Wesley 1979.

[7] See references quoted in [Eells and Elworthy] and [Elworthy].

[8] As usual one identifies T_bM and R^n and thinks of z either as $z: T \to R^n$ such that $z(t_b) = 0$, or $z: T \to T_bM$ such that $z(t_b) = b$. The metric on T_bM and R^n is $g(b)$.

[9] This notation gives the erroneous feeling that $\delta Y(t,x)$ is a small increment; it is used nevertheless for its obvious convenience.

CORRESPONDENCE RULES AND PATH INTEGRALS

Maurice M. Mizrahi
Center for Naval Analyses of
the University of Rochester
1401 Wilson Boulevard
Arlington, Va. 22209, USA

Abstract. A path-integral representation is constructed for propagators corresponding to quantum Hamiltonian operators obtained from classical Hamiltonians by an arbitrary rule of correspondence. Each rule yields a unique way of defining the path integral in the context of a formalism which does not require a limiting process. This formalism is more reliable than the usual time-slicing (lattice) definition in that all the expressions it entails are well-defined for computational purposes and it allows the explicit evaluation of large classes of path integrals. Direct substitution in the Schrödinger equation shows that there are no restrictions (such as Hermiticity or time-independence) on the Hamiltonian operator. Examples are given.

I. INTRODUCTION

The purpose of this paper is to propose a solution to the following problem: Given an arbitrary classical Hamiltonian $H_c(p,q,t)$ and an arbitrary rule of correspondence which enables one to derive a quantum Hamiltonian operator $\underset{\sim}{H}(\underset{\sim}{P},\underset{\sim}{Q},t)$ (Hermitian or not) from H_c, find a path-integral representation for the propagator $K(q_b,t_b;q_a,t_a)$ corresponding to $\underset{\sim}{H}$ which (1) takes proper account of the correspondence rule and (2) does not involve a limiting process (i.e. a "skeletonization of the path" or "time-slicing" technique or "polygonal-path" technique) in its definition. The latter requirement purports to avoid the many ambiguities inherent in this process and to enable one to actually compute path integrals, rather than simply exhibit formal expressions.

In 1975 we showed [1], by time-slicing and Weyl transform techniques, that a formal path-integral expression in phase space can be written for the propagator, where the Weyl transform of the Hamiltonian operator takes the place of the classical Hamiltonian in the action functional. Subsequent work by Cohen [2] and Dowker [3] showed that formal path integrals can be obtained where an arbitrary transform of the Hamiltonian operator replaces the classical Hamiltonian in the action functional. In all the foregoing papers, however, work stopped when a "formal" path integral was obtained, leaving open the problems of evaluation of the path integrals, substitution of the path-integral expression in the Schrödinger equation for verification, and justification of some possibly ambiguous limits inherent in the time-slicing approach. This paper will address the above problems by proposing an alternative approach, and supersedes reference 1.

The starting points are the general framework for path integration in phase space without limiting procedure introduced in reference 4 and Cohen's mappings between correspondence rules and ordinary functions [5]. It will be shown that an infinite series representing the propagator (where each term is a path integral) satisfies the Schrödinger equation with arbitrary $\underset{\sim}{H}$ and the boundary condition, provided a consistent well-defined algorithm is used to properly take the correspondence rule into account.

It must be stressed that since the resulting propagator, in its path-integral form, is found to satisfy the Schrödinger equation and the boundary condition in the most general possible case, no apologies need be made for any lack of mathematical rigor which may be encountered in the derivation of this propagator.

For simplicity of exposition, only path integrals representations with respect to the free-particle measure will be considered here. The more general measures introduced in reference 4, which allow a semiclassical expansion of the propagator, will be treated elsewhere.[1] We work in one dimension to simplify the discussion, although the results can be readily generalized to n dimensions. All integrals are over R^s for suitable s, unless otherwise specified.

II. THE PATH-INTEGRAL FORMALISM

There are many difficulties and ambiguities associated with the time-slicing approach to path integrals. Some of these are:

(i) Key existence and uniqueness theorems on the limit (when the time interval Δt between successive points on the lattice goes to 0) are lacking,

(ii) When terms of order higher than 1 in Δt should be retained (Edwards and Gulyaev [6] showed that in some cases they must),

(iii) Whether position- (q-) dependent terms should be evaluated at the subdivision point q_j, the midpoint $(q_j + q_{j+1})/2$, or other points, a question closely related to factor ordering,

(iv) When the p_j integrals should be carried out before the q_j integrals in phase-space path-integrals (different results are obtained if the order is changed),

(v) Evaluation of the integrals is cumbersome and lengthy.

This is not to say that these problems cannot be tackled within the time-slicing approach, but their handling is usually tricky, ad-hoc, cookbook-style and frequently hinges on knowing the answer in advance. What is needed is a formalism where the basic

---

[1] However, if one is interested only in perturbation series in α of the propagator corresponding to $\underset{\sim}{H} = g(t)\underset{\sim}{P}^2/2m + f(t)\underset{\sim}{Q}^2/2 + k(t)(\underset{\sim}{P}\underset{\sim}{Q} + \underset{\sim}{Q}\underset{\sim}{P})/2 + \alpha \underset{\sim}{H}_1(\underset{\sim}{P},\underset{\sim}{Q},t)$, then the method described applies directly.

objects are clearly defined, and from which results can be obtained in a straightforward, natural manner.

The formalism for constructing phase-space path integrals without resorting to a limiting procedure with ambiguous epsilonics was introduced in reference 4 and only a brief description will be given here. A similar formalism obtained by different methods was described in reference 7. The general approach consists of defining what plays the role of a measure in phase space by its Fourier transform, which is a simple closed-form expression. For example, the normalized free-particle measure w(p,q) in phase space $\mathcal{P}$, corresponding to

$$dw(p,q) \sim \frac{1}{K_o}\left[\frac{dpdq}{2\pi\hbar}\right]\exp\left\{\frac{i}{\hbar}\int_{t_a}^{t_b}\left[p(t)\dot{q}(t) - \frac{p^2(t)}{2m}\right]dt\right\}, \tag{1}$$

can be shown [4] to have, as its Fourier transform:

$$\begin{aligned} Fw(\mu,\nu) = \exp\Big\{ & -i\langle\mu,\bar{q}\rangle - i\langle\nu,\bar{p}\rangle - \frac{i\hbar}{2}\int_T\int_T G_{ab}(t,t')d\mu(t)d\mu(t') \\ & - i\hbar\int_T\int_T \bar{G}(t,t')d\mu(t)d\nu(t') - \frac{i\hbar}{2}\int_T\int_T G_p(t,t')d\nu(t)d\nu(t')\Big\}, \end{aligned} \tag{2}$$

where:

(1) $\mathcal{P} \equiv \Big\{[p(t),\ q(t)]$ on $T \equiv [t_a,t_b] \,|\, q(t_a) = q_a,\ q(t_b) = q_b,\ p(t)$ unrestricted$\Big\}$, (3)

(2) $$K_o = \left(\frac{m}{2\pi i\hbar T}\right)^{1/2}\exp\left[\frac{im(q_b-q_a)^2}{2\hbar T}\right] \tag{4}$$

is the free-particle propagator, with $T \equiv t_b - t_a$,

(3) $[\bar{q}(t),\bar{p}(t)]$ is the average free-particle path:

$$\bar{q}(t) = [q_b(t-t_a) + q_a(t_b-t)]/T \tag{5}$$

$$\bar{p}(t) = m(q_b-q_a)/T, \tag{6}$$

(4) The G functions are the free-particle covariances:

$$G_{ab}(t,t') = [(t'-t_a)(t_b-t)Y(t-t') + (t-t_a)(t_b-t')Y(t'-t)]/mT, \tag{7}$$

$$\bar{G}(t,t') = [(t_b-t)Y(t-t') - (t-t_a)Y(t'-t)]/T, \tag{8}$$

$$G_p(t,t') = -m/T \tag{9}$$

[$Y(x)$ being 1 when $x > 0$ and 0 otherwise],

This is done to split off the free-particle part so we can use the free-particle measure $w(p,q)$. Although the propagator will be expressed as a power-series in k, our aim is not a perturbation expansion. Rather, it is the manner in which the ordering of the factors in $\underset{\sim}{H}_1$ is taken into account in evaluating the path integrals, so that the propagator K thereby obtained satisfies the Schrödinger equation associated with $\underset{\sim}{H}$ to all orders in k. Thus, k should be regarded as simply a "bookkeeping" parameter. Second, we define a function $H_{no}(p,q,t)$, obtained by normal-ordering the operator $\underset{\sim}{H}_1$ and then replacing $\underset{\sim}{P}$ by p and $\underset{\sim}{Q}$ by q. For example (since $\underset{\sim}{Q}\underset{\sim}{P}-\underset{\sim}{P}\underset{\sim}{Q} = i\hbar$):

$$H_c = p^2/2m + kpq^2 \tag{35}$$

$$\underset{\sim}{H} = \underset{\sim}{P}^2/2m + k\underset{\sim}{Q}\underset{\sim}{P}\underset{\sim}{Q} = \underset{\sim}{P}^2/2m + k(\underset{\sim}{Q}^2\underset{\sim}{P} - i\hbar\underset{\sim}{Q}) \tag{36}$$

$$H_{no} = q^2 p - i\hbar q. \tag{37}$$

Our main result is expressed in the following theorem.

Theorem

The propagator $K(q_b,t_b;q_a,t_a)$[1], or probability amplitude that a particle at position q_a at time t_a will be at position q_b at time t_b, for an arbitrary Hamiltonian operator $\underset{\sim}{H}$ [which we write as $\underset{\sim}{P}^2/2m + k\underset{\sim}{H}_1(\underset{\sim}{P},\underset{\sim}{Q},t)$, $\underset{\sim}{H}_1$ being arbitrary], can be written as a phase space path integral as follows:

$$K \doteq K_o \int_P dw(p,q)\, \exp\left\{\frac{-ik}{\hbar}\int_T H_{no}\,[p(t),q(t),t]\,dt\right\} \tag{38}$$

$$\doteq K_o \left\{1 + \sum_{j=1}^{\infty}\left(\frac{-ik}{\hbar}\right)^j \frac{1}{j!}\int_P dw(p,q)\left[\int_T H_{no}(p(t),q(t),t)dt\right]^j\right\} \tag{39}$$

$$\doteq K_o \left\{1 + \sum_{j=1}^{\infty}\left(\frac{-ik}{\hbar}\right)^j \frac{1}{j!}\int_P dw(p,q)\int_T dt_1\, H_{no}[p(t_1),q(t_1),t_1] \ldots \right.$$

$$\left. \ldots\int_T dt_j\, H_{no}\,[p(t_j),q(t_j),t_j]\right\} \tag{40}$$

---

[1]Note that the propagator $K(q_b,t_b;q_a,t_a)$ is generally not equal to the matrix element $\langle q_b,t_b|q_a,t_a\rangle$. For example, for non-Hermitian time-independent Hamiltonians, K is given by $\langle q_b|\exp[-i(t_b-t_a)\underset{\sim}{H}/\hbar]|q_a\rangle$, but $\langle q_b,t_b|q_a,t_a\rangle$ is equal to $\langle q_b|[\exp(-it_b\underset{\sim}{H}/\hbar)]\,[\exp(it_a\underset{\sim}{H}^\dagger/\hbar)]|q_a\rangle$, which is different from K.

$$\equiv K_o \left\{ 1 + \sum_{j=1}^{\infty} \left(\frac{-ik}{\hbar}\right)^j \frac{1}{j!} \int_{T^j} dt_1 \ldots dt_j \right.$$

$$\times \lim_{(t_1 - t_1') \to 0^+} \cdots \lim_{(t_j - t_j') \to 0^+} \int_P dw\,(p,q)$$

$$\left. \times H_{no}[p(t_1'),q(t_1),t_1] \ldots H_{no}\,[p(t_j'),q(t_j),t_j] \right\}, \tag{41}$$

where

(1) K_o is the free-particle propagator (4), $w(p,q)$ is the free-particle measure defined by (1) and (2)ff, and $H_{no}(p,q,t)$ is a function obtained by normal-ordering $\underset{\sim}{H}_1$ (i.e. $\underset{\sim}{Q}$ before $\underset{\sim}{P}$), then replacing $\underset{\sim}{Q}$ by q and $\underset{\sim}{P}$ by p;

(2) The dot over the equal sign in the first three equations, (38)-(40), denotes a formal, as yet undefined, relation. The equations are undefined because as soon as one interchanges the path integral over P and the time integral over T^j, one is faced with the path integral $\int_P dw(p,q)H_{no}(p(t),q(t),t)$. Since the integrand contains terms coupling p and q at the same time t, this path integral is undefined because, as explained earlier, the pq correlation function, $\bar{G}(t,t')$ in (8), is undefined at $t = t'$. However, if we replace $\int_P dw(p,q)H_{no}(p(t),q(t),t)$ by $\lim_{(t-t')\to 0+} \int_P dw(p,q)H_{no}(p(t'),q(t),t)$, the resulting expression is well-defined. This is done in the last equation, (41), which gives an unambiguous path-integral representation of K directly tied to the ordering of the factors in the quantum operator $\underset{\sim}{H}_1$;

(3) It is not necessary that the H function in (38)-(41) be H_{no}. However, if another is chosen, the "time" limits in (41) will be different. There is a close connection between the function chosen to replace H_{no} and the type of time limits which will give a well-defined, correct expression for the propagator K. This will be proved and discussed in lemma 1 below;

(4) The propagator (41) satisfies the Schrödinger equation and the boundary condition:

$$\left[-\frac{\hbar^2}{2m} \frac{\partial^2}{\partial q_b{}^2} + k\underset{\sim}{H}_1\left(-i\hbar \frac{\partial}{\partial q_b}, q_b, t_b\right) - i\hbar \frac{\partial}{\partial t_b} \right] K(q_b,t_b;q_a,t_a) = 0 \tag{42a}$$

$$K(q_b,t_b;q_a,t_b) = \delta(q_b - q_a), \tag{42b}$$

to all orders in k;

(5) The path integral in (41) can be evaluated in case a perturbation series in k is sought. The result, proved later in the paper, can be expressed in terms

(5) μ and ν are elements of M, the space of bounded measures on the time interval T; $\langle\mu,q\rangle \equiv \int_T q(t)d\mu(t)$ if μ is induced by a function, $\langle\delta_t,q\rangle \equiv q(t)$ if μ is δ_t, the "delta-function" measure at t.

The most general Gaussian measure, which absorbs all the quadratic terms [i.e. not only $p^2/2m$ but $g(t)p^2/2m + f(t)q^2/2 + k(t)pq$] in either the Hamiltonian or the action functional expanded about the classical path, was constructed in reference 4, but the free-particle measure will be sufficient for our purpose here. w is a true phase-space measure: it does not entail performing separate, successive path integrals in configuration space and momentum space.

This definition enables one to carry out path integrals of cylindrical functionals, i.e., functionals which depend on only a finite number of terms of the form $\langle\mu,q\rangle$ or $\langle\nu,p\rangle$, by converting them into finite-dimensional ordinary integrals. The result is the following fundamental integral [in a form slightly different from that given in reference 4, equation (97)]:

$$\int_P F(\langle\mu_1 q\rangle,\ldots,\langle\mu_n,q\rangle,\langle\nu_1,p\rangle,\ldots,\langle\nu_m,p\rangle)\; dw\;(p,q)$$
$$= \int_{\mathbb{R}^{n+m}} \frac{F(y)\,dy}{[(2\pi i\hbar)^{m+n}\det A]^{1/2}} \exp\left\{\frac{i}{2\hbar}\sum_{i,j=1}^{n+m} (A^{-1})_{ij}(y_i-a_i)(y_j-a_j)\right\} \tag{10}$$

where

$$y \equiv (u_1,\ldots,u_n,\; v_1,\ldots,v_m) \tag{11}$$

$$dy \equiv du_1\ldots du_n dv_1\ldots dv_m \tag{12}$$

$$a \equiv (\langle\mu_1,\bar{q}\rangle,\ldots,\langle\mu_n,\bar{q}\rangle,\langle\nu_1,\bar{p}\rangle,\ldots,\langle\nu_m,\bar{p}\rangle) \tag{13}$$

$$A \equiv \begin{pmatrix} W & C \\ \tilde{C} & V \end{pmatrix} \qquad [(n+m)\times(n+m)] \tag{14}$$

$$W_{ij} \equiv \int_T\int_T G_{ab}(t,t')\,d\mu_i(t)\,d\mu_j(t') \qquad (n\times n) \tag{15}$$

$$C_{ij} \equiv \tilde{C}_{ji} \equiv \int_T\int_T \bar{G}(t,t')\,d\mu_i(t)\,d\nu_j(t') \qquad (n\times m) \tag{16}$$

$$V_{ij} \equiv \int_T\int_T G_p(t,t')\,d\nu_i(t)\,d\nu_j(t') \qquad (m\times m) \tag{17}$$

In the above formula, w is not restricted to being the free-particle measure, but can be the most general quadratic measure mentioned earlier. Of particular interest are an expression for the Fourier transform of w:

$$\int_P \left\{ \exp[-i\langle\mu,q\rangle - i\langle\nu,p\rangle] \right\} dw\,(p,q) = \mathcal{F}w(\mu,\nu) \tag{18}$$

and the special case where $F(x_1,\ldots,x_k) = x_1\ldots x_k$, which yields the generalized moments formula in phase space [an extension of the one given in reference 8, equation (65), for configuration space]:

$$\int_P \langle\mu_1,q\rangle\ldots\langle\mu_n,q\rangle\langle\nu_1,p\rangle\ldots\langle\nu_m,p\rangle\ dw\,(p,q)$$

$$= i^{n+m}\ \mathcal{H}_{n+m}\left(\frac{-ic_1}{2},\ldots,\frac{-ic_{n+m}}{2}\right) \tag{19}$$

where

$$c_k \equiv \begin{cases} \langle\mu_k,\bar{q}\rangle & \text{for } k = 1,\ldots,n \\ \langle\nu_k,\bar{p}\rangle & \text{for } k = n+1,\ldots,n+m \end{cases} \tag{20}$$

and $\mathcal{H}_{n+m}$ is the generalized Hermite polynomial of order $n+m$ and matrix $i\hbar A/2$, defined in references 8 and 9. The proof of (19) is similar to the one given in reference 7. The examples shown below, for the case where all μs and νs are δ functions, reveal the general pattern:

$$\int_P q(t)dw(p,q) = \bar{q}(t) \tag{21}$$

$$\int_P q(t)p(t')dw(p,q) = \bar{q}(t)\bar{p}(t') + i\hbar\bar{G}(t,t') \tag{22}$$

$$\int_P q(t)q(t')dw(p,q) = \bar{q}(t)\bar{q}(t') + i\hbar G_{ab}(t,t') \tag{23}$$

$$\int_P p(t)p(t')dw(p,q) = \bar{p}(t)\bar{p}(t') + i\hbar G_p(t,t') \tag{24}$$

$$\int_P q(t_1)q(t_2)p(t_3)dw(p,q) = \bar{q}(t_1)\bar{q}(t_2)\bar{p}(t_3)$$

$$+ i\hbar\bar{G}(t_2,t_3)\bar{q}(t_1) + i\hbar\bar{G}(t_1,t_3)\bar{q}(t_2) + i\hbar G_{ab}(t_1,t_2)\bar{p}(t_3) \tag{25}$$

$$\int_P q(t_1)q(t_2)p(t_3)p(t_4)dw(p,q) = \bar{q}(t_1)\bar{q}(t_2)\bar{p}(t_3)\bar{p}(t_4)$$
$$+ i\hbar\bar{G}(t_1,t_4)\bar{q}(t_2)\bar{p}(t_3) + i\hbar\bar{G}(t_2,t_4)\bar{q}(t_1)\bar{p}(t_3)$$
$$+ i\hbar G_p(t_3,t_4)\bar{q}(t_1)\bar{q}(t_2) + i\hbar G_{ab}(t_1,t_2)\bar{p}(t_3)\bar{p}(t_4) \qquad (26)$$
$$+ i\hbar\bar{G}(t_2,t_3)\bar{q}(t_1)\bar{p}(t_4) + i\hbar\bar{G}(t_1,t_3)\bar{q}(t_2)\bar{p}(t_4)$$
$$+ (i\hbar)^2 G_{ab}(t_1,t_2)G_p(t_3,t_4) + (i\hbar)^2\bar{G}(t_2,t_3)\bar{G}(t_1,t_4)$$
$$+ (i\hbar)^2\bar{G}(t_1,t_3)\bar{G}(t_2,t_4).$$

It will be noted that the pq correlation function, $\bar{G}(t,t')$, is discontinuous across the diagonal $t = t'$, and its jump there is of magnitude 1:

$$\left[\lim_{(t-t')\to 0^+} - \lim_{(t'-t)\to 0^+}\right]\bar{G}(t,t') = 1 \qquad (27)$$

[see, e.g., (8) for the free-particle case]. Thus, $\mathcal{F}w(\delta_t,\delta_t)$ is not defined. This indefiniteness, which occurs in a natural manner as one builds the measure [4], stems from the non-commutativity of $\underset{\sim}{P}$ and $\underset{\sim}{Q}$, and will give us the flexibility we need to take various correspondence rules into account.

III. THE CORRESPONDENCE RULES

We will consider the most general quantum Hamiltonian operator $\underset{\sim}{H}(\underset{\sim}{P},\underset{\sim}{Q},t)$ which can be derived from a classical Hamiltonian $H_c(p,q,t)$. The form most convenient for our purposes is that given by Cohen [5]:

$$\underset{\sim}{H} = (2\pi\hbar)^{-2}\int dpdqdudv\ F(u,v)H_c(p,q,t) \quad \exp\left\{(i/\hbar)\ [(q-\underset{\sim}{Q})u + (p-\underset{\sim}{P})v]\right\}, \qquad (28)$$

where $F(u,v)$ is the transformation function. This scheme for correspondence rules was also used in reference 10 to determine the range of validity of a commonly-used formula for the WKB approximation of the propagator $K(q_b,t_b;q_a,t_a)$. Each F uniquely determines the correspondence rule. For instance, $F = 1$, $F = \cos(uv/2\hbar)$ and $F = (uv/2\hbar)^{-1}\sin(uv/2\hbar)$ give the Weyl, symmetrized, and Born-Jordan ordering schemes, respectively. For real H_c, $\underset{\sim}{H}$ is Hermitian iff $F^*(-u,-v) = F(u,v)$, a condition we do not impose here. Also, we must have $F(0,v) = F(u,0) = 1$ to insure that $f(\underset{\sim}{P})$ and $g(\underset{\sim}{Q})$ correspond to $f(p)$ and $g(q)$. The transform can be inverted to yield:

$$H_c(p,q,t) = (2\pi\hbar)^{-1}\int dq'dp'dq''\exp\{(i/\hbar)\ [q'p + p'(q-q'')]\}$$
$$\times F^{-1}(q',p') \langle q'' - q'/2|\underset{\sim}{H}|q'' + q'/2\rangle, \qquad (29)$$

and finally, a transformation function F can be deduced from the knowledge of the Hamiltonian H_c and its transform $\underset{\sim}{H}$:

$$F(u,v)\int dpdq\ H_c(p,q,t)\exp[i(qu+pv)/\hbar] = (2\pi\hbar)\,tr(e^{i(\underset{\sim}{Q}u+\underset{\sim}{P}v)/\hbar}\underset{\sim}{H}), \tag{30}$$

where $tr\underset{\sim}{A} \equiv \int <q|\underset{\sim}{A}|q>dq = \int <p|\underset{\sim}{A}|p>dp$. This formula can be proved by left-multiplying both sides of (28) by $\exp[i(\underset{\sim}{Q}u' + \underset{\sim}{P}v')/\hbar]$, taking traces and inserting complete sets of states after using equation (32).

Note that:

(1) the F which relates a given $\underset{\sim}{H}$ with a given H_c is usually not unique. For example, for $\underset{\sim}{H} = f(\underset{\sim}{Q})$ and $H_c = f(q)$, (30) gives:

$$F(u,v)\delta(v)\int dqf(q)e^{iqu/\hbar} = \int dqf(q)e^{iqu/\hbar}\ . \tag{31}$$

Thus, any F such that $F(u,0) = 1$ will do. Similarly, one finds that pq is mapped into $(\underset{\sim}{P}\underset{\sim}{Q} + \underset{\sim}{Q}\underset{\sim}{P})/2$ by any $F = F(uv/\hbar)$ such that $F(0) = 1$ and $F'(0) = 0$. However, when the ordering of the factors of the generic $p^m q^n$ is given for <u>all</u> m and n, then the F function is unique, modulo a test function $\phi(x)$ all of whose derivatives are 0 at 0. For example, given that the most general ordering of $H \equiv p^m q^n$ is $\underset{\sim}{H} = \Sigma_{j=0}^{n} a_{jn}\, \underset{\sim}{Q}^j \underset{\sim}{P}^m \underset{\sim}{Q}^{n-j}$, with $\Sigma_{j=0}^{n} a_{jn} = 1$, the corresponding F depends on u and v through the combination $uv/\hbar \equiv x$, and can be calculated [using (30)] to be:

$$F(x) = e^{-ix/2} \lim_{n\to\infty} \sum_{j=0}^{n} \frac{a_{jn}}{n!} \sum_{\ell=0}^{j} \binom{j}{\ell} (n-\ell)!(ix)^{\ell}$$

(modulo a test function). This formula can be inverted to yield the coefficients a_{jn} in terms of F:

$$a_{jn} = \frac{(-1)^j}{2\pi}\binom{n}{j}\int_{-\infty}^{\infty} k^j(1+k)^{n-j}dk\int_{-\infty}^{\infty} F(x)e^{ix(k+1/2)}dx.$$

Thus, we have the following correspondence table:

F	a_{jn}	F	a_{jn}
1	$\binom{n}{j}/2^n$	$\exp(-ix/2)$	$\begin{cases}1 & \text{for } j = 0\\ 0 & \text{otherwise}\end{cases}$
$\cos(x/2)$	$\begin{cases}1/2 & \text{for } j=0,n\\ 0 & \text{otherwise}\end{cases}$	$\exp(ix/2)$	$\begin{cases}1 & \text{for } j = n\\ 0 & \text{otherwise}\end{cases}$
$\dfrac{\sin(x/2)}{(x/2)}$	$\dfrac{1}{n+1}$		

Hermiticity is expressed by the condition $a_{n-j,n} = a^{*}_{jn}$.

(2) A correspondence rule is not limited to factor ordering. For example, the (Hermitian) correspondence

$$p^2 \rightarrow e^{i\underset{\sim}{Q}/a}\, \underset{\sim}{P}^2\, e^{-i\underset{\sim}{Q}/a},$$

where a is some fundamental length, is not obtained by factor-ordering (since there are no qs in the classical function). A corresponding F, obtained using (30), is

$$F(u,v) = \exp(iv/a).$$

(3) Formula (30) proves that Cohen's scheme is exhaustive: each correspondence between a classical function and a quantum operator, whether a factor-ordering scheme (F is then a function of $uv/\hbar$ alone) or not, is represented by at least one function F.

It will be more convenient in certain cases to put $\underset{\sim}{H}$ in (28) in its normal-ordered form ($\underset{\sim}{Q}$ before $\underset{\sim}{P}$). For this, one uses the Baker-Campbell-Hausdorff formula, $e^{\underset{\sim}{A}+\underset{\sim}{B}} = e^{\underset{\sim}{A}} e^{\underset{\sim}{B}} e^{-[\underset{\sim}{A},\underset{\sim}{B}]/2}$, valid for all $\underset{\sim}{A}$ and $\underset{\sim}{B}$ which commute with $[\underset{\sim}{A},\underset{\sim}{B}]$. It gives:

$$e^{-i(\underset{\sim}{Q}u+\underset{\sim}{P}v)/\hbar} = e^{-i\underset{\sim}{Q}u/\hbar}\, e^{-i\underset{\sim}{P}v/\hbar}\, e^{iuv/2\hbar} . \tag{32}$$

Since $\underset{\sim}{H}$ will be applied to functions of the endpoint q_b, $\underset{\sim}{Q}$ is represented by q_b and $\underset{\sim}{P}$ by $-i\hbar\partial/\partial q_b$. Thus, the most general operator $\underset{\sim}{H}$ derived from H_c is, in its normal-ordered form,

$$\underset{\sim}{H} = (2\pi\hbar)^{-2}\int dpdqdudvF(u,v)\ H_c(p,q,t) \quad \exp\left[\frac{iu}{\hbar}(q-q_b+\frac{v}{2})\right]\exp\left(\frac{ipv}{\hbar}\right)\exp\left(-v\frac{\partial}{\partial q_b}\right) . \tag{33}$$

Consequently, when $\underset{\sim}{H}$ is applied to $f(q_b)$, the result is the right-hand side of (33) with $\exp(-v\partial/\partial q_b)$ replaced by $f(q_b - v)$. Expressions (28) and (33) for $\underset{\sim}{H}$ will both be used.

IV. THE PROPAGATOR BY PATH INTEGRALS, FOR ARBITRARY HAMILTONIANS

With the two foregoing tools in hand, we can proceed to write a path-integral representation for the propagator corresponding to an arbitrary Hamiltonian operator. First, we write the latter in the form:

$$\underset{\sim}{H} = \frac{\underset{\sim}{P}^2}{2m} + k\, \underset{\sim}{H}_1\ (\underset{\sim}{P},\underset{\sim}{Q},t). \tag{34}$$

of only the classical Hamiltonian H_1 and the correspondence function F. It is:

$$K = K_o\left[1 + \sum_{j=1}^{\infty} k^j \alpha_j\right], \tag{43}$$

where α_j is displayed in (70) and (71) below.

The proof of the theorem will consist in showing, by recurrence, that (41) satisfies (42) by using both (28) and (33) to relate the classical and quantum Hamiltonians. First, we give a simple illustration of the theorem.

Example

Calculate, to first order in k, the propagator corresponding to the following Hamiltonian:

$$\underset{\sim}{H} = \underset{\sim}{P}^2/2m + k\left[\alpha \underset{\sim}{P}\underset{\sim}{Q}^2 + \beta \underset{\sim}{Q}\underset{\sim}{P}\underset{\sim}{Q} + (1-\alpha-\beta)\underset{\sim}{Q}^2\underset{\sim}{P}\right], \tag{44}$$

(which represents all the possible $\underset{\sim}{H}$s corresponding to $H_c = p^2/2m + kpq^2$).[1]

Answer. The normal-ordered $\underset{\sim}{H}$ is:

$$\underset{\sim}{H} = \underset{\sim}{P}^2/2m + k[\underset{\sim}{Q}^2\underset{\sim}{P} - i\hbar(\beta+2\alpha)\underset{\sim}{Q}]. \tag{45}$$

Using (41), we have:

$$K = K_o\left\{1 - \frac{ik}{\hbar}\int_T dt\left[\lim_{t-t'\to 0^+}\int_P q^2(t)p(t')dw(p,q) - i\hbar(\beta+2\alpha)\int_P q(t)dw(p,q)\right]\right\} \tag{46}$$

$$= K_o\left[1 - \frac{ikm}{3\hbar}(q_b^3 - q_a^3) + \frac{kT}{2}(q_b+q_a)(1-\beta-2\alpha)\right], \tag{47}$$

where we have used (25), (21), and (5)-(9). It can be directly verified that (47) satisfies the Schrödinger equation to first order in k, i.e., that

$$\left\{-(\hbar^2/2m)\frac{\partial^2}{\partial q_b^{\,2}} - i\hbar k\left[q_b^{\,2}\frac{\partial}{\partial q_b} + (\beta-2\alpha)q_b\right] - i\hbar\frac{\partial}{\partial t_b}\right\}K = 0(k^2) \tag{48}$$

---

[1] Let us mention in passing that one F-function which fulfills this correspondence is $F(u,v) = \exp[iuv(1-\beta-2\alpha)/2\hbar]$. It was derived as follows. (30) yields $i\hbar\delta''(u)\delta'(v)N(u,v) + (\beta+2\alpha)\delta(v)\delta'(u) = 0$, where $N(u,v) \equiv F(u,v) - \exp(-iuv/2\hbar)$. By using $f(x)\delta^{(n)}(x) = \Sigma_{j=0}^{n} f^{(j)}(0)\binom{n}{j}(-1)^j\delta^{(j)}(x)$ and equating coefficients, we conclude that any F such that $2i\hbar N_{,12}(0,0) = -(\beta+2\alpha)$ and $N(0,0) = N_{,1}(0,0) = N_{,11}(0,0) = N_{,2}(0,0) = N_{,112}(0,0) = 0$ will do. The F we chose is a convenient one which satisfies these conditions.

together with $\lim_{t_b \to t_a} K = \delta(q_b - q_a)$.

Note that we could have obtained the same result without normal-ordering first, as stated earlier. If we consider the function obtained by replacing $\tilde{Q}$ and $\tilde{P}$ by q and p in (44), order the times in the sequence suggested by (44), then take successive coincidence limits, we get:

$$K = K_o \left\{ 1 - \frac{ik}{\hbar} \int_T dt \left[\alpha \lim_{t-t'\to 0^+} \int_P p(t) q^2(t') dw(p,q) \right.\right.$$

$$+ \beta \lim_{t-t'\to 0^+} \lim_{t'-t''\to 0^+} \int_P q(t) p(t') q(t'') dw(p,q)$$

$$\left.\left. + (1-\alpha-\beta) \lim_{t-t'\to 0^+} \int_P q^2(t) p(t') dw(p,q) \right] \right\}, \tag{49}$$

which also yields the correct result (47). The general proof of this flexibility is found in lemma 1 below.

Proof of theorem

Lemma 1. For all phase-space functionals $F[q,p]$ which do not contain the path (p,q) evaluated at either t or t', we have:

$$\lim_{t-t'\to 0^+} \int_P F[q,p] [q(t)p(t') - q(t')p(t) - i\hbar] \, dw(p,q) = 0 \tag{50}$$

Proof. Consider the measure

$$dw_{tt'}(p,q) \equiv [q(t)p(t') - q(t')p(t) - i\hbar] \, dw(p,q). \tag{51}$$

Its Fourier transform is

$$F_{w_{tt'}}(\mu,\nu) = \int_P e^{-i\langle\mu,q\rangle - i\langle\nu,p\rangle} dw_{tt'}(p,q) \tag{52}$$

To evaluate it, we proceed as follows:

$$\int_P q(t) p(t') \, e^{-i\langle\mu,q\rangle - i\langle\nu,p\rangle} dw(p,q)$$

$$= \frac{-\partial^2}{\partial\lambda\partial\sigma} \int_P e^{-i\langle\mu+\lambda\delta_t, q\rangle - i\langle\nu+\sigma\delta_{t'}, p\rangle} dw(p,q) \Big|_{\lambda=\sigma=0}$$

$$= -\frac{\partial^2}{\partial\lambda\partial\sigma} Fw(\mu+\lambda\delta_t, \nu+\sigma\delta_{t'}) \Big|_{\lambda=\sigma=0}$$

$$= Fw(\mu,\nu)\left[\frac{i\hbar}{2}G_{ab}(t,t') + \frac{i\hbar}{2}G_p(t,t') + i\hbar\bar{G}(t,t') + \xi(\mu,\nu;t,t')\right], \tag{53}$$

where

$$\xi(\mu,\nu;t,t') \equiv \left[i\bar{q}(t) + \frac{i\hbar}{2}\int_T G_{ab}(t,s')d\mu(s') + \frac{i\hbar}{2}\int_T G_p(t,s')d\nu(s') + i\hbar\int_T \bar{G}(t,s')d\nu(s')\right] \times \left[i\bar{p}(t') + \frac{i\hbar}{2}\int_T G_{ab}(s,t')d\mu(s) + \frac{i\hbar}{2}\int_T G_p(s,t') d\nu(s) + i\hbar\int_T \bar{G}(s,t')d\mu(s)\right] \tag{54}$$

and (2) was used. Therefore,

$$Fw_{tt'}(\mu,\nu) = Fw(\mu,\nu)\left\{-i\hbar + \frac{i\hbar}{2}\left[G_{ab}(t,t') - G_{ab}(t',t)\right] + \frac{i\hbar}{2}\left[G_p(t,t') - G_p(t',t)\right] + i\hbar\left[\bar{G}(t,t') - \bar{G}(t',t)\right] + \xi(\mu,\nu;t,t') - \xi(\mu,\nu\,;t',t)\right\}. \tag{55}$$

Since G_{ab}, G_p, and ξ are continuous across the diagonal $t = t'$ (provided μ and ν are different from δ_t and $\delta_{t'}$), and since $\bar{G}$ has a jump of magnitude 1 there [equation (27)], we conclude that

$$\lim_{t-t'\to 0^+} Fw_{tt'}(\mu,\nu) = 0. \tag{56}$$

Consequently, the measure $w_{tt'}(p,q)$ is effectively the zero measure in the limit $(t-t')\to 0^+$, provided (1) this limit is taken <u>after</u> a path integral with respect to $w_{tt'}(p,q)$ is performed [(50) is obviously false if the limit is taken before the path integral is done], and (2) the integrand does not contain p or q evaluated at t or t' [if it does, then $\xi(\mu,\nu;t,t')$ is no longer continuous across the diagonal $t = t'$]. Q.E.D.

This lemma insures that the various path integrals obtained by changing the form of the given Hamiltonian operator (by repeated use of the commutation relation $\underset{\sim\sim}{QP}-\underset{\sim\sim}{PQ} = i\hbar$) will all yield the same result. This was illustrated in the example above where the two path integrals (46) and (49), corresponding to the same Hamiltonian operator written in two different forms (45) and (44), gave the same correct answer.

Therefore, it is sufficient to prove the theorem for the normal-ordered form of $\underset{\sim}{H}_1$, i.e., with H_{no}.

<u>Lemma 2.</u>

$$\left(\int_{t_a}^{t_b} f(t)dt\right)^n = \int_{T^n} dt_1 \ldots dt_n f(t_1)\ldots f(t_n)$$

$$= n!\int_{t_a}^{t_b} dt_1 \int_{t_a}^{t_1} dt_2 \ldots \int_{t_a}^{t_{n-2}} dt_{n-1} \int_{t_a}^{t_{n-1}} dt_n f(t_1)\ldots f(t_n). \tag{57}$$

Proof. The lemma is true for $n = 1$. Assume it is true for $n = k - 1$. Consider $F_k(s) \equiv \left(\int_{t_a}^{s} f(x)dx\right)^k$. Then

$$\dot{F}_k(s) = kF_{k-1}(s)f(s) = kf(s)(k-1)! \int_{t_a}^{s} dt_1 \int_{t_a}^{t_1} dt_2 \ldots \int_{t_a}^{t_{k-2}} dt_{k-1} f(t_1)\ldots f(t_{k-1}). \tag{58}$$

Integrating from t_a to t_b with respect to s gives

$$F_k(t_b) = k!\int_{t_a}^{t_b} ds \int_{t_a}^{s} dt_1 \int_{t_a}^{t_1} dt_2 \ldots \int_{t_a}^{t_{k-2}} dt_{k-1} f(t_1)\ldots f(t_{k-1}) f(s). \tag{59}$$

Changing variables: $s = t_1$, $t_1 = t_2$, ..., $t_{k-1} = t_k$, proves that the formula is true for $n = k$. Therefore the formula is proved true by recurrence. Q.E.D.

Proof to first order in k. The theorem will be proved by recurrence. Thus, we must first prove that (41) satisfies the Schrödinger equation (42) to first order in k. The propagator to first order in k is:

$$K = K_o\left[1 - \frac{ik}{\hbar}\int_T dt \lim_{t-t'\to 0^+} \int_P dw(p,q) H_{no}[p(t'),q(t),t]\right]. \tag{60}$$

Using the correspondence rule (28), along with (32) to put $\underset{\sim}{H}_1$ in normal-ordered form, we can write:

$$H_{no}\left[p(t'),q(t),t\right] = (2\pi\hbar)^{-2}\int dpdqdudv\; F(u,v)H_1(p,q,t) \quad \exp\Big\{(i/\hbar)(qu+pv+uv/2)$$
$$- iuq(t)/\hbar - ivp(t')/\hbar\Big\}. \tag{61}$$

Substituting (61) in (60) reveals that the path integral is a particularly simple one, namely the Fourier transform of w at $(u\delta_t/\hbar,\; v\delta_{t'}/\hbar)$. It can be evaluated using (2) and (5)-(9), and the result, after the limit,

$$K = K_o[1+k\alpha_1], \tag{62}$$

where

$$\alpha_1 \equiv -\frac{i}{\hbar(2\pi\hbar)^2}\int dpdqdudvF(u,v)\int_T dt\; H_1(p,q,t) \quad \exp\Big\{(i/\hbar)(qu+pv+uv/2)$$

$$- (i/\hbar T)[uq_b(t-t_a) + uq_a(t_b-t) + mv(q_b-q_a) + u^2(t-t_a)(t_b-t)/2m + uv(t_b-t) - mv^2/2]\Big\} . \tag{63}$$

We must show that

$$\left[-\frac{\hbar^2}{2m}\frac{\partial^2}{\partial q_b{}^2} + k\underset{\sim}{H}_1\left(q_b,-i\hbar\frac{\partial}{\partial q_b},t_b\right) - i\hbar\frac{\partial}{\partial t_b}\right]\left[K_o(1+k\alpha_1)\right] = O(k^2). \tag{64}$$

Since K_o satisfies the free-particle Schrödinger equation, we must simply show that the coefficient of k is zero, i.e., that

$$\left[-\frac{\hbar^2}{2m}\frac{\partial^2}{\partial q_b{}^2} - i\hbar\frac{\partial}{\partial t_b}\right](K_o\alpha_1) + \underset{\sim}{H}_1K_o = 0. \tag{65}$$

For $\underset{\sim}{H}_1$ we will use (33), its normal-ordered form in terms of the classical function H_1. Thus,

$$\underset{\sim}{H}_1K_o = (2\pi\hbar)^{-2}\int dpdqdudvF(u,v)H_1(p,q,t_b)\left\{\exp\left[\frac{iu}{\hbar}(q-q_b+\frac{v}{2}) + \frac{ipv}{\hbar}\right]\right\}\left(\frac{m}{2\pi i\hbar T}\right)^{1/2}\exp\left[\frac{im}{2\hbar T}(q_b-v-q_a)^2\right]. \tag{66}$$

The remainder of the proof is tedious and straightforward. Differentiations with respect to q_b and t_b are performed under the integral sign, assuming interchangeability, using Leibnitz's rule, $(\partial/\partial t_b)\int_{t_a}^{t_b}\phi(t,t_b)dt = \phi(t_b,t_b) + \int_{t_a}^{t_b}[\partial\phi(t,t_b)/\partial t_b]dt$, where needed. The $\underset{\sim}{H}_1K_o$ term cancels the term equivalent to $\phi(t_b,t_b)$. Upon collecting terms, the integrand vanishes, and the theorem is established to first order in k for arbitrary Hamiltonian operators.

Proof to any order in k. Using lemma 2 [equation (57)] along with (61), we can write K in (41) as:

$$K = K_o[1+\sum_{j=1}^{\infty} k^j\alpha_j] , \tag{67}$$

where

$$\alpha_j \equiv \left(\frac{-i}{\hbar}\right)^j\int_{t_a}^{t_b}dt_1\int_{t_a}^{t_1}dt_2\cdots\int_{t_a}^{t_{j-2}}dt_{j-1}\int_{t_a}^{t_{j-1}}dt_j(2\pi\hbar)^{-2j}\int dq_1\ldots dq_j$$

$$\times\, dp_1\ldots dp_j du_1\ldots du_j dv_1\ldots dv_j\; F(u_1,v_1)\ldots F(u_j,v_j)\quad H_1(p_1,q_1,t_1)\ldots H_1(p_j,q_j,t_j)$$

$$\times \exp\left[i\sum_{s=1}^{j}(q_su_s+p_sv_s+u_sv_s/2)/\hbar\right]\lim_{t_1-t'_1\to 0^+}\cdots\lim_{t_j-t'_j\to 0^+}\int_P dw(p,q)$$

$$\times \exp\{(-i/\hbar)[u_1 q(t_1) + \ldots + u_j q(t_j) + v_1 p(t_1') + \ldots + v_j p(t_j')]\} \,. \tag{68}$$

The path integral above is readily recognized as being

$$Fw\left[(u_1\delta_{t_1}+\ldots+u_j\delta_{t_j})/\hbar, (v_1\delta_{t_1'}+\ldots+v_j\delta_{t_j'})/\hbar\right], \tag{69}$$

which can be evaluated, by use of (2) and (5)-(9). By virtue of lemma 2, the times entering the integral are now ordered $(t_b \geq t_1 \geq t_2 \geq \ldots \geq t_j \geq t_a)$. The result is:

$$\alpha_j = \left(\frac{-i}{\hbar}\right)^j \int_{t_a}^{t_b} dt_1 \int_{t_a}^{t_1} dt_2 \ldots \int_{t_a}^{t_{j-2}} dt_{j-1} \int_{t_a}^{t_{j-1}} dt_j (2\pi\hbar)^{-2j}$$

$$\times \int dq_1 \ldots dq_j dp_1 \ldots dp_j du_1 \ldots du_j dv_1 \ldots dv_j \quad F(u_1, v_1) \ldots F(u_j, v_j) H_1(p_1, q_1, t_1)$$

$$\ldots H_1(p_j, q_j, t_j) E_j , \tag{70}$$

where

$$E_j \equiv \exp\left\{(i/\hbar)\sum_{s=1}^{j}(q_s u_s + p_s v_s + u_s v_s/2) - (i/\hbar T)\sum_{r=1}^{j}[u_r(t_r - t_a)q_b + u_r(t_b - t_r)q_a\right.$$

$$+ m(q_b - q_a)v_r] - (i/\hbar T)\sum_{\substack{r,s=1\\ r\geq s}}^{j}[u_r u_s (t_r - t_a)(t_b - t_s)/m - m v_r v_s](1 - \delta_{rs}/2) - (i/\hbar T)$$

$$\left.\times\left[\sum_{\substack{r,s=1\\ r\leq s}}^{j} u_r v_s (t_b - t_r) - \sum_{\substack{s=1,r=2\\ r>s}}^{j} u_r v_s (t_r - t_a)\right]\right\} . \tag{71}$$

We must now show that if

$$K_n \equiv K_o\left[1 + \sum_{j=1}^{n} k^j \alpha_j\right] \tag{72}$$

satisfies the Schrödinger equation to nth order in k, then K_{n+1} satisfies it to (n+1)th order in k, i.e.

$$\left(-\frac{\hbar^2}{2m}\frac{\partial^2}{\partial q_b^2} + k\underset{\sim}{H}_1 - i\hbar\frac{\partial}{\partial t_b}\right)\left[K_o(1 + \sum_{j=1}^{n+1} k^j \alpha_j)\right] = O(k^{n+2}). \tag{73}$$

This can be shown to be true if and only if the coefficient of k^{n+1} is 0, i.e., iff

$$\left(-\frac{\hbar^2}{2m}\frac{\partial^2}{\partial q_b^2} - i\hbar\frac{\partial}{\partial t_b}\right)(K_o\alpha_{n+1}) + \underset{\sim}{H}_1\,(K_o\alpha_n) = 0. \tag{74}$$

In the k-expansion of the left side of (73), the vanishing of the 0th order term results from the Schrödinger equation for K_o, the vanishing of the coefficient of k results from (65), and the vanishing of the coefficients of k^2 to k^{n+1} is what (74) (from n = 2 to n + 1) proves. (74) readily reduces to

$$\underset{\sim}{H}_1(K_o\alpha_n) - i\hbar K_o \frac{\partial\alpha_{n+1}}{\partial t_b} - \frac{i\hbar K_o(q_b-q_a)}{T}\frac{\partial\alpha_{n+1}}{\partial q_b} - \frac{\hbar^2}{2m}K_o\frac{\partial^2\alpha_{n+1}}{\partial q_b^{\,2}} = 0. \tag{75}$$

We will calculate each term in (75) separately and show that they cancel each other out. Using (33) to represent $\underset{\sim}{H}_1$, as we did earlier, we have:

$$\underset{\sim}{H}_1(K_o\,\alpha_n) = (2\pi\hbar)^{-2}K_o\int dpdqdudvF(u,v)H_1(p,q,t_b)\quad \exp\Big\{(i/\hbar)[uq+pv+uv/2 - uq_b$$

$$- mv^2/2T - mv(q_b-q_a)/T]\Big\}\quad (-i/\hbar)^n\int_{t_a}^{t_b}dt_1\int_{t_a}^{t_1}dt_2\ldots\int_{t_a}^{t_{n-1}}dt_n(2\pi\hbar)^{-2}$$

$$\int dq_1\ldots dq_n dp_1\ldots dp_n\quad du_1\ldots du_n dv_1\ldots dv_n F(u_1,v_1)\ldots F(u_n,v_n)$$

$$H_1(p_1,q_1,t_1)\ldots H_1(p_n,q_n,t_n)\exp\Big\{(i/\hbar)\sum_{s=1}^{n}(q_su_s+p_sv_s+u_sv_s/2)$$

$$-(i/\hbar T)\sum_{r=1}^{n}[u_r(t_r-t_a)(q_b-v) + u_r(t_b-t_r)q_a + m(q_b-v-q_a)v_r]$$

$$-(i/\hbar T)\sum_{\substack{r,s=1\\ r\geq s}}^{n}[u_ru_s(t_r-t_a)(t_b-t_s)/m - mv_rv_s](1-\delta_{rs}/2) - (i/\hbar T)$$

$$\times\Big[\sum_{\substack{r,s=1\\ r\leq s}}^{n}u_rv_s(t_b-t_r) - \sum_{\substack{s=1,r=2\\ r>s}}^{n}u_rv_s(t_r-t_a)\Big]\Big\}\;. \tag{76}$$

The time-derivative term in (75), $-i\hbar K_o\partial\alpha_{n+1}/\partial t_b$, can be written as A + B, where B is the derivative of the integrand and A evaluates the integrand at t_b. Thus,

$$A = -i\hbar K_o(-i/\hbar)^{n+1}\int_{t_a}^{t_b}dt_2\ldots\int_{t_a}^{t_n}dt_{n+1}(2\pi\hbar)^{-2(n+1)}\quad\int dq_1\ldots dq_{n+1}dp_1\ldots dp_{n+1}$$

$$du_1\ldots du_{n+1}dv_1\ldots dv_{n+1}\;F(u_1,v_1)\ldots F(u_{n+1},v_{n+1})\;H_1(p_1,q_1,t_b)H_1(p_2,q_2,t_2)$$

$$\ldots H_1(p_{n+1},q_{n+1},t_{n+1})\;\exp\Big\{(i/\hbar)\sum_{s=1}^{n+1}(q_su_s+p_sv_s+u_sv_s/2) - (i/\hbar T)\Big[u_1Tq_b$$

$$+ m(q_b-q_a)v_1 + \sum_{r=2}^{n+1}[u_r(t_r-t_a)q_b + u_r(t_b-t_r)q_a + m(q_b-q_a)v_r]\Big]$$

$$-(i/\hbar T)\left[-mv_1{}^2/2 - mv_1\sum_{r=2}^{n+1} v_r + \sum_{\substack{r,s=2\\ r\geq s}}^{n+1}[u_r u_s(t_r-t_a)(t_b-t_s)/m - mv_r v_s](1-\delta_{rs}/2)\right]$$

$$-(i/\hbar T)\left[\sum_{\substack{r,s=2\\ r\leq s}}^{n+1} u_r v_s(t_b-t_r) - v_1\sum_{r=2}^{n+1} u_r(t_r-t_a) - \sum_{\substack{s=2,r=3\\ r>s}}^{n+1} u_r v_s(t_r-t_a)\right]\Bigg\} \qquad (77)$$

The t_b dependence of the integrand of $K_o\alpha_{n+1}$ is of the form $\exp[at_b/(t_b-t_a)]$. Therefore,

$$B = -i\hbar K_o(-i/\hbar)^{n+1}\int_{t_a}^{t_b} dt_1\cdots\int_{t_a}^{t_n} dt_{n+1}(2\pi\hbar)^{-2(n+1)}\int dq_1\cdots dq_{n+1}dp_1\cdots dp_{n+1}du_1\cdots$$

$$du_{n+1}dv_1\cdots dv_{n+1}\; F(u_1,v_1)\cdots F(u_{n+1},v_{n+1})\; H_1(p_1,q_1,t_1)\cdots H_1(p_{n+1},q_{n+1},t_{n+1})$$

$$E_{n+1}(-i/\hbar T^2)\Bigg\{Tq_a\sum_{r=1}^{n+1} u_r + \sum_{\substack{r,s=1\\ r\geq s}}^{n+1} u_r u_s(t_r-t_a)T(1-\delta_{rs}/2)/m + T\sum_{\substack{r,s=1\\ r\leq s}}^{n+1} u_r v_s - \sum_{r=1}^{n+1}$$

$$[u_r(t_r-t_a)q_b + u_r(t_b-t_r)q_a + m(q_b-q_a)v_r] - \sum_{\substack{r,s=1\\ r\geq s}}^{n+1}[u_r u_s(t_r-t_a)(t_b-t_s)/m$$

$$-mv_r v_s](1-\delta_{rs}/2) - \sum_{\substack{r,s=1\\ r\leq s}}^{n+1} u_r v_s(t_b-t_r) + \sum_{\substack{s=1,r=2\\ r>s}}^{n+1} u_r v_s(t_r-t_a)\Bigg\}. \qquad (78)$$

The q_b dependence of the integrand of α_{n+1} is of the simple form $\exp(aq_b)$. Therefore,

$$-\frac{i\hbar}{T}K_o(q_b-q_a)\frac{\partial\alpha_{n+1}}{\partial q_b} - \frac{\hbar^2}{2m}K_o\frac{\partial^2\alpha_{n+1}}{\partial q_b{}^2} = i\hbar K_o(-i/\hbar)^{n+1}\int_{t_a}^{t_b} dt_1\cdots\int_{t_a}^{t_n} dt_{n+1}(2\pi\hbar)^{-2(n+1)}$$

$$\times\int dq_1\cdots dq_{n+1}dp_1\cdots dp_{n+1}du_1\cdots du_{n+1}dv_1\cdots dv_{n+1}F(u_1,v_1)\cdots F(u_{n+1},v_{n+1})$$

$$\times H_1(p_1,q_1,t_1)\cdots H_1(p_{n+1},q_{n+1},t_{n+1})E_{n+1}(-i/\hbar T^2)\Bigg\{-(q_b-q_a)\sum_{r=1}^{n+1}[u_r(t_r-t_a) + mv_r]$$

$$+\frac{1}{2m}\left[\sum_{r=1}^{n+1}\Big(u_r(t_r-t_a) + mv_r\Big)\right]^2\Bigg\}, \qquad (79)$$

where E_{n+1} is defined in (71).

From (76) and (77) one can show that $A + \underset{\sim}{H}_1(K_o\alpha_n) = 0$. For this, it is sufficient to make the following changes of variable: in (77), $t_i = t_{i-1}$ for $i = 2$ to $n+1$; $C_1 = C_{n+1}$ and $C_i = C_{i-1}$ for $i = 2$ to $n+1$, where C denotes u,v,p, or q; in (76), $C = C_{n+1}$, where C denotes u,v,p, or q. It is also seen, by rearranging the terms inside the curly brackets of (78), that B added to either side of (79) gives zero.

The boundary condition (42b) is satisfied, as can be seen from the expression (43) for K. Indeed, K_o satisfies (42b) and $\lim_{t_b \to t_a} \alpha_j = 0$ as can be seen from (71).

This completes the proof of the theorem. Q.E.D.

Conjecture

As discussed in reference 4, section IVB, a semiclassical expansion (in powers of $\hbar$) of a propagator which allows such an expansion [see reference 9] yields:

$$K = K_{WKB} \int_{P_o} dw(p_o, q_o) \exp\{(-i/\hbar)\Omega[p_o, q_o]\} \tag{80}$$

where K_{WKB} is the semiclassical approximation to K, P_o is the same as P except that $q_a = q_b = 0$, $(p_o, q_o) \epsilon P_o$,

$$\Omega[p_o, q_o] \equiv \sum_{n=3}^{\infty} \sum_{k=0}^{n} \frac{n!}{(n-k)!k!} \int_T dt \left[\frac{\partial^n H_c}{\partial p^k \partial q^{n-k}} \right]_{\substack{q=q_c \\ p=p_c}} (t)\, p_o^k(t) q_o^{n-k}(t), \tag{81}$$

and w absorbs the full quadratic part of the expansion of the action functional about the classical path (q_c, p_c). However, if one expands the exponential and attempts to carry out the path integral before the time integral, the indefiniteness discussed before appears again. One conjectured answer, which remains to be verified, is that the $p_o^k(t) q_o^{n-k}(t)$ term should be "time-ordered" (in the manner discussed in this paper) with the same correspondence rule as the original Hamiltonian operator of the problem. This topic will be the subject of a follow-up study.

V. CONCLUSION

The object of this paper was to show that any Hamiltonian operator is amenable to a path-integral treatment, by providing an unambiguous, computationally viable formalism and taking proper account of the correspondence rule leading from the classical function to the quantum operator.

REFERENCES

1. Maurice M. Mizrahi, "The Weyl Correspondence and Path Integrals", J. Math. Phys. 16(1975), 2201-6.

2. L. Cohen, "Correspondence Rules and Path Integrals", J. Math. Phys. 17(1976), 597-8.

3. J.S. Dowker, "Path Integrals and Ordering Rules", J. Math. Phys. 17(1976), 1873-4.

4. Maurice M. Mizrahi, "Phase Space Path Integrals, Without Limiting Procedure", J. Math. Phys. 19(1978), 298-307.

5. L. Cohen, "Generalized Phase-Space Distribution Functions", J. Math. Phys. 7(1966), 781-6.

6. S.F. Edwards and Y.V. Gulyaev, Proc. Roy. Soc. A279(1964), 299.

7. C. DeWitt-Morette, A. Maheshwari, and B. Nelson, "Path Integration in Phase Space", Gen. Rel. and Grav. 8(1977), 581-93.

8. Maurice M. Mizrahi, "On Path Integral Solutions of the Schrödinger Equation, Without Limiting Procedure", J. Math. Phys. 17(1976), 566-75.

9. Maurice M. Mizrahi, "Generalized Hermite Polynomials", J. Comp. and Appl. Math. 1(1975), 273-7.

10. Maurice M. Mizrahi, "On the Semiclassical Expansion in Quantum Mechanics for Arbitrary Hamiltonians", J. Math. Phys. 18(1977), 786-90.

Feynman-type integrals defined in terms of general cylindrical approximations

Jan Tarski

Institut für Theoretische Physik
Technische Universität Clausthal
3392 Clausthal-Zellerfeld, F.R.Germany

Abstract.—A Feynman-type integral over an abstract Hilbert space is defined in terms of approximations which are determined by finite-dimensional projections. One obtains on this basis a theory that appears to be an attractive alternative to other approaches. The usual specialization to nonrelativistic path integrals is discussed, and phase-space integrals as well as integrals for free fields are considered briefly.

1. Introduction.

In this article we present an alternative definition of the Feynman (path) integral and develop a few consequences. This definition resembles that of Itô [1] (cf. also [2]), insofar as both describe an integral of the form $\int \mathcal{D}(\xi) \times e^{\frac{1}{2}i\kappa\langle\xi,\xi\rangle}F(\xi)$ over an (abstract) real Hilbert space. However, our definition has allowed a theory which is slightly simpler and broader than that based on Itô's definition.

Our definition depends on cylindrical approximations to the integral, and we adapt a few ingredients from the approach of Friedrichs and Shapiro to integration over a Hilbert space [3], [4]. Our definition appears rather natural, but it contains some ad hoc elements, and many variations are possible. We do not discuss these possibilities here.

However, we discuss different forms of the integral, which are characterized by weight factors other than $e^{\frac{1}{2}i\kappa\langle\xi,\xi\rangle}$. Such integrals can describe e.g. quantized fields.—We may note in this connection that the usual path integral for nonrelativistic quantum mechanics corresponds to the form given above, with the Hilbert space determined by

$$\langle\dot{\eta},\dot{\eta}\rangle = \int_0^t d\tau \sum_{j=1}^n (d\eta^j/d\tau)^2 \quad \text{and} \quad \eta(0) = 0 \quad , \qquad (1.1)$$

for the case of a particle on R^n.

We should also say that the object of this investigation are Feynman integrals and their mathematical structure, but not their applications. The reader should not be surprised therefore, that this article contains largely the adaptations of available proofs into a new framework. In particular, we give an (incomplete) discussion of rather general potentials, by adapting some methods used by Nelson [5].

Section 2 includes the definition, and secs. 3 and 4 describe simple consequences concerning integrability and interchange of limits. In sec. 5 we investigate path integral representations for Schrödinger Green's functions. In secs. 6 and 7 we discuss some other forms of Feynman-type integrals. A few proofs are relegated to the appendix. We include also a supplement, to rectify some shortcomings of a previous paper (on Feynman integrals for quantized fields).

An addendum follows the appendix, and is devoted to Green's functions for the case of bounded potentials.

The author expresses his thanks to Drs. K. Gawędzki and A. Uhlmann for discussions related to the supplement. Furthermore, he presented preliminary versions of the material that follows at various seminars, especially at the University of Colorado in Boulder. This article was completed at T.U. Clausthal. The author thanks Professors W.E. Brittin and H.D. Doebner for hospitality at the respective institutions.

2. The definition.

We fix a parameter κ (occurring in $e^{\frac{1}{2}i\kappa\langle\xi,\xi\rangle}$) such that $\mathrm{Im}\,\kappa \geqslant 0$, $\kappa \neq 0$. The dependence of the integrals on κ often will not be shown explicitly.

We will refer to the following set of projections over a real Hilbert space $\mathcal{H}$,

$$\mathrm{Pr} := \{(P\colon \mathcal{H}\to\mathcal{H})\colon\ P \text{ linear},\ P^2 = P = P^*,\ \dim P < \infty\}. \qquad (2.1)$$

Given $P \in \mathrm{Pr}$ with $\dim P = q$, we select an orthonormal basis $x_1, x_2, \ldots$ such that $P\mathcal{H}$ is spanned by $\{x_1,\ldots,x_q\}$. Each $\xi \in \mathcal{H}$ can then be written as $(\xi^1, \xi^2, \ldots)$ with

$$P\xi = (\xi^1, \ldots, \xi^q, 0, 0, \ldots) \qquad (2.2)$$

Now let $\alpha \in \mathcal{H}$, let $b \in C^1$ satisfy $\mathrm{Re}\, b > 0$, and let $F\colon \mathcal{H}\to C^1$ be the function to be integrated. (The necessary regularity conditions on F are implied by the subsequent formulas.) We set, in the sense of L_1-convergence,

$$I_P^{b,\alpha}(F) = c\int d^qu\, e^{-\frac{1}{2}b\langle u-P\alpha,\ u-P\alpha\rangle}\, e^{\frac{1}{2}i\kappa\langle u,u\rangle} F(u^1,\ldots,u^q,0,0,\ldots)\ , \qquad (2.3a)$$

where c is defined by

$$I_P^{b,0}(1) = 1 \ . \qquad (2.3b)$$

Explicitly,

$$c = [(b - i\kappa)/2\pi]^{\frac{1}{2}q}\ , \qquad (2.4a)$$

with

$$-\tfrac{1}{2}\pi < \arg(b - i\kappa)^{\frac{1}{2}} < \tfrac{1}{2}\pi\ . \qquad (2.4b)$$

We next introduce a set of sequences $\{P_j\}$,

$$\hat{\mathcal{Q}} := \{\{P_j\}\colon\ P_k \in \mathrm{Pr} \text{ and } P_{k+1} \geqslant P_k \text{ for } \forall k,\ \lim_{j\to\infty} P_j = 1\}. \qquad (2.5)$$

If $\{P_j\} = \Pi \in \hat{\mathcal{Q}}$, we set

$$I_{\Pi}^{b,\alpha}(F) = \lim_{j\to\infty} I_{P_j}^{b,\alpha}(F) \quad , \tag{2.6}$$

and if $\mathcal{P} \subseteq \hat{\mathcal{Q}}$ is a family of sequences such that

$$\Pi', \Pi'' \in \mathcal{P} \Rightarrow I_{\Pi'}^{b,\alpha}(F) = I_{\Pi''}^{b,\alpha}(F) \quad , \tag{2.7}$$

then we denote by $I_{\mathcal{P}}^{b,\alpha}(F)$ the common value. We next set

$$I_{\Pi}(F) = \lim_{b\to 0} I_{\Pi}^{b,\alpha}(F) \ , \qquad I_{\mathcal{P}}(F) = \lim_{b\to 0} I_{\mathcal{P}}^{b,\alpha}(F) \quad , \tag{2.8}$$

with the provisions that (a) there be a unique nontangential limit as $b\to 0$, and that (b) the limit be independent of α. We comment below on these provisions. Now, if the last limit exists, we say that F is (Feynman) integrable with reference to the family $\mathcal{P}$.

The polygonal approximations as described in [6] and elsewhere correspond roughly to the above procedure and to a particular choice of families $\mathcal{P}$. These families are meaningful only for spaces of paths η such as in (1.1). However, we want to give a more general definition of the Feynman integral. As it is sometimes inconvenient to require integrability with reference to the maximal family $\hat{\mathcal{Q}}$, we proceed as follows.

Definition 1. A family $\mathcal{P} \subseteq \hat{\mathcal{Q}}$ is called determining if it is nonempty, and if

$$\{P_j\} \in \mathcal{P} \ , \quad \{Q_j\} \in \hat{\mathcal{Q}} \ , \quad \text{and} \quad P_k \leq Q_k \quad \text{for } \forall k \Rightarrow \{Q_j\} \in \mathcal{P} \quad . \tag{2.9}$$

In other words, $\mathcal{P}$ must contain all sequences which are, in a sense, sufficiently large.

Lemma 2. Let $\mathcal{P}$ and $\mathcal{P}'$ be determining. Then: (a) $\mathcal{P} \cap \mathcal{P}'$ is determining. (b) If $I_{\mathcal{P}}(F)$, $I_{\mathcal{P}'}(F)$ are both defined, then

$$I_{\mathcal{P}}(F) = I_{\mathcal{P}'}(F) \quad . \tag{2.10}$$

We see that by restricting ourselves to determining families, we obtain integrals which are uniquely determined.

Proof: For (a): Let $\{P_j\} \in \mathcal{P}$, $\{P_j'\} \in \mathcal{P}'$. Then one sees that

$$\{P_j \vee P_j'\} \in \mathcal{P} \cap \mathcal{P}' \quad , \tag{2.11}$$

where $P \vee Q$ denotes the l.u.b. of P and Q in Pr. Thus $\mathcal{P} \cap \mathcal{P}'$ is nonempty. Moreover, (2.9) is clearly valid for $\mathcal{P} \cap \mathcal{P}'$. For (b): $I_{\mathcal{P}}(F) = I_{\mathcal{P} \cap \mathcal{P}'}(F) = I_{\mathcal{P}'}(F)$. □

Definition 3. If $I_{\mathcal{P}}(F)$ exists for a determining family $\mathcal{P}$, then this quantity is called the Feynman integral of F (with the weight $e^{\frac{1}{2}i\kappa\langle\xi,\xi\rangle}$).

We will denote such integrals by $I(F)$, and also in the way given before, so that

$$I(F) = \int \mathcal{D}(\xi) e^{\frac{1}{2} i \kappa \langle \xi, \xi \rangle} F(\xi) \quad . \tag{2.12}$$

The reference family is suppressed in such notation. However, sometimes it is necessary to know the family in order to justify the manipulations.

In the sequel we will refer to the determining families $\mathcal{Q}(P)$ which are defined in terms of projections $P \in Pr$ in the following way:

$$\mathcal{Q}(P) := \left\{ \{P_j\} \in \hat{\mathcal{Q}} : \; P_k \geqslant P \;\text{ for } \forall k \right\} \quad . \tag{2.13}$$

In particular, $\mathcal{Q}(0) = \hat{\mathcal{Q}}$.

When we say in this article that F is integrable, we refer to the existence of $I(F)$, unless another kind of integral is explicitly stated.

Let us return to the limits $b \to 0$. It is clear that if $I_{P_j}^{b,\alpha}(F)$ is defined for $\operatorname{Re} b > 0$ near $b = 0$ (explicitly, if it is defined on $N_0 \cap \{b \in C^1 : \operatorname{Re} b > 0\}$, where N_0 is any neighborhood of 0), then it is analytic in b when $\operatorname{Re} b > 0$. We recall, moreover, that a limit of analytic functions is analytic, if the elements of the sequence are uniformly bounded (theorem of Vitali, [5] [7] [8]). In the general case, the limit function may have disconnected domains of analyticity [7].

If $I_{\pi}^{b,\alpha}(F)$ is analytic in b when $\operatorname{Re} b > 0$, then the limit $b \to 0$ may be taken in an arbitrary way from within a sector $|\arg b| < \frac{1}{2}\pi - \delta$, $\delta > 0$ [8], and such a limit is called nontangential. In the examples that we consider the $I_{\pi}^{b,\alpha}(F)$ are indeed analytic. However, if a given $I_{\pi}^{b,\alpha}(f)$ is not analytic, then the existence of a unique nontangential limit has to be regarded as a separate assumption.

Let us also comment on the condition that $\lim_{b \to 0} I_{\pi}^{b,\alpha}(F)$ be independent of α. This condition is used in particular to prove the translational invariance of the integral (proposition 4). However, this condition has been awkward to establish for various functions, and a surprisingly large part of this paper is devoted to the ensuing complications. At the same time, we have not found an example which would show an α-dependent limit. A closer examination of the α-dependence should be worthwhile. (We call α sometimes a shift vector.)

Finally, we should like to cite an example of $I_{\pi}(f)$ where one has dependence on the sequence π. We refer to proposition 6 below and to the accompanying discussion. If the operator which occurs there is not of trace class, then its trace is given by a series which is at best conditionally convergent. In such cases the value of $I_{\pi}^{b,\alpha}(f)$ may depend on the chosen orthonormal basis, or equivalently, on the chosen sequence π.

3. Elementary integrability properties.

Most of the results of this section have also been deduced on the basis of Itô's definition, see [1] [2].

We take $\mathcal{H}$ and κ as before. The projections P, P_j', etc. will always be elements of Pr. If $\alpha \in \mathcal{H}$, $\alpha \neq 0$, then we denote by P_α the element of Pr such that

$$P_\alpha \zeta = \alpha\langle\alpha,\zeta\rangle/\langle\alpha,\alpha\rangle \quad \text{for } \forall\zeta\in\mathcal{H} \quad . \tag{3.1}$$

Proposition 4. Let F be integrable (with reference to a determining family $\mathcal{P}$), let $\beta \in \mathcal{H}$, and let R be an orthogonal transformation of $\mathcal{H}$. Then $F(R^{-1}\cdot)$ and $e^{-i\kappa\langle\cdot,\beta\rangle}F(\cdot-\beta)$ are also integrable (with reference to the determining families $R^{-1}\mathcal{P}R$ and $\mathcal{P}$, respectively), and

$$\int\mathcal{D}(\xi)\; e^{\frac{1}{2}i\kappa\langle\xi,\xi\rangle}\,F(\xi) = \int\mathcal{D}(\xi)\; e^{\frac{1}{2}i\kappa\langle\xi,\xi\rangle}F(R^{-1}\xi) \tag{3.2}$$

$$= e^{\frac{1}{2}i\kappa\langle\beta,\beta\rangle}\int\mathcal{D}(\xi)\; e^{\frac{1}{2}i\kappa\langle\xi,\xi\rangle}\; e^{-i\kappa\langle\xi,\beta\rangle}\;F(\xi-\beta) \quad . \tag{3.3}$$

The meaning of $R^{-1}\mathcal{P}R$ is evident.—In (3.3) the exponentials may be combined into $\exp(\frac{1}{2}i\kappa\langle\xi-\beta,\xi-\beta\rangle)$. Equations (3.2)-(3.3) can therefore be interpreted as

$$\mathcal{D}(R\xi) = \mathcal{D}(\xi) = \mathcal{D}(\xi+\beta) \quad . \tag{3.4}$$

Proof: For (3.2): We first observe that

$$P \leq P' \iff RPR^{-1} \leq RP'R^{-1} \quad . \tag{3.5}$$

From this and from the assumption that $\mathcal{P}$ is determining it follows that $R^{-1}\times\mathcal{P}R$ is also determining. We now take $\{P_j\}\in\mathcal{P}$ and use the rotational invariance of the approximating integrals to conclude:

$$I_{P_j}^{\,b,R\alpha}(F) = I_{R^{-1}P_jR}^{\,b,\alpha}(F(R^{-1}\cdot)) \quad . \tag{3.6}$$

Taking $\lim_{b\to 0}\lim_{j\to\infty}$ in (3.6) yields (3.2) (and $R^{-1}\mathcal{P}R$ is the new reference family).—For (3.3): The change of variable $\xi\to\xi-\beta$ in the approximating integrals yields

$$I_{P_j}^{\,b,\alpha-\beta}(F) = e^{\frac{1}{2}i\kappa\langle P_j\beta,\,P_j\beta\rangle}\; I_{P_j}^{\,b,\alpha}(e^{-i\kappa\langle\cdot,\beta\rangle}F(\cdot-\beta)). \tag{3.7}$$

We apply $\lim_{b\to 0}\lim_{j\to\infty}$ as before, and obtain (3.3).

Proposition 5. Let μ be a complex Borel measure on $\mathcal{H}$ of bounded absolute variation. Then the function

$$f(\xi) = \int d\mu(\chi)\;\exp(i\langle\chi,\xi\rangle) \tag{3.8a}$$

is integrable with reference to $\hat{\mathcal{Q}}$, and

$$I(f) = \int d\mu(\chi)\;\exp\left(\tfrac{1}{2}(i\kappa)^{-1}\langle\chi,\chi\rangle\right) \quad . \tag{3.8b}$$

In particular, if we take for μ the δ-measure at the origin, then we see that the function $f(\xi)=1$ is integrable, and

$$I(1) = 1 \quad . \tag{3.9}$$

Proof: In the approximating integrals we may interchange the order of integrations, and then a routine evaluation yields

$$I_{P_j}^{b,\alpha}(f) = \exp\left(\frac{1}{2}\frac{ib\kappa}{i\kappa - b}\langle\alpha, P_j\alpha\rangle\right)\int d\mu(\chi)$$
$$\times \exp\left[\tfrac{1}{2}(i\kappa - b)^{-1}\langle P_j\chi, P_j(\chi - 2ib\alpha)\rangle\right] \quad . \qquad (3.10)$$

The vector $P_j(\chi - \ldots)$ lies in $\mathcal{H} + i\mathcal{H}$, and the indicated scalar product is linear in the second argument. Since $\mathrm{Re}(i\kappa - b) < 0$, it follows that we have a damped Gaussian in the integrand. With some arithmetic one can establish the following bound. Let us write $\kappa = \kappa_1 + i\kappa_2$, $b = b_1 + ib_2$, with $b_1 \geqslant 0$, $\kappa_2 \geqslant 0$, and let us suppose that $\kappa_1 \neq 0$. (The case $\kappa_1 = 0$, $\kappa_2 > 0$ is similar but simpler.) We restrict b further by $|b_2| \leqslant \frac{1}{2}|\kappa_1|$. Then

$$\left|\exp\left[\tfrac{1}{2}(i\kappa - b)^{-1}\langle P_j\chi, P_j(\chi - 2ib\alpha)\rangle\right]\right| \leqslant \exp\left(4|\kappa_1|\langle\alpha,\alpha\rangle\right) \quad . \qquad (3.11)$$

This bound is independent of P_j and of b. Therefore, in view of the hypothesis on μ and of the bounded convergence theorem, we may take the limits $j \to \infty$ and $b \to 0$ in (3.10) inside the integral sign. □

We also note: The approximations $I_{P_j}^{b,\alpha}(f)$ are analytic in b and have a uniform bound implied by (3.11). Thus the limit as $j \to \infty$ is also analytic in b (at least if $|b_2| < \frac{1}{2}|\kappa_1|$; cf. Vitali's theorem, sec. 2).

The next proposition is an adaptation of a familiar result on Gaussian integrals of polynomials [3], [9]. A polynomial function is a sum of monomials. Of particular interest are those of even rank, and such a monomial of rank $2n$ can be expressed as

$$m_{2n}(\xi) = \langle \xi\otimes\ldots\otimes\xi,\; M_{2n}(\xi\otimes\ldots\otimes\xi)\rangle \quad , \qquad (3.12)$$

where $M_{2n}: \mathcal{H}_{sym}^{\otimes n} \to \mathcal{H}_{sym}^{\otimes n}$. (Here "sym" refers to symmetric tensors.)

Proposition 6. If M_{2n} in (3.12) is of trace class, then the monomial m_{2n} is integrable with reference to $\hat{\mathcal{Q}}$, and

$$I(m_{2n}) = (-1/i\kappa)^n \cdot 1\cdot 3\cdot \ldots \cdot(2n-1)\ \mathrm{tr}\, M_{2n} \quad . \qquad (3.13)$$

This proposition is proved in the appendix. We remark that if we set $\alpha = 0$ in the approximating integrals, then (3.13) follows readily from loc. cit.

The condition that M_{2n} be of trace class ensures that the trace does not depend on the approximating sequence of projections [10]. Cf. the remarks at the end of sec. 2.

If a monomial of odd rank is integrable, then its integral is necessarily zero. Indeed, we may then set $\alpha = 0$, and each approximating integral will vanish by symmetry. Finding a general criterion for integrability of monomials of odd rank remains, however, an open problem. It appears that such a criterion has not been established also for positive-definite Gaussian integrals. (We note that monomials which are cylinder functions are integrable. See the remarks following proposition 9.)

Proposition 7. Let $L: \mathcal{H} \to \mathcal{H}$ be a symmetric trace-class operator and let κ be real. Moreover, let κ and $\kappa + L$ be both >0 or both <0. Then

$$g(\xi) = e^{\frac{1}{2}i\langle \xi, L\xi\rangle} \Rightarrow I(g) = [\det(1 + \kappa^{-1}L)]^{-\frac{1}{2}} , \qquad (3.14)$$

where $\det(1+\kappa^{-1}L) > 0$ and $I(g) > 0$. The reference family is $\hat{\mathcal{Q}}$.

The determinant can be expressed in terms of the eigenvalues $l_1, l_2, \ldots$ of L as follows:

$$\det(1 + \kappa^{-1}L) = \prod_{j=1}^{\infty} (1 + \kappa^{-1} l_j) \quad . \qquad (3.15)$$

The hypothesis implies that each factor is >0, and the condition $\sum |l_j| < \infty$ implies convergence (to a nonzero value): $0 < \det(1+\kappa^{-1}L) < \infty$. The assumptions of reality, or of positivity of $1+\kappa^{-1}L$, can be relaxed, but then the determination of the square root is less direct.

On the other hand, it is essential to have L of trace class in order to have absolute convergence and independence of the approximating sequence of projections (as in proposition 6).

Proof: The approximating integrals $I_{P_j}{}^{b,\alpha}(g)$ can be done in closed form by standard methods (see e.g. [2], Appendix (a); we refer to $\mathcal{H} + i\mathcal{H}$, as before):

$$I_{P_j}{}^{b,\alpha}(g) = (\det[1 - i(b - i\kappa)^{-1}P_jLP_j])^{-\frac{1}{2}} \exp(-\tfrac{1}{2}b\langle P_j\alpha, P_j\alpha\rangle)$$
$$\times \exp(\tfrac{1}{2}b^2\langle P_j\alpha, [(b - i\kappa - iL)\restriction P_j\mathcal{H}]^{-1} P_j\alpha\rangle) . \qquad (3.16)$$

Since λL is of trace class (here λ is a scalar), the determinant approaches a limit as $j\to\infty$ which is independent of the chosen sequence $\{P_j\}$, and the result depends continuously on λ [11]. Explicitly,

$$\det[1 - i(b - i\kappa)^{-1}P_jLP_j] \xrightarrow[j\to\infty]{} \det[1 - i(b - i\kappa)^{-1}L] \xrightarrow[b\to 0]{} \det(1 + \kappa^{-1}L). \quad (3.17)$$

Note that a judicious choice of the sequence $\{P_j\}$ will yield eq. (3.15), which shows (as above) that $\det(..) > 0$. The choice of the square root can then be justified by examining one-dimensional eigen-subspaces of L.

We return to (3.16) and consider the last scalar product there. Let us show that it approaches finite limits as $j\to\infty$ and $b\to 0$. We extend the domain of the operator $[(b - \ldots]^{-1}$ of (3.16) from $P_j\mathcal{H}$ to $\mathcal{H}$ without changing the value of the scalar product, as follows:

$$\langle P_j\alpha, [(b - ..)\restriction P_j\mathcal{H}]^{-1} P_j\alpha\rangle = \langle P_j\alpha, (b - i\kappa - iP_jLP_j)^{-1}P_j\alpha\rangle . \qquad (3.18a)$$

Now, it follows from the hypothesis that $\kappa + L$ is bounded from below, and therefore, that $(\kappa + L)^{-1}$ is bounded (cf. [12], p. 268). Therefore theorem 1.5 of [12], p. 429, implies that

$$\text{s-lim}_{b\to 0}\ \text{s-lim}_{j\to\infty} (b - i\kappa - iP_jLP_j)^{-1} = i(\kappa + L)^{-1} \quad . \tag{3.18b}$$

Since also s-lim $P_j\alpha = \alpha$, we conclude that

$$\lim_{b\to 0} \lim_{j\to\infty} \langle P_j\alpha, (\ldots)^{-1}P_j\alpha\rangle = i\langle \alpha, (\kappa + L)^{-1}\alpha\rangle \quad . \tag{3.18c}$$

This is a finite quantity, and so the last exponent of (3.16) $\to 0$ as $b\to 0$. The same conclusion holds for the other exponent. Therefore (3.16)-(3.17) yield in the limit the desired answer (3.14). □

One can also show that $\lim_{j\to\infty} I_{P_j}{}^{b,\alpha}(g)$ is analytic in b when Re $b>0$, by considering separately the three factors in (3.16). Analyticity is evident for the first exponent, and for the second follows from known properties of resolvents. (See [12], p. 367. The resolvent in question is $(b - i\kappa - iL)^{-1}$.) Finally, the approximating determinants are analytic in b. If b is restricted to a compact subset of $\{\text{Re } b>0\}$, then these can be bounded uniformly by $\sup_j \times \det(1+\lambda_0 P_j|L|P_j)$, where $|L| = (L^*L)^{\frac{1}{2}}$ and λ_0 is a suitable number >0. The sequence $\{\det(1+\lambda_0 P_j|L|P_j)\}$ is convergent and therefore bounded. Thus by Vitali's theorem the limiting determinant is also analytic, and our assertion follows.

The next proposition gives a form of Fubini's theorem.

Proposition 8. Let $\mathcal{H} = \mathcal{H}_1 \dot{+} \mathcal{H}_2$, where $\mathcal{H}_1$ is finite-dimensional. Let $f(\xi_1,\xi_2) = f_1(\xi_1)f_2(\xi_2)$, $\xi_j \in \mathcal{H}_j$, with $I^{(\mathcal{H}_j)}(f_j)$ both defined. Then f is integrable over $\mathcal{H}$ (with reference to $\mathcal{P}$ given below), and

$$I(f) = I^{(\mathcal{H}_1)}(f_1)\ I^{(\mathcal{H}_2)}(f_2) \quad . \tag{3.19}$$

Proof: Let $P_{(1)} \in \Pr(\mathcal{H})$ project onto $\mathcal{H}_1$. Let f_2 be integrable with reference to a determining family $\mathcal{P}^{(\mathcal{H}_2)}$, and consider

$$\mathcal{P} := \left\{\{P_j\} \in \hat{\mathcal{Q}}^{(\mathcal{H})}:\ P_k \geq P_{(1)} \text{ for } \forall k, \text{ and } \{P_j \restriction \mathcal{H}_2\} \in \mathcal{P}^{(\mathcal{H}_2)}\right\}, \tag{3.20}$$

which is determining. Note that $\mathcal{P} \subseteq \mathcal{Q}^{(\mathcal{H})}(P_{(1)})$. For $\{P_j\} \in \mathcal{P}$ the approximating integrals factorize, and the limits $j\to\infty$ and $b\to 0$ exist for the factors, hence for the product. □

Proposition 9. Let F be a cylinder function on $\mathcal{H}$, satisfying $F(\xi) = F(P\xi)$ for some $P \in \Pr$, and set $\dim P = q$. Then the following are equivalent:
(1) F is (Feynman) integrable over $R^q = P\mathcal{H}$.
(2) F is integrable over $\mathcal{H}$, with reference to $\hat{\mathcal{Q}}$.
(3) F is integrable over $\mathcal{H}$, with reference to some determining family $\mathcal{P}$.
Furthermore, in case of integrability, $I^{(R^q)}(F) = I^{(\mathcal{H})}(F)$. If also $\alpha \in P\mathcal{H}$, then $I^{(R^q)}{}_{P=1}{}^{b,\alpha}(F) = I^{(\mathcal{H})}{}_{\hat{\mathcal{Q}}}{}^{b,\alpha}(F)$, where meaning of the symbols is evident.

We say that such an F has $P\mathcal{H}$ as its base.

Note that (1), proposition 8, and eq. (3.9) imply directly that F is integrable over $\mathcal{H}$ with reference to $\mathcal{Q}(P)$, but not the stronger assertion in (2). The equivalence of (1) and (2) will be useful for us in an example in sec. 4.

Proof: In case of integrability, the first equation follows from (3.9) and (3.19). Next, the part (2)$\Rightarrow$(3) is immediate. For (3)$\Rightarrow$(1): We take $\{P_j\} \in \mathcal{P}$ and consider $\{P_j \vee P\} \in \mathcal{P}$. Then the approximating integrals factorize. The integrals over $((P_j \vee P) - P)\mathcal{H}$ are trivial, and the assumed existence of $\lim_{b\to 0} \times \lim_{j\to\infty}$ for the products of integrals implies the existence of the corresponding limits for the nontrivial factors. The part (1)$\Rightarrow$(2) and the second equation are proved in the appendix. (□)

Examples of integrable functions over R^q can be found in [13] (where a stronger criterion of integrability was adopted). For instance, functions which are restrictions to R^q of entire functions of order <2 are integrable. In particular, polynomials are integrable.

We also include for reference:

Proposition 10. Let F be one of the functions f, m_{2n}, or g of propositions 5-7, or an integrable cylinder function. Then $I_{\hat{\mathcal{E}}}^{b,\alpha}(F)$ is analytic in b when $\mathrm{Re}\, b > 0$.

Proof: The case of an integrable cylinder function follows from analyticity of $I^{(R^q)}{}_{P=1}{}^{b,\alpha}(F)$ and from the second equality of proposition 9. (See also the remarks in sec. 2.) If $\alpha \notin P\mathcal{H}$, we can consider F as a function based on $(P \vee P_\alpha)$. Next, the case of g was discussed fully above. The case of m_{2n} follows from the proof in the appendix, where $I_{\hat{\mathcal{E}}}^{b,\alpha}(m_{2n})$ is shown to be a rational function of b. Finally, the case of f was covered previously, under the additional restriction $|\mathrm{Im}\, b| < \frac{1}{2}|\mathrm{Re}\, \kappa|$. For other subsets of $\{\mathrm{Re}\, b > 0\}$ bounds analogous to (3.11) can be obtained, and these imply analyticity. (Cf. the discussion of g.) □

4. Examples of interchange of limits.

The general problem of justifying interchanges of Feynman integration with other limits does not appear easy. However, we can give two examples where such interchanges can be easily justified.

The first example relates to integration by parts and to the action principle, and is adapted from [2]. It can be motivated by the following heuristic manipulations. We start with

$$\int \mathcal{D}(\xi)\, e^{\frac{1}{2}i\kappa\langle\xi,\xi\rangle} F(\xi) = \int \mathcal{D}(\xi)\, e^{\frac{1}{2}i\kappa\langle\xi + a\zeta,\, \xi + a\zeta\rangle} F(\xi + a\zeta) \quad , \qquad (4.1)$$

cf. (3.3) (here $a \in R^1$). We apply $(d/da)_{a=0}$ and interchange this with the integration:

$$0 = \int \mathcal{D}(\xi)\, e^{\frac{1}{2}i\kappa\langle\xi + a\zeta,\, \xi + a\zeta\rangle} \Big[(i\kappa\langle\xi,\zeta\rangle + i\kappa a\langle\zeta,\zeta\rangle)\, F(\xi + a\zeta) + $$
$$+ (d/da)F(\xi + a\zeta)\Big]\Big|_{a=0}$$
$$= \int \mathcal{D}(\xi)\, e^{\frac{1}{2}i\kappa\langle\xi,\xi\rangle} \Big[i\kappa\langle\xi,\zeta\rangle F(\xi) + D_\zeta F(\xi)\Big] \quad , \qquad (4.2)$$

where D_ζ denotes the Gâteau derivative.

Proposition 11. Equation (4.2) is valid if F, $\langle\cdot,\zeta\rangle F$, and $D_\zeta F$ are all integrable.

Proof: Let the three families of reference be $\mathcal{P}_1$, $\mathcal{P}_2$, $\mathcal{P}_3$, let $\mathcal{P} = \mathcal{Q}(P_\zeta)\cap\mathcal{P}_1\cap\mathcal{P}_2\cap\mathcal{P}_3$, and let $\{P_j\}\in\mathcal{P}$. In the approximating integrals we change variables and apply $(d/da)_{a=0}$ to two expressions. In view of the assumed integrability, we may interchange d/da with integration in the approximating integrals. Therefore, upon letting $q = \dim P_j$,

$$\frac{d}{da}\, I_{P_j}{}^{b,-a\zeta}(f)\Big|_{a=0} = \left(\frac{b-i\kappa}{2\pi}\right)^{\frac{1}{2}q}\int d^q u\left[\frac{d}{da}\, e^{-\frac{1}{2}b\langle u-a\zeta,\, u-a\zeta\rangle}\right]_{a=0} e^{\frac{1}{2}i\kappa\langle u,u\rangle}F(u) =$$

$$= \left(\frac{b-i\kappa}{2\pi}\right)^{\frac{1}{2}q}\int d^q u\, e^{-\frac{1}{2}b\langle u,\,u\rangle}\left[\frac{d}{da}\, e^{\frac{1}{2}i\kappa\langle u+a\zeta,\, u+a\zeta\rangle}F(u+a\zeta)\right]_{a=0} \quad . \tag{4.3a}$$

Evaluation of the derivatives yields

$$b\, I_{P_j}{}^{b,0}(\langle\cdot,\zeta\rangle F) = I_{P_j}{}^{b,0}(i\kappa\langle\cdot,\zeta\rangle F + D_\zeta F) \quad . \tag{4.3b}$$

Taking $\lim_{b\to 0}\lim_{j\to\infty}$ now yields (4.2). We need to consider only $\alpha = 0$ in (4.3b), since the assumed integrability implies independence of α. □

We remark that this proof did not make use of the relation (4.1).

The next example refers to the path integral and to the scalar product $\langle\dot\eta,\dot\eta\rangle$, as in (1.1). The time interval will be $[0,t]$. For brevity in writing we take the case of a particle on R^1.

It is significant that we can express the value $\eta(T)$ as a scalar product, as follows,

$$\theta_T(\tau) := \min(\tau,T) \Rightarrow \eta(T) = \int_0^T d\tau\,\dot\eta(\tau) = \langle\dot\theta_T,\dot\eta\rangle \quad . \tag{4.4}$$

Consequently, if F is (Feynman) integrable over R^1, then $F(\eta(T))$ is integrable over $\mathcal{H}$, cf. proposition 9.

Lemma 12. Let F be (Feynman) integrable over R^1, let $0<T\leq t$, and let

$$g(T,\kappa) = \int_{\eta(0)=0}\mathcal{D}(\eta)\; e^{\frac{1}{2}i\kappa\langle\dot\eta,\dot\eta\rangle}\, F(\eta(T)) \quad . \tag{4.5}$$

Then g depends on T and κ through the combination T/κ.

Proof: We observe that $\langle\dot\theta_T,\dot\theta_T\rangle = \theta_T(T) = T$, and so

$$\theta_T{}^0 := T^{-\frac{1}{2}}\theta_T \qquad \text{satisfies} \qquad \langle\dot\theta_T{}^0,\, \dot\theta_T{}^0\rangle = 1 \quad . \tag{4.6}$$

We make a change of variable in (4.5) and use the rotational invariance property of proposition 4, as follows (the restriction to $\eta(0) = 0$ is to be understood):

$$\int\mathcal{D}(\eta)\; e^{\frac{1}{2}i\kappa\langle\dot\eta,\dot\eta\rangle}F(\eta(T)) = \int\mathcal{D}(\eta)\; e^{\frac{1}{2}i\kappa\langle\dot\eta,\dot\eta\rangle}F(\langle\dot\theta_T{}^0,\, T^{\frac{1}{2}}\dot\eta\rangle) =$$

$$= \int\mathcal{D}(\eta)\; e^{\frac{1}{2}i(\kappa/T)\langle\dot\eta,\dot\eta\rangle}F(\langle\dot\theta_T{}^0,\dot\eta\rangle) = \int\mathcal{D}(\eta)\; e^{\frac{1}{2}i(\kappa/T)\langle\dot\eta,\dot\eta\rangle}F(\langle\beta,\dot\eta\rangle) \quad , \tag{4.7}$$

for any $\beta\in\mathcal{H}$ of unit norm. □

<u>Proposition 13.</u> Let F be integrable over R^1 (as in lemma 12) and be such that the integral $g(T/\kappa)$ is a bounded and measurable function when T varies over $[s,t]$, where $s>0$. We assume, moreover, that $\int_s^t dT\, F((T))$ is integrable over $\mathcal{H}$, with reference to $\hat{\mathfrak{a}}$. Then

$$\int_{\eta(0)=0} \mathcal{D}(\eta)\, e^{\frac{1}{2}i\kappa\langle\dot\eta,\dot\eta\rangle} \int_s^t dT\, F(\eta(T)) = \int_s^t dT \int_{\eta(0)=0} \mathcal{D}(\eta)\, e^{\frac{1}{2}i\langle\dot\eta,\dot\eta\rangle} F(\eta(T)) \; . \tag{4.8}$$

The strong hypothesis enables us to give a rather short proof. We expect, however, that integrability of $\int dT\, F$ could be established by additional but similar arguments.

Expressions resembling those in (4.8) (but differing in details) occur in perturbation expansions of Schrödinger Green's functions. In the case $F(T) = T^{2N}$ where N is a positive integer (this corresponds to an anharmonic oscillator), (3.13) yields the following evaluation for the two members of (4.8):

$$(-1/i\kappa)^N \cdot 1\cdot 3\cdot \ldots \cdot (2N-1) \int_s^t dT\, T^N \; . \tag{4.9}$$

<u>Proof</u>: Let $\Pi = \{P_j\} \in \hat{\hat{\mathfrak{a}}}$ (a particular form of the P_j's will be specified below), and we take $\alpha = 0$. In view of proposition 9, we may also use Π for approximating the Feynman integrals on the r.h.s. of (4.8), so that

$$\text{r.h.s.} = \int_s^t dT\, \lim_{b\to 0} \lim_{j\to\infty} I_{P_j}{}^{b,0}(F(\eta(T))) \; . \tag{4.10}$$

Interchanging $\int dT$ with the two limits will give the l.h.s., and the problem is to justify these interchanges. We will use the notation $F_T(\eta) = F(\eta(T))$.

First of all, an extension of lemma 12 (and previous observations) imply that $I_{\hat{\mathfrak{a}}}{}^{b,0}(F_T)$ is an analytic function of $T/(b-i\kappa)$ when $\mathrm{Re}(b-i\kappa)>0$. Since this function remains bounded when $b\to 0$, $T\in[s,t]$, it must also be bounded on each set of a system which allows nontangential limits. This conclusion and the bounded convergence theorem allow interchanging $\int dT$ with $\lim_{b\to 0}$.

For justifying the second interchange we use a technique due to Friedrichs and Shapiro ([3], chapter V of notes). We consider θ_s, $\theta_s{}^0$ as in (4.4), (4.6), we write Q_s for P_{θ_s}, and we want to evaluate $I_{Q_s}{}^{b,0}(F_T)$. The projection Q_s entails a replacement of $\eta(T)$, as follows:

$$\eta(T) = T^{\frac{1}{2}}\langle\dot\eta,\dot\theta_T{}^0\rangle = \langle\dot\eta,\dot\theta_T\rangle \to \langle Q_s\dot\eta,\dot\theta_T\rangle = \langle\dot\eta,\dot\theta_s{}^0\rangle\,\langle\dot\theta_s{}^0,\dot\theta_T\rangle = s^{\frac{1}{2}}\langle\dot\eta,\dot\theta_s{}^0\rangle \; . \tag{4.11}$$

Therefore the approximating integrals of F_T reduce to one-dimensional ones in the following way:

$$I^{(R^1)}{}_{P=1}{}^{b,0}(F_T) = I_{\hat{\mathfrak{a}}}{}^{b,0}(F_T) = [(b-i\kappa)/2\pi]^{\frac{1}{2}} \int_{-\infty}^{\infty} du\, e^{-\frac{1}{2}(b-i\kappa)u^2} F(T^{\frac{1}{2}}u) \; , \tag{4.12a}$$

$$I_{Q_s}{}^{b,0}(F_T) = [(b-i\kappa)/2\pi]^{\frac{1}{2}} \int_{-\infty}^{\infty} du\, e^{-\frac{1}{2}(b-i\kappa)u^2} F(s^{\frac{1}{2}}u) \; . \tag{4.12b}$$

Let us now suppose that each P_j projects onto the linear span of $\{\theta_s, \theta_{\tau_1},\ldots,\theta_{\tau_{n_j}}\}$ for some $\tau_1,\ldots,\tau_{n_j}$. Short calculations show that $I_{P_j}{}^{b,0}(F_T)$ will be such as in (4.12), except that the argument of F

will be $\sigma^{\frac{1}{2}}u$ where $s \leq \sigma \leq T$. By changing the variable of integration we may obtain a bound independent of T and j,

$$\left|I_{P_j}{}^{b,0}(F_T)\right| \leq \left|(b - i\kappa)/2\pi\right|^{\frac{1}{2}} \frac{1}{s^{1/2}} \int_{-\infty}^{\infty} dv \left| e^{-\frac{1}{2}(b-i\kappa)v^2/t} F(v)\right| \quad , \qquad (4.13)$$

and this bound allows the desired interchange.

5. Representation of Green's functions.

In this section we investigate in our framework the usual expression for the Schrödinger Green's function:

$$G(t;y,x) = \int_{\eta(o)=o} \mathcal{D}(\eta)\ e^{\frac{1}{2}im\langle\dot\eta,\dot\eta\rangle} \exp\left[-i\int_o^t d\tau\ V(\eta(\tau)+x)\right] \delta(\eta(t)+x-y) \quad . \qquad (5.1a)$$

The scalar product $\langle\dot\eta,\dot\eta\rangle$ and the space $\mathcal{H}$ of paths are as in (1.1), for a particle on R^n, and the time interval is $[0,t]$. Note that if we used paths taking the value $x \neq 0$ at $\tau = 0$, then such paths would not form a linear space. —In general G is a distribution, which defines an integral operator, to be denoted also by G. Convergence of the path integral in (5.1a) refers therefore to convergence of the integrated expression,

$$\langle\chi,\ G\psi\rangle = \int d^n y\ d^n x\ \chi^*(y)\ \psi(x)\ G(t;y,x) \quad , \qquad (5.1b)$$

where $\chi, \psi \in L_2$.

We make a preliminary observation. Equation (4.4) can be extended to n dimensions by the vectorial relation,

$$\eta(t) = \int_o^t d\tau\ \dot\theta_t(\tau)\ \dot\eta(\tau) \quad . \qquad (5.2a)$$

Let P' project onto $\{v\theta_t:\ v \in R^n\}$, or explicitly, let

$$P' := \text{g.l.b.}\left\{P'' \in Pr:\quad P''(v\theta_t) = v\theta_t \quad \text{for} \quad \forall v \in R^n\right\} \quad . \qquad (5.2b)$$

The δ-function in (5.1a) can be conveniently handled by utilizing an integral over $(1-P')\mathcal{H}$, as in [1], or by restricting the families of reference to $\mathcal{Q}(P')$ (or to its subfamilies). We will adopt the second alternative.

Our first conclusion corresponds to the main results of [1].

Proposition 14. Let V be one of the following potentials, for a particle on R^1:
(a) $V(x) = \int d\lambda(y)\ e^{iyx}$ where $\int d|\lambda|(y) < \infty$, (b) $V(x) = cx$ where $c \in R^1$,
(c) $V(x) = \frac{1}{2}kx^2$, where $k \geq 0$ and $t < \pi(m/k)^{\frac{1}{2}}$.
Then eq. (5.1) are valid. The path integral converges with reference to $\mathcal{Q}(\theta_t)$.

Outline of proof: Integrability of the relevant functions (over $\mathcal{H}$) follows from propositions 5 and 7. The verification that one gets indeed the Green's functions is the same as in [1], since this verification depends only on the evaluations (3.8) and (3.15). (□)

We made the restrictions to R^1 and in (c) to $t < \pi(m/k)^{\frac{1}{2}}$, since these restrictions were made in [1], and we cited the calculations of loc. cit. in the proof. The measure λ in (a) can be complex.

We now turn to the case of more general potentials, where our immediate problem (that of proving convergence of the path integral) remains unsolved. We follow in part the article of Nelson [5], and we extend his results by including the arbitrary shift vector $\alpha \in \mathcal{H}$. We give also a detailed treatment of the passage to the limit $b \to 0$. However, complications develop when we try to incorporate sequences of projections $\{P_j\}$ into the analysis.

To start with, we take the above path integral with $b > 0$ and α, and with m positive imaginary, so that $m' := m/i > 0$. We set

$$G_{b,m'}(t;y,x) := \int_{\eta(o)=o} \mathcal{D}(\eta)\, e^{-\frac{1}{2}b\langle\dot{\eta}-\dot{\alpha},\dot{\eta}-\dot{\alpha}\rangle}\, e^{-\frac{1}{2}m'\langle\dot{\eta},\dot{\eta}\rangle}$$
$$\times \exp\left[-i \int_o^t d\tau\, V(\eta(\tau) + x)\right] \delta(\eta(t) + x - y) \quad , \tag{5.3a}$$

$$\langle \chi, G_{b,m'}\psi\rangle = \int d^n y\, d^n x\, \chi^*(y)\, \psi(x)\, G_{b,m'}(t;y,x) \quad . \tag{5.3b}$$

This functional integral can be interpreted in the sense of sec. 2, and also as a (conditional) Wiener integral. Both interpretations yield the same result for a wide class of potentials. We will consider this integral to be the Wiener integral. In this case the term $\int d\tau\, \dot{\eta}\dot{\alpha}$ in the exponent should be interpreted through a translation of $\mathcal{H}$, which transforms the measure into an equivalent one [14].

An examination of short-time approximations to (5.3a) suggests that

$$\partial G_{b,m'}/\partial t = \left[\tfrac{1}{2}(b + m')^{-1}(\nabla - b\dot{\alpha})^2 - \tfrac{1}{2}b\dot{\alpha}(\tau)^2 - iV(y)\right] G_{b,m'} \quad , \tag{5.4a}$$

$$\lim_{t \searrow 0} G(t;y,x) = \delta(y - x) \quad . \tag{5.4b}$$

However, it appear that the available results concerning integrals like $G_{b,m'}$ $\chi(t..)$ (see e.g. [5]) do not cover the case of $G_{b,m'}$ itself. Since we will use eq. (5.4) only as a convenient heuristic background, we did not try to prove them.

We want to adapt Trotter's approximations to the time evolution indicated by (5.3)-(5.4). We subdivide $[0,t]$ into ν subintervals by the division points $t/\nu, \ldots, (\nu - 1)t/\nu$, and set

$$\alpha_j := (\nu/t)\int_{(j-1)t/\nu}^{jt/\nu} d\tau\, \dot{\alpha}(\tau) , \quad A_j := \tfrac{1}{2}(b + m')^{-1}(\nabla - b\alpha_j)^2 \quad . \tag{5.5}$$

Here $\alpha_j \in R^n$, and we suppressed the dependence on ν of α_j and of A_j. Since α_j results from applying a projection to α, we have the bounds

$$\sum_{j=1}^{\nu} \alpha_j^{\,2} \leq \int_o^t d\tau\, \dot{\alpha}^2 = \langle\dot{\alpha},\dot{\alpha}\rangle \quad , \qquad \sum|\alpha_j| \leq (\nu\Sigma\alpha_j^{\,2})^{\frac{1}{2}} \quad , \tag{5.6}$$

where we use also Schwartz' inequality. The spectrum of A_j is the set

$$\mathrm{Sp}\, A_j = \left\{\tfrac{1}{2}(b + m')^{-1}(ik - b\alpha_j)^2 \ : \ k \in R^n\right\} \quad . \tag{5.7}$$

We now allow b and m' to be complex, $b = b_1 + ib_2$ and $m' = m_1' + im_2'$, and a short computation gives

$$\mathrm{Re}\left[(ik - b\alpha_j)^2/2(b + m')\right] \leq C_j(b,m') := \tfrac{1}{2} b_1{}^2\alpha_j{}^2 \,|b + m'|^{-2}$$
$$\times\left[(b_1 + m_1') + (b_2 + m_2')^2(b_1 + m_1')^{-1}\right] \quad . \tag{5.8}$$

Therefore for $\mathrm{Re}(b + m') > 0$, the following operator is dissipative:

$$B_j(b,m') := A_j - C_j \Rightarrow \| e^{B_j(b,m')t/\nu} \| \leq 1 \quad . \tag{5.9}$$

We now construct the approximations,

$$G_{b,m'}{}^{(\nu)} := e^{\sum_j C_j t/\nu}\left[e^{-iVt/\nu} e^{B_\nu t/\nu} \cdots e^{-iVt/\nu} e^{B_1 t/\nu}\right] e^{-\frac{1}{2}b\sum_j \alpha_j{}^2} \quad , \tag{5.10a}$$

and for $\chi, \psi \in L_2$,

$$F^{(\nu)}(b,m') := \left\langle \chi,\ G_{b,m'}{}^{(\nu)}\psi \right\rangle \quad . \tag{5.10b}$$

We return to the case $b + m' > 0$. The $G_{b,m'}{}^{(\nu)}$ and $F^{(\nu)}(b,m')$ can be expressed in terms of Wiener integrals. These integrals can be obtained from that in (5.3a) by replacing $\int d\tau\, V$ by Riemann sums, and by replacing α by $\alpha^{(\nu)}$, defined as follows: $\alpha^{(\nu)}(0) = 0$, $\alpha^{(\nu)}$ is continuous, and $\dot{\alpha}^{(\nu)} = \chi_j$ for $(j-1)t/\nu < \tau < jt/\nu$. The bounded convergence theorem then implies

$$\lim_{\nu\to\infty} F^{(\nu)}(b,m') = \left\langle \chi,\ G_{b,m'}\psi \right\rangle \quad . \tag{5.11}$$

See [5] for further details. This equation is valid for V as in the following proposition. For "capacity zero" see [15]. The terms $\int d\tau\, \dot{\eta}\dot{\alpha}^{(\nu)}$ should be eliminated by translation in $\mathcal{H}$, cf. above.

<u>Proposition 15</u>. Let V be continuous on the complement of a set of capacity zero. Then the scalar product $\langle \chi,\ G_{b,m'}\psi \rangle$ extends to a function of b and m' which is holomorphic when $\mathrm{Re}(b + m') > 0$.

We will denote this holomorphic function also by $\langle \chi,\ G_{b,m'}\psi \rangle$.

<u>Proof</u>: The approximations $F^{(\nu)}(b,m')$ are holomorphic when $\mathrm{Re}(b + m') > 0$. (Cf. [5] [7], and e.g. [16] for Hartogs' theorem.) They are also uniformly bounded by $\exp\left[\sum_j C_j(b,m')t/\nu\right]$. [Cf. (5.10). The sums $\sum C_j/\nu$ are uniformly bounded, in view of (5.6).] Moreover, the $F^{(\nu)}$ converge on the product of positive real half-axes by (5.11). Therefore, by applying Vitali's theorem to each variable in turn, and in each order, one deduces convergence to a limit function holomorphic in each variable. Hartogs' theorem completes the proof. □

The boundary behavior of $\langle \chi,\ G_{b,m'}\psi \rangle$ for some potentails is given in the following proposition, which we prove in the appendix:

<u>Proposition 16</u>. Let V be continuous on the complement of a set of capacity zero, and let it be such that $H = H_0 + V$ is self-adjoint (where $H_0 = -\nabla^2/2m$). Then

$$\lim_{b\to 0} \left\langle \chi,\ G_{b,m/i}\psi \right\rangle = \left\langle \chi,\ e^{-itH}\psi \right\rangle \quad , \tag{5.12}$$

where the limit is to be nontangential.

(When $\mathrm{Re}\, b > 0$, the point $m' = m/i$ is in the region of analyticity.)

The restriction to those $H_0 + V$ which are self-adjoint is due to our method of proof (which is based on the techniques of [5]). We expect that this proposition can be extended to essentially self-adjoint operators.

This proposition describes a limiting behavior of operators of evolution. We refer to [17], especially to lemma 6.3 and to theorem 8.2, for other results about such limits.

We should now like to relate $\langle \chi, G\psi\rangle$ of (5.1b) to $\langle \chi, e^{-itH}\psi\rangle$. This requires that we consider also sequences $\{P_j\}$ and limits as $j\to\infty$. For $\{P_j\}\in \mathcal{Q}(P')$, cf. (5.2), the approximating integrals are analytic when $\mathrm{Re}(b+m')>0$, and one has convergence as $j\to\infty$ when $b+m'>0$ (by the bounded convergence theorem, as before). However, this convergence might not extend to the complex region, in particular to small values of $\mathrm{Re}(b+m')$. It would extend if the approximations were uniformly bounded, but such a bound has not been established.

We can reduce our problem in the following way. We observe that the approximations $F^{(\nu)}(b,m')$ of (5.10) come from finite-dimensional integrals, which are the result of replacing $\int d\tau V$ by a Riemann sum $\sum V_k$ (and α by $\alpha^{(\nu)}$). But for such a finite-dimensional integral, one obtains the expected limit for any $\{P_j\}\in\mathcal{Q}(P')$, by proposition 9. We conclude:

<u>Lemma 17</u>. Let V be continuous on the complement of a set of capacity zero. If $\mathrm{Re}(b+m')>0$ and $\{P_j\}\in\mathcal{Q}(P')$, then

$$\lim_{\nu\to\infty}\lim_{j\to\infty}\langle \chi, I_{P_j}{}^{b,\alpha}(\exp(-i\textstyle\sum_{k=1}^{\nu} V_k)\delta)\psi\rangle = \langle \chi, G_{b,m'}\psi\rangle \quad . \qquad (5.13)$$

One now easily sees that the validity of interchange of limits in (5.13) is sufficient, and in essence necessary, for the convergence of the Feynman integral in $\langle\chi, G\psi\rangle$ and for the equality of this expression to $\langle \chi, e^{-itH}\psi\rangle$. (This depends on proposition 16 and hence on self-adjointness. The necessity should be qualified with respect to some minor details, so we wrote "in essence".)

<u>6. A path integral over phase space</u>.

Typically, a path integral over phase space has the form

$$\int_{q(o)=o}\mathcal{D}(q)\,\mathcal{D}(p)\; e^{i\langle p,\dot q\rangle}\; e^{-i(2\kappa)^{-1}\langle p,p\rangle}\; F(p,q) \quad . \qquad (6.1)$$

(For the familiar applications one takes κ real, $=m$.) We have here two Hilbert spaces $\mathcal{H}_p$ and $\mathcal{H}_q$ with the respective scalar products,

$$\langle p,p\rangle = \int_o^t d\tau \sum_{j=1}^n (p^j(\tau))^2 \;, \qquad \langle q,q\rangle_{\mathcal{H}_q} := \langle \dot q,\dot q\rangle = \int_o^t d\tau \sum_{j=1}^n (dq^j/d\tau)^2 . \quad (6.2)$$

Both spaces consist of functions over $[0,t]$, and there is a natural isomorphism between $\mathcal{H}_p$ and $\mathcal{H}_q$, where $p\leftrightarrow q$ if and only if $p(\tau)=\dot q(\tau)$ (modulo sets of measure zero) and $q(0)=0$. We presupposed this isomorphism in writing $\langle p,\dot q\rangle$ in (6.1).

The integral in (6.1) is discussed in [2], [18] on the basis of definitions that are modeled on Itô's.

We want to define this integral in a way which would be close to our definition 3, by using projections over $\mathcal{H}_p \dot{+} \mathcal{H}_q$. However, the two component spaces do not occur in a symmetric way, and it appears preferable to modify the handling of families of reference somewhat.

Definition 18. The integral in (6.1) is a Feynman-type integral over $\mathcal{H}_p \dot{+} \mathcal{H}_q$, and is defined as in definition 3 except for the following modifications:
(1) In place of $e^{\frac{1}{2}i\kappa\langle\zeta,\zeta\rangle}$ we use the weight $W(p,q) = \exp[i\langle p,q\rangle - i(2\kappa)^{-1}\langle p,p\rangle]$.
(2) Determining families are not used. Therefore the family of reference must be specified for each integrand.

For this integral it is natural to consider sequences $\{P_j\}$ such that

$$P_j = P_{(p)j} \dot{+} P_{(q)j}, \quad \{P_{(p)j}\} \in \hat{\mathcal{Q}}(\mathcal{H}_p), \quad \{P_{(q)j}\} \in \hat{\mathcal{Q}}(\mathcal{H}_q) \quad . \tag{6.3}$$

A family of such sequences cannot be determining in the sense of definition 1.

For the approximating integrals we use the weights $W(P_{(p)j}p, P_{(q)j}q)$ and the normalization $I_{P_j}^{b,0}(1) = 1$. The explicit form of the normalizing constant in (2.4a) does not apply.

Various conclusions in secs. 3 and 4, and in [2] and [18], should have their analogues for the integral that we just defined. We do not discuss these possibilities. However, path integrals over phase space give rise to some questions which we did not encounter before, and which we will consider briefly.

In some examples of interest the function F depends on q only. This is the case e.g. when the integral (6.1) is used to represent the Green's function, and then F is the same integrand as in (5.1a). The phase-space integrals of such functions $F(q)$ may be regarded as iterated integrals, and then integration over p yields directly the expected integral of F over $\mathcal{H}_q$. These integrals may also be regarded as integrals directly over $\mathcal{H}_p \dot{+} \mathcal{H}_q$. We expect that in typical cases the two kinds of integrals are equivalent. However, a slight complication arises.

Let us consider an approximating integral for which $P_{(p)} \geqslant P_{(q)}$, with reference to the natural isomorphism. After doing the trivial p-integration, we obtain the following convergence factor:

$$\exp\left[-\tfrac{1}{2}(b + (i/\kappa))^{-1}\langle u + iba_p, u - iba_p\rangle - \tfrac{1}{2}\, b\langle u - a_q, u - a_q\rangle\right] , \tag{6.4a}$$

where

$$\alpha = \alpha_p + \alpha_q , \qquad a_q = P_{(q)}\alpha_q , \qquad a_p = P_{(p)}\alpha_p \quad . \tag{6.4b}$$

The occurrence of b as a coefficient of a_p means that we do not recover the exact form of an approximating integral for integrating F over $\mathcal{H}_q$. We believe that this additional b-dependence of the effective shift vector α should not affect integrability. However, it does not seem easy to establish a general conclusion of this kind.

We make two further comments about (6.4). First: The condition $P_{(p)} \geqslant P_{(q)}$ introduced a certain simplification. Otherwise (6.4a) would be somewhat more complicated. Second: In (6.4a) a part of the exponent takes the form $-\frac{1}{2}\langle u, u\rangle$ $\chi(i\kappa + b')$, where b' approaches zero with b. If b and κ are real, then b' is complex. This is one reason for considering complex b in the construction of the Feynman integral given in sec. 2.

Finally, we give one simple lemma, whose proof is contained in the above discussion.

Lemma 19. If $P_j = P_{(p)j} \dotplus P_{(q)j}$ and $P_{(p)j} \geqslant P_{(q)j}$ for each j, then

$$\lim_{b\to 0}\lim_{j\to\infty} I^{(\mathcal{H}_p \dotplus \mathcal{H}_q)}{}^{b,0}_{P_j}(F(q)) = \lim_{b\to 0}\lim_{j\to\infty} I^{(\mathcal{H}_q)}{}^{b,0}_{P(q)j}(F). \tag{6.5}$$

Here the existence of either member implies that of the other and the equality.

7. A path integral for the free field.

Such a path integral (one can also use the term, history integral) for the free scalar relativistic field was studied in [19] on the basis of a definition that was adapted from that of Itô. The integral in question was put into the schematic form,

$$\bar{I}(F) = \int \mathcal{D}(\eta) \exp\left[\tfrac{1}{2}\langle \eta, (iB - C)\eta\rangle\right] F(\eta) \quad , \tag{7.1}$$

where η ranges over a real Hilbert space $\mathcal{H}$, B and C are bounded symmetric operators on $\mathcal{H}$, and $C \geqslant 0$. (Of course, $(iB - C)\eta \in \mathcal{H} + i\mathcal{H}$.)

The quadratic form $[\ldots]$ has the interpretation of $iS_\varepsilon^{(o)}$ where $S_\varepsilon^{(o)}$ is the action, with a small increment $i\varepsilon$, $\varepsilon > 0$, included:

$$S_\varepsilon^{(o)}(\eta) = \tfrac{1}{2}\int d^4k\, |\tilde{\eta}(k)|^2 \left[(k^o)^2 - \underset{\sim}{k}^2 - (m^2 - i\varepsilon)\right] \quad . \tag{7.2}$$

For B and C to be bounded, we select the following scalar product, which depends on the chosen Lorentz frame and on a constant $K^2 > 0$:

$$\langle \zeta, \eta\rangle_L := \int d^4k\, \tilde{\zeta}(-k)\, \tilde{\eta}(k) \left[(k^o)^2 + \underset{\sim}{k}^2 + K^2\right] \quad . \tag{7.3}$$

We remark that by dropping $\underset{\sim}{k}^2$ and $d^3\underset{\sim}{k}$ in (7.2)-(7.3) we can describe the harmonic oscillator, cf. [19].

We proceed to study such integrals by utilizing projections, as before. We first consider the general form (7.1), independently of (7.2)-(7.3).

Definition 20. The integral $\bar{I}(F)$ in (7.1) is defined as in definition 3, except for the following modification: In place of $e^{\frac{1}{2}i\kappa\langle\xi,\xi\rangle}$ we use the weight $W(\eta) = \exp \frac{1}{2}\langle \eta, (iB - C)\eta\rangle$, where B and C are bounded symmetric linear operators $\mathcal{H}\to\mathcal{H}$, with $C \geqslant 0$ and with $(iB - C)\eta \neq 0$ for $\eta \neq 0$.

In the approximating integrals we use the weights $W(P_j\eta)$ and the normalization $\bar{I}_{P_j}{}^{b,0}(1) = 1$, as before.

As an immediate consequence of this definition, we have:

Proposition 21. We assume that $\bar{I}(F)$ converges, with reference to a determining family $\mathcal{P}$. Then:

(a) In the following one has convergence on the r.h.s. with reference to $\mathcal{P}$, and the equality, for $\beta \in \mathcal{H}$:

$$\bar{I}(F) = \int \mathcal{D}(\eta)\; e^{\frac{i}{2}\langle \eta-\beta,\,(iB-C)(\eta-\beta)\rangle} F(\eta-\beta) \quad . \tag{7.4}$$

(b) Let $A:\mathcal{H}\to\mathcal{H}$ be a linear transformation such that for $\forall \eta \in \mathcal{H}$,

$$\langle A\eta, A\eta\rangle = \langle \eta,\eta\rangle \;, \qquad \langle A\eta,(iB-C)A\eta\rangle = \langle \eta,(iB-C)\eta\rangle \;.$$

Then $F(A\cdot)$ is integrable with reference to $A\mathcal{P}A^{-1}$, and $\bar{I}(F) = \bar{I}(F(A\cdot))$.

The proof is like that of proposition 4. (□)

The next proposition makes the same assertion as theorem 9 of [19], but for the new definition of the integral. We remark that for the scalar field, as in (7.2), the inverse $(iB-C)^{-1}$ is unbounded. However, if $\underset{\sim}{k}^2$ and $d^3\underset{\sim}{k}$ are eliminated from (7.2)-(7.3), so that we describe a harmonic oscillator, then this inverse is bounded.

Proposition 22. Let $\mathcal{H}' \subseteq \mathcal{H}$ be a subspace on which $(iB-C)^{-1}$ is bounded, and let μ be a Borel measure on $\mathcal{H}'$ of bounded absolute variation and of bounded support. Then

$$f(\eta) = \int_{\mathcal{H}'} d\mu(\chi)\; e^{i\langle\chi,\eta\rangle} \Rightarrow \bar{I}(f) = \int_{\mathcal{H}'} d\mu(\chi)\; \exp\left[\tfrac{1}{2}\langle \chi,(iB-C)^{-1}\chi\rangle\right]. \tag{7.5}$$

The integral $\bar{I}(F)$ converges with reference to $\hat{\mathcal{Q}}$.

Outline of proof: The proof is similar to that of the cited theorem, see appendix A of loc. cit. However, some additional steps are needed in the beginning. We start with an evaluation: for $f_\chi(\eta) = e^{i\langle\chi,\eta\rangle}$,

$$I_{P_j}{}^{b,\alpha}(f_\chi) = e^{-\frac{1}{2}b\langle\alpha,P_j\alpha\rangle} \exp \tfrac{i}{2}\langle P_j(\chi+ib\alpha),\,\left[(-b+iB-C)\restriction P_j\mathcal{H}\right]^{-1} \times P_j(\chi-ib\alpha)\rangle\} \quad . \tag{7.6}$$

We next proceed as in eq. (3.18a-b), namely, we extend the definition of the operator $[(-b+i\ldots]^{-1}$ to $\mathcal{H}$ and use a standard result about resolvents to obtain $(-b+iB-C)^{-1}$. We arrive in this way at an evaluation as in eq. (A.4) of loc. cit., except that b replaces the unbounded operator $(nT)^{-1}$ (where T is of trace class, symmetric, and >0). The proof continues as in loc. cit., but some simplifications might be possible. (□)

The present situation can in fact be contrasted with that of loc. cit. by means of an example depending on lemma A4 of loc. cit. This lemma implies that for $\chi \in D(iB-C)^{-1}$, and with T as above,

$$\text{s-lim}_{b\to 0}\,(-b+iB-C)^{-1}\chi = (iB-C)^{-1}\chi \quad , \tag{7.7a}$$

but only

$$\text{w-lim}_{b\to 0}\,(-bT^{-1}+iB-C)^{-1}\chi = (iB-C)^{-1}\chi \quad . \tag{7.7b}$$

In the next proposition we return to the free fields.

Proposition 23. We specialize the integral so that $\langle\eta,(iB-C)\eta\rangle_L = iS_\varepsilon^{(0)}(\eta)$, as in (7.2)-(7.3), and we impose an additional condition: The integral (7.1) must be independent of the Lorentz frame chosen to define $\langle\zeta,\eta\rangle_L$. We denote the integral so limited by $\check{I}$.

(a) We assume that F is integrable (for $\check{I}$). Let g be a transformation of the Poincaré group, including reflections. Then $F(\eta(g^{-1}\cdot))$ is also integrable, and $\check{I}(F) = \check{I}(F(\eta(g^{-1}\cdot)))$.

(b) Proposition 22 remains valid for the integral $\check{I}$.

In part (a), the reference families could be easily given in terms of the family for F.

Proof: For (b), one needs to check that $\check{I}(f)$ is independent of the chosen Lorentz frame, and this is evident. For (a): The effect of a Lorentz transformation can be compensated by using a transformed scalar product and by utilizing the assumed independence of the Lorentz frame. For the reflections of space and/or time in a particular Lorentz frame, and for translation of space-time, the conclusion follows from (b) of proposition 21. Note that under translations, $\tilde{\eta}(k) \to e^{ika}\tilde{\eta}(k)$. □

We make a final remark. In the construction of the integral $\bar{I}(F)$, the scalar product $\langle \eta, \eta \rangle$ is only an auxiliary object. We expect therefore that (b) of proposition 21 should be valid without the hypothesis $\langle \eta,\eta\rangle = \langle A\eta, A\eta\rangle$. If this is so, then the additional condition for $\check{I}$, i.e. that it be independent of the Lorentz frame, will be automatically fulfilled.

Appendix. Some proofs.

Proof of proposition 6. Let $P \in Pr$ with $\dim P = q$, let $\beta = P\alpha$, and we consider β as a variable vector in $P\mathcal{U}$. Standard evaluations and the rule

$$(d/dv)\ e^{\frac{1}{2}av^2} = e^{\frac{1}{2}av^2}((d/dv) + av) \tag{A.1}$$

yield, with $m_{2n}^{(P)}(u^1,\dots,u^q) := m_{2n}(u^1,\dots,u^q,0,0,\dots)$,

$$e^{\frac{1}{2}b\langle\beta,\beta\rangle}\, I_{P_j}^{\ b,\beta}(m_{2n}) = \left[(b - i\kappa)/2\pi\right]^{-q} \int d^q u\ e^{-\frac{1}{2}(b-i\kappa)\langle u,u\rangle}\ e^{b\langle u,\beta\rangle} \times m_{2n}^{(P)}(u^1,\dots,u^q) \tag{A.2a}$$

$$= m_{2n}^{(P)}\left(\frac{1}{b}\frac{\partial}{\partial\beta^1},\dots,\frac{1}{b}\frac{\partial}{\partial\beta^q}\right) e^{\frac{1}{2}\langle\beta,\beta\rangle b^2/(b-i\kappa)} \tag{A.2b}$$

$$= e^{\frac{1}{2}\langle\beta,\beta\rangle b^2/(b-i\kappa)}\ m_{2n}^{(P)}\left(\frac{1}{b}\frac{\partial}{\partial\beta^1} + \frac{b}{b-i\kappa}\beta^1,\dots,\frac{1}{b}\frac{\partial}{\partial\beta^q} + \frac{b}{b-i\kappa}\beta^q\right)\cdot 1 . \tag{A.2c}$$

We expand the last-mentioned function $m_{2n}^{(P)}((1/b)\dots\beta^q)$ and we examine the derivatives. The terms can be arranged as follows,

$$m_{2n}^{(P)}(\tfrac{1}{b}\dots\beta^q)\cdot 1 = (b - i\kappa)^{-n}\left[r_0^{(P)} + \frac{b^2}{b-i\kappa} r_2^{(P)}(\beta) + \dots + \frac{b^{2n}}{(b-i\kappa)^n} r_{2n}^{(P)}(\beta)\right], \tag{A.3}$$

where $r_k^{(P)}(\beta)$ is a polynomial of order k in the components of β.

Let $\{P_j\} \in \hat{\mathcal{Q}}$. We observe that

$$(b - i\kappa)^{-n} \lim_{j\to\infty} r_0^{(P_j)} = \lim_{j\to\infty} I_{P_j}^{\ b,0}(m_{2n}) . \tag{A.4}$$

The limit on the l.h.s. is $1\cdot 3\cdot \ldots \cdot(2n-1)$ tr M, for any $\{P_j\}\in\hat{\mathcal{Q}}$. (See e.g. the discussion in [3], chapter VII of notes.) We will show below that the terms $r_k^{(P_j)}$ remain finite as $j\to\infty$. Therefore, if we let $b\to 0$ (after letting $j\to\infty$), we will obtain the desired conclusion (3.13) from (A.2c)-(A.4).

Let us consider the term $r_2^{(P_j)}(P_j\alpha)$. We replace $e^{b\langle u,\beta\rangle}$ by $e^{b\langle u,P_j\alpha\rangle}$ in (A.2a) and expand, and $\frac{1}{2}b^2\langle u,P_j\alpha\rangle$ will yield the contribution of order 2 in $P_j\alpha$. In view of (A.2c) and (A.3), this contribution is

$$b^2(b-i\kappa)^{-(n+1)}\left[\tfrac{1}{2}r_0^{(P_j)}\langle P_j\alpha, P_j\alpha\rangle + r_2^{(P_j)}(P_j\alpha)\right] \quad . \tag{A.5}$$

Let us recall the following well-known formula, for ν a positive integer and $\mathrm{Re}\, c>0$ (this is in fact a special case of (3.13)):

$$(c/2\pi)^{\frac{1}{2}}\int dw\, e^{-\frac{1}{2}cw^2}\, w^{2\nu} = c^{-\nu}\left[1\cdot 3\cdot \ldots \cdot(2\nu-1)\right] \quad . \tag{A.6}$$

In particular, the coefficient of $c^{-\nu}$ on the r.h.s. is >0.

Now, M must be the difference of two positive semidefinite trace-class operators [10], and it suffices to consider the case $M\geq 0$. Then the operator associated with $b^2\langle u,P_j\alpha\rangle^2 m_{2n}(P_j)$ is also ≥ 0, and in (A.5) (as in (A.6)) $[\ldots]\geq 0$. Moreover, as $j\to\infty$, the successive approximations to tr M and to $[\ldots]$ in (A.5) are nondecreasing.

We may now suppose $P_j\alpha=(v_j,0,0,\ldots)$. Since m_{2n} has terms in u^1 to powers at most $2n$, $\langle u,P_j\alpha\rangle^2 m_{2n}$ has u^1 to powers $\leq 2n+2$. One then obtains integrals such as in (A.6) with $w=u^1$ and with successive integers ν, say $\nu=N$ (for $r_0^{(P_j)}$) and $\nu=N+1$ (for $r_2^{(P_j)}$), where $N\leq n$. It follows that the corresponding integrals differ by a factor $\leq(2n+1)$. Therefore

$$0\leq \tfrac{1}{2}v_j^2 r_0^{(P_j)} + r_2^{(P_j)}((v_j,0,0,..))\leq \tfrac{1}{2}(2n+1)v_j^2 r_0^{(P_j)} \quad . \tag{A.7}$$

Since $r_0^{(P_j)}$ tends to a finite limit as $j\to\infty$, we conclude that $r_2^{(P_j)}((v_j,0,\ldots))$ does likewise (while $v_j\to\|\alpha\|$ as $j\to\infty$). A similar argument clearly applies to all the other $r_{2k}^{(P_j)}(\alpha)$. □

<u>Proof of proposition 9</u>. The part $(1)\Rightarrow(2)$ remains to be shown. We first suppose that $\alpha\in P\mathcal{H}$ (and assume integrability over R^q), and will show that $I^{(R^q)}{}_{P=1}^{b,\alpha}(F) = I^{(\mathcal{H})}{}_{\hat{\mathcal{Q}}}^{b,\alpha}(F)$. The proof depends on a result of Friedrichs and Shapiro ([3], chapter V of notes), which states in particular that if f is a cylinder function with an s-dimensional base, and if $f\in L_1(d^s u\, e^{-\langle u,u\rangle/2\sigma})$ for some $\sigma>0$, then for $\forall\{P_j\}\in\hat{\mathcal{Q}}$,

$$\lim_{j\to\infty}\int\mathcal{D}(\xi)\, e^{-\langle\xi,\xi\rangle/2\sigma}|f(P_j\xi)-f(\xi)| = 0 \quad . \tag{A.8}$$

(To interpret the functional integral, note that for $\forall j$ the integral reduces to a finite-dimensional one.) The condition $F\in L_1(\ldots)$ is of course fulfilled for $\forall\sigma$ with $\mathrm{Re}\,\sigma>0$, if F is Feynman-integrable over R^s. In <u>loc. cit.</u> it is also shown how (A.8) may be extended to complex σ, and our proof amounts to a rephrasing of the argument given there.

By integrability hypothesis on R^q, we have $F\in L_1(d^q u\, e^{-\frac{1}{2}b\langle u,u\rangle})$, and also $F \in L_1(d^q u\, e^{-\frac{1}{4}b\langle u,u\rangle})$. Since $e^{\langle u,\alpha\rangle}\, e^{-\frac{1}{4}b\langle u,u\rangle}$ is bounded, we conclude that

$$f := F\, e^{b\langle\cdot,\alpha\rangle} \in L_1(d^q u\, e^{-\frac{1}{2}b\langle u,u\rangle}) \quad . \tag{A.9}$$

For time being we will work with f and will ignore the dependence of the integrals on α.

Let $b = b_1 + ib_2$, $\kappa = \kappa_1 + i\kappa_2$, with $b_1 > 0$ and $\kappa_2 \geqslant 0$, and let $\{P_j\} \in \hat{\mathcal{Q}}$ be arbitrary. Let $r_j = \dim P_j \vee P$ and $u = (P_j \vee P)\xi$. Then

$$D_j := \left| I^{(\mathcal{H})}{}_{P_j}{}^{b,0}(f) - I^{(R^q)}{}_{P=1}{}^{b,0}(f) \right| = \left| \left[(b - i\kappa)/2\pi\right]^{\frac{1}{2}r_j} \int d^{r_j}u \right.$$
$$\left. \times\, e^{-\frac{1}{2}(b-i\kappa)\langle u,u\rangle}\left[f(P_j u) - f(u)\right] \right| \quad . \tag{A.10}$$

Note that integration over redundant variables, for $f(P_j u)$ and for $f(u)$ separately, gives unity. We have the following bound,

$$D_j \leqslant \left[|b - i\kappa|/(b_1 + \kappa_2)\right]^{\frac{1}{2}r_j} \Big\{ \left[(b_1 + \kappa_2)/2\pi\right]^{\frac{1}{2}r_j} \int d^{r_j}u\, e^{-\frac{1}{2}(b_1+\kappa_2)\langle u,u\rangle}$$
$$\times \left| f(P_j u) - f(u) \right| \Big\} \quad . \tag{A.11}$$

As $j \to \infty$, then $\{\ldots\} \to 0$ by (A.8), but its coefficient $\to \infty$. We resolve this difficulty by exploiting the fact, as in <u>loc. cit.</u>, that $f(P_j u) = f(P'P_j u)$ where P' projects onto the range of $P_j P$ (and $P' \in Pr$). Therefore $\dim P' \leq q$ and $\dim P' \vee P \leq 2q$.

We now write

$$(P' \vee P)\mathcal{H} =: \mathcal{H}' \quad \text{with } \dim \mathcal{H}' \leqslant 2q, \qquad (P_j \vee P)\mathcal{H} = \mathcal{H}' \dot{+} \mathcal{H}'' \quad , \tag{A.12}$$

and obtain

$$D_j = \left| \int_{\mathcal{H}' \dot{+} \mathcal{H}''} d^{r_j}u \ldots \right| = \left| \int_{\mathcal{H}'} \ldots \right| \left| \int_{\mathcal{H}''} \ldots \right| = \left| \int_{\mathcal{H}'} \ldots \right| \quad , \tag{A.13}$$

since the integral over $\mathcal{H}''$ yields unity. (The normalizing factors are to be included in the integrals of (A.13).) We may now apply an estimate as in (A.11) to $\left|\int_{\mathcal{H}'} \ldots\right|$, and the first factor on the r.h.s. will be bounded by $(|b - i\kappa|/ /(b_1 + \kappa_2))^{2q}$. The second factor $\to 0$ when $j \to \infty$, as before, and $D_j \to 0$.

Upon setting $f = Fe^{b\langle\cdot,\alpha\rangle}$ with $\alpha \in P\mathcal{H}$, there follows $I^{(R^q)}{}_{P=1}{}^{b,\alpha}(F) = I^{(\mathcal{H})}{}_{\hat{\mathcal{Q}}}{}^{b,\alpha}(F)$. Finally, if $\alpha \notin P\mathcal{H}$, we may again consider $f = Fe^{b\langle\cdot,\alpha\rangle}$, but this function has a $(q+1)$-dimensional base. We obtain then $I^{(R^{q+1})}{}_{P=1}{}^{b,\alpha}(F) = I^{(\mathcal{H})}{}_{\hat{\mathcal{Q}}}{}^{b,\alpha}(F)$, and also

$$\lim_{b\to 0} I^{(R^{q+1})}{}_{P=1}{}^{b,\alpha}(F) = I^{(R^q)}(F) = I^{(\mathcal{H})}(F) \quad , \tag{A.14}$$

since the redundant integration on R^{q+1} gives a factor which $\to 1$ as $b \to 0$. (Cf. the part $(3) \Rightarrow (1)$ of the proof.) □

<u>Proof of proposition 16</u>. In this proof the scalar product $\langle\dot{\gamma},\dot{\gamma}\rangle$ and the space $\mathcal{H}$ are not relevant except for $\langle\dot{x},\dot{x}\rangle$, and otherwise all operators, norms, etc. refer to the space $L_2(R^n)$ or to its subsets. We use the entities introduced in sec. 5 and the techniques of [5], appendix B.

We observe that continuity in b as $b \to 0$ is manifest for $G_{b,m'}^{(\nu)}$ and for $F^{(\nu)}(b,m')$, and $\lim_{\nu\to\infty}\lim_{b\to 0} F^{(\nu)}(b,m/i)$ gives the desired answer

$\langle \chi, e^{-itH}\psi \rangle$ by Trotter's formula. (Cf. loc. cit. Also, in view of analyticity, the nontangential limit is unique if it exists.) We will justify the interchange of limits, and this will prove the proposition.

By hypothesis H is self-adjoint, hence its domain $D(H)$ is closed and is a Banach space in the norm $\|\psi\|_H := \|H\psi\| + \|\psi\|$. The operators ∂_k and H_o from $D(H)$ to L_2 are everywhere finite and hence have finite H-norms,

$$\|H_o\|_H := \sup \|H_o\psi\| / \|\psi\|_H < \infty , \quad \text{also} \quad \|\partial_k\|_H < \infty . \tag{A.15}$$

Next, the operators $G_{b,m'}^{(\nu)}$ are constructed in terms of exponentials of the A_j. We introduce the factorization,

$$\exp\left[(b + m/i)^{-1}(t/\nu)A_j\right] = e^{-iH_o t/\nu}\, e^{(bt/\nu)K_j(b)} , \tag{A.16}$$

where $K_j(b)$ has terms proportional to α_j^2, to $\alpha_j \cdot \nabla$, and to ∇^2 (and is regular in b). The arguments of loc. cit. and the bounds of (A.15), together with those of (5.8)-(5.9), show that for small $|b|$,

$$(\nu/tb)\left[\exp\left((bt/\nu)K_j(b)\right) - 1\right] - K_j(b) \tag{A.17a}$$

is a uniformly bounded set of operators on $D(H)$, which tend to zero strongly as $bt/\nu \to 0$. It follows (loc. cit. and [12], p. 151) that for $\psi \in D(H)$,

$$(bt/\nu)^{-1}\left\|\left[e^{(bt/\nu)K_j(b)} - 1\right]\psi\right\| = \|K_j(b)\psi\| + \delta_1(bt/\nu) , \tag{A.17b}$$

where δ_1 is a quantity such that $\delta_1(bt/\nu)(bt/\nu)^{-1} \to 0$ as $bt/\nu \to 0$ (with $\mathrm{Re}\, b > 0$), and this convergence is uniform over any compact set of vectors ψ in $D(H)$. We also recall from [5] that

$$\left\|\left[e^{-iVt/\nu}\, e^{-iH_o t/\nu} - e^{-i(H_o + V)t/\nu}\right]\psi\right\| = \delta_2(t/\nu) , \tag{A.18}$$

where similarly $\delta_2(t/\nu)(t/\nu)^{-1} \to 0$ as $t/\nu \to 0$, uniformly over any compact set of ψ's.

Now, let $\varphi \in D(H)$ be arbitrary, and the following set of vectors is compact: $\{e^{-isH} : 0 \le s \le t\}$. Consider the following rearrangement, with reference to (5.10a),

$$\begin{aligned} G_{b,m/i}^{(\nu)} - (e^{-iHt/\nu})^{\nu} &= \left[e^{-iVt/\nu}\, e^{-iH_o t/\nu} - e^{-iHt/\nu}\right] e^{-iHt(\nu-1)/\nu} + \\ &+ e^{-iVt/\nu}\, e^{-iH_o t/\nu}\left[e^{(bt/\nu)K_\nu(b)} - 1\right] e^{-iHt(\nu-1)/\nu} + \\ &+ e^{-iVt/\nu} e^{-iH_o t/\nu} e^{(bt/\nu)K_\nu(b)}\left[e^{-iVt/\nu} e^{-iH_o t/\nu} - e^{-iHt/\nu}\right] e^{-iHt(\nu-2)/\nu} + \\ &+ \dots + e^{-iVt/\nu} e^{-iH_o t/\nu} e^{(bt/\nu)K_\nu(b)} \dots \left[e^{(bt/\nu)K_1(b)} - 1\right] , \end{aligned} \tag{A.19}$$

which implies

$$\begin{aligned} \left\|\left[G_{b,m/i}^{(\nu)} - e^{-iHt}\right]\varphi\right\|_H &\le \delta_2(t/\nu) + |b|\, t\nu^{-1}\|K_\nu e^{-iHt(\nu-1)/\nu}\varphi\|_H + \\ &+ \delta_1(bt/\nu) + \dots + |b|\, t\nu^{-1}\|K_1\varphi\|_H + \delta_1(bt/\nu) \\ &\le \nu\cdot\delta_2(t/\nu) + \nu\cdot\delta_1(bt/\nu) + |b|\, t\nu^{-1}\|\varphi\|_H \textstyle\sum_{j=1}^{\nu}\|K_j(b)\|_H , \end{aligned} \tag{A.20}$$

since $\|\varphi\|_H = \|e^{-iHs}\varphi\|_H$. The first two terms $\to 0$ as $\nu \to \infty$, and the last term has a bound proportional to b and independent of ν. To establish the last assertion, we use in particular (5.6) to deduce

$$\nu^{-1}\sum\|\alpha_j{}^2\|_H \leq (\text{const.})\nu^{-1}, \qquad \nu^{-1}\sum\|\alpha_j\cdot\nabla\|_H \leq (\text{const.})\nu^{-\frac{1}{2}},$$
$$\nu^{-1}\sum\|\nabla^2\|_H \leq (\text{const.}) \quad . \tag{A.21}$$

The foregoing bounds now imply the existence of a double limit, and also allow interchanges of limits. In particular,

$$\lim_{b\to 0}\langle\chi, G_{b,m/i}\varphi\rangle = \lim_{b\to 0,\nu\to\infty}\langle\chi, G_{b,m/i}^{(\nu)}\varphi\rangle = \langle\chi, e^{-itH}\varphi\rangle \quad . \tag{A.22}$$

Finally, since the operators in (A.22) are bounded, we conclude that these limit relations remain valid if $\varphi\in D(H)$ is replaced by any $\psi\in L_2$.

Addendum. Bounded potentials.

In this addendum we establish the path-integral representation of the Schrödinger Green's functions for the case of potentials which are bounded, and continuous as in sec. 5. (We completed this addendum too late for correlating the main part of the text with the present material.)

The results here included extend the class of potentials for which the representation in question has been proved (i.e. in the framework of the present article). This extension is of interest in particular for the following reason: The potentials which we specified before, i.e. those in proposition 14, lead to path integrals which can be done essentially in closed form. The potentials that we are about to consider cannot be handled in this way, and our proof depends on the general analysis of sec. 5.

We recall that one kind of problem that we encountered in sec. 5 was to show that a sequence of approximations is uniformly bounded. This circumstance motivates the following approach. We start with (5.3a) for a particle on R^n, and let b and m' be real with $b+m'>0$. We introduce and evaluate:

$$G_{b,m'}(t;y,x)|_{V=0} =: G_{b,m'}^{(o)}(t;y,x) = [(b+m')/2\pi t]^{\frac{1}{2}n}\, e^{-(b+m')(y-x)^2/2t}$$
$$\times\exp[(b/t^{\frac{1}{2}})\langle y-x,\alpha(t)\rangle]\exp[-\tfrac{1}{2}b\langle\dot\alpha,\dot\alpha\rangle + \tfrac{1}{2}b^2(b+m')^{-1}\langle\dot\alpha_o,\dot\alpha_o\rangle] \quad , \tag{B.1}$$

where we used the same symbol $\langle\cdot,\cdot\rangle$ for the scalar product in R^n and in $\mathcal{H}$, and where (cf. (5.2b))

$$\alpha_o := (1-P')\alpha, \quad \alpha_o(\tau) = \alpha(t)(\tau/t) - \alpha(\tau) \quad . \tag{B.2}$$

One sees that

$$|G_{b,m'}(t;y,x)| \leq G_{b,m'}^{(o)}(t;y,x) \quad , \tag{B.3}$$

but we have not succeeded in extending this to a useful bound when b or m' is complex and when projections are introduced. However, if V is bounded, $|V|\leq M$, then we can expand $\exp(-i\int d\tau V)$ in a convergent series, which yields a weaker estimate:

$$|\exp(-i\int d\tau\, V)| \leq \sum(k!)^{-1}(\int d\tau|V|)^k \leq e^{tM} \quad , \tag{B.4a}$$

$$|G_{b,m'}(t;y,x)| \leq e^{tM}\, G_{b,m'}^{(o)}(t;y,x) \quad . \tag{B.4b}$$

(Here V may be complex, but we must have b and m' as before.) The last inequality can be adapted to the relevant approximating integrals, as follows:

Proposition 24. Let $V:R^n \to C^1$ be bounded, $|V| \leq M$, and measurable. Let $Re(b+m')>0$ and $P \geq P'$ (see (5.2b)). Then

$$\left| I_P^{b,\alpha}(\exp(-i\int_0^t d\tau V)\delta)\right| \leq \left[|b+m'|/(b_1+m_1')\right]^{\frac{1}{2}}$$
$$\times \exp\left\{tM\left[|b+m'|/(b_1+m_1')\right]^{\frac{1}{2}(n+1)}\right\} \sup_{P\in Pr} G_{b_1,m'_1}{}^{(o)(P\alpha)}(t;y,x), \quad (B.5)$$

where the integrand on the l.h.s. is given in detail in (5.3a), where $b=b_1+ib_2$, $m'=m_1'+im_2'$, and where the dependence on the shift vector in $G_{b_1,m'_1}{}^{(o)}$ is explicitly shown.

The factor $\sup_{P\in Pr}(\ldots)$ in (B.5) is finite, as one sees from (B.1).

Proof: We expand the integrand on the l.h.s. and examine a given term:

$$I_P^{b,\alpha}((\int d\tau\, V)^k\delta) = \int d\tau_1\ldots d\tau_k\, I_P^{b,\alpha}(V(\eta(\tau_1))\ldots V(\eta(\tau_k))\delta) \quad . \quad (B.6)$$

The expression $V(\eta(\tau_1))\ldots\delta$ is a cylinder function with an $n(k+1)$-dimensional base (we are ignoring distribution-theoretic subtleties). According to the discussion which precedes (A.12), we may find a projection $P''\in Pr$ such that $\dim P'' =: q'' \leq n(k+1)+1$ and such that for $\forall \eta\in\mathcal{H}$,

$$V((P''\eta)(\tau_1))\ldots\delta((P''\eta)(t)..) = V((P\eta)(\tau_1))\ldots\delta((P\eta)(t)..) \quad \text{and}$$
$$e^{b\langle\dot\eta,(P''\alpha)^{\cdot}\rangle} = e^{b\langle\dot\eta,(P\alpha)^{\cdot}\rangle} \quad . \quad (B.7)$$

As in (A.13), the integral over $P\mathcal{H}$ then factorizes into integrals over $P''\mathcal{H}$ and over $(P-P'')\mathcal{H}$. The latter integral (with normalizing factor included) equals unity.

For estimating the integral over $P''\mathcal{H}$, we proceed as follows (here $P''\eta = u$, and we use these designations interchangeably):

$$\left|I_P^{b,\alpha}((\int d\tau V)^k\delta)\right| = \left|I_{P''}^{b,\alpha}(\ldots)\right| = \left[|b+m'|/(b_1+m_1')\right]^{\frac{1}{2}q''}\int d\tau_1\ldots d\tau_k$$
$$\times\left[(b_1+m_1')/2\pi\right]^{\frac{1}{2}q''}\int d^{q''}u\, e^{-\frac{1}{2}(b+m')\langle u,u\rangle}\, e^{b\langle u,(P''\alpha)^{\cdot}\rangle}\, e^{-\frac{1}{2}b\langle\dot\alpha,(P''\alpha)^{\cdot}\rangle}$$
$$\times\delta((P''\eta)(t)-(y-x))\, V((P''\eta)(\tau_1))\ldots V((P''\eta)(\tau_k))$$
$$\leq \left|\frac{b+m'}{b_1+m_1'}\right|^{\frac{1}{2}}\left[tM\left(\frac{b+m'}{b_1+m_1'}\right)^{\frac{1}{2}(n+1)}\right]^k\left(\frac{b_1+m'_1}{2\pi}\right)^{\frac{1}{2}q''}\int d^{q''}u\, e^{-\frac{1}{2}(b_1+m'_1)\langle u,u\rangle}$$
$$e^{b_1\langle u,(P''\alpha)^{\cdot}\rangle}\, e^{-\frac{1}{2}b_1\langle\dot\alpha,(P''\alpha)^{\cdot}\rangle}\,\delta((P''\eta)(t)-(y-x)) \quad . \quad (B.8)$$

The last integral (with the normalization factor) is just $G_{b_1,m'_1}{}^{(o)(P\alpha)}$, since P'' must satisfy $P''\alpha = P\alpha$, $(P''\eta)(t) = (P\eta)(t) = \eta(t)$.

Note that each $(P''\eta)(\tau_j)$ is continuous in τ_j. This follows from (5.2a) and from the fact that elements of $\mathcal{H}$ are continuous. Since V is by hypothesis measurable on R^n, it remains such on $P''\mathcal{H}$ (for a fixed τ_j). Consequently all the integrals in (B.8) are defined.

Now, summing over k in (B.8) and taking $\sup_{P\in Pr}$ yields (B.5). □

Theorem 25. Let $V:R^n \to R^1$ be bounded, and continuous on the complement of a set of capacity zero. Then eq. (5.1) are valid, and the path integrals converge with reference to $\mathcal{A}(P')$.

Proof: A set of capacity zero has measure zero (cf. [5] [15]), so the preceding proposition applies, and the approximating integrals for $G_{b,m'}(t;y,x)$ (or for $\langle\chi, G_{b,m'}\psi\rangle$) have a uniform bound on compact subsets of $\{Re(b+m')>0\}$. They converge therefore to $G_{b,m'}(t;y,x)$ (or to $\langle\chi, G_{b,m'}\psi\rangle$ respectively); see the discussion which precedes lemma 17. Next, for V as specified, $H_o + V$ is self-adjoint. Therefore by proposition 16, the limit $b \to 0$ yields the desired Green's function. □

References.

1. K. Itô, in: Proc. 5th Berkeley symposium on mathematical statistics and probability, vol. II part I (Univ. California Press, Berkeley, 1966), p. 145.
2. J. Tarski, in: Complex analysis and its applications, vol. III (IAEA, Vienna, 1976), p. 193.
3. K.O. Friedrichs and H.N. Shapiro, Proc. Nat. Acad. Sci. (USA) 43, 336 (1957); K.O. Friedrichs, H.N. Shapiro, et al., Integration of functionals (Courant Institute, New York University, seminar notes, 1957).
4. J. Tarski, in: IV winter school of theoretical physics, vol. I (Acta Universitatis Wratislaviensis No. 88, Wrocław, 1968), p. 42.
5. E. Nelson, J. Math. Phys. 5, 332 (1964).
6. A. Truman, J. Math. Phys. 17, 1852 (1976) and in these proceedings.
7. E. Hille and R.S. Phillips, Functional analysis and semigroups, 2nd edition (American Mathematical Society, Providence, R.I. 1957; Colloquium Publications vol. XXXI).
8. I.I. Priwalow, Randeigenschaften analytischer Funktionen, 2nd edition (VEB Deutscher Verlag der Wissenschaften, Berlin, 1956), pp. 18-19.
9. L. Gross, Trans. Amer. Math. Soc. 105, 372 (1962).
10. J. von Neumann, Mathematische Grundlagen der Quantenmechanik (Springer-Verlag, Berlin-Heidelberg-New York, reprinted 1968), Kap. II, Sek. 11.
11. D. Shale, Trans. Amer. Math. Soc. 103, 149 (1962).
12. T. Kato, Perturbation theory for linear operators (Springer-Verlag, Berlin--Heidelberg-New York, 1966).
13. D. Buchholz and J. Tarski, Ann. Inst. H. Poincaré A-24, 323 (1976).
14. I.E. Segal, Trans. Amer. Math. Soc. 88, 12 (1958).
15. J.L. Doob, Trans. Amer. Math. Soc. 77, 327 (1954).
16. L. Hörmander, An introduction to complex analysis (D. Van Nostrand Company, Inc., Princeton, N.J. 1966), pp. 27-29.
17. Yu.L. Daletskii, Russian Mathematical Surveys 17, No. 5, 1 (1962).
18. K. Gawedzki, Reports on Math. Phys. 6, 327 (1974).
19. J. Tarski, Ann. Inst. H. Poincaré A-15, 107 (1971).

Supplement.

We include some supplementary remarks for the article [19] on path integrals for fields. We continue to denote the integral there defined by $\hat{I}_\infty(f)$.

First: The following proposition belongs naturally in loc. cit., but was not stated there explicitly.

Proposition S. We consider the case where $\frac{1}{2}\langle\eta, (iB-C)\eta\rangle = iS_\varepsilon^{(0)}(\eta)$ (as in (7.2)-(7.3) of the present article). Let F be such that $\hat{I}_\infty(F)$ exists, and let g be a transformation of the Poincaré group, including reflections. Then $\hat{I}_\infty(F) = \hat{I}_\infty(F(\eta(g^{-1}\cdot)))$.

Outline of proof: Only the case where g is a Lorentz transformation Λ needs some discussion. (Otherwise the remarks of sec. 7 continue to apply.) Since the definition of $\hat{I}_\infty$ is based on a fixed noninvariant scalar product, we have to compensate for Λ by adjusting the covariance operator T which defines the approximating integrals. Now, Λ induces the operator M_Λ: $\eta(k) \to \eta(\Lambda^{-1}k) =: \check{\eta}(k)$. This operator is bounded, like B and C.

We next consider approximating integrals of $F(\eta)$, we replace there $F(\eta)$ by $F(\check{\eta})$, and we bring the integrals to the original form, with $\check{\eta}$ as the new variable of integration. The procedure is routine. In particular, the covariance operator T has to be replaced by $M_\Lambda T M_\Lambda^*$. (□)

Second: An example is given in loc. cit., p. 115, to show that one may have $R_1 \geq R_0 \geq 0$ and $R_1^2 \not\geq R_0^2$. We should therefore like to note that general results related to such situations are presented in: K. Löwner, Math. Z. 38, 177 (1934) and E. Heinz, Math. Ann. 123, 415 (1951).

Third: We should like to correct certain details of [19], especially the proof of the second part of lemma A4 (p. 134). It turns out that by a minor rearrangement of manipulations a slightly stronger result than that given in loc. cit. can be proved.—The author is indebted to Dr. K. Gawedzki for having found the error in the published proof.

We give the corrections in French, since the cited article was so written. Above eq. (A.8b) (p. 134) should be: Si $\chi \in D(S_\infty)$, alors [omit: et ... ∞]
In eq. (A.12) and above should be:

...l'on commence avec (A.9) et avec une variante de (A.7c):

$$\langle \zeta, S_n\chi\rangle = \langle (S^*)^{-1}S_n^*\zeta, S_\infty\chi\rangle \tag{A.12a}$$

$$= \langle T^{-\frac{1}{2}}(1-uM^{*-1})^{-1}T^{\frac{1}{2}}\zeta, S_\infty\chi\rangle. \tag{A.12b}$$

On p. 135 line 1 should be: ... à la limite $u\to 0$, l'opérateur $T^{-\frac{1}{2}}$ étant fermé.

In the proof of theorem 9 (p. 135) there is no need now to mention T_o'' (and the prime on T_o' can be dropped).

Furthermore: In the statement of lemma A5 (p. 135), the following sentence should be omitted: Prenons $T \in \mathfrak{J} \ldots \mathcal{H}_o$.

On p. 116 line 4 from bottom, replace: 230, 234 by: 230-234 .

- SECTION VI -

BOUNDS ON THE EUCLIDEAN FUNCTIONAL DETERMINANT

H. HOGREVE
R. SCHRADER
R. SEILER

Institut für Theoretische Physik
Freie Universität Berlin

Consider the effective Lagrangian $\mathcal{L}(A)$ for the electromagnetic field A, or more generally a Yang-Mills field, interacting with a scalar field ϕ respectively a spinor field Ψ. $\mathcal{L}$ is defined by

$$e^{-\mathcal{L}(A)} = \frac{\int d\phi\, d\phi^* \exp - \phi^* (p+A)^2 \phi}{\int d\phi\, d\phi^* \exp - \phi^* p^2 \phi}$$

respectively

$$e^{-\mathcal{L}(A)} = \frac{\int d\Psi\, d\Psi^* \exp - \Psi^* (\not{p} + \not{A}) \Psi}{\int d\Psi\, d\Psi^* \exp - \Psi^* \not{p} \Psi} .$$

A theory is called dynamically stable if the effective Lagrangian is positive. Due to the identity

$$2\mathcal{L}(A) = - \log \frac{\det p^2}{\det (p+A)^2} \tag{1}$$

respectively

$$2\mathcal{L}(A) = - \log \frac{\det (\not{p} + \not{A})^2}{\det \not{p}^2} \tag{2}$$

stability is tantamount with the two inequalities on the Euclidean functional determinants

$$\frac{\det(p+A)^2}{\det p^2} \geqslant 1 \tag{3}$$

$$\frac{\det(\not p+\not A)^2}{\det \not p^2} \leqslant 1 \quad . \tag{4}$$

From (1) and (2) follows the well known and remarkable fact that $\mathcal{L}$ has an interpretation in non relativistic quantum mechanics with the one particle Hamiltonian $H_D = (p+eA)^2$ respectively $H_p = (\not p + e\not A)^2$.

The proper definition of the functional determinants require renormalization. In the case of a boson field ϕ interacting with an external Yang-Mills field we have used a ζ -function renormalization [1] . Since our present aim is to elucidate the mechanism responsible for the inequalities (3) and (4) , we circumvent renormalization as follows : a possible start for a proper definition of functional determinants is the formal identity

$$\log \frac{\det(p+eA)^2}{\det p^2} = -\int_0^\infty \frac{d\beta}{\beta} \, \text{Trace} \left(e^{-\beta H_D} - e^{-\beta H_0} \right) \tag{5}$$

respectively

$$\log \frac{\det(\not p+e\not A)^2}{\det \not p^2} = -\int_0^\infty \frac{d\beta}{\beta} \, \text{Trace} \left(e^{-\beta H_p} - e^{-\beta H_0 \otimes 1} \right) \tag{6}$$

where $H_0 = p^2$ and 1 is the identity on spin space. The right hand side has an obvious interpretation as an integral over a partition function or rather a partition function density $\exp -\beta H_D(x,x)$ respectively $\text{Trace}_s \exp -\beta H_p(x,x)$. Trace_s denotes the partial trace over spin space and $\exp -\beta H(x,y)$ the kernel of the operator $\exp H$. Since generally neither the x-space integral nor the β -integral in (6), (7), exists as it stands, we rather ask the question whether the following inequalities corresponding to (3) and (4) hold :

$$\exp -\beta H_D(x,x) \leq \exp -\beta H_0(x,x) \tag{3'}$$

respectively

$$\text{Trace}_s \exp -\beta H_p(x,x) \geqslant \text{Trace}_s \exp -\beta H_0 \otimes 1(x,x) . \tag{4'}$$

Formally they imply (3) respectively (4). Alternatively we add to all Hamiltonians a positive potential $\mathcal{U}$, strongly growing at infinity, so as to make the corresponding partition functions well defined,

$$Z(\beta, H_D+\mathcal{U}) = \int dx \exp -\beta (H_D + \mathcal{U})(x,x) \tag{7}$$

$$Z(\beta, H_p+\mathcal{U}\otimes 1) = \int dx \, \mathrm{Trace}_S \exp -\beta(H_p+\mathcal{U}\otimes 1)(x,x) . \tag{8}$$

In this case the inequalities considered are

$$Z(\beta, H_D+\mathcal{U}) \leq Z(\beta, H_0+\mathcal{U}) \tag{3"}$$

respectively

$$Z(\beta, H_p+\mathcal{U}\otimes 1) \geqslant Z(\beta, H_0+\mathcal{U}\otimes 1) . \tag{4"}$$

Again the inequalities imply formally (3) respectively (4).

The inequalities (3) and (4) have a straightforward interpretation as a diamagnetic and paramagnetic property of the corresponding Hamiltonians. Since formally the determinant is the product of all the eigenvalues the inequalities tell us that the spectrum of H_D and H_p is lifted and lowered respectively due to the presence of A .

The diamagnetic inequality (3) holds in great generality [1] that is to say for arbitrary Yang-Mills fields and a proper renormalization of the functional determinant. The key elements of the proof are Kato's inequality [2] for the kernels of $\exp -\beta H_D$ and $\exp -\beta H_0$; saying that

$$|\exp -\beta H_D(x,y)| \leq \exp -\beta H_0(x,y) \tag{9}$$

and Seeley's definition of the trace of an operator [3].

There is however no general proof of the paramagnetic inequality (4) , although some further progress has been made since the delivery of this talk [4]. The conjecture of the paramagnetic inequality is based on the following facts [5] :

1. The inequality (4") can be shown to be correct to order $\hbar^2$ and e^2 for arbitrary Yang-Mills potentials.

2. In one dimension the inequality holds provided the determinant is properly defined [5].

3. In the two dimensional case with an external electromagnetic field - the Schwinger model - the determinant can be computed explicitly. Inequality (4) is satisfied.

4. In three dimensions with a constant magnetic field strength B and an external harmonic potential U the partition function is monotonically increasing in the modulus of B .

5. We have checked inequality (4') for the case of constant electromagnetic field strength $F_{\mu\nu} = \partial_\mu A_\nu - \partial_\nu A_\mu$ in four dimensions by an explicit calculation.

Rather than giving a proof of all the above results we shall discuss in more detail only the last mentioned case of constant field strength in four dimensions. It shows what we think is the important mechanism for paramagnetism.

The kernel of $\exp-\beta H_P$ can be computed explicitly [6] . This is even more than we need to check (4'). Because F is a constant matrix the Hamiltonian H_P is the sum of two commuting terms,

$$H_P = H_D \otimes 1 - \frac{e}{2} 1 \otimes F\Sigma. \qquad (10)$$

H_D acts on the spatial and $F\Sigma$ on the spinorial part of the wave function. Hence the kernel of H_P factorizes into two terms

$$\exp-\beta H_P(x,y) = \exp-\beta H_D(x,y) \otimes \exp \tfrac{1}{2}\beta e F\Sigma .$$

The explicit form of the first term is

$$\exp-\beta H_D(x,y) = \frac{1}{(4\pi\beta)^2} e^{-\frac{1}{4\beta}(x-y).\beta eF \operatorname{cotg} e\beta F.(x-y)} \cdot e^{-ie\beta\int_x^y dz\, A(z)} \cdot e^{-L(\beta)} .$$

where we used the abbreviation

$$L(\beta) = \frac{1}{2} \operatorname{Trace} \log \frac{\sin e\beta F}{e\beta F} .$$

The line integral goes over the straight line from x to y. The formulae can be simplified if we introduce the eigenvalues $\pm\alpha_\pm$ of iF. Then

$$e^{-L(\beta)} = \frac{e\beta\alpha_+}{\sinh e\beta\alpha_+} \cdot \frac{e\beta\alpha_-}{\sinh e\beta\alpha_-} \leq 1 \tag{11}$$

and

$$\mathrm{Trace}_s\, e^{\frac{1}{2}\beta e F\Sigma} = 4 \cosh e\beta\alpha_+ \cosh e\beta\alpha_- \geq 1. \tag{12}$$

Hence the partial trace of $\exp-\beta H_p$ is simply

$$\mathrm{Trace}_s\, e^{-\beta H_p}(x,y) = \frac{1}{(4\pi\beta)^2}\, e^{-\frac{1}{4\beta}(x-y).\beta e F \coth \beta e F.(x-y)} \\ e^{-ie\beta\int_x^y dz\, A(z)} \cdot 4\, \frac{e\beta\alpha_+}{\mathrm{Tgh}\, e\beta\alpha_+}\, \frac{e\beta\alpha_-}{\mathrm{Tgh}\, e\beta\alpha_-}\,. \tag{13}$$

Now we are ready to analyze the case of constant field strength in more detail. Obviously it follows from the last equation that

$$\mathrm{Trace}_s\, e^{-\beta H_p}(x,x) = \mathrm{Trace}\; e^{-\beta H_0\otimes 1} \cdot \frac{e\beta\alpha_+}{\mathrm{Tgh}\, e\beta\alpha_+}\, \frac{e\beta\alpha_-}{\mathrm{Tgh}\, e\beta\alpha_-}$$

$$\mathrm{Trace}\; e^{-\beta H_0\otimes 1} = \frac{1}{(4\pi\beta)^2} \cdot 4.$$

Hence the paramagnetic inequality (4') holds because $x/\mathrm{Tgh}\,x \geq 1$ for every real x.

We should like to argue that the mechanism responsible for paramagnetism is rather subtle. For $\mathrm{Trace}_s \exp-\beta H_p$ can be written as a product of three terms

$$\mathrm{Trace}_s\, e^{-\beta H_p}(x,x) = \mathrm{Trace}_s\, e^{-\beta H_0\otimes 1}(x,x)\, e^{-L(\beta)}\, e^{\frac{1}{2}\beta e F\Sigma} \tag{14}$$

and (4') holds if the left hand side is larger than the first term on the right. Due to (11) the second term $\exp-L(\beta)$ is smaller than one. (This is a special case of Kato's inequality (9)). Its influence is diamagnetic and in fact it results from the diamagnetic scalar term H_D in H_p (9). Furthermore the last term in (13) is larger than one (11).(This is a special case of the

Bogoliubov-Peierls inequality). Its influence is paramagnetic and it reflects the paramagnetic spin term in H_P (10).

Finally we should like to add two remarks on paramagnetism for an arbitrary external electromagnetic field A and a scalar potential $\mathcal{U}$:

1. As before the Hamiltonian splits into two terms, a scalar term and a spin term,

$$H_P + \mathcal{U}\otimes 1 = (H_D + \mathcal{U})\otimes 1 - \tfrac{1}{2}e\, 1\otimes F\Sigma .$$

The influence of the last term is always paramagnetic,

$$Z(\beta, H_P + \mathcal{U}\otimes 1) \geq Z(\beta, (H_D + \mathcal{U})\otimes 1)$$

because of the Bogoliubov-Peierls inequality [7]. Furthermore the effect of the electromagnetic field A is always diamagnetic,

$$Z(\beta, (H_D + \mathcal{U})\otimes 1) \leq Z(\beta, (H_0 + \mathcal{U})\otimes 1)$$

due to Kato's inequality [2]. The general problem is to show that the paramagnetic effect always wins over the diamagnetic one.

2. The paramagnetic inequality (4') would be easy to prove if there were an inverted Kato inequality for $\exp -\beta H_P$ and $\exp -\beta(H_0\otimes 1)$. Generally such an inequality does not hold. The formula (13) for the case of constant field strength proves that, however, close to the diagonal of the kernel, such an inequality does hold for this special case. A generalization of this statement for arbitrary external fields A would suffice to prove paramagnetism.

-:-:-:-:-:-

NOTES AND REFERENCES

[1] R. SCHRADER and R. SEILER, A Uniform Lower Bound on the Renormalized Euclidean Functional Determinant, to appear in Commun.math.Phys.

[2] H. HESS, R. SCHRADER, D.A. UHLENBROCK, Kato's Inequality and Spectral Distribution of Laplace Operators on Compact Riemannian Manifolds, to appear in Journ.Diff.Geom.

T. KATO, Schrödinger Operators with Singular Potentials, Israel J. Math. 13, 135-148 (1972).

B. SIMON, An abstract Kato's Inequality for Generators of Positivity Preserving

Semigroups, Indiana University Math. Journal 26, 1067 (1977) ; Kato's Inequality and the Comparison of Semigroups, submitted to Journ. Funct. Anal.

H. HESS, R. SCHRADER, D.A. UHLENBROCK, Domination of Semigroups and Generalization of Kato's Inequality, Proceedings of the Conference on Mathematical Physics, Rome 1977, Springer Lecture Notes in Physics, and Duke Mathematical Journal 44, 893 (1977).

B. SIMON, Universal Diamagnetism of Spinless Bose Systems, Phys.Rev.Letters 36, 1083 (1976).

[3] R.T. SEELEY, Complex Powers of an Elliptic Operator, Amer. Math. Soc. Proc. Sympos. Pure Math. 10, 288-307 (1967).

Analytic Extension of the Trace associated with Elliptic Boundary Problems, Amer. J. Math. 91, 963-983 (1969).

[4] Using Fermion functional integration and a lattice cutoff, a proof of the paramagnetic inequality (4) was given by D. Bridges, J. Fröhlich and E. Seiler ; On the Construction of Quantized Gauge Fields. I. General Results ; IHES Preprint June 1978. The proof works as yet up to dimension three.

A particular case of our conjecture has been demonstrated by E. Lieb, private communication. He has shown for the case of a constant magnetic field and an arbitrary potential V in three dimensions that the groundstate is lowered by the magnetic field. In fact, this particular case was the subject of an independent conjecture by I. Herbst, J. Avron and B. Simon in their article "Schrödinger Operators with Magnetic Fields. I, General Interactions ; to appear in Duke Math. Journ. 1978, Princeton University Preprint, where also the argument by Lieb can be found.

Lieb's argument can be extended to the case of nonconstant magnetic fields, see J. Avron and R. Seiler, Paramagnetism for Nonrelativistic Electrons and Euclidean Massless Dirac Particles, Preprint 78/P.1038, CNRS Marseille.

[5] H. HOGREVE, R. SCHRADER and R. SEILER ; A Conjecture on the Spinor Functional Determinant, to appear in Nucl. Phys. B, FUB-HEP Preprint, April 1978.

[6] W. HEISENBERG, H. EULER ; Folgerungen aus der Diracschen Theorie des Positrons, Z. Physik 98, 714-732 (1936).

J. SCHWINGER ; On Gauge Invariance and Vacuum Polarization, Phys. Rev. 82, 664-679 (1951).

V. WEISSKOPF, über die Electrodynamik des Vakuums auf Grund der Quantentheorie des Elektrons ; Kongelige Danske Videnskabernes Selskas, Mathematisk-fysike Meddelelser XIV, N° 6 (1936).

V. FOCK, Physik.Z.Sowjetunion 12, 404 (1937).

[7] See e.g. H. ARAKI and E. LIEB, Entropy Inequalities, Commun.math.Phys. 18, 160-170 (1970).

APPLICATION OF PATH INTEGRALS TO NON-PERTURBATIVE STUDY OF MASSIVE YANG-MILLS THEORY

A.A. Slavnov*
Service de Physique Théorique
C.E.A.-Saclay, BP n°2, 91190 Gif-sur-Yvette, France
(*On leave from Steclov Mathematical Institute, Moscow, USRR)

1. INTRODUCTION

The only self-consistent way of gauge invariant description of massive vector fields is provided at present by the Higgs mechanism[1,2]. This mechanism is based on the ad hoc introduction of scalar fields which appear as fundamental elementary particles together with leptons, quarks and Yang-Mills mesons. Contrary to the Yang-Mills interaction the Higgs mechanism has no clear geometrical meaning and introduction of fundamental scalar fields for several reasons does not seem to be very attractive possibility.

There exist an alternative possibility of the gauge invariant description of massive vector mesons which uses chiral fields, i.e. the fields with the values in some non-linear manifold. This possibility was known long ago (see e.g. the review[3]), and was used also for the construction of models of weak and electromagnetic interactions[4]. The main drawback of these models was the lack of a self-consistent approximation scheme. Non-renormalizable ultraviolet divergencies do not allow to use for these models ordinary perturbation theory.

In this paper a non-perturbative definition of the massive Yang-Mills theory is proposed which allows self-consistent calculation of the S-matrix. The spectrum of the model differs crucially from the spectrum one would expect on the basis of a naive perturbation theory. It includes together with massive vector fields also massive scalars. The resulting theory is actually equivalent to the Higgs-Kibble model giving in some sense dynamical explanation of the Higgs mechanism. From the point of view of the original fundamental Lagrangian which includes only massive vector fields the Higgs meson may be considered as a bound state. Being considered in the space including as asymptotic states also Higgs scalars, the model may be treated perturbatively.

In the second section the gauge invariant formulation of the massive Yang-Mills theory is described, which uses chiral fields. In the third section a non-perturbative definition of the path integral over the chiral fields using $(N)^{-1}$ expansion is given. In the fourth section it is shown that this integral satisfies reasonable equation of motion. In the fifth section the equivalence of the theory of the Higgs-Kibble model is shown.

2. GAUGE INVARIANT FORMULATION OF THE MASSIVE YANG-MILLS THEORY

Let $A_\nu(x)$ be a vector field with the values in the Lie algebra of a gauge group SU(N). A complex scalar field $\varphi(x)$ belongs to the fundamental representation of SU(N) and satisfies the condition

$$\varphi^+\varphi = \frac{N}{\gamma} \tag{1}$$

$A_\mu(x)$ may be presented as a matrix in the adjoint representation

$$A_\mu(x) = A^a_\mu(x)\, t^a \tag{2}$$

where the generators t^a are normalized as follows

$$tr(t^a . t^b) = -2\delta^{ab} \tag{3}$$

Postulating invariance with respect to the gauge transformations

$$A^a_\mu(x) \to A^a_\mu(x) - t^{abc}\,[A^b_\mu(x)\,\varepsilon^c(x)] + \partial_\mu\,\varepsilon^a(x) \tag{4}$$

$$\begin{aligned} \varphi(x) &\to \varphi(x) + i\varepsilon^a(x)\,\lambda^a\,\varphi(x) \\ \varphi^+(x) &\to \varphi^+(x) - i\varepsilon^a(x)\,\varphi^+(x)\,\lambda^a \end{aligned} \tag{5}$$

where λ^a are the generators of the fundamental representation one gets a gauge invariant Lagrangian

$$\mathcal{L} = -(\partial_\mu\varphi^+ - i\varphi^+\lambda^a A^a_\mu)(\partial_\mu\varphi + i\,\lambda^a A^a_\mu\varphi) + \mathcal{L}_{YM} \tag{6}$$

where

$$\mathcal{L}_{YM} = -\frac{1}{8g^2}\, tr\,\{F_{\mu\nu}F_{\mu\nu}\} \tag{7}$$

$$F_{\mu\nu} = \partial_\nu A_\mu - \partial_\mu A_\nu + [A_\mu, A_\nu] \tag{8}$$

The fields φ determine the directions in the internal "charge" space. Using gauge freedom one can choose the direction in such a way that

$$\varphi_i = \delta_{iN}\sqrt{\frac{N}{\gamma}} \tag{9}$$

In this gauge scalar fields are completely eliminated and the first term in the formula (6) reduces to

$$\frac{N}{\gamma}\,(\lambda^a\lambda^b)_{NN}\,A^a_\mu(x)\,A^b_\mu(x) \tag{10}$$

providing some of the vector mesons with the non-zero masses. In particular for the case of SU(2) gauge group when $\lambda^a = \tau^a/2$, all vector mesons acquire equal non-zero masses.

Therefore in the framework of a classical fiels theory the Lagrangian (6) describes only vector fields, scalar fields being purely gauge degrees of freedom.

Another possible gauge which we shall exploit in the following is the Lorentz gauge

$$\partial_\mu A_\mu(x) = 0 \tag{11}$$

Now we should define a quantum theory corresponding to the classical Lagrangian (6). Using the well-known procedure one can write the Green function generating functional in the Lorentz gauge

$$Z_{YM} = N^{-1} \int \exp\left\{i \int \mathcal{L} dx + s.t.\right\} \delta(\partial_\mu A_\mu)\ \Delta(A)\ d\varphi^+\ d\varphi\ dA \tag{12}$$

where $\Delta(A)$ is the Faddeev-Popov determinant and integration is performed over the fields satisfying the condition (1). s.t. means source terms which shall not be specified for the moment. One can think for example that the sources are introduced for some gauge invariant functions.

Usually the expression (12) is understood in the framework of the renormalized perturbation theory. Namely, the integrand in the eq.(12) is presented as a formal series in the coupling constant. Then all the integrals become gaussian and may be given a precise meaning[5)], provided that intermediate ultraviolet regularization is also introduced. Finally, allowing the masses, charges etc entering the Lagrangian to depend on the regularization parameter (ε), i.e. introducing counterterms, one can define limiting procedure $\varepsilon \to 0$, giving finite Green functions satisfying all necessary conditions.

To apply this procedure to the Lagrangian (6) one must firstly solve explicitly the eq.(1) :

$$\mathrm{Re}\,\varphi_N = \sqrt{N\gamma^{-1} - \sum_{i=1}^{N-1} |\varphi_i|^2 - (\mathrm{Im}\,\varphi_N)^2} = \sqrt{\frac{N}{\gamma}} \left[1 - \frac{\gamma}{2N} \sum |\varphi_i|^2 + \ldots\right] \tag{13}$$

and substitute this solution into eq.(6). However the resulting Lagrangian cannot be presented as a finite polynomial, and one cannot define limiting procedure $\varepsilon \to 0$ by fixing a finite number of parameters. Therefore the perturbative definition of the functional (12) makes no sense, and some other definition must be used.

What are the physical and mathematical requirements to be satisfied by the quantum Green functions ? Obviously they must be defined as a distribution and must satisfy the conditions of Lorentz and gauge invariance, unitarity and causality. They must also satisfy the equations which generalize the classical equations of motion in the domain where this generalization is unambiguous.

Below we shall show explicitly how to construct the Green functions satisfying all these conditions. We adopt straightforward renormalization point of view. Some intermediate regularization preserving the invariance properties of the theory is assumed. For example one can use dimensional regularization which makes all the Feynman diagrams, generated by the functional (12), finite as long as $\varepsilon = 4-n$ (n is the dimension of the space-time) is different from zero. Then redefining the parameters entering the original Lagrangian, i.e. introducing the counterterms depending on ε, we shall be able to perform the limit $\varepsilon \to 0$.

The difficulties in the calculation of the integral (12) are due to the presence of φ-dependent terms which being treated perturbatively are essentially non-polynomial. So first of all we shall show how the integral over the fields φ may be defined.

3. NON-PERTURBATIVE DEFINITION OF THE INTEGRAL OVER φ

In this section we shall consider the integral

$$Z(A_\mu) = \int \exp\left\{i\int\left[(\partial_\mu\varphi^+ - i\varphi^+\lambda^a A_\mu^a)(\partial_\mu\varphi + i\lambda^a A_\mu^a\varphi)\right] dx\right\} \prod_x d\varphi \tag{14}$$

where the integration is performed over the fields, satisfying the constraint (1). As we have already mentioned perturbative definition of this integral fails, leading to non-renormalizable theory.

Instead one can try to define the integral (14) using N^{-1} expansion. We shall use the particular form of N^{-1} expansion described in ref.6).

The integral (14) may be written in the following convenient form

$$Z(A_\mu) = \int \exp\left\{i\int\left[|\partial_\mu\varphi^+ - i\varphi^+\lambda^a A_\mu^a|^2 - \sigma(x)(\varphi^+\varphi - \frac{N}{\gamma})\right] dx\right\} \prod_x d\sigma\, d\varphi^+\, d\varphi \tag{15}$$

where the integration over the Lagrange multiplier $\sigma(x)$ takes into account explicitly the constraint (1). An invariant regularization (for example dimensional) is always understood.

Assuming that one may interchange the integrations over φ and σ we calculate the Gaussian integral over φ and get

$$Z(A_\mu) = \int \exp\left\{i \operatorname{Tr}\ln\widetilde{K} + i\frac{N}{\gamma}\int \sigma(x)\, dx\right\} \prod_x d\sigma \tag{16}$$

where the operator $\widetilde{K}$ is defined as follows

$$\widetilde{K}\varphi = \left\{\Box + \sigma(x) - 2\lambda^a A_\mu^a(x)\left(\partial_\mu^x + i\frac{\lambda^b}{2}A_\mu^b(x)\right)\right\}\varphi(x) \tag{17}$$

and Tr means the trace operation which includes also continuous variables, i.e. integration over x .

The integral over $\sigma(x)$ will be calculated by expanding the effective action in the exponent (16) around the stationary point. Differentiating the effective action with respect to $\sigma(x)$ one gets the stability condition

$$N\, D^\varepsilon(0,\sigma_o) = \frac{iN}{\gamma} \tag{18}$$

where $D^\varepsilon(x,\sigma_o)$ stands for the Green function of the operator $(\Box + \sigma_o)$. Index ε means that this operator is regularized. When regularization is removed $D(0,\sigma_o)$ has a singularity ε^{-1}. Therefore to make sense of the eq.(18) in the limit when $\varepsilon \to 0$, one should renormalize the constant γ : $\gamma = \gamma_R \cdot \varepsilon$.

Shifting the fields $\sigma(x)$ to the stationary point $\sigma \to \sigma + \sigma_o$ we get

$$Z(A_\mu) = \int \exp\left\{i \operatorname{Tr} \ln K + \frac{iN}{\gamma} \int \sigma(x)\, dx\right\} \prod_x d\sigma \tag{19}$$

where

$$K\varphi = \left[\Box + \sigma_o + \sigma(x) - 2\lambda^a A^a_\mu(x) \left(\partial^x_\mu + i \frac{\lambda^b}{2} A^b_\mu(x)\right)\right] \varphi(x) \quad . \tag{20}$$

The expression in the exponent (19) may be presented as a sum of one-loop diagrams. When $\varepsilon \to 0$ some of them become divergent. Superficially divergent diagrams are shown in Fig.1.

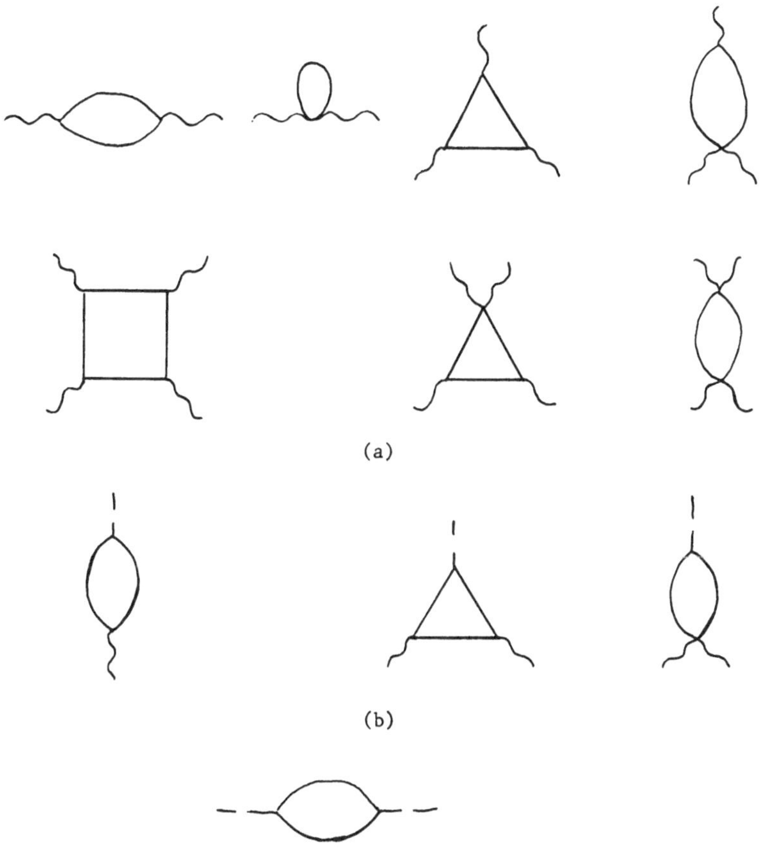

(a) (b) (c)

Fig.1 - Wavy line means the propagator of A_μ , dotted line means the propagator of σ , solid line the propagator of φ .

Due to the gauge invariance of regularization procedure the divergent (in the limit $\varepsilon \to 0$) parts of the diagrams (a) form a gauge invariant structure, i.e. are proportional to Yang-Mills Lagrangian

$$- c_1 \, \varepsilon^{-1} \mathcal{L}_{YM}(x) \tag{21}$$

The diagrams (b) have logarithmic superficial divergencies. But due to Lorentz and gauge invariance this divergence is actually absent. Indeed the corresponding counterterm

$$c \, \varepsilon^{-1} \, \sigma(x) \, A_\mu^2(x) \tag{22}$$

is not gauge invariant. Therefore $c = 0$.

Finally the diagram (c) contributes the term

$$\sim \varepsilon^{-1} \, \sigma^2(x) \qquad .$$

The divergencies (21) may be eliminated by introducing in the original Lagrangian (6) as a counterterm the same expression with the opposite sign. We cannot however in the same way to eliminate the term (23) because the corresponding counterterm $\sim \sigma^2(x)$ would change the structure of the original Lagrangian. Instead we shall calculate the integral (19) perturbatively in N^{-1} , keeping $\varepsilon \neq 0$, and show that in the final expression one can put $\varepsilon = 0$ without introducing new counterterms. The propagator of the ε-field is defined by the quadratic part of Eq.(20) which is

$$G^{-1}(p) = -N \int \frac{e^{-ipx}}{(2\pi)^4} \, D^\varepsilon(x,\sigma_o) \, D^\varepsilon(x,\sigma_o) \, dx = N(\text{const.}\varepsilon^{-1} + f(p)) \, , \tag{24}$$

where $f(p)$ increases logarithmically at large p .

The expansion of the effective Lagrangian (20) generates non-local vertices of the type $\sigma^n A_\mu^m$. Alternatively one can rescale the fields σ : $\sigma \to N^{-1/2} \sigma$. Then the σ-propagator becomes of zero order, but the vertices acquire the factor $N^{-n/2}$.

The leading order terms in N^{-1} are given by the eq.(19) with $\sigma(x) = 0$, i.e.

$$Z_o^\varepsilon(A_\mu) = \exp \left\{ i \, \mathrm{Tr} \ln K_o^\varepsilon + i c_1 \varepsilon^{-1} \int \mathcal{L}_{YM} \, dx \right\} \tag{25}$$

where K_o denotes the operator K, given by the eq.(20) for $\sigma(x) = 0$. This expression admits the limiting process $\varepsilon \to 0$ giving renormalized Green function generating functional in the leading order in N^{-1}. Now we shall show that all higher order corrections to eq.(25) vanish in the limit $\varepsilon \to 0$, and therefore (25) is an exact answer.

Instead of working with the non-local vertices, it is more convenient to go back and introduce again explicitly the φ-propagators. Then the diagram technique will include the following elements. The propagator of σ-field is given by the eq. (24). The propagator of φ-field

$$D^\varepsilon(x,\sigma_o) = \frac{1}{i(2\pi)^4} \int e^{-i\bar{\bar{k}}x} \, \frac{1}{\sigma_o - \bar{\bar{k}}^2} \, d\bar{\bar{k}} \qquad . \tag{26}$$

The interaction vertices are

$$\varphi^{+}\varphi\,\sigma \quad , \quad 2i\,\partial_{\mu}\varphi^{+}\lambda^{a}A^{a}_{\mu} \quad , \quad \varphi^{+}\lambda^{a}\lambda^{b}\varphi A^{a}_{\mu}A^{b}_{\mu} \tag{27}$$

One loop insertions into the σ-lines should be omitted because they are already taken into account by (24). Tadpoles containing one φ-loop are also already taken into account due to the stability condition (18).

Let us consider firstly N^{-1} corrections. The corresponding diagram contain one internal σ-line. Due to the structure of the σ-propagator (24) every finite integral containing this propagator vanishes in the limit $\varepsilon \to 0$. Therefore it is sufficient to consider only divergent (in the limit $\varepsilon \to 0$) diagrams. Superficially divergent diagrams shown in Fig.2.

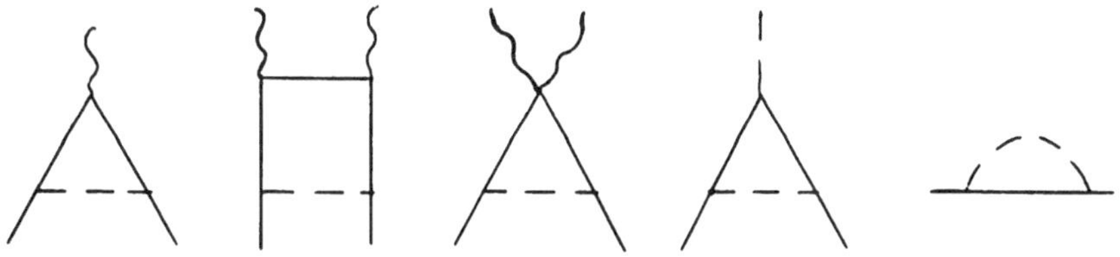

a b c d

Fig.2

The divergent parts of the diagrams (a), (b) at most may produce gauge invariant counter-terms leading to finite renormalization of φ-wave function. The diagram (c) produces the term

$$N^{-1}\,\{\mathrm{const}\,\sigma\varphi^{2} + O(\varepsilon)\} \tag{28}$$

which leads to the finite renormalization of the vertex present in the original effective action.

Finally diagram (d) produces term

$$N^{-1}\,\{\mathrm{const}\,\varphi^{2} + O(\varepsilon)\} \tag{29}$$

which may be compensated by the finite renormalization of the linear term $\frac{N}{\gamma}\sigma(x)$. All other diagrams contribute terms of order ε or higher.

Therefore in the limit $\varepsilon \to 0$, N^{-1} corrections produce only non-essential finite renormalization. At the order N^{-2} there exist superficially divergent diagrams of the same structure and two new diagrams shown in Fig.3.

Fig. 3

The diagram (a) contributes the term

$$N^{-2}\ \{\text{const}\ \varepsilon(\varphi^+\varphi)^2 + O(\varepsilon^2)\} \tag{30}$$

and the diagram (b) after the introduction of one-loop counter-term (28) contributes

$$N^{-2}\ \{\text{const}\ \varepsilon\sigma^2 + O(\varepsilon^2)\} \tag{31}$$

Two-loop diagrams having the same structure as the diagrams shown in Fig.2, after the introduction of one-loop counter-terms contribute the terms of order ε or higher. So all N^{-2} corrections also vanish in the limit $\varepsilon \to 0$.

In the higher orders superficially divergent diagrams have the same structure as those shown in Figs.2, 3 and for the same reason contribute the terms vanishing in the limit $\varepsilon \to 0$.

Therefore the lowest order expression (25) indeed gives up to non-essential renormalization an exact answer for the integral (15) in the limit $\varepsilon \to 0$.

This expression may be written in the form

$$Z_R = \int \exp\left\{i \int \left[|\partial_\mu\varphi^+ - i\varphi^+\lambda^a A^a_\mu|^2 - \sigma^R_o\ \varphi^+\varphi + c_1\varepsilon^{-1}\mathcal{L}_{YM}\right] dx\right\} \prod_x d\varphi^+\, d\varphi \tag{32}$$

where the integration is performed over all the fields φ^+, φ . This result agrees completely with the statement[7)] that the four dimensional non-linear σ-model is equivalent to the model of free scalar fields. For $A_\mu = 0$ the Lagrangian in the exponent (15) is simply the Lagrangian of the non-linear σ-model. Our final result(32) clearly corresponds for $A_\mu = 0$ to the free field theory.

One sees also that N^{-1} expansion is actually not necessary, because we need only zero order terms. Instead one can use the expansion in the parameter ε , as have been done in our paper[8)].

4. EQUATIONS OF MOTION FOR THE GENERATING FUNCTIONAL

In this section we shall show that the Green functions generated by the functional (32) satisfy the equations of motion which follows from the corresponding classical equations of motion in the domain where these equations are defined unambiguously. We shall consider in details the case of SU(2) gauge group, when these equations have a particularly simple form.

For this purpose it is convenient to rewrite the functional (15) in the alternative equivalent form

$$Z = \int \exp\left\{-i\,\frac{2}{\gamma}\int tr(A_\mu - L_\mu)^2\,dx\right\}\,d\Omega \tag{33}$$

Here the chiral currents L_μ are defined by the formula

$$L_\mu = \partial_\mu \Omega\Omega^{-1} \tag{34}$$

Ω is an element of the gauge group.

The equivalence is most easily seen by noticing that the expression (33) is gauge invariant with respect to transformations

$$A_\mu \to \tilde{\Omega} A_\mu \tilde{\Omega}^{-1} + \partial_\mu \tilde{\Omega}\tilde{\Omega}^{-1} \quad , \qquad \Omega \to \tilde{\Omega}\Omega \tag{35}$$

and in a special gauge $L_\mu = 0$ coincides with the eq.(14) in the gauge (9).

The classical action which stands in the exponent (33) generates the equation of motion

$$(\partial_\mu A_\mu - \partial_\mu L_\mu) + [A_\mu , L_\mu] = 0 \tag{36}$$

which is nothing but the current conservation law in the presence of the external field $A_\mu(x)$.

The second equation satisfied by the chiral currents L_μ is a Maurer-Cartan equation

$$\partial_\mu L_\nu - \partial_\nu L_\mu - [L_\mu , L_\nu] = 0 \tag{37}$$

which states simply that L_μ is a pure gauge.

In quantum theory one would expect analogous equations for the Green functions

$$< T\left\{\partial_\mu A_\mu(x) - \partial_\mu L_\mu(x) + \left[A_\mu(x) , L_\mu(x)\right]\right\} L_{\mu_1}(x_1) \ldots L_{\mu_n}(x_n)> = 0 \tag{38}$$

$$< T\left\{\partial_\mu L_\nu(x) - \partial_\nu L_\mu(x) - \left[L_\mu(x) , L_\nu(x)\right]\right\} L_{\mu_1}(x_1) \ldots L_{\mu_n}(x_n)> = 0 \ . \tag{39}$$

However these equations are not well defined when the arguments of $L_{\mu_i}(x_i)$ coincide. If the Green functions

$$< T \; L_{\mu_1}(x_1) \ldots L_{\mu_n}(x_n) > \tag{40}$$

are defined as distributions by means of the renormalization procedure, then the first equation (38) acquires a precise meaning. But the equation (39) which includes the product of the operators L_μ at the same point x (i.e. a composite operator) still needs an independent definition. In other words in quantum theory the equation (39) is defined up to the terms proportional to $\delta(x-x_i)$. The particular form of these terms depends on the specific renormalization procedure used for the calculation of the functional (14).

Below we shall verify that the Green functions generated by eq.(32) in the limit $\varepsilon \to 0$ satisfy the Eqs.(38) and (39) for $x \neq x_i$. For this purpose we rewrite these equations in a functional form introducing the generating functional for the connected Green functions

$$W = - i \ln Z \tag{41}$$

In terms of the functional W these equations look as follows

$$\partial^x_\mu \frac{\delta W}{\delta A_\mu(x)} - \left[A_\mu(x) \, , \, \frac{\delta W}{\delta A_\mu(x)}\right] = 0 \tag{42}$$

$$\begin{aligned} &\partial_\mu \left(\frac{\delta W}{\delta A_\nu(x)} + iA_\nu(x)\right) - \partial_\nu \left(\frac{\delta W}{\delta A_\mu(x)} + iA_\mu(x)\right) = \\ &= - \frac{\gamma}{2} \frac{\delta^2 W}{[\delta A_\mu(x) \, , \, \delta A_\mu(x)]} - i \frac{\gamma}{2} \left[\left(\frac{\delta W}{\delta A_\mu(x)} + iA_\mu(x)\right), \left(\frac{\delta W}{\delta A_\nu(x)} + iA_\nu(x)\right)\right] \end{aligned} \tag{43}$$

The equation (42) is nothing but the Ward identity and is evidently satisfied due to the manifest gauge invariance of the functional (32).

To check the equation (43) we shall use the formulae

$$\begin{aligned} \frac{\delta W}{\delta A^a_\mu(x)} &= \frac{\delta}{\delta A^a_\mu(x)} (\mathrm{Tr} \ln K_o) = \mathrm{Tr} \left[\frac{\delta K_o}{\delta A^a_\mu(x)} K_o^{-1}\right] \\ &= \mathrm{tr} \left[(-A^a_\mu(x) + i\tau^a \partial^x_\mu) \, K_o^{-1}(x,y)\Big|_{y=x}\right] \end{aligned} \tag{44}$$

$$\begin{aligned} \varepsilon^{abc} \frac{\delta^2 W}{\delta A^b_\mu(x) \, \delta A^c_\nu(x)} &= \varepsilon^{abc} \, \mathrm{Tr} \left[\frac{\delta^2 K_o}{\delta A^b_\mu(x) \, \delta A^c_\nu(x)} K_o^{-1} - \frac{\delta K_o}{\delta A^b_\mu(x)} K_o^{-1} \frac{\delta K_o}{\delta A^c_\nu(x)} K_o^{-1}\right] \\ &= \varepsilon^{abc} \, \mathrm{tr} \left[(-A^b_\mu(x) + i\tau^b \partial^x_\mu) \, K_o^{-1}(x,y)\Big|_{y=x} (-A^c_\nu(x) + i\tau^c \partial^x_\nu) \, K_o^{-1}(x,y)\Big|_{y=x}\right] \end{aligned} \tag{45}$$

Being interested in the Green functions for $x \neq x_i$ we can drop in these formulae the terms proportional to $A_\mu(x)$ because they contribute the terms $\sim \delta(x-x_i)$

So we must check the equation

$$\partial^x_\mu \left[\mathrm{tr}\ \tau^a \partial^x_\nu K_o^{-1}(x,y)\Big|_{y=x}\right] - \partial^x_\nu \left[\mathrm{tr}\ \tau^a \partial^x_\mu K_o^{-1}(x,y)\Big|_{y=x}\right] =$$

$$= \frac{i\gamma}{2}\,\varepsilon^{abc}\left[\mathrm{tr}\left\{ \tau^b \partial^x_\mu K_o^{-1}(x,y)\Big|_{y=x}\ \tau^c \partial^x_\nu K_o^{-1}(x,y)\Big|_{y=x}\right\} - i\,\mathrm{tr}\left\{ \tau^b \partial^x_\mu K_o^{-1}(x,y)\Big|_{y=x}\right\}\ \mathrm{tr}\left\{ \tau^c \partial^x_\nu K_o^{-1}(x,y)\Big|_{y=x}\right\}\right] \tag{46}$$

in the limit $\varepsilon \to 0$, keeping in mind that for the moment we are interested only in the Green functions generated by K_o for $x \neq x_i$.

The left hand side represents the sum of one-loop diagrams which are evidently finite when $x \neq x_i$. The second term in the r.h.s. represents the product of one-loop diagrams which are also finite for $x_i = x$ and therefore being multiplied by γ , where $\gamma = \varepsilon . \gamma_R$, vanishes in the limit $\varepsilon \to 0$. The first term at the r.h.s. of eq.(46) generates contracted diagrams shown in Fig.4.

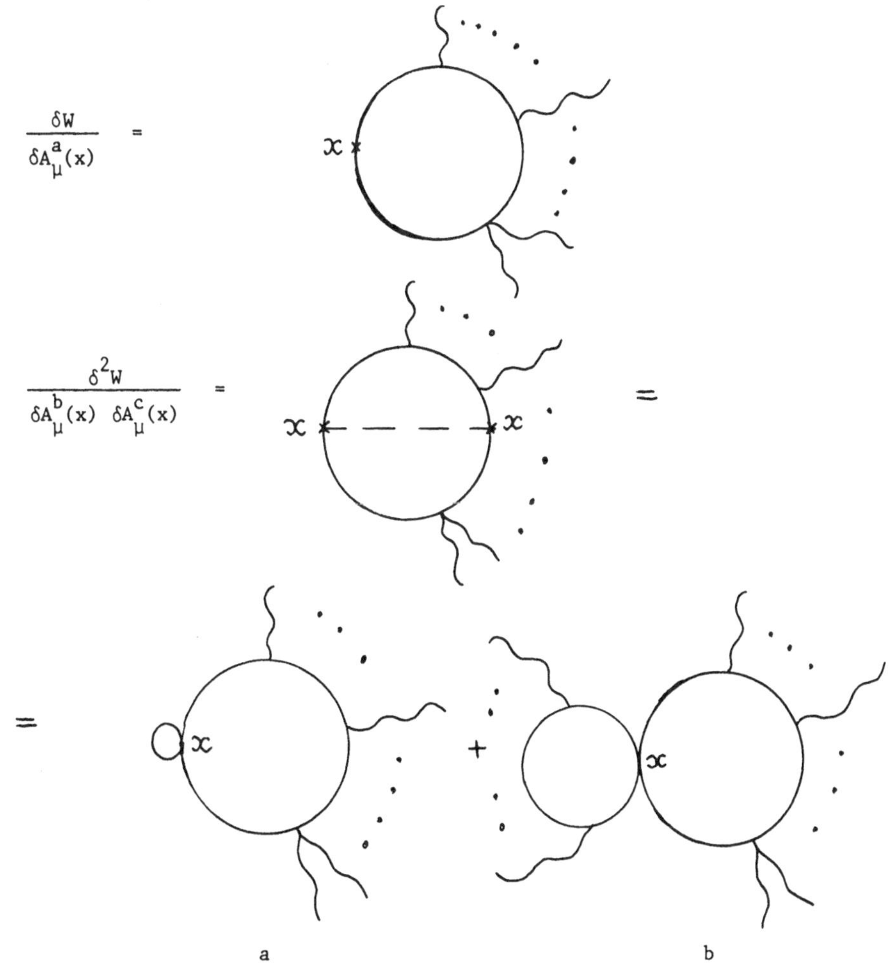

Fig. 4

All these diagrams except for the diagrams (a), corresponding to the contraction of two adjacent vertices are finite for $x \neq x_i$ and therefore do not contribute in the limit $\varepsilon \to 0$. The only diagrams which should be taken into account are the diagrams (a). That means that the only relevant rerms in the r.h.s. of eq.(46) are those where at least one of the operators $K_o^{-1}(x,y)$ is replaced by the free Green function $D^{\varepsilon}(x-y)$. Simple algebraic transformations give for the r.h.s. of eq.(46) in the limit $\varepsilon \to 0$ the expression

$$\lim_{\varepsilon \to 0} \int - i\gamma D^{\varepsilon}(0,\sigma_o) \; \partial_\mu^x \, \mathrm{tr}\, \tau^a \partial_\nu^x K_o^{-1}(x,y)\Big|_{y=x} - (\nu \leftrightarrow \mu)_i \tag{47}$$

Due to the stability condition (18) this expression is equal to the l.h.s. of eq. (46). So the functional (32) indeed satisfies the quantum analogue of the classical equations of motion (36), (37) in the domain where these equations are defined unambiguously.

5. EQUIVALENCE TO THE HIGGS-KIBBLE MODEL

Up to now we considered the Yang-Mills field A_μ as an external field. As long as the propagators of A_μ are kept regularized all our considerations are equally applicable to the complete generating functional (12) which therefore can be written in the form

$$Z_{YM} = N^{-1} \int \exp\left\{ i \int \left[(1 + c_1 \varepsilon^1) \mathcal{L}_{YM} + \left|\partial_\mu \varphi^+ - i\varphi^+ \frac{\tau^a}{2} A_\mu^a\right|^2 - \right.\right.$$
$$\left.\left. - \sigma_o^R(\varphi^+\varphi) + \mathrm{s.t.} \right] dx \right\} \delta(\partial_\mu A_\mu)\, \Delta(A) \prod_x dA\, d\varphi^+\, d\varphi \tag{48}$$

where the integration is performed over all fields φ^+, φ .

The equation (48) is equivalent to the eq.(12) if one regularize separately integrals over φ^+, φ and over A_μ and removes the regularization of φ-integrals, keeping the integral over A_μ regularized. In what follows we shall always assume that the intermediate regularization is performed in this way, i.e. the regularization of the integral over A is removed at the very end. Of course one may use another prescription, e.g. to remove regularization of A_μ and φ-integrals simultaneously, but it makes the analyses much more complicated.

The integration over A_μ leads to two modifications. Due to the presence of loop diagrams including internal A_μ lines the additional counter-terms renormalizing the wave functions, the mass and the charge are necessary. Introduction of these counter-terms changes nothing in the previous arguments.

However apart from these counter-terms we need also the counter-term $h(\varphi^+\varphi)^2$ to make the functional (33) finite. Let us show how to incorporate this new vertex in pur scheme.

The original functional (12) does not change if we add to the action the term

$h(\varphi^+\varphi)^2$ because in this formula the integration is performed over the fields satisfying the condition $\varphi^+\varphi = 2\gamma^{-1}$. The normalized functional (12) may be re-written in the following form

$$Z_{YM} = N^{-1}\int \exp\left\{i\int\left[\mathcal{L}_{YM} + |\partial_\mu\varphi^+ - i\varphi^+\frac{\tau^a}{2}A^a_\mu|^2 - \sigma(x)(\varphi^+\varphi - 2\gamma^{-1})\right.\right.$$

$$\left.\left. - \sigma_o(\varphi^+\varphi) - h(\varphi^+\tau^a\varphi)\,\chi^a + \frac{1}{2}\chi^a\chi^a + \text{s.t.}\right]dx\right\}\,\delta(\partial_\mu A_\mu)\,\Delta(A)\,\prod_x d\sigma\, d\varphi\, dA \qquad (49)$$

Here we introduced the auxiliary field χ^a with a point-like propagator to make the Lagrangian quadratic in φ , and used the identity

$$(\varphi^+\tau^a\varphi)(\varphi^+\tau^a\varphi) = (\varphi^+\varphi)^2 \qquad . \qquad (50)$$

The integration over φ^+ , φ may be performed exactly in the same way as before, giving

$$Z(A_\mu,\chi) = \int \exp\left\{i\ \mathrm{Tr}\ln K_\chi + \frac{2i}{\gamma}\int\sigma(x)\ dx\right\}\prod_x d\sigma \qquad (51)$$

where

$$K_\chi\ \varphi = \left[\Box + \sigma_o + \sigma(x) + h\tau^a\chi^a - \frac{A_\mu^2}{2} + i\tau^a A^a\partial^x_\mu\right]\varphi(x) \qquad . \qquad (52)$$

As before the integrand in (51) may be presented as a sum of one-loop diagrams, where the additional vertex $h\tau^a\chi^a$ is present. The only new diagram which is superficially divergent when $\varepsilon\to 0$ is the diagram with two external χ-lines. This new divergence is eliminated by renormalization of the χ-wave function, i.e. by a counter-term $\sim\varepsilon^{-1}\chi^2$. After that the integration over σ-field may be performed in a complete analogy with the discussion in the previous section. The final result in the limit $\varepsilon\to 0$ looks as follows

$$Z(A_\mu,\chi) = \exp\left\{i\ \mathrm{Tr}\ln K^o_\chi + \int\left[c_1\varepsilon^{-1}\mathcal{L}_{YM} + c_2\varepsilon^{-1}(\chi^a\chi^a)\right]dx\right\} \qquad (53)$$

where K^o_χ is given by the equation (54) with $\sigma(x) = 0$.

(This particularly simple structure of $\mathrm{Tr}\ln K_\chi$ is due to our choice of the auxiliary field interaction in the form $(\varphi^+\tau^a\varphi)\chi^a$ and not in the isoscalar form $(\varphi^+\varphi)\chi$. Otherwise $\mathrm{Tr}\ln K_\chi$ would also contain divergent diagram with one σ and one χ-line leading to σ-χ mixing.)

Summarizing we see that after all necessary renormalizations the generating functional (12) for the massive Yang-Mills theory can be presented in the form

$$Z_{YM} = N^{-1}\int\exp\left\{i\int\left[\mathcal{L}_R(A_\mu,\varphi) - \sigma_o\varphi^+\varphi - h_R(\varphi^+\tau^a\varphi)\chi^a + \frac{1}{2}(1 + c_2\varepsilon^{-1})\chi^a\chi^a\right]dx\right.$$

$$\left.\delta(\partial_\mu A_\mu)\,\Delta(A_\mu)\prod_x d\chi^a\, d\varphi^+\, d\varphi\, dA\right. \qquad , \qquad (54)$$

where $\mathcal{L}_R(A_\mu,\varphi)$ denotes the gauge invariant renormalized Lagrangian for the massless

Yang-Mills field interacting with the scalar fields φ . The integration in the equation (54) is performed over all the fields φ^+ , φ without any constraints.

The last point to be discussed concerns the value of the parameter σ_o . This parameter is defined by the equation

$$\frac{2}{(2\pi)^4} \int \frac{1}{\sigma_o - \bar{\bar{k}}_o^2 + \vec{\bar{k}}^2} d^4\bar{k} = \frac{2i}{\gamma_R \varepsilon} \qquad (55)$$

where again the regularization is understood. This equation has a solution

$$\sigma_o = m^2 - i0 \qquad (56)$$

Indeed, substituting (56) into eq.(55) one can pass to the Euclidean space

$$\frac{1}{(2\pi)^4} \int \frac{1}{m^2 + \bar{\bar{k}}_o^2 + \vec{\bar{k}}^2} d\bar{\bar{k}} = (\gamma\varepsilon)^{-1} \qquad (57)$$

and solve this equation explicitly.

But if (56) is a solution of eq.(55), then

$$\sigma_o = -(m^2 - i0) \qquad (58)$$

is also a solution. Substituting (58) into eq.(55), one can firstly integrate over the space angles φ, θ and then rotate the contour of integration in the $|\vec{K}|$ plane. This leads to the equation

$$-\frac{2}{(2\pi)^4} \left(\int \frac{1}{-m^2 - \bar{\bar{k}}_o^2 - \vec{\bar{k}}^2} d\bar{\bar{k}} \right) = \frac{2}{\gamma_R \varepsilon} \qquad (59)$$

which coincides with the equation (57).

So for the mass term in the functional (54) we have two possibilities

$$\sigma_o = m^2 > 0 \quad , \quad \text{or} \quad \sigma_o = -m^2 < 0 \quad . \qquad (60)$$

The first one corresponds to the massless Yang-Mills field interacting with the scalar particles having the mass $m = \sqrt{\sigma_o}$. The second possibility is just the Higgs-Kibble model. Shifting the scalar fields to zero expectation values we have massive Yang-Mills field interacting with Higgs scalars. The second possibility corresponds to a lower ground state energy and therefore is stable. The massless Yang-Mills theory corresponds to the unstable solution and therefore is not realized in the nature.

6. DISCUSSION

We have shown that starting with the classical massive Yang-Mills Lagrangian which contains only vector fields one comes to a quantum theory with quite a different mass spectrum. The final theory includes apart from massive vector fields also massive scalar. The mass of this scalar depends non-analytically on the coupling constant and clearly cannot be obtained in perturbation theory. If however we include this scalar into the asymptotic space and fix its mass as an independent parameter the theory may be treated perturbatively in the coupling constant f and is equivalent to the Higgs-Kibble model. Contrary to the ordinary approach in our scheme the Higgs model and massless Yang-Mills theory correspond to the same fundamental Lagrangian, and Higgs model is favored by nature as having a lower ground state energy.

In our approach the limit $M \to 0$ may be performed. The resulting theory is not just massless Yang-Mills theory, but includes also massless scalar particles. It would be interesting to check if the resulting model being a limiting case of a sensible theory exhibits a reasonable infrared behaviour.

The method described above is clearly applicable to the case of other representations of scalar fields leading to essentially the same conclusion.

REFERENCES

1 - Higgs P.W. *Phys. Rev. Lett.* 13, 508 (1964) ; *Phys. Rev.* 145, 1156 (1966).

2 - Kibble T.W.B. *Phys. Rev.* 155, 1554 (1967).

3 - Slavnov A.A. *Theor. and Math. Phys.* 10, 305 (1972).

4 - Faddeev L.D. *Doklady Acad. Sci. USSR* 210, 807 (1973).

5 - Slavnov A.A. *Theor. and Math. Phys.* 22, 177 (1975).

6 - Arefyeva I.Ya. *Theor. and Math. Phys.* 36, 24 (1978).
Arefyeva I.Ya. et al., Preprint LOMI,E-1-1978 .

7 - Brézin E., Zinn-Justin J., *Phys. Rev.* D14, 2615 (1976).

8 - Slavnov A.A., Preprint Saclay DPh-T/78-45 (1978).

$F = {*}F$, A REVIEW

by

J. MADORE

J.L. RICHARD

R. STORA *

Centre de Physique Théorique du CNRS, Marseille

* On leave of absence at CERN, Theory Division.

I - INTRODUCTION

Among the classical Euclidean gauge field configurations which are relevant - may be not the most relevant ones - to the structure of gauge theories, are the so-called instanton configurations, solutions of the reduced Yang-Mills equations

$$F = \pm * F \tag{1}$$

which realize the absolute minima of the Euclidean action

$$\mathcal{A} = \int_{S^4} (F , *F) \tag{2}$$

for given topological number

$$k = \frac{1}{4\pi^2} \int_{S^4} (F , F) . \tag{3}$$

It is not known whether regularity on the one-point compactification S^4 of Euclidean space E_4 is necessary to insure finiteness of the action. Besides, no finite action solution of the full Yang-Mills equation

$$\nabla * F = 0 \tag{4}$$

is known, other than those of the reduced equations which have now been all constructed by M.F. Atiyah, V.G. Drinfeld, N.J. Hitchin, Yu. I. Manin (ADHM) [1], thus completing the class discovered by G. 't Hooft and followers [2] of dimensionality $5k + 4$ (for the SU2 gauge group, $k > 1$) up to the right dimensionality $8k - 3$ [3] (for the SU2 gauge group) which can be both obtained by physical and mathematical arguments.

The development of this work has been rather fast since the discovery of the $k = 1$ solutions for SU2 by Belavin, Polyakov, Schwarz, Tyupkin [4], through the formulation of the 't Hooft, Jackiw, Rebbi Ansatz, the counting of the solutions by Jackiw, Nohl, Rebbi, Schwarz, Atiyah, Hitchin, Singer [3], and the solution of the equivalent problem of algebraic geometry set up by Atiyah and Ward [5]. It has gradually slipped off the hands of physicists into those of mathematicians. The mathematical techniques which have been used are not very well spread among physicists : they range from the index theory [3], [6], which, together with simple positivity arguments allow to solve all dimensionality questions which come up (starting with "8k - 3") to reasonably simple algebraic geometry that can mostly be found in textbooks at the graduate level [7]. Actually it is not totally unlikely that the construction might be completely reduced to harmonic theory on S^4 .

These notes will attempt to give a description of the ADHM construction [1] in some details. The reader whose mathematical background needs to be refreshed on some or other standard chapter of mathematics will be helped by accurate references to available textbooks or lecture notes.

Section II is entitled "Isomorphism and Vanishing Theorems". It is divided into three subsections :

a) bundles over $S^4 \sim P_1(H)$ on which there is given a connection with self-dual curvature are shown to be in a one-to-one correspondence with "special" algebraic bundles over $P_3(\mathbb{C})$ from which the connection can be reconstructed.

b) A technical section is devoted to establishing the equivalence between the cohomology spaces of $P_3(\mathbb{C})$ with values in the associated "special" algebraic sheaves and some (finite dimensional) spaces of spinor fields over S^4 . Čech cohomology is used throughout.

c) Vanishing theorems follow from simple positivity arguments.

Section III is devoted to the characterization of "special" algebraic bundles and sheaves over $P_3(\mathbb{C})$:

a) A reminder is given about holomorphic differential forms over $P_3(\mathbb{C})$

b) The Koszul resolution introduced by Drinfeld and Manin for special algebraic bundles is exhibited and reduced with the help of the previously established vanishing theorems. Special bundles are thereby characterized in terms of a pair of finite dimensional vector spaces which can be interpreted in terms of spinor fields over S^4 , and a linear mapping between them which depends linearly on $x \in E_4$.

Section IV concludes the analysis with the construction of connections over S^4 with self-dual curvature.

Although this construction has been performed for all classical groups some of the detailed calculations still need to be improved. Very little is known about the topology of the solution manifold and its parametrization [8] . For this reason, and in order to keep the notation lighter, we shall deal with the SU2 gauge group and only make a few occasional remarks about other classical groups.

II - ISOMORPHISM AND VANISHING THEOREMS.

a) Bundles over $S^4 \sim P_1(H)$ and special algebraic bundles over $P_3(\mathbb{C})$

The one-point compactification of E^4 will be identified with the quaternionic projective line

$$P_1(H) = (y,y' \mid y,y' \sim yq,\ y'q) \quad ,\ y\ ,\ y'\ ,\ q \in H \tag{5}$$

H is the field of quaternions $(H \ni x = x^4 + i\vec{x}.\vec{\sigma}\ ;\ (x^4,\vec{x}) \in E^4)$, $P_1(H)$ is covered by two charts :

$$\begin{array}{lll} y \neq 0 & y'\,y^{-1} = x \in H & (\Omega_0) \\ y' \neq 0 & y\,y'^{-1} = \xi \in H & (\Omega_\infty) \end{array} \tag{6}$$

In the intersection of the two charts :

$$\Omega_0 \cap \Omega_\infty \equiv \Omega_{0\infty}\ ,\quad x = \xi^{-1} \tag{7}$$

A complex vector bundle $\mathcal{E}$ over $P_1(H)$ with structure group G is defined by a transition function $g_{0\infty}(x)$ with values in G , defined in $\Omega_{0\infty}$.

A connection on $\mathcal{E}$ can be described as a pair of $\mathcal{G}$-valued ($\mathcal{G}$: Lie algebra of G) one-forms α_0 , α_∞ defined on Ω_0 , Ω_∞ respectively, with

$$\alpha_0 = g_{0\infty}^{-1}\ \alpha_\infty\, g_{0\infty} + g_{0\infty}^{-1}\, d\, g_{0\infty} \qquad \text{in } \Omega_{0\infty}$$

Accordingly, the curvature 2-forms

$$\mathcal{F}_{\substack{0\\ \infty}} = d\,\alpha_{\substack{0\\ \infty}} + \tfrac{1}{2}[\alpha_{\substack{0\\ \infty}}, \alpha_{\substack{0\\ \infty}}] \tag{8}$$

fit according to

$$\mathcal{F}_0 = g_{0\infty}^{-1}\ \mathcal{F}_\infty\, g_{0\infty} \tag{9}$$

We now consider the complex projective 3-plane $P_3(\mathbb{C})$ as fibered over $P_1(H)$ according to :

$$\begin{array}{ll} P_3(\mathbb{C}) \ni z = \begin{pmatrix} u \\ v \end{pmatrix} & u,\ v \in \mathbb{C}^2 \\ v = xu \quad u \in P_1(\mathbb{C}) & \text{above } \Omega_0 \end{array} \tag{10}$$

$$u = \xi \, v \qquad v \in P_1(\mathbb{C}) \qquad \text{above } \Omega_\infty$$

$P_3(\mathbb{C})$ can be covered by the four open sets analytically isomorphic with $\mathbb{C}^3$:

$$\begin{array}{llll} u_1 \neq 0 , & u_2 \neq 0 & \text{fibered over } \Omega_0 , & \text{denoted } \Omega_0^+ , \Omega_0^- \\ v_1 \neq 0 , & v_2 \neq 0 & \text{fibered over } \Omega_\infty , & \text{denoted } \Omega_\infty^+ , \Omega_\infty^- \end{array} \tag{11}$$

The fiber above x is the projective "real" line ($\sim P_1(\mathbb{C})$) $v = xu$; conversely, $z = (u,v)$ projects on $x \sim (q(u) , q(v))$ with

$$q(u) = (u, -i\sigma_2 u^*) , \tag{12}$$

the quaternion associated with u .

Reality is understood under the antilinear involution induced on $P_3(\mathbb{C})$ by

$$\begin{pmatrix} u \\ v \end{pmatrix} \longrightarrow \begin{pmatrix} -i\sigma_2 u^* \\ -i\sigma_2 v^* \end{pmatrix} = \tau_2 K \begin{pmatrix} u \\ v \end{pmatrix} \tag{13}$$

Any bundle $\mathcal{E}$ over $P_1(H)$ can be pulled back to a bundle E over $P_3(\mathbb{C})$ through performing the change of variables

$$x = q(v)\, q(u)^{-1} \tag{14}$$

Similarly, a connection $\mathcal{A}$ on $\mathcal{E}$ can be pulled back to a connection a on E. Let now $\mathcal{A}$ give rise to a self-dual curvature

$$\mathcal{F} = * \mathcal{F}$$

where the $*$ (duality operation) is defined with respect to any metric conformal to the flat metric e.g. the $O(5)$ - invariant metric

$$\frac{dx^2}{(1+x^2)^2} = \frac{d\xi^2}{(1+\xi^2)^2} \tag{15}$$

Then one has :

<u>Theorem 1</u> (Atiyah, Hitchin, Singer, Ward) [3],[5]

Let $\mathcal{E}$ be a complex $SU(2)$ - bundle over $P_1(H)$, E its pull back over $P_3(\mathbb{C})$. Let $\mathcal{A}$ be a connection on $\mathcal{E}$ with self-dual curvature. Then E possesses a complex analytic structure which is uniquely associated with the conformal structure of $P_1(H)$. E is holomorphically trivial on the real lines (x) of $P_3(\mathbb{C})$. If $\mathcal{A}$ is compatible with a hermitean structure on the fibers of $\mathcal{E}$, a is the unique type $(1,0)$ connection compatible with the induced hermitean structure on E .

Proof [9] : (G = SU 2)

$\mathcal{F} = *\mathcal{F}$ is the integrability condition for the almost complex structure on E : $J = J_V \oplus J_H$ where J_V is the almost complex structure along the fibers inherited from their complex structure, J_H is the lift on the horizontal subspace defined by $\mathfrak{a}$ of the almost complex structure defined by the complex analytic structure of $P_3(\mathbb{C})$. Computational details can be found e.g. in [9] . The Newlander-Nirenberg theorem can be used in a somewhat simplified form, due to the linear structure in the fiber. In terms of the x , u , $\bar{u}$ (resp. ξ , v , $\bar{v}$) variables, the Ward equations

$$\partial_{\bar{u}} \mathcal{A} = 0 \quad , \quad u \overset{\circ}{\nabla}_x \mathcal{A} = 0$$
$$\text{resp.} \quad \partial_{\bar{v}} \mathcal{A} = 0 \quad , \quad v \overset{\infty}{\nabla}_\xi \mathcal{A} = 0 \tag{16}$$

are integrable. One can furthermore choose $\mathcal{A}$ with determinant 1. Since any complex analytic bundle over $\mathbb{C}^3$ is trivial, the Ward equation can be integrated in $\Omega_0^\pm$ (resp. $\Omega_\infty^\pm$), thus providing the change of variables along the fiber

$$\eta \longrightarrow \tilde{\eta} = \overset{\circ}{\mathcal{A}}{}^{\pm}(x,u)\,\eta$$
$$\text{resp.} \quad \eta \longrightarrow \tilde{\eta} = \overset{\infty}{\mathcal{A}}{}^{\pm}(\xi,v)\,\eta \tag{17}$$

which trivializes E over $\Omega_0^\pm$, $\Omega_\infty^\pm$ in such a way that the corresponding transition functions are holomorphic. In particular, for fixed x (resp. ξ), the transition function splits into two factors holomorphic in u (resp. v) within $\Omega_0^+ \cap \Omega_0^- = \Omega_0^{+-}$ (resp. Ω_∞^{+-}), which expresses the holomorphic triviality of E on real lines.

If $\mathfrak{a}$ is compatible with the metric h on $\mathcal{E}$, so is a . In terms of the $\tilde{\eta}$ variables, the gauge transformed $\tilde{a}$ of a through the gauge transformation $\mathcal{A}^{-1}$ is then [9] [10] <u>the</u> type (1,0) canonical connection compatible with the metric

$$\tilde{h} = \mathcal{A}^{+\,-1} \mathcal{A}^{-1} \tag{18}$$

From there, follows Ward's procedure to recover $\mathfrak{a}$ [5]. Recovering $\mathfrak{a}$ with value in SU2 stems from the fact that $\mathcal{A}$ can be chosen such that

$$\mathcal{A}^{\pm}_{\substack{0\\ \infty}}(x,u) = \sigma_2\, \bar{\mathcal{A}}^{\mp}_{\substack{0\\ \infty}}(x,\sigma_2\bar{u})\,\sigma_2 \left(= \sigma_2\, \mathcal{A}^{\pm\,-1T}_{\substack{0\\ \infty}}\sigma_2\right) \tag{19}$$

which expresses the holomorphic antilinear isomorphism

$$\check{E} \simeq (\tau_2 K)^* \bar{E} \overset{\sigma_2}{\simeq} E \qquad (20)$$

where $\check{E}$ denotes the dual of E, $\bar{E}$ its complex conjugate, $(\tau_2 K)^*$ the transformation induced by the involution $\tau_2 K$.

E is thus a holomorphic bundle over $P_3(\mathbb{C})$, and, therefore, algebraic [7]. By construction it has vanishing first Chern number and second Chern number equal to -k, and is thus isomorphic with its dual $\check{E}$ (through conjugation by σ_2, since its transition functions can be chosen in $SL(2,\mathbb{C})$). The associated sheaf of germs of holomorphic sections will be denoted by the symbole $\underline{E}$.

In the sequel, it will be analyzed in terms of some of the cohomology groups of $P_3(\mathbb{C})$ with values in the associated sheaf and its twisted versions $\underline{E}(n) = \underline{E} \otimes \mathcal{O}(n)$ where $\mathcal{O}(n)$ is the sheaf of germs of holomorphic functions homogeneous of degree n ($n \in Z$) associated with the nth power of the Hopf line bundle L i.e. with transition functions $g_{ij}([z]) = \left(\frac{z_i}{z_j}\right)^n$ from $\Omega_i : z_i \neq 0$ to $\Omega_j : z_j \neq 0$. We shall also need to consider $\underline{E} \otimes \Omega^p$ where Ω^p is the sheaf of germs of holomorphic p-forms over $P_3(\mathbb{C})$. The cohomology groups are the Dolbeaut cohomology groups [10] ($\bar{\partial}$ cohomology) which can conveniently be analyzed in terms of the Čech cohomology relative to the covering $\Omega_0^\pm$, $\Omega_\infty^\pm$ of $P_3(\mathbb{C})$ [10]. They will be shown to be isomorphic with finite dimensional spaces of spinor fields over $P_1(H)$ fulfilling massless Dirac-like equations with minimal coupling to the gauge field α.

b) Cohomology Spaces of $P_3(\mathbb{C})$.

As will be explained later we shall only need $H^0(P_3, \underline{E}(\ell))$, $H^1(P_3, \underline{E}(\ell))$ (for some values of ℓ), $H^1(P_3, \underline{E} \otimes \Omega^1)$ where Ω^1 denotes the sheaf of germs of holomorphic one-forms ($\omega = \Sigma\, a_i dz_i$, $a_i \in \mathcal{O}(-1)$, $\Sigma\, z_i a_i = 0$).

We shall consistently use the standard covering of P_3 by $\Omega_0^\pm$, $\Omega_\infty^\pm$ and respectively the variables (x,u), (ξ, v) which describe the fibration of $\Omega_0^\pm$ over Ω_0, resp. of $\Omega_\infty^\pm$ over Ω_∞. We shall furthermore express the cocycle conditions in terms of the $\sim$ fiber variables obtained through operating on the initial variables with $\mathcal{A}(x,u)$ resp. $\mathcal{A}(\xi, v)$ which expresses the trivialization of E on real lines. Most of the following arguments are adapted from R. Penrose and coworkers [11].

i) $H^0(P_3(\mathbb{C}), \underline{E}(\ell))$

ia) For $\ell < 0$, $H^0(P_3, \underline{E}(\ell))$ is represented in terms of the (x,u), resp.

(ξ, v) variables by $\tilde{f}_0^{\pm}(x,u)$, $\tilde{f}^{\pm}(\xi,v)$ homogeneous of degree $-\ell$ in u , resp. v , fitting through the proper transition functions. In particular $\tilde{f}_0^{+}(x,u) = \tilde{f}_0^{-}(x,u)$; $\tilde{f}_\infty^{+}(\xi,v) = \tilde{f}_\infty^{-}(\xi,v)$. Since $H^\circ(P_1, \mathcal{O}(-\ell)) = 0$, $\tilde{f}_0^{\pm} = \tilde{f}_\infty^{\pm} = 0$:

$$\boxed{H^\circ(P_3, \underline{E}(-\ell)) = 0} \tag{ia}$$

ib) <u>For $\ell = 0$</u> , we get a pair

$$\begin{aligned} \tilde{f}_0^{+}(x,u) &= \tilde{f}_0^{-}(x,u) = \tilde{f}_0(x) \\ \tilde{f}_\infty^{+}(\xi,v) &= \tilde{f}_\infty^{-}(\xi,v) = \tilde{f}_\infty(\xi) \end{aligned} \tag{29}$$

fitting through

$$\tilde{f}_0(x) = g_{0\infty}(x)\, \tilde{f}_\infty(\xi) \tag{30}$$

The holomorphy condition yields

$$(\nabla_A{}^{\dot{A}} \tilde{f}_0)(x) = 0 \qquad (\nabla_{\dot{A}}{}^{A} \tilde{f}_\infty)(\xi) = 0 \tag{31}$$

It then follows that

$$(\mathcal{F}^{\dot{A}\dot{B}} \tilde{f}_0)(x) = 0 \qquad (\mathcal{F}^{AB} \tilde{f}_\infty)(\xi) = 0 \tag{31}$$

Thus if the connection $(\mathcal{a}_0 , \mathcal{a}_\infty)$ is irreducible,

$$\tilde{f}_0(x) = 0 \qquad \tilde{f}_\infty(\xi) = 0 \tag{32}$$

$$\boxed{H^\circ(P_3, \underline{E}) = 0 \quad \text{if } \mathcal{a} \text{ is irreducible}} \tag{ib}$$

ic) <u>For $\ell > 0$</u> , $H^\circ(P_3, \underline{E}(\ell))$ is represented by $\tilde{f}_0^{\pm}(x,u)$, $\tilde{f}_\infty(\xi,v)$ homogeneous of degree ℓ . Writing

$$\begin{aligned} \tilde{f}_0^{\pm}(x,u) &= u^{A_1} \cdots u^{A_\ell}\, \tilde{f}_{0\, A_1 \cdots A_\ell}(x) \\ \tilde{f}_\infty(\xi,v) &= v^{\dot{A}_1} \cdots v^{\dot{A}_\ell}\, \tilde{f}_{\infty\, \dot{A}_1 \cdots \dot{A}_\ell}(\xi) \end{aligned} \tag{33}$$

The fitting condition reads

$$\tilde{f}_{0\, A_1 \cdots A_\ell}(x) = x^{\dot{A}_1}{}_{A_1} \cdots x^{\dot{A}_\ell}{}_{A_\ell}\, g_{0\infty}\, \tilde{f}_\infty(\xi)_{\dot{A}_1 \cdots \dot{A}_\ell} \tag{34}$$

The holomorphy condition reads

$$\nabla_{(A}{}^{\dot{A}} \tilde{f}_{0 A_1 \dots A_\ell)}(x) = 0 \qquad \nabla_{(\dot{A}}{}^{A} \tilde{f}_{\infty \dot{A}_1 \dots \dot{A}_\ell)}(\xi) = 0 \tag{35}$$

which are some kinds of generalized twistor equations. The dimensionality of $H^\circ(P_3, \underline{E}(\ell))$ combined with that of $H^1(P_3, \underline{E}(\ell))$ occurs in the evaluation of the Euler Poincaré characteristics of $\underline{E}(\ell)$ [12] which is the same as the index of the AHS complex [3] after tensorization. We shall come back to this later when we give the spinor description of $\underline{E}(\ell)$ [13].

ii) $H^1(P_3(\mathbb{C}), \underline{E}(\ell))$

We shall consider three cases : $\ell < -2$, $\ell = -2$; $\ell > -2$.

iia) $H^1(P_3 \underline{E}(-\ell-2)$, $\ell > 0$

We write the cocycle conditions as follows

$$\begin{aligned}
&\tilde{f}^{+-}_{00} - \tilde{f}^{++}_{0\infty} + \tilde{f}^{-+}_{0\infty} = 0 && \text{in } \Omega^{+-+}_{00\infty} \\
&\tilde{f}^{+-}_{00} - \tilde{f}^{+-}_{0\infty} + \tilde{f}^{--}_{0\infty} = 0 && \text{in } \Omega^{+--}_{00\infty} \\
&\tilde{f}^{+-}_{\infty\infty} - \tilde{f}^{++}_{\infty 0} + \tilde{f}^{-+}_{\infty 0} = 0 && \text{in } \Omega^{+-+}_{\infty\infty 0} \\
&\tilde{f}^{+-}_{\infty\infty} - \tilde{f}^{+-}_{\infty 0} + \tilde{f}^{--}_{\infty 0} = 0 && \text{in } \Omega^{+--}_{\infty\infty 0}
\end{aligned} \tag{36}$$

recalling that the usual antisymmetry of the f_{ij}'s has to be twisted through the relevant transition function. Now [11] multiply the first pair of equations through $u_A u_{A_1} \dots u_{A_\ell}$, the second pair through $v_{\dot{A}} v_{\dot{A}_1} \dots v_{\dot{A}_\ell}$, and denote the corresponding functions by adding subscripts $[A] = (A, A_1 \dots A_\ell)$, $[\dot{A}] = (\dot{A}, \dot{A}_1 \dots \dot{A}_\ell)$. All components are now homoegeneous of degree -1 for all x, resp. ξ.
We can thus perform the splitting [11]

$$\begin{aligned}
\tilde{f}^{+-}_{00\,[A]} &= \tilde{f}^{+}_{0\,[A]} - \tilde{f}^{-}_{0\,[A]} \\
\tilde{f}^{+-}_{\infty\infty\,[\dot{A}]} &= \tilde{f}^{+}_{\infty\,[\dot{A}]} - \tilde{f}^{-}_{\infty\,[\dot{A}]}
\end{aligned} \tag{37}$$

as indicated by the vanishing of $H^1(P_1, \mathcal{O}(-1))$. This splitting is unique in view of the vanishing of $H^\circ(P_1, \mathcal{O}(-1))$. Combining the cocycle condition with this splitting yields in $\Omega^+_0 \cap \Omega^-_0 \cap \Omega^+_\infty \cap \Omega^-_\infty$:

$$\tilde{f}^{++}_{0\infty[A]} + g_{0\infty}\, \xi_{[A]}^{[\dot{A}]} \tilde{f}^{+}_{\infty[\dot{A}]} - \tilde{f}^{+}_{0[A]} = \tilde{f}^{+-}_{0\infty[A]} + g_{0\infty}\, \xi_{[A]}^{[\dot{A}]} \tilde{f}^{-}_{\infty[\dot{A}]} - \tilde{f}^{+}_{0[A]}$$

$$= \tilde{f}^{-+}_{0\infty[A]} + g_{0\infty}\, \xi_{[A]}^{[\dot{A}]} \tilde{f}^{-}_{\infty[\dot{A}]} - \tilde{f}^{+}_{\infty[A]} = \tilde{f}^{--}_{0\infty[A]} + g_{0\infty}\, \xi_{[A]}^{[\dot{A}]} f^{-}_{\infty[\dot{A}]} - \tilde{f}^{-}_{0[A]} \qquad (38)$$

$$\left(\xi_{[A]}^{[\dot{A}]} = \xi_A^{\ \dot{A}}\, \xi_{A_1}^{\ \dot{A}_1} \cdots \xi_{A_\ell}^{\ \dot{A}_\ell} \right)$$

which therefore can be continued into $(\Omega_0^+ \cap \Omega_\infty^+) \cup (\Omega_0^+ \cap \Omega_\infty^-) \cup (\Omega_0^- \cap \Omega_\infty^+) \cup (\Omega_0^- \cap \Omega_\infty^-)$ i.e. all of P_3 except for the lines $x = 0$, $\xi = 0$. Therefore, for all x , $x \neq 0$, $\xi \neq 0$, by the vanishing of $H^\circ(P_1, \mathcal{O}(-1))$, all the above combinations vanish wherever they are defined (C^∞ in x (resp. ξ), holomorphic in u , (resp. v)).

Saturating Eqs. (37) by u^A , resp. $v^{\dot{A}}$ yields [11] :

$$u^A \tilde{f}^{+}_{0(AA_1\cdots A_\ell)} = u^A \tilde{f}^{-}_{0(AA_1\cdots A_\ell)} = \varphi_{0A_1\cdots A_\ell}(x)$$

$$v^{\dot{A}} \tilde{f}^{+}_{\infty(\dot{A}A_1\cdots \dot{A}_\ell)} = v^{\dot{A}} \tilde{f}^{-}_{\infty(AA_1\cdots A_\ell)} = \varphi_{\infty\dot{A}_1\cdots\dot{A}_\ell} \qquad (39)$$

by $H^\circ(P_1, \mathcal{O}) = \mathbb{C}$.

with

$$\varphi_{0A_1\cdots A_\ell}(x) = \xi^2\, \xi_{A_1}^{\ \dot{A}_1} \cdots \xi_{A_\ell}^{\ \dot{A}_\ell}\, g_{0\infty}\, \varphi_{\infty\dot{A}_1\cdots\dot{A}_\ell}(\xi) \qquad (40)$$

Note the ξ^2 factor which comes from saturation by u^A since $u^A \xi_A^{\ \dot{A}} = \xi^2 v^{\dot{A}}$.

Furthermore, the holomorphy of f^{+-}_{00} , $f^{+-}_{\infty\infty}$ yields [11] through saturation by $\nabla^{A_1}{}_{\dot{B}}$, $\nabla_B{}^{\dot{A}_1}$

$$\overset{\circ}{\nabla}{}^{A_1}{}_{\dot{B}}\, \tilde{f}^{+}_{0(AA_1\cdots A_\ell)} = \overset{\circ}{\nabla}{}^{A_1}{}_{\dot{B}}\, \tilde{f}^{-}_{0(AA_1\cdots A_\ell)} = 0$$

$$\overset{\infty}{\nabla}{}_B{}^{\dot{A}_1}\, \tilde{f}^{+}_{\infty(\dot{A}\dot{A}_1\cdots A_\ell)} = \overset{\infty}{\nabla}{}_B{}^{\dot{A}_1}\, \tilde{f}^{-}_{\infty(\dot{A}\dot{A}_1\cdots A_\ell)} = 0 \qquad (41)$$

since $H^\circ(P_1, \mathcal{O}(-1)) = 0$. Hence [11] , contracting with u^A resp. $v^{\dot{A}}$ yields :

$$\overset{\circ}{\nabla}{}^{A_1}{}_{\dot{B}}\, \varphi_{0\,A_1 \cdots A_\ell}(x) = 0$$

$$\overset{\infty}{\nabla}{}^{\dot{A}_1}{}_{B}\, \varphi_{\infty\,\dot{A}_1 \cdots \dot{A}_\ell}(\xi) = 0 \tag{42}$$

Putting

$$\varphi_0(x) = \frac{\widetilde{\varphi}_0}{(1+x^2)^{1+\frac{\ell}{2}}} \qquad \varphi_\infty(\xi) = \frac{\widetilde{\varphi}_\infty}{(1+\xi^2)^{1+\frac{\ell}{2}}} \tag{43}$$

we have

$$\widetilde{\varphi}_{0\,A_1 \cdots A_\ell}(x) = \hat{\xi}_{A_1}{}^{\dot{A}_1} \cdots \hat{\xi}_{A_\ell}{}^{\dot{A}_\ell}\, g_{0\infty}\, \varphi_{\infty\,\dot{A}_1 \cdots \dot{A}_\ell}(\xi) \tag{44}$$

which characterize the pair $\widetilde{\varphi}_0$, $\widetilde{\varphi}_\infty$ as defining a spinor field on $P_1(H)$.

Using the Fierz identities for the $\widetilde{\sigma}$, $\underset{\sim}{\sigma}$ matrices together with the specific form for the spin connection derived from the invariant metric Eq. (15), one finds :

$$\left(\underset{\sim}{\sigma}^\mu\, \overset{\circ}{\nabla}{}^{cov.}_\mu\, \widetilde{\varphi}_0\right)_{\dot{A}_1, A_2 \cdots A_\ell} = 0$$

$$\left(\widetilde{\sigma}^\mu\, \overset{\infty}{\nabla}{}^{cov.}_\mu\, \widetilde{\varphi}_\infty\right)_{A_1, \dot{A}_2 \cdots \dot{A}_\ell} = 0 \tag{45}$$

where

$$\overset{\circ}{\nabla}{}^{cov.}_\mu = \partial_\mu + \overset{\circ}{\alpha}_\mu + \Gamma_\mu^- \qquad \overset{\infty}{\nabla}{}^{cov.}_\mu = \partial_\mu + \overset{\infty}{\alpha}_\mu + \Gamma_\mu^+$$

$$\Gamma_\mu^\pm = \sum_{p=1}^{\ell} 1\!\!1 \otimes \cdots \otimes \Gamma_\mu^{(p)\pm} \otimes \cdots \otimes 1\!\!1 \tag{46}$$

$$\Gamma_\mu^{(p)\pm} = -i\, \vec{\eta}^{\,\pm}_{\mu\nu} \cdot \vec{\sigma}^{(p)} \frac{x^\nu}{1+x^2}$$

The corresponding iterated equations follow

$$\left[\overset{\circ}{\underset{\infty}{\nabla}}{}^{cov.}_\mu\, \overset{\circ}{\underset{\infty}{\nabla}}{}^\mu_{cov.} - 4(\ell+2)\right] \widetilde{\varphi}_{\overset{\circ}{\infty}} = 0 \tag{47}$$

which implies the non-existence of non-trivial solutions, since $\nabla^{cov.}_{\mu}\nabla^{\mu}_{cov.}$ is a negative operator on $P_1(H)$. The last term in Eq. (47) is a curvature term. We can thus solve Eq.(39) in their domain of definition :

$$\tilde{f}^{\pm}_{o(AA_1\cdots A_\ell)} = u_A u_{A_1} \cdots u_{A_\ell} \tilde{f}^{\pm}_o(x,u)$$
$$\tilde{f}^{\pm}_{\infty(\dot{A}\dot{A}_1\cdots\dot{A}_\ell)} = v_{\dot{A}} v_{\dot{A}_1} \cdots v_{\dot{A}_\ell} \tilde{f}^{\pm}_{\infty}(\xi,v) \qquad (48)$$

From the Dirac equation, Eq. (41), it follows that $f^{\pm}_o(x,u)$, $f^{\pm}_{\infty}(\xi,v)$ define holomorphic functions. Hence $f^{\pm\pm}_{\binom{o}{\infty}\binom{o}{\infty}}$ defines a coboundary. Thus,

$$\boxed{H^1(P_3,\underline{E}(-\ell-2)) = 0 \qquad \ell>0} \qquad \text{(iia)}$$

iib) $\underline{H^1(P_3,\underline{E}(-2))}$. For $\ell = 0$, the previous construction [11] goes through defining

$$\varphi_o(x) = \xi^2 g_{o\infty}\, \varphi_{\infty}(\xi) \qquad (49)$$

from

$$u^{\dot{A}} \tilde{f}^{+}_{o\,A}(x,u) = u^{\dot{A}} \tilde{f}^{-}_{o\,A}(x,u) = \varphi_o(x)$$
$$v^{\dot{A}} \tilde{f}^{+}_{\infty\,\dot{A}}(\xi,v) = v^{\dot{A}} \tilde{f}^{-}_{\infty\,\dot{A}}(\xi,v) = \varphi_{\infty}(\xi) \qquad (49')$$

Holomorphy of the cocycle implies

$$\nabla^{A}{}_{\dot{B}} \tilde{f}^{+}_{oA} = \nabla^{A}{}_{\dot{B}} \tilde{f}^{-}_{oA} = 0 \qquad \nabla^{\dot{A}}{}_{B} \tilde{f}^{+}_{\infty\dot{A}} = \nabla^{\dot{A}}{}_{B} \tilde{f}^{-}_{\infty\dot{A}} = 0 \qquad (50)$$

Saturating through $u^{B}\nabla_{B}^{\dot{B}}$, $v^{\dot{B}}\nabla_{\dot{B}}^{B}$ yields

$$\Box\varphi_o = 0 \qquad \Box\varphi_{\infty} = 0 \qquad (51)$$

Thus, putting

$$\varphi_o = \frac{\tilde{\varphi}_o}{1+x^2} \qquad \varphi_{\infty} = \frac{\tilde{\varphi}_{\infty}}{1+\xi^2} \qquad (52)$$

so that

$$\tilde{\varphi}_o(x) = \tilde{\varphi}_{\infty}(\xi) \qquad \text{in } \Omega_o \cap \Omega_{\infty} \qquad (53)$$

we get

$$(\Box^{cov} - 8)\,\tilde{\varphi} = 0 \qquad (54)$$

where

$$\Box^{Cov.} = (1+x^2)^2 \Box - 4(1+x^2)\vec{x}.\vec{\nabla} \qquad (55)$$

is the scalar Laplacian for the metric Eq.(15). Hence $\tilde{\varphi} = 0$ and, as before, one deduces :

$$\boxed{H^1(P_3, \underline{E}(-2)) = 0} \qquad \text{(iib)}$$

Actually this is the main result since it follows by an argument of algebraic geometry that it entails the vanishing of $H^1(P_3, \underline{E}(-2-\ell))$ for all $\ell \geqslant 0$ [14].

iic) $H^1(P_3, \underline{E}(-2+\ell))$ $\ell > 0$

Since $H^1(P_1, -2+\ell) = 0$, we can right away split :

$$\tilde{f}_{oo}^{+-} = \tilde{f}_o^{+} - \tilde{f}_o^{-}$$
$$\tilde{f}_{\infty\infty}^{+-} = \tilde{f}_\infty^{+} - \tilde{f}_\infty^{-} \qquad (56)$$

Now, holomorphy implies

$$u^A \nabla_A{}^{\dot{B}} \tilde{f}_o^{+} = u^A \nabla_A{}^{\dot{B}} \tilde{f}_o^{-} = u^{A_1} \cdots u^{A_{\ell-1}} \varphi_{o A_1 \cdots A_{\ell-1},}{}^{\dot{B}}(x)$$
$$v^{\dot{A}} \nabla_{\dot{A}}{}^{B} \tilde{f}_\infty^{+} = v^{\dot{A}} \nabla_{\dot{A}}{}^{B} \tilde{f}_\infty^{-} = v^{\dot{A}_1} \cdots v^{\dot{A}_{\ell-1}} \varphi_{\infty \dot{A}_1 \cdots \dot{A}_{\ell-1},}{}^{B}(\xi) \qquad (57)$$

where, in the right hand side, we have written out sections of $H^\circ(P_1, \mathcal{O}(\ell-1))$. $\tilde{f}_o^{\pm}$ resp. $\tilde{f}_\infty^{\pm}$ are ambiguous up to an element of $H^\circ(P_1, \mathcal{O}(\ell-2))$, i.e.

$\tilde{f}_o^{\pm}$ is ambiguous up to $u^{A_1} \cdots u^{A_{\ell-2}} \Delta_{o A_1 \cdots A_{\ell-2}}(x)$

$\tilde{f}_\infty^{\pm}$ is ambiguous up to $v^{\dot{A}_1} \cdots v^{\dot{A}_{\ell-2}} \Delta_{\infty \dot{A}_1 \cdots \dot{A}_{\ell-2}}(\xi)$

Hence,

$\varphi_{o A_1 \cdots A_{\ell-1},}{}^{\dot{B}}(x)$ is ambiguous up to $\nabla_{(A_{\ell-1}}{}^{\dot{B}} \Delta_{o A_1 \cdots A_{\ell-2})}$

$\varphi_{\infty \dot{A}_1 \cdots \dot{A}_{\ell-1},}{}^{B}(\xi)$ is ambiguous up to $\nabla_{(\dot{A}_{\ell-1}}{}^{B} \Delta_{\infty \dot{A}_1 \cdots \dot{A}_{\ell-2})}$

Applying $u_B \nabla_{\dot{B}}{}^{B}$ resp. $v_{\dot{B}} \nabla_B{}^{\dot{B}}$ to Eq. 57 yields

$$\nabla_{\dot{B}(B}\, \varphi_{0\, A_1 \cdots A_{\ell-1})}^{\dot{B}} = \nabla_{B(\dot{B}}\, \varphi_{\infty}{}^{B}{}_{\dot{A}_1 \cdots \dot{A}_{\ell-1})} = 0 \tag{58}$$

The fitting conditions are derived from Eq.(36), which implies the existence of an element of $H^\circ(P_1, \mathcal{O}(\ell-2))$ for all x, $x \neq 0$, $\xi \neq 0$:

$$u^{A_1} \cdots u^{A_{\ell-2}}\, \Delta_{0\, A_1 \cdots A_{\ell-2}} = v^{\dot{A}_1} \cdots v^{\dot{A}_{\ell-2}}\, \Delta_{\infty\, \dot{A}_1 \cdots \dot{A}_{\ell-2}} \tag{59}$$

by applying $u^A \nabla_A{}^{\dot{B}}$, resp. $v^{\dot{A}} \nabla_{\dot{A}}{}^{B}$, one gets

$$\begin{aligned} u^{A_1} \cdots u^{A_{\ell-1}}\, \varphi_{0\, A_1 \cdots A_{\ell-1}}{}^{\dot{B}}(x) = {} & \xi^2 v^{\dot{A}_1} \cdots v^{\dot{A}_{\ell-1}}\, g_{0\infty}\, \varphi_{\infty\, \dot{A}_1 \cdots \dot{A}_{\ell-1}\, B} \\ & + u^{A_{\ell-1}} \nabla_{A_{\ell-1}}{}^{\dot{B}}\, u^{A_1} \cdots u^{A_{\ell-2}}\, \Delta_{0\, A_1 \cdots A_{\ell-2}} \end{aligned} \tag{60}$$

i.e.

$$\begin{aligned} \varphi_{0\, A_1 \cdots A_{\ell-1}}{}^{\dot{B}} = {} & \xi^2\, \xi_{A_1}{}^{\dot{A}_1} \cdots \xi_{A_{\ell-1}}{}^{\dot{A}_{\ell-1}}\, \varphi_{\infty\, \dot{A}_1 \cdots \dot{A}_{\ell-1}}{}^{B}\, \xi_B^{\dot{B}} \\ & + \nabla_{(A_{\ell-1}}{}^{\dot{B}}\, \Delta_{0\, A_1 \cdots A_{\ell-2})} \end{aligned} \tag{61}$$

For $\ell = 1$, i.e. $H^1(P_3, \underline{E}(-1))$, we have the pair

$$\varphi_{0\dot{B}} = \xi^2\, \xi^{B}{}_{\dot{B}}\, g_{0\infty}\, \varphi_{\infty B} \tag{62}$$

with the Dirac equation

$$\nabla^{\dot{B}}{}_{A}\, \varphi_{0\dot{B}} = 0 \qquad \nabla^{B}{}_{\dot{A}}\, \varphi_{\infty B} = 0 \tag{63}$$

and no gauge ambiguity. One knows that this space is k-dimensional since the adjoint system which we have encountered in the description of $H^1(P_3, E(-3))$ has no solution.

For $\ell = 2$, i.e. $H^1(P_3, \underline{E})$, we encounter the AHS complex whose index is 2k-2 (instead of 8k-3 because, here, the adjoint representation is replaced by the fundamental representation. In general, the dimensionality is given by the index of the AHS complex tensored by $V^{-\otimes \ell-2}$.

We have finally to show that the spaces we have just described are isomorphic with the cohomology group of P_3 under investigation. Starting with a solution of Eq.(58), construct $f_0^{\dot{B}} = u^{A_1} \cdots u^{A_{\ell-1}}\, \varphi_{0\, A_1 \cdots A_{\ell-1}}{}^{\dot{B}}$, $f_\infty{}^{B} = v^{\dot{A}_1} \cdots v^{\dot{A}_{\ell-1}}\, \varphi_{\infty\, \dot{A}_1 \cdots \dot{A}_{\ell-1}}{}^{B}$, restrict them to $\Omega_0^\pm$, $\Omega_\infty^\pm$ respectively.

From Eq. (58), we deduce that, considered as 1-forms, they are annihilated by the $\bar{\partial}$ -operator, hence they are of the form $u^{A} \nabla_{A\dot{B}} \tilde{f}_{0}^{\pm}$, $v^{\dot{A}} \nabla_{\dot{A}B} \tilde{f}_{\infty}^{\pm}$
Now, $f_{0\infty}^{\pm\pm} = f_{0}^{\pm} - f_{\infty}^{\pm} - \Delta$ are holomorphic and provide a cocycle. Furthermore if $(\varphi_{0}, \varphi_{\infty})$ is a gauge ambiguity, it is easy to see that this cocycle is a coboundary.

iii) $H^{1}(P_{3}, \underline{E} \otimes \Omega^{1})$

We start with holomorphic forms $\omega^{\pm\pm}_{\substack{0\,0\\ \infty\,\infty}}$, i.e. objects of the type

$$\omega = U_{A}\, du^{A} + V_{\dot{A}}\, dv^{\dot{A}} \tag{64}$$

where

$$\Omega = (U_{A} , V_{\dot{A}}) \tag{65}$$

are homogeneous of degree -1 subject to the condition

$$U_{A}\, u^{A} + V_{\dot{A}}\, v^{\dot{A}} = 0 \tag{66}$$

As is usual [11] we shall split

$$\begin{aligned} \omega^{+-}_{00} &= \overset{0}{U}{}^{+-}_{A}\, du^{A} + \overset{0}{V}{}^{+-}_{\dot{A}}\, dv^{\dot{A}} = \omega^{+}_{0} - \omega^{-}_{0} \\ \omega^{+-}_{\infty\infty} &= \overset{\infty}{U}{}^{+-}_{A}\, du^{A} + \overset{\infty}{V}{}^{+-}_{\dot{A}}\, dv^{\dot{A}} = \omega^{+}_{\infty} - \omega^{-}_{\infty} \end{aligned} \tag{67}$$

Going over to the variables x , u , resp. ξ , v , holomorphy insures :

$$\begin{aligned} u^{B} \nabla_{B}{}^{\dot{B}} \overset{0}{\Omega}{}^{+} &= u^{B} \nabla_{B}{}^{\dot{B}} \overset{0}{\Omega}{}^{-} = \overset{0}{\Omega}{}^{\dot{B}}(x) \\ - v^{\dot{B}} \nabla_{\dot{B}}{}^{B} \overset{\infty}{\Omega}{}^{+} &= - v^{\dot{B}} \nabla_{\dot{B}}{}^{B} \overset{\infty}{\Omega}{}^{-} = \overset{\infty}{\Omega}{}^{B}(\xi) \end{aligned} \tag{68}$$

Iterating the operation $u^{C} \nabla_{C\dot{B}}$ resp. $-v^{\dot{C}} \nabla_{\dot{C}B}$ yields

$$\nabla^{C}{}_{\dot{B}} \overset{0}{\Omega}{}^{\dot{B}} = 0 \qquad \nabla^{\dot{C}}{}_{B} \overset{\infty}{\Omega}{}^{B} = 0 \tag{69}$$

Condition (66) yields :

$$\begin{aligned} \overset{0}{U}{}^{+}_{A}\, u^{A} + \overset{0}{V}{}^{+}_{\dot{A}}\, v^{\dot{A}} &= \overset{0}{U}{}^{-}_{A}\, u^{A} + \overset{0}{V}{}^{-}_{\dot{A}}\, v^{\dot{A}} = \overset{0}{\varphi} \\ \overset{\infty}{U}{}^{+}_{A}\, u^{A} + \overset{\infty}{V}{}^{+}_{\dot{A}}\, v^{\dot{A}} &= \overset{\infty}{U}{}^{-}_{A}\, u^{A} + \overset{\infty}{V}{}^{-}_{\dot{A}}\, v^{\dot{A}} = \overset{\infty}{\varphi} \end{aligned} \tag{70}$$

Applying $u^B \nabla_B{}^{\dot B}$, resp. $-v^{\dot B} \nabla_{\dot B}{}^{B}$ to Eq.(70) yields

$$\nabla_B{}^{\dot B} \overset{o}{\varphi} = \overset{o}{U}_B{}^{\dot B} + \overset{o}{V}_{\dot A}{}^{\dot B} x^{\dot A}{}_B$$
$$-\nabla_{\dot B}{}^{B} \overset{\infty}{\varphi} = \overset{\infty}{U}_A{}^{B} \xi^A{}_{\dot B} + \overset{\infty}{V}_{\dot B}{}^{B} \tag{71}$$

The fitting laws provided by Eq.(36) read :

$$\overset{o}{\Omega}{}^{\dot B}(x) = \xi^2 \xi_A{}^{\dot B} g_{o\infty} \overset{\infty}{\Omega}{}^A(\xi)$$
$$\overset{o}{\varphi}(x) = g_{o\infty} \overset{\infty}{\varphi}(\xi) \tag{72}$$

From Eq.(71), one can eliminate $\overset{o}{U}_B{}^{\dot B}$ $\overset{\infty}{V}_{\dot B}{}^{B}$. Taking into account the Dirac equation yields

$$\square \overset{o}{\varphi} = 2 \overset{o}{V}_{\dot A}{}^{\dot A}$$
$$\square \overset{\infty}{\varphi} = 2 \overset{\infty}{U}_A{}^{A} \tag{73}$$

Conversely, given $(\overset{o}{\varphi}, \overset{o}{V}_{\dot A}{}^{\dot B})$ $(\overset{\infty}{\varphi}, \overset{\infty}{U}_A{}^{B})$ fulfilling the coupled Dirac and Laplace equations (71, 73) allows to reconstruct $\overset{o}{U}_B{}^{\dot B}$, $\overset{\infty}{V}_{\dot B}{}^{B}$ fulfilling the Dirac equation.

Eq.(71) characterizes $H^1(P_3, \underline{E} \otimes \Omega^1)$ as the space of sections of E" whose covariant derivatives are sums of products of solutions of the Dirac equation and of the twistor equation $(1, x^{\dot A}{}_B)$" (cf. ref. [1]). The dimensionality of this space can be shown to be 2k+2 (in the case of SU2), using some algebraic geometry [15] .

It is however of some interest to exhibit a simpler characterization : Since $(\overset{o}{\varphi}, \overset{\infty}{\varphi})$ defines a scalar field on $P_1(H)$, we may rewrite Eq.(73) in invariant terms :

$$(\square^{cov.} - 8)\overset{o}{\varphi} = 2(1+x^2)^2 \left(\overset{o}{V}_{\dot A}{}^{\dot A} - \frac{x_{\dot B}{}^{B}}{1+x^2} \nabla_B{}^{\dot B} \overset{o}{\varphi} - 2\frac{\epsilon_{\dot A}{}^{\dot A}}{(1+x^2)^2} \overset{o}{\varphi} \right)$$
$$(\square^{cov.} - 8)\overset{\infty}{\varphi} = 2(1+\xi^2)^2 \left(\overset{\infty}{U}_A{}^{A} - \frac{\xi_B{}^{\dot B}}{1+\xi^2} \nabla_{\dot B}{}^{B} \overset{\infty}{\varphi} - 2\frac{\epsilon_A{}^{A}}{(1+\xi^2)^2} \overset{\infty}{\varphi} \right) \tag{74}$$

Now eliminating $\overset{o}{U}_B{}^{\dot B}$, $\overset{\infty}{V}_{\dot B}{}^{B}$ from the fitting laws (72) , we get :

$$\xi^2 \xi_A{}^{\dot B} g_{o\infty} \overset{\infty}{U}_B{}^{A} = \nabla_B{}^{\dot B} \overset{o}{\varphi} - \overset{o}{V}_{\dot A}{}^{\dot B} x^{\dot A}{}_B \tag{75}$$

i.e.

$$x^2 \overset{\circ}{V}_{\dot{A}}{}^{\dot{B}} - x^{B}{}_{\dot{A}} \nabla_B{}^{\dot{B}} \overset{\circ}{\varphi} = - \xi^2 \xi_A{}^{\dot{B}} g_{0\infty} \overset{\infty}{U}_B{}^A x^B{}_{\dot{A}} \tag{75'}$$

or, equivalently

$$\begin{aligned} \overset{\circ}{V}_{\dot{A}}{}^{\dot{B}} &= \xi^2 x^B{}_{\dot{A}} \nabla_B{}^{\dot{B}} \overset{\circ}{\varphi} - \xi^2 \xi^B{}_{\dot{A}} \xi_A{}^{\dot{B}} g_{0\infty} \overset{\infty}{U}_B{}^A \\ &= \xi^B{}_{\dot{A}} \xi_A{}^{\dot{B}} g_{0\infty} \left(\xi_{\dot{C}}{}^A \nabla^{\dot{C}}{}_B \overset{\infty}{\varphi} - \xi^2 \overset{\infty}{U}_B{}^A \right) \end{aligned} \tag{75''}$$

Combining Eq.(75', 75") yields :

$$\begin{aligned} (1+x^2) \overset{\circ}{V}_{\dot{A}}{}^{\dot{B}} - x^B{}_{\dot{A}} \nabla_B{}^{\dot{B}} \overset{\circ}{\varphi} = - \xi^B{}_{\dot{A}} \xi_A{}^{\dot{B}} g_{0\infty} \Big[& (1+\xi^2) \overset{\infty}{U}_B{}^A \\ & - \xi_{\dot{C}}{}^A \nabla^{\dot{C}}{}_B \overset{\infty}{\varphi} \Big] \end{aligned} \tag{76}$$

In terms of

$$\begin{aligned} \overset{\circ}{W}_{\dot{A}}{}^{\dot{B}} &= \overset{\circ}{V}_{\dot{A}}{}^{\dot{B}} - \frac{x^B{}_{\dot{A}}}{1+x^2} \nabla_B{}^{\dot{B}} \overset{\circ}{\varphi} - 2 \frac{\epsilon_{\dot{A}}{}^{\dot{B}}}{(1+x^2)^2} \overset{\circ}{\varphi} \\ \overset{\infty}{W}_A{}^B &= - \left(\overset{\infty}{U}_A{}^B - \frac{\xi_{\dot{B}}{}^B}{1+\xi^2} \nabla^{\dot{B}}{}_A \overset{\infty}{\varphi} - 2 \frac{\epsilon_A{}^B}{(1+\xi^2)^2} \overset{\infty}{\varphi} \right) \end{aligned} \tag{77}$$

we have

$$\begin{aligned} \overset{\circ}{W}_{\dot{A}}{}^{\dot{B}} &= \xi^2 \xi^B{}_{\dot{A}} \xi_A{}^{\dot{B}} g_{0\infty} \overset{\infty}{W}_B{}^A \\ (\Box_{sph} - 8) \overset{\circ}{\varphi} &= 2 (1+x^2)^2 \overset{\circ}{W}_{\dot{A}}{}^{\dot{A}} \\ (\Box_{sph} - 8) \overset{\infty}{\varphi} &= 2 (1+\xi^2)^2 \overset{\infty}{W}_A{}^A \end{aligned} \tag{78}$$

The symmetric part of ($\overset{\circ}{W}_{\dot{A}\dot{B}}$, $\overset{\infty}{W}_{AB}$) behaves as a self-dual two-form ; ($(1+x^2)^2 \overset{\circ}{W}_{\dot{A}}{}^{\dot{A}}$, $(1+\xi^2)^2 \overset{\infty}{W}_A{}^A$) as a scalar.

$$\tilde{\overset{\circ}{W}}_{\dot{A}\dot{B}} = (1+x^2)^{-2} \overset{\circ}{W}_{\dot{A}\dot{B}} \quad , \quad \tilde{\overset{\infty}{W}}_{AB} = (1+\xi^2)^{-2} \overset{\infty}{W}_{AB} \tag{79}$$

behave as a bispinor :

$$\tilde{\overset{\circ}{W}}_{\dot{A}\dot{B}} = \hat{\xi}^A{}_{\dot{A}} \hat{\xi}^B{}_{\dot{B}} \tilde{\overset{\infty}{W}}_{AB} \tag{80}$$

It is straightforward but slightly lengthy to see that

$$\overset{\circ}{\nabla}{}^{cov}{}_C{}^{\dot{A}} \tilde{\overset{\circ}{W}}_{\dot{A}\dot{B}} = 0 \qquad \overset{\infty}{\nabla}{}^{cov}{}_{\dot{C}}{}^A \tilde{\overset{\infty}{W}}_{AB} = 0 \tag{81}$$

where ($\tilde{\overset{o}{\nabla}}^{cov}$, $\tilde{\overset{\infty}{\nabla}}^{cov}$) are the Dirac operators for bispinors. Sticking to the W's, one gets in terms of self-dual forms and scalars an analogue of the transposed Schwarz [3] elliptic system which is equivalent to the AHS complex. The transposed system has no solution by the usual positivity argument involving the bispinor Laplacian $\tilde{\nabla}^{cov\,C}_{\dot{A}}\;\tilde{\nabla}^{cov\,\dot{A}}_{C}$. The dimensionality of the space of solutions is thus obtained from the AHS dimensionality 2k-2 by changing k into -k to take into account the interchange of dotted and undotted indices, and an overall - sign to take into account the interchange of the system with its adjoint. Since $\overset{o}{\varphi}$ is uniquely determined in terms of W , by virtue of the vanishing of $H^1(P_3, \underline{E}(-2)) = 0$ (($\Box_{sph} - 8) \varphi = 0 \Rightarrow \varphi = 0$), it is easy to see that

$$\boxed{\begin{array}{l} H^1(E \otimes \Omega^1) \approx \tilde{\overset{o}{W}}_{\dot{A}\dot{B}} , \tilde{\overset{\infty}{W}}_{AB} ; \\ \qquad \tilde{\overset{o}{W}}_{\dot{A}\dot{B}} = \hat{X}_{\dot{A}}{}^{A}\, \hat{X}_{\dot{B}}{}^{B}\, \tilde{\overset{\infty}{W}}_{AB} \\ \qquad \tilde{\overset{o}{\nabla}}^{cov.\,\dot{A}}_{C}\, \tilde{\overset{o}{W}}_{\dot{A}\dot{B}} = 0 \quad \tilde{\overset{\infty}{\nabla}}^{cov.\,A}_{\dot{C}}\, \tilde{\overset{\infty}{W}}_{AB} = 0 \\ \dim H^1(E \otimes \Omega^1) = 2k+2 \end{array}} \qquad \text{(iii)}$$

M.F. Atiyah [16] has given an argument based on algebraic geometry which shows why the vanishing of $H^1(P_3, \underline{E}(-2))$ allows to simplify the description of $H^1(P_3, \underline{E} \otimes \Omega^1)$ as indicated here.

III - SPECIAL ALGEBRAIC BUNDLES AND SHEAVES OVER $P_3(\mathbb{C})$.

Whereas the previous section was devoted to the differential geometry part of the construction, this one deals with the algebraic geometry which simplifies, due to the vanishing theorems we have seen in Section II.

The first step is to write down a proper resolution for $\underline{E}$, the sheaf of germs of holomorphic sections of E . The "Koszul resolution used by Drinfeld and Manin [1] makes use of holomorphic forms as well as the twisting sheaves $\mathcal{O}(\ell)$.

The first paragraph will thus be devoted to a reminder about holomorphic forms over $P_n(\mathbb{C})$. The Koszul resolution is then written down and cut into short pieces which, thanks to the vanishing theorems, characterize $\underline{E}$ as the defect of exactness of a short three term complex involving two finite dimensional spaces : $H^1(P_3, E(-1))$ and its dual, both of dimension k, and $H^1(P_3, E \otimes \Omega^1)$, of dimension $2k+2$.

a) Holomorphic forms on $P_n(\mathbb{C})$

We shall only be interested in the case $n = 3$ but it is shorter to treat the general case. A holomorphic p-form on $P_n(\mathbb{C})$ is defined by

$$\omega^p = \sum \omega_{i_1 \dots i_p} dz_{i_1} \wedge \dots dz_{i_p} \tag{82}$$

where $z = (z_1 \dots z_{n+1})$ are the homogeneous coordinates for P_n, $\omega_{i_1 \dots i_p}$ is antisymmetric, holomorphic, homogeneous of degree $(-p)$ and fulfills

$$\omega_{i_1 \dots i_p} z_{i_p} = 0 \tag{83}$$

This is easy to check by taking locally inhomogeneous coordinates and express them in terms of homogeneous ones.

Furthermore, Eq. (83) implies

$$\omega^p_{i_1 \dots i_p} = \sum \omega^{p-1}_{i_1 \dots i_{p-1}} \wedge z_{i_p} \tag{84}$$

for some ω^{p-1} in any holomorphy domain, and therefore at the level of germs. This means one has the exact sequence of sheaves :

$$0 \longrightarrow \Omega^p \longrightarrow \mathcal{O}(-p)^{\oplus \binom{n+1}{p}} \longrightarrow \Omega^{p-1} \longrightarrow 0 \tag{85}$$

In particular, for $p = n+1$ $\Omega^{n-1} \simeq \mathcal{O}(-(n+1))$. This sequence can of course be tensorized through $\underline{E}(\ell)$ and the long cohomology sequence yields a number of vanishings and isomorphisms provided the vanishing theorems of the previous sections are used. We recall the long cohomology sequence :

$$\begin{aligned} 0 \longrightarrow H^\circ(P_3, \underline{E}(\ell) \otimes \Omega^p) &\longrightarrow \bigoplus_{\binom{n+1}{p}} H^\circ(P_3, \underline{E}(-p+\ell)) \longrightarrow H^\circ(P_3, \underline{E}(+\ell) \otimes \Omega^{p-1}) \\ &\longrightarrow H^1(P_3, \underline{E}(\ell) \otimes \Omega^p) \longrightarrow \cdots \end{aligned} \tag{86}$$

The statement of the Serre duality [7], [10], [12] takes a very simple form :

$$\check{H}^p(P_n, \underline{E}(\ell) \otimes \Omega^q) \simeq H^{n-p}(P_n, \check{\underline{E}}(-\ell) \otimes \Omega^{n-q}) \tag{87}$$

b) The Koszul Resolution.

One considers two copies of $P_3(\mathbb{C})$, with homogeneous coordinates $[z]$, $[\zeta]$, and the diagonal $\Delta : \{z, \zeta, \underline{z} = \lambda \underline{\zeta} \quad \lambda \neq 0\}$. Let $\mathcal{O}_\Delta$ be the structure sheaf of Δ, i.e. the sheaf of holomorphic functions on Δ (holomorphic functions in a neighbourhood of a point of Δ modulo those which vanish on Δ). Then, one has the following exact sequence of sheaves :

$$0 \longrightarrow \mathcal{O}_z(-2) \otimes \Omega^3_\zeta(2) \xrightarrow{z\cdot\frac{\partial}{\partial\zeta}} \mathcal{O}_z(-1) \otimes \Omega^2_\zeta(1) \xrightarrow{z\cdot\frac{\partial}{\partial\zeta}} \mathcal{O}_z \otimes \Omega^1_\zeta \xrightarrow{z\cdot\frac{\partial}{\partial\zeta}} \mathcal{O}_z(1) \otimes \mathcal{O}_\zeta(-1) \longrightarrow \mathcal{O}_\Delta \rightarrow 0 \tag{88}$$

The germs for $\mathcal{O}_z(-p) \otimes \Omega^{p+1}(p)$ are of the form

$$\sum \omega^{(p+1)}_{i_1 \cdots i_{p+1}}(z, \zeta)\, d\zeta_{i_1} \wedge \cdots d\zeta_{i_{p+1}}$$

where the coefficients are homogeneous of degree $-p$ in z, $-(p+1) + p = 1$ in ζ. Going from one step to the next means evaluating the $p+1$ - form ω^{p+1} on the vector field $\sum z_i \frac{\partial}{\partial \zeta_i}$ i.e. replacing $d\zeta_i$ by z_i once thus decreasing the degree of the form, keeping the homogeneity degree in ζ fixed, increasing the homogeneity degree in z by one unit. That this complex is a consequence of the antisymmetry of forms, and, at the last step of the fact that the functions $\sum z_i g_i(z, \zeta)$ with g_i homogeneous of degree 0 in z, -1 in ζ, $\sum \zeta_i g_i(z, \zeta) = 0$ are just those which vanish on Δ $(z_i/\zeta_i = \lambda)$.

The exactness means that, at the level of germs,

$$\sum \omega^{p+1}_{i_1 \cdots i_{p+1}}(z, \zeta)\, z_{i_{p+1}} = 0 \longrightarrow \omega^{p+1}_{i_1 \cdots i_{p+1}}(z, \zeta) = \omega^{p+2}_{i_1 \cdots i_{p+2}}(z, \zeta)\, z_{i_{p+2}}$$

(cf. eqs. (83), (84)), and, at the last step is just the definition of $\mathcal{O}_\Delta$.

We now split this resolution into three term pieces : after tensorizing through $\underline{E}_\zeta$

$$(\alpha): \quad 0 \longrightarrow \mathcal{O}_z(-2) \otimes \Omega^3_\zeta(2) \otimes \underline{E}_\zeta \longrightarrow \mathcal{O}_z(-1) \otimes \Omega^2_\zeta(1) \otimes \underline{E}_\zeta \longrightarrow \alpha \longrightarrow 0$$

$$\Omega^3_\zeta(2) \simeq \mathcal{O}_\zeta(-2) \tag{89}$$

$$(\alpha\beta): \quad 0 \longrightarrow \alpha \longrightarrow \mathcal{O}_z \otimes \Omega^1_\zeta \otimes \underline{E}_\zeta \longrightarrow \beta \longrightarrow 0$$

$$(\beta):\ 0 \longrightarrow \beta \longrightarrow \mathcal{O}_z(1) \otimes \mathcal{O}_\zeta(-1) \otimes \underline{E}_\zeta \longrightarrow \mathcal{O}_\Delta \otimes \underline{E}_\zeta \longrightarrow 0 \tag{89}$$

We now write down the long sequences [7] corresponding to projecting onto the copy of P_3 labelled by the variables z, which involves cohomology groups of the ζ factors on the inverse image of a point z, namely $P_3(\mathbb{C})$, whose specification will be omitted

$$\begin{aligned}
(\alpha)_\ell:\ 0 &\longrightarrow \mathcal{O}(-2) \otimes \underset{\substack{\wr\wr\\ 0}}{\overset{}{H^0(\underline{E}(-2))}}\ (1) \longrightarrow \mathcal{O}(-1) \otimes H^0(\underline{E} \otimes \Omega^2(1)) \\
&\longrightarrow \alpha^0 \longrightarrow \mathcal{O}(-2) \otimes \underset{\substack{\wr\wr\\ 0}}{H^1(\underline{E}(-2))}\ (1) \longrightarrow \mathcal{O}(-1) \otimes H^1(\underline{E} \otimes \Omega^2(1)) \\
&\longrightarrow \alpha^1 \longrightarrow \mathcal{O}(-2) \otimes \underset{\substack{\wr\wr\\ 0}}{H^2(\underline{E}(-2))}\ (2) \longrightarrow \mathcal{O}(-1) \otimes H^2(\underline{E} \otimes \Omega^2(1)) \\
&\longrightarrow \alpha^2 \longrightarrow \mathcal{O}(-2) \otimes \underset{(2)\ \simeq 0}{H^3(\underline{E}(-2))} \longrightarrow \mathcal{O}(-1) \otimes H^3(\underline{E} \otimes \Omega^2(1)) \longrightarrow \alpha^3 \longrightarrow 0
\end{aligned} \tag{90}$$

We use ① the vanishing theorems

② the Serre duality [7], [10], [12] together with the isomorphism of $\underline{E}$ with its dual $\check{\underline{E}}$, Eq. (20)

$$H^2(\underline{E}(-2)) \simeq \check{H}^1(\underline{E}(-2)) = 0$$

$$H^3(\underline{E}(-2)) \simeq \check{H}^0(\underline{E}(-2)) = 0$$

There follows :

$$\alpha^p \simeq \mathcal{O}(-1) \otimes H^p(\underline{E} \otimes \Omega^2(1))$$

Similarly

$$\begin{aligned}
(\alpha\beta)_\ell:\ 0 &\longrightarrow \alpha^0 \longrightarrow \mathcal{O} \otimes H^0(\underline{E} \otimes \Omega^1) \longrightarrow \beta^0 \\
&\longrightarrow \alpha^1 \longrightarrow \mathcal{O} \otimes H^1(\underline{E} \otimes \Omega^1) \longrightarrow \beta^1 \\
&\longrightarrow \alpha^2 \longrightarrow \mathcal{O} \otimes H^2(\underline{E} \otimes \Omega^1) \longrightarrow \beta^2 \\
&\longrightarrow \alpha^3 \longrightarrow \mathcal{O} \otimes H^3(\underline{E} \otimes \Omega^1) \longrightarrow \beta^3 \longrightarrow 0
\end{aligned} \tag{91}$$

$$\begin{aligned}(\beta)_\ell :\ 0 &\longrightarrow \beta^0 \longrightarrow \mathcal{O}(1) \otimes H^0(\underline{E}(-1)) \longrightarrow \underline{E}\\ &\longrightarrow \beta^1 \longrightarrow \mathcal{O}(1) \otimes H^1(\underline{E}(-1)) \longrightarrow \underline{0}\\ &\longrightarrow \beta^2 \longrightarrow \mathcal{O}(1) \otimes H^2(\underline{E}(-1)) \longrightarrow \underline{0}\\ &\longrightarrow \beta^3 \longrightarrow \mathcal{O}(1) \otimes H^3(\underline{E}(-1)) \longrightarrow \underline{0}\end{aligned} \tag{92}$$

In $(\beta)_\ell$ $\underline{E}$ and $\underline{0}$ appear because $\mathcal{O}_\Delta \otimes E_\zeta$ is $\underline{E}$ concentrated on the diagonal. The inverse image of a neighbourhood U_z of z picks up contributions from the diagonal alone. The H° term yields $\underline{E}$ whereas the subsequent $H^i(p_z^{-1} U_z, E_\zeta \otimes \mathcal{O}_\Delta) \simeq H^i(U_z, \underline{E}) = 0$. Thus

$$\beta^i \simeq \mathcal{O}(1) \otimes H^i(\underline{E}(-1)) \tag{93}$$

So, from the Serre duality and vanishing :

$$\beta^0 = \beta^2 = \beta^3 = 0 \tag{94}$$

We thus have the following, which combines $(\beta)_\ell$ and $(\alpha\beta)_\ell$ described by simple and double arrows respectively

$$H^0(\underline{E} \otimes \Omega^2(1)) \overset{②}{=} H^0(\underline{E} \otimes \Omega^1) \overset{①}{=} 0 \tag{95}$$

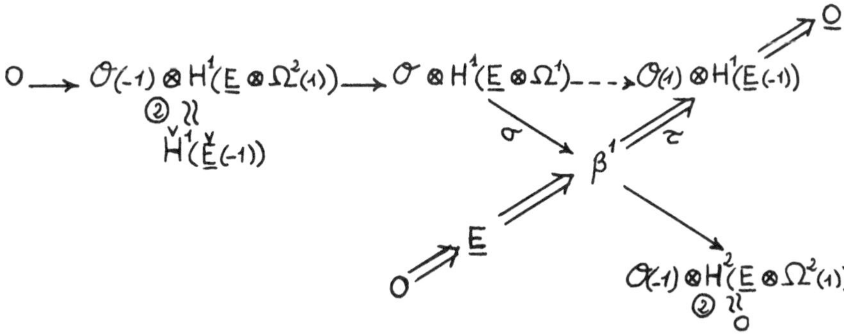

(96)

$$① \quad 0 \longrightarrow \Omega^1 \longrightarrow \mathcal{O}(-1)^{\oplus 4} \longrightarrow \mathcal{O} \longrightarrow 0 \tag{97}$$

hence

$$0 \longrightarrow H^0(\underline{E} \otimes \Omega^1) \longrightarrow \underset{\approx 0}{H^0(E(-1))^{\oplus 4}} \tag{98}$$

(2) From

$$0 \longrightarrow \underline{E} \otimes \Omega^3(1) \longrightarrow \underline{E} \otimes \mathcal{O}(-2)^{\oplus 4} \longrightarrow \underline{E} \otimes \Omega^2(1) \longrightarrow 0 \qquad (99)$$

$$\underline{E} \otimes \Omega^3(1) \simeq \underline{E}(-3)$$

we get

$$H^\circ(\underline{E} \otimes \Omega^2(1)) = 0 \qquad (100)$$

$$\cdots\ H^1(\underline{E}(-2))^{\oplus 4} \longrightarrow H^1(\underline{E} \otimes \Omega^2(1)) \longrightarrow H^2(\underline{E}(-3)) \longrightarrow H^2(\underline{E}(-2))^{\oplus 4} \qquad (101)$$

$$H^1(\underline{E}(-2)) \simeq 0, \quad H^2(\underline{E}(-3)) \simeq \check{H}^1(\underline{\check{E}}(-1)), \quad H^2(\underline{E}(-2)) \simeq \check{H}^1(\underline{\check{E}}(-2))$$

Thus

$$H^1(\underline{E} \otimes \Omega^2(1)) \simeq \check{H}^1(\underline{\check{E}}(-1)) \qquad (k \ \text{dimensional}) \qquad (102)$$

Next :

$$H^2(\underline{E}(-2))^{\oplus 4} \longrightarrow H^2(\underline{E} \otimes \Omega^2(1)) \longrightarrow H^3(\underline{E}(-3)) \qquad (103)$$

$$H^2(\underline{E}(-2)) \simeq \check{H}^1(\underline{\check{E}}(-2)) = 0, \quad H^3(\underline{E}(-3)) \simeq \check{H}^\circ(\underline{\check{E}}(-1)) = 0$$

Thus

$$H^2(\underline{E} \otimes \Omega^2(1)) = 0 \qquad (104)$$

3) Let now $\dashrightarrow^{j} = \tau \circ \sigma$ (with σ into β^1 and τ out of β^1), $\quad j = \tau \circ \sigma$

$$0 \longrightarrow \mathcal{O}(-1) \otimes \check{H}^1(\underline{\check{E}}(-1)) \longrightarrow \mathcal{O} \otimes H^1(E \otimes \Omega^1) \dashrightarrow \mathcal{O}(1) \otimes H^1(E(-1)) \Rightarrow 0 \qquad (105)$$

is a complex :

$\dashrightarrow$ goes through β^1 and thus starting from $\mathcal{O}(-1) \otimes \check{H}^1(\underline{\check{E}}(-1))$ goes to zero in β^1 after two steps and thus to zero in $\mathcal{O}(1) \otimes H^1(\underline{E}(-1))$. Next starting from $\mathcal{O} \otimes H^1(\underline{E} \otimes \Omega^1)$ goes surjectively on β^1 and thus surjectively on $\mathcal{O}(1) \otimes H^1(\underline{E}(-1))$. Finally, $\underline{E}$ is the defect of exactness of this complex : Im i = ker σ by the exactness of β_ℓ ; on the other hand $x \in$ ker j $\Leftrightarrow \tau \circ \sigma\, x = 0$, i.e. $\sigma x \in$ ker $\tau \simeq \underline{E}$, thus $\underline{E} \simeq$ ker j / ker σ $\sim$ ker j/Im i :

$$E = \ker j \,/\, \operatorname{Im} i \qquad (106)$$

The Final Step

The sheaf situation goes over to the bundle situation : E is the defect of exactness of the complex

$$0 \longrightarrow L \otimes \check{H}^1(\check{\underline{E}}(-1)) \xrightarrow{i_E} P_3 \times H^1(\underline{E} \otimes \Omega^1) \xrightarrow{j_E} \check{L} \otimes H^1(\underline{E}(-1)) \longrightarrow 0 \quad (107)$$

where L denotes the Hopf line bundle over P_3 , i_E is defined as a section of Hom $(L \otimes \check{H}^1(\check{\underline{E}}(-1)), P_3 \times H^1(\underline{E} \otimes \Omega^1)) \approx \check{L} \otimes H^1(\underline{E}(-1)) \otimes H^1(\underline{E} \otimes \Omega^1)$,

i.e. a matrix of rank k for all z :

$$A(z) = \sum_i A_i z_i \quad (108)$$

where the A_i's are linear mappings from $\check{H}^1(\check{\underline{E}}(-1))$ to $H^1(\underline{E} \otimes \Omega^1)$.

Now $\check{E}$ is defined by :

$$0 \longrightarrow L \otimes \check{H}^1(\underline{E}(-1)) \xrightarrow{i_{\check{E}}} P_3 \times H^1(\check{\underline{E}} \otimes \Omega^1) \xrightarrow{j_{\check{E}}} \check{L} \otimes \check{H}^1(\underline{E}(-1)) \longrightarrow 0 \quad (109)$$

and also by the complex dual to (107)

$$0 \longrightarrow L \otimes \check{H}^1(\underline{E}(-1)) \xrightarrow{j_E^T} P_3 \times \check{H}^1(\underline{E} \otimes \Omega^1) \xrightarrow{i_E^T} \check{L} \otimes \check{H}^1(E(-1)) \longrightarrow 0 \quad (110)$$

It follows that $H^1(\check{E} \otimes \Omega^1)$ and $\check{H}^1(E \otimes \Omega^1)$ are isomorphic and can thus be identified because of the isomorphism between E and $\check{E}$, Eq. (20). The duality pairing between $H^1(E \otimes \Omega^1)$ and $H^1(\check{E} \otimes \Omega^1)$ thus defines a quadratic form on $H^1(E \otimes \Omega^1)$ [17] which, expressed in terms of the representative spinors reads

$$\begin{aligned} J(\varphi, \psi) &= \int_{P_1(H)} < \check{\varphi}^{\dot{A}\dot{B}} \psi_{\dot{A}\dot{B}} > d\tau = \int_{P_1(H)} \varphi^{\dot{A}\dot{B}} \sigma_2 \psi_{\dot{A}\dot{B}} \, d\tau \\ &= - J(\psi, \varphi) \end{aligned} \quad (111)$$

which realizes the isomorphism :

$$H^1(E \otimes \Omega^1) \overset{J}{\simeq} \check{H}^1(E \otimes \Omega^1) \quad (112)$$

There follows

$$j_E = j_{\check{E}} \, J \quad (113)$$

and, from

$$j_E \circ i_E = 0$$

one deduces

$$i_E^T \, J \, i_E = 0 \tag{114}$$

Finally, the antilinear isomorphism Eq. (20), results into the antilinear isomorphism on $H^1(E \otimes \Omega^1)$:

$$\psi \longrightarrow \psi^c : \psi_{\dot{A}\dot{B}} \longrightarrow \sigma_2 \overline{\psi}^{\dot{A}\dot{B}} \quad ; \quad \psi^{cc} = -\psi \tag{115}$$

such that

$$J(\varphi^c, \psi) = \int_{P_1(H)} \overline{\varphi}_{\dot{A}\dot{B}} \, \psi_{\dot{A}\dot{B}} \, d\tau \tag{116}$$

which defines a hermitean positive definite form. On $H^1(E(-1))$, the corresponding conjugation is

$$\varphi \longrightarrow \varphi^c : \varphi_{\dot{A}} \longrightarrow \sigma_2 \overline{\varphi}^{\dot{A}} \qquad \varphi^{cc} = \varphi \tag{117}$$

and one must have the reality property :

$$\left(i_E(z)\varphi\right)^c = i_E(\tau_2 \bar{z}) \, \varphi^c \tag{118}$$

The isotropy property Eq.(114) and reality property Eq.(118) when expressed in terms of A_i (Eq.(108)), together with the injectivity property characterize the linear algebra data of ref. [1].

IV - GAUGE FIELDS WITH SELF-DUAL CURVATURE.

This section will be rather brief since the actual construction of the gauge fields has been reported in many places [9] , [18].

Following the description given in ref. [1], the situation can be summarized as follows : Let

$$\begin{aligned} W &= \check{H}^1(\check{E}(-1)) \simeq \mathbb{C}^k \\ V &= H^1(E \otimes \Omega^1) \simeq \mathbb{C}^{2k+2} = \bigoplus_1^{k+1} \mathcal{V}_p \underset{(\text{Eq.12})}{\approx} H^{k+1} \end{aligned} \tag{119}$$

$$J = \bigoplus_{1}^{k+1} (-i\sigma_2)_p$$

$$w \in W \qquad w^c = \bar{w} \tag{119}$$

$$v \in V \quad v^c = J\bar{v} = \bigoplus_{1}^{k+1} (-i\sigma_2 \bar{v}_p)$$

$$(v_1^c , J v_2) = (\bar{v}_1, v_2)$$

The reality condition allows to express $A(z)$ as a quaternionic $k \times (k+1)$ matrix.
The isotropy condition then reads

$$\sum_{i,j=1}^{k} [u^T i\sigma_2 (A_j^+ + x^+ B_j^+) w_j][(A_i + B_i x) w_i \, u] = 0 \tag{120}$$

$\forall w$, u , x , where we have substituted

$$[z] = [u , xu]$$

There follows that

$$M_{ji}(x) \equiv (A_j^+ + x^+ B_j^+).(A_i + B_i x)$$
$$= M_{ij}(x) = M_{ij}^{\;\;+}(x) \tag{121}$$

is a symmetric real matrix.

The injectivity property is equivalent to

$$\det \| M_{ij}(x) \| \neq 0 \qquad \forall x \text{ (including } \infty \text{)} \tag{122}$$

The matrix $A(z)$ is defined up to the linear equivalence

$$A(z) \simeq U\, A(z)\, R$$
$$U \in U(k+1, H) \tag{123}$$
$$R \in GL(k,R)$$

which realizes the equivalence of basis in W , V compatible with the reality properties.

E can be trivialized over the straight line x by choosing as a representative of the defining quotient $U_z^{\perp} \cap \; U_{\sigma_2 \bar{z}}^{\perp}$ which can be seen to be independent of the chosen z. The fiber at x can thus be described as the quaternionic line :

$$\begin{aligned} &\mathcal{w}(x)\, q \in H^{k+1} \qquad q \in H \\ &\mathcal{w}^{+}(x).\left[A_i + B_i x\right] = 0 \qquad i = 1 \ldots k \qquad (124) \\ &\mathcal{w}^{+}(x).\mathcal{w}(x) = 1 \end{aligned}$$

The connection is induced by the trivial connection on $P_1(H) \times H^{k+1}$ of which $\mathcal{E}$ is a subundle :

$$a(x) = \frac{1}{i}\; \mathcal{w}^{+}(x)\; d\,\mathcal{w}(x) \qquad (125)$$

The calculation of $w(x)$ is greatly simplified by the fact that $M(x)$ (Eq. (121)) is a scalar matrix so that the quaternionic orthogonalization procedure to be performed really boils down to scalars.

For details including the count of parameters, and remarkable expressions for Green functions in the instanton field and expressions for the zero modes of the Dirac equation, we refer to [1] , [18].

Acknowledgements

We wish to thank M.F. Atiyah, W. Barth, A. Douady, V.G. Drinfeld, K. Hulek, R. Hartshorne, N.J. Hitchin, Yu. I. Manin, J.H. Rawnsley, A. Schwartz, I.M. Singer, J.L. Verdier, for keeping us informed of their work prior to publication as well as for useful discussions and correspondence on various mathematical questions ; H. Grosse, S. Ferrara and V. Glaser for remarks concerning conformal invariant field equations ; E.F. Corrigan and P. Goddard for private communications concerning the construction of instanton fields from the linear algebra data. Special thanks are due to E. Arbarello who introduced one of us (R.S.) to some techniques of algebraic geometry during a visit at the Mathematical Institute Castelnuovo, in Roma.

- REFERENCES AND FOOTNOTES -

[1] M.F. ATIYAH, V.G. DRINFELD, N.J. HITCHIN, Yu.I. MANIN,
Phys. Lett. 65A (1978), 185

V.G. DRINFELD, Yu.I. MANIN,
- Uspehi Mat. Nauk 33 : 3 (1978), 241
- Funct. An. Appl. 12 : 2 (1978), 81
- Preprint ITEP N 72 (1978)
- Preprints Moscow University and Steklov Institute, May 1978,
Commun.math.Phys. 63, 177 (1978)

W. BARTH, K. HULEK, Manuscripta Mathematica 25, 323 (1978)

[2] G. 't HOOFT, unpublished

R. JACKIW, C. NOHL, C. REBBI,
Phys. Rev. D 15 , 1642 (1977)

[3] R. JACKIW, C. REBBI,
Phys. Rev. Lett. B 67, 189 (1977)

A.S. SCHWARTZ,
Phys. Lett. B 67, 172 (1977)

Commun.math.Phys., to appear

M.F. ATIYAH, N.J. HITCHIN, I.M. SINGER,
Proc. Nat. Acad. Sci. USA, 74 (1977)

Proc. Lond. Math. Soc. (1978)

[4] A. BELAVIN, A. POLYAKOV, A.S. SCHWARZ, Y. TYUPKIN,
Phys. Lett. B 59, 85 (1975)

[5] R.S. WARD,
Phys. Lett. A 61, 81 (1977)

M.F. ATIYAH, R.S. WARD,
Commun.math.Phys. 55 (1977), 117

[6] P.B. GILKEY,
The Index Theorem and the Heat Equation. Mathematical Lecture Series, 4,
Publish or Perish, Boston (1974)

R. PALAIS,
Seminar on the Atiyah-Singer Index Theorem, Ann. Math. Stud. n° 57,
Princeton University Press, Princeton (1965)

H. CARTAN, L. SCHWARTZ, Ed.,
Séminaire Henri CARTAN, E.N.S. Paris, 1963-1964, Secrétariat Mathématique,
11, rue Pierre Curie, 75005 Paris

[7] R. HARTSHORNE,
Algebraic Geometry, Graduate Texts in Math. 52 , Springer Verlag,
New York (1977) ; see, in particular, p. 250.

[8] R. HARTSHORNE,
Commun.math. Phys. 59, 1 (1978)

[9] Ref.[5], [3]
- R. STORA in International School of Mathematical Physics, Ettore Majorana,
Erice Sicily 1977, Invariant Wave Equations, G. Velo, A.S. Wightman Ed.,
Springer Lecture Notes in Physics, Vol. 73, Berlin (1978)

- J. MADORE, J.L. RICHARD, R. STORA in Meeting on Solitons, Instantons and Turbulence, Les Houches 1978, E. Brezin, J.L. Gervais Ed., Physics Reports, to be published.

[10] R.O. WELLS,
Differential Analysis on Complex Manifolds, Prentice Hall Series in Modern Analysis (1973)

[11] "Twistor News Letters"

[12] F. HIRZEBRUCH,
Topological Methods in Algebraic Geometry, Third Edition, Springer Verlag, New York (1966).

[13] That the Euler characteristics of $E(\ell)$ is the index of an AHS complex was noticed by Yu. M. Malyuta (Kiev) in November 1977. This author however did not show why this is so.

See also R. HARTSHORNE, Stable Vector Bundles of Rank 2 over P_3, Berkeley preprint. We wish to thank I.M. Singer for pointing out this reference.

[14] The derivation of $H^1(P_3, \underline{E}(-2)) = 0$ has been given by several authors, see ref. [1], Dolbeaut cohomology is usually preferred rather than Čech cohomology (see e.g. J.H. Rawnsley : "Differential Geometry of Instantons", "On the Atiyah Hitchin Drinfeld Manin vanishing theorem", Dublin Institute for Advanced Study preprint). A spectral sequence argument has been given by J.L. Verdier in Séminaire ENS 1977-1978, A. Douady, J.L. Verdier Ed. to be published in Astérisque, and communication in "Journées sur les champs de Yang et Mills", S.M.F. May 24-26, (1978). One can prove [1] by recursion that if $H^\circ(P_3, \underline{E}(-1)) = 0$, $H^1(P_3, \underline{E}(-2)) = 0$, $H^\circ(P_3, \underline{E}(-\ell)) = H^1(P_3, \underline{E}(-\ell-1)) = 0$, for $\ell \geqslant 1$ by using the fact that on a two plane $P_2 \subset P_3$ containing a real line, $\underline{E}$ is trivial on almost all straightlines in P_2 , hence $H^\circ(P_3, E_{|P_2}(-\ell)) = 0$ $\ell \leqslant -1$. Actually, $H^\circ(P_3, \underline{E}(-1)) = 0$ follows from $H^\circ(P_3, \underline{E}) = 0$.

Let P_2 be the equation of the two plane (a section of $\mathcal{O}(1)$) ,

$$0 \longrightarrow \mathcal{O}(-1) \xrightarrow{P_2} \mathcal{O} \longrightarrow \mathcal{O}/P_2\,\mathcal{O}(-1) = \mathcal{O}_{P_2} \longrightarrow 0$$

where $\mathcal{O}_{P_2}$ is the sheaf of germs of functions holomorphic in P_2 .

It follows :

$$0 \longrightarrow \underline{E}(-1-\ell) \longrightarrow \underline{E}(-\ell) \longrightarrow E_{|P_2}(-\ell) \longrightarrow 0$$

Hence

$$0 \longrightarrow H^0(P_3, \underline{E}(-1-\ell)) \longrightarrow H^0(P_3, \underline{E}(-\ell)) \longrightarrow H^0(P_3, E_{|P_2}(-\ell))$$
$$\wr\wr$$
$$H^0(P_2, \mathcal{O}(-\ell))^{\oplus 2}$$

$$\longrightarrow H^1(P_3, E(-\ell)) \longrightarrow H^1(P_3, \underline{E}(-\ell))$$

from which the recursion is established.

[15] N.J. HITCHIN,

Communication presented at the "Journées sur les Champs de Yang et Mills, Société Mathématique de France, Paris, May 24, 25, 26, 1978, and to be published :

write

$$0 \longrightarrow \Omega^1 \longrightarrow \mathcal{O}(-1)^{\oplus 4} \longrightarrow \mathcal{O} \longrightarrow 0$$

tensorize with E :

$$0 \longrightarrow H^0(E \otimes \Omega^1) \longrightarrow H^0(\underline{E}(-1)) \longrightarrow H^0(E) \longrightarrow H^1(E \otimes \Omega^1) \longrightarrow H^1(E(-1))^{\oplus 4} \longrightarrow H^1(E)$$
$$\underset{(a)}{\wr\wr\ 0} \quad \Leftarrow \quad \wr\wr\ 0 \qquad \wr\wr\ 0 \qquad\qquad \dim = 4k \qquad \dim = 2k-2$$
$$\longrightarrow H^2(E \otimes \Omega^1)$$
$$\wr\wr\ 0\ (b)$$

$$H^2(E \otimes \Omega^1) \approx \check{H}^1(\check{E} \otimes \Omega^2)$$

$$0 \longrightarrow \Omega^2 \longrightarrow \mathcal{O}(-2)^{\oplus 6} \longrightarrow \Omega^1 \longrightarrow 0$$
$$\wr\wr$$
$$\mathcal{O}(-4)$$

tensorize with $\check{E}$:

$$H^2(\check{\underline{E}}(-2))^{\oplus 6} \longrightarrow H^2(\check{\underline{E}} \otimes \Omega^1) \longrightarrow H^3(\check{\underline{E}} \otimes \Omega^2)$$
$$\wr\wr \qquad\qquad \wr\wr \qquad\qquad \wr\wr$$
$$H^1(\underline{E}(-2)) \approx 0 \quad \Longrightarrow (b)\ 0 \qquad \check{H}^0(\underline{E} \otimes \Omega^1) \overset{(a)}{\approx} 0$$

[16] M.F. ATIYAH,

Pisa Lectures and Private Communication. Consider :

$$0 \longrightarrow N \longrightarrow T \xrightarrow{Az} \mathcal{O}(2) \longrightarrow 0$$

where T is the tangent bundle, $A = -A^T$ a non degenerate correlation :

$$N \sim \left\{ \sum z_i \frac{\partial}{\partial z_i} \in T \;\middle|\; (z, Az) = 0 \right.$$

The dual sequence

$$0 \longrightarrow \mathcal{O}(-2) \xrightarrow{Az} \Omega^1 \longrightarrow N^* \longrightarrow 0$$

gives rise after tensorization by E to :

$$H^1(E(-2)) \longrightarrow H^1(E\otimes\Omega^1) \longrightarrow H^1(E\otimes N^*) \longrightarrow H^2(E(-2))$$
$$H^1(E(-2)) \cong 0 \quad \Rightarrow \approx \quad H^2(E(-2)) \cong^* H^1(E^*(-2)) = 0$$

Hence $H'(E\otimes\Omega^1) \approx H^1(E(-1)\otimes N^*(1))$

Now, $N^*(1) = V^{\circ *}$ dual and thus isomorphic with the null correlation sub-bundle V° of $T(-1)$ which is the $P^3(\mathbb{C})$ version of the $k = 1$ instanton. The computation of $H^1(E)$ holds with however the extra index pertaining to the fiber of $\overset{\circ}{V}{}^*$ and the connection $\mathcal{a}$ altered by the corresponding connection for $\overset{\circ}{V}{}^*$ which is nothing else than the spinor connection for the invariant metric on $P_1(H)$!
A complete description of V° can be found in W. Barth, Math. Ann. 226, 125 (1977). See also ref. [12] , p. 166

[17] This can also be seen directly from algebraic geometry and Serre duality, [7], [10], [12] :

$$H^1(\underline{E}\otimes\Omega^1) \simeq H^2(\check{E}\otimes\Omega^2) \qquad \text{(Serre)}$$

From the sequence

$$0 \longrightarrow \check{E}\otimes\Omega^2 \longrightarrow \check{E}(-2)^{\oplus 6} \longrightarrow \check{E}\otimes\Omega^1 \longrightarrow 0$$

it follows

$$H^1(\check{E}(-2))^{\oplus 6} \longrightarrow H^1(\check{E}\otimes\Omega^1) \longrightarrow H^2(\check{E}\otimes\Omega^2) \longrightarrow H^2(\check{E}(-2))^{\oplus 6}$$
$$H^1(\check{E}(-2)) \cong 0 \quad \Rightarrow \approx \quad H^2(\check{E}(-2)) \cong 0$$

see e.g. J.L. Verdier in Séminaire ENS 1977-1978, A. Douady, J.L. Verdier, Ed.

[18] E.F. CORRIGAN, D.B. FAIRLIE, P. GODDARD, S. TEMPLETON,
Nucl. Phys. B 140 (1978), 31-44

N.H. CHRIST, E.J. WEINBERG, N.K. STANTON,
Columbia Preprint (1978)

R.J. CREWTHER,
CERN-TH 2522 (1978)

C. MEYERS, M. de ROO,
CERN-TH Aug. 1978, Nucl. Phys. to be published

QUARTIC OSCILLATOR*

R. Balian
Service de Physique Théorique, CEA-Saclay, BP n°2, 91190 Gif-sur-Yvette, France

G. Parisi
I.N.F.N., Frascati, Italy

A. Voros+
Service de Physique Théorique, CEA-Saclay, BP n°2, 91190 Gif-sur-Yvette, France

Abstract : On the example of the semi-classical expansion for the levels of the quartic oscillator $-(d^2/dq^2)+q^4$, we show how the complex WKB method provides information about the singularities of the Borel transform of the semi-classical series. In this problem there occurs a tunneling effect between complex turning points, by which those singularities generate exponentially small, yet detectable, corrections to the energy levels.

Résumé : Sur l'exemple du développement semi-classique des niveaux de l'oscillateur quartique $-(d^2/dq^2)+q^4$, nous montrons comment la méthode BKW complexe renseigne sur les singularités de la transformée de Borel de la série semi-classique. Dans ce problème, les singularités engendrent, par un effet tunnel entre points tournants complexes, des corrections aux niveaux d'énergie qui sont exponentiellement petites et cependant détectables.

One of the oldest recipes to make numerical sense out of an asymptotic (divergent) expansion like $\sum_{k=0}^{\infty} F_k x^{-k}$ for a function F(x) $(x\to\infty)$ is, for any x large but fixed, to sum the series up to the point $k=K$ where the general term $F_k x^{-k}$ attains its minimum modulus as a function of k [2]. The resulting sum $F^*(x) = \sum_{k=0}^{K} F_k x^{-k}$ is taken as an estimate for the exact value F(x), and the first neglected term allegedly provides an estimate for the error : $|F(x) - F^*(x)| \lesssim \varepsilon = |F_{K+1} x^{-K-1}|$.

Such a procedure relies on a faith supported by experience, but it has no rigorous derivation from the general definition of an asymptotic series. In some cases, like the Stirling series for $\log\Gamma(x)$ (x real), it can indeed be justified. But we intend to show that many interesting series exhibit a different type of behaviour ; on one hand they may lead to a systematic discrepancy $F(x) - F^*(x)$ much larger than the "apparent uncertainty" ε measuring the minimum fluctuation of the sequence of partial sums. On the other hand, the asymptotic representation of F(x) may contain additional terms that are *subdominant*, i.e. exponentially small when compared to all terms

---

* Dedicated to the centennial of the instanton[1]

+ Member of CNRS

of the initial *(dominant)* series. These new terms are asymptotically negligible and seem ill-defined mathematically, but their numerical contributions must be retained as it precisely compensates for the observed discrepancy from the dominant series. We have been surprised not to find this elementary observation in classic textbooks [2,3] (it was noted independently of us by J. Zinn-Justin[4]).

In general, this raises the following questions about the asymptotic representation $\Sigma F_k x^{-k}$ for a given function F(x) :

- how to define subdominant contributions if and when they exist ? (possibly by analytic continuation to regions in x whose they become dominant, but this may not always be explicitly feasible).

- when is a subdominant contribution numerically relevant ? The answer to this question results from a competition as $x \to \infty$ between the size of the subdominant series (essentially of its leading term) and the size of the *smallest term* of the dominant series, which is governed by the *large order behaviour* of its coefficients.

- how do subdominant contributions influence the currently used Borel summation procedure ?

Following the approach we introduced in ref.[5], we shall relate these questions to the study of the analyticity properties of a Fourier-Laplace transform of the function F(x) . We shall first treat the more or less explicit and illustrative example of expansions generated by the saddle-point method, but our main application will be the study of the semi-classical expansion for the levels of the quartic oscillator.

1. The case of saddle-point expansions.

As a typical example, we shall study asymptotic expansions generated by the method of steepest descent[3]. We consider an integral along a complex path without endpoints :

$$F(x) = \int e^{-x\,\varphi(u)}\,du \tag{1.1}$$

where $x > 0$, and $\varphi(u)$ is an analytic function that makes the integral (1.1) convergent and that has only isolated simple critical points (or saddle-points) where $\varphi'(u) = 0$, $\varphi''(u) \neq 0$. In the saddle-point method, the integration path is first distorted to a path of steepest descent. Then for $x \to +\infty$, each saddle-point u_j encountered by this new path formally produces an independent asymptotic contribution to (1.1) of the form $e^{-x\,\varphi(u_j)}\left(\sum_{k=0}^{\infty} F_k^{(j)}\, x^{-k-1/2}\right)$. Assume for simplicity that $u_o = 0$ is a saddle-point with $\varphi(0) = 0$, and that all other saddle-points u_j satisfy $\mathrm{Re}\,\varphi(u_j) > 0$). Then :

$$F(x) \sim \sum_{k=0}^{\infty} F_k^{(0)}\, x^{-k-1/2} + \sum_j e^{-x\,\varphi(u_j)} \left(\sum_k F_k^{(j)}\, x^{-k-1/2}\right) \tag{1.2}$$

We are precisely in the case of a dominant series (the first one, produced by the saddle-point $u = 0$) supplemented by exponentially small contributions. But (1.2) is up to now a formal expression that requires a suitable interpretation.

One method to understand and to build the expression (1.2) is to take $\varphi(u) = s$ as the integration variable. This transforms (1.1) into :

$$F(x) = \int_C e^{-xs} \rho(s)\, ds \tag{1.3}$$

Thus, F is the Laplace transform along a suitable path C of the function $\rho(s) = \left.\frac{du}{d\varphi}\right|_{\varphi=s}$. $\rho(s)$ is multivalued, the critical values $s_j = \varphi(u_j)$ are its branch points (of the $(s-s_j)^{-1/2}$ type). When we take the path of steepest descent in (1.1), the contour C, which lies on the Riemann surface of $\rho(s)$, gets pushed as far to the right as permitted by the branch points (Fig.1).

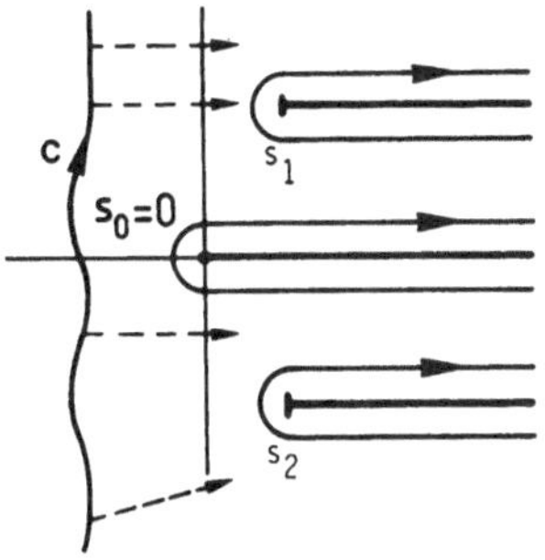

Fig.1

By expanding $\rho(s) = \sum_{n=-1}^{\infty} \frac{\rho_{n/2}^{(0)} s^{n/2}}{2\Gamma(\frac{n}{2}+1)}$ around $s_o = 0$, we then find the coefficients of the dominant expansion as :

$$F_k^{(0)} = \rho_{k-1/2}^{(0)} \tag{1.4}$$

This means that

$$\Delta_o \rho(s) = \rho(s+i0) - \rho(s-i0) = \sum_0^{\infty} \frac{\rho_{k-1/2}^{(0)} s_+^{k-1/2}}{\Gamma(k+1/2)} , \tag{1.5}$$

the *discontinuity* of ρ across the cut from $s = 0$, is a *Borel transform* of the dominant series (s_+^α is the function $s_+^\alpha \equiv s^\alpha$ for $s > 0$, $s_+^\alpha \equiv 0$ for $s < 0$ [6]). The same analysis applies to the other cuts : the series multiplying $e^{-x\,\varphi(u_j)}$ in (1.2) results from expanding the discontinuity $\Delta_j\rho$ of $\rho(s)$ across the cut from $s_j = \varphi(u_j)$. Finally the function F itself can be reconstructed from the integral representation

$$F(x) = \int_0^\infty e^{-xs} \Delta_o\rho(s)\, ds + \sum_j \int_{s_j}^\infty e^{-xs} \Delta_j\, \rho(s)\, ds \tag{1.6}$$

This exact formula shows how each contribution to F(x), whether dominant or subdominant, exists independently of the others.

We now want to compare the subdominant terms (at least the largest ones) in (1.2) with the size ε of the smallest term $F_K^{(0)} x^{-K-1/2}$ in the dominant series. Its rank K satisfies

$$|F^{(0)}_{K+1} / F^{(0)}_{K}| \sim x \qquad (1.7)$$

and should be large for large x . Thus, K and ε are asymptotically governed by the *large order behaviour* of $F_k^{(0)}$, itself controlled by the radius of convergence $|s_\varepsilon|$ of the series (1.5) : here s_ε is the singularity of ρ(s) closest to the origin and lying on either of the two sheets associated with the branch point s = 0 . From the relations (1.4) and $|\rho^{(0)}_{k-1/2}| / \Gamma(k+\frac{1}{2}) \propto |s_\varepsilon|^{-k}$, we get :

$$|F_k^{(0)}| \propto (k+\frac{1}{2}) \; |s_\varepsilon|^{-k} \quad . \qquad (1.8)$$

Returning to Eq.(1.7), we approximately find

$$K \sim x|s_\varepsilon| \quad \text{and} \quad \varepsilon \propto \exp(-x|s_\varepsilon|) \quad . \qquad (1.9)$$

By comparison, the largest subdominant contribution to (1.2) is produced by the *leftmost* singularity s_δ encountered (besides s = 0) as the contour C is pushed to the right. The leading contribution of s_δ is $\delta \propto \exp(-xs_\delta)$. Although δ is exponentially small since $\mathrm{Re}\, s_\delta > 0$, we can distinguish two generic situations :

(i) $|s_\varepsilon| < \mathrm{Re}\, s_\delta$. Subdominant terms can be neglected in the numerical scheme of the introduction (they nevertheless *remain in the Borel summation formula* (1.6)). In the extreme case $\mathrm{Re}\, s_\delta = +\infty$, there are strictly no subdominant contributions, as when the function F(x) is Borel summable in the usual sense.

(ii) $|s_\varepsilon| > \mathrm{Re}\, s_\delta$. A systematic deviation is introduced by the dominant term, which should be corrected by including the subdominant contribution δ. This correction becomes relatively more and more important as $x \to \infty$, since $|\delta/\varepsilon| \to \infty$. A frequent occurrence of this case is when the *same* branch point plays both roles : $s_\varepsilon = s_\delta$, and $\mathrm{Im}\, s_\delta \neq 0$ (as for the Bessel function $J_o(z)$ when Re z > 0 [5]). Another example is the integral (1.1) taken on the real axis with $\varphi(u) = 36u^2 - 20u^3 + 3u^4$: the subdominant saddle-point is u = 3 ($s_\delta = 27$), whereas the saddle-point u = 2 ($s_\varepsilon = 32$) controls the large order behaviour (Fig.2, showing the cuts and the contour C in this case, demonstrates that some realistic cases may be harder to describe than the ideal situation pictured on Fig.1).

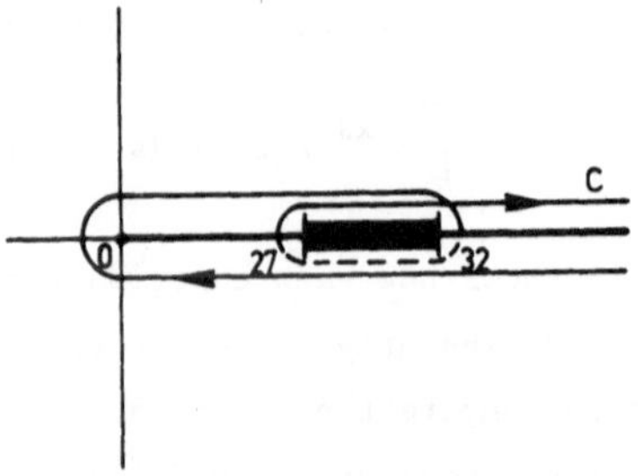

Fig.2

A refinement of Eq.(1.8) has become very popular with the recent advent of instanton calculations[7] : it consists in computing a *large order expansion* for the coefficients F_k , in the form

$$F_k \underset{k\to\infty}{\sim} \Gamma(k+\alpha)\, r^{-k} \left(d_o + \frac{d_1}{k} + \frac{d_2}{k^2} + \dots\right)$$

with all constants to be determined. This question is actually answered by a hundred year-old theorem of Darboux[1], discussed in ref.[3] : if an analytic function $f(s) = \sum_{k=0}^{\infty} f_k\, s^k/k!$ has its nearest singularity at s_ε with a local expansion $f(s) \sim \sum_j c_{\ell_j} (s-s_\varepsilon)^{\ell_j}$ ($\{\ell_j\}$ is an increasing sequence of real numbers, and we place the cut for the functions z^ℓ on $[0,\infty)$), then the dominant large order behaviour of the Taylor coefficients f_k is :

$$f_k \underset{k\to\infty}{\sim} -\pi^{-1}\, s_\varepsilon^{-k} \sum_j c_{\ell_j} \sin(\pi\ell_j)\, \Gamma(\ell_j+1)\, (-s_\varepsilon)^{\ell_j}\, \Gamma(k-\ell_j) \tag{1.10}$$

In terms of the *discontinuity* of f across its cut :

$$\Delta f(s) = f_+(s) - f_-(s) = \sum_{\ell_j} f^{(\varepsilon)}_{\ell_j} (s-s_\varepsilon)_+^{\ell_j} / \Gamma(\ell_j+1) \tag{1.11}$$

$\left(f^{(\varepsilon)}_{\ell_j} = (1-e^{2\pi i\ell_j})\, \Gamma(\ell_j+1)\, c_{\ell_j}\right)$, Eq.(1.10) becomes :

$$f_k \sim (2i\pi)^{-1}\, s_\varepsilon^{-k} \sum_j f^{(\varepsilon)}_{\ell_j} (-s_\varepsilon)^{\ell_j}\, e^{-i\pi\ell_j}\, \Gamma(k-\ell_j) \tag{1.12}$$

which has the advantage of allowing positive integral powers ℓ_j in the discontinuity ; for negative integers, $(s-s_\varepsilon)_+^{\ell_j} / \Gamma(\ell_j+1)$ in Eq.(1.11) must now be interpreted as the generalized function $\delta^{(-\ell_j-1)}(s-s_\varepsilon)$ [6].

From Eq.(1.4) we see that to get the large order behavior of $F_k^{(0)}$ we must apply Darboux's theorem to $f(s) = \Delta_o\, \rho(s)$. The fact that this function admits not a Taylor but a Puiseux series is only a slight complication : we must separate the odd and even parts of ρ as a function of $s^{1/2}$. The result is that if $\rho(s) \sim \sum_{n=-1}^{\infty} \rho^{(\varepsilon)}_{n/2} (s-s_\varepsilon)^{n/2} / 2\Gamma(\frac{n}{2}+1)$ around $s = s_\varepsilon$, the coefficients of the series (1.5) have the large order expansion :

$$\rho^{(0)}_{k-1/2} \sim (2i\pi)^{-1}\, s_\varepsilon^{-(k-1/2)} \sum_{\ell=0}^{\infty} \rho^{(\varepsilon)}_{\ell-1/2} (-s_\varepsilon)^{\ell-1/2}\, e^{-i\pi(\ell-1/2)}\, \Gamma(k-\ell) \tag{1.13}$$

where the singularity s_ε is now located on the *Riemann surface* of $\rho(s)$: Eq.(1.13) is then unambiguous as long as the two cuts (from $s = 0$ and $s = s_\varepsilon$) stay disjoint (if there are several branch points at the same distance $|s_\varepsilon|$, their contributions add up).

This example (1.1) suggests the following conclusion, probably connected with

Dingle's interpretation of asymptotic expansions [3] : the best understanding of a general function F(x) is achieved by expressing it as a Laplace transform of a ramified analytic function $\rho(s)$, the integral being taken along some path on the Riemann surface $\mathcal{S}$ of $\rho(s)$. Some branch points of $\rho(s)$, depending on the global structure of $\mathcal{S}$, then contribute to the asymptotic expressions of F(x) through the expansions of the corresponding discontinuities of $\rho(s)$. In turn, the large order behaviour of each such contribution is a reflection of the first orders in the corresponding expansion at (a) neighboring branch point(s) also specified by the global structure of $\mathcal{S}$. The discontinuity expansions can often be computed perturbatively ; moreover if we can fully resum the discontinuities from their power series, and either express the Laplace integral for F(x) in terms of the discontinuities alone as in Fig.1, or recover $\rho(s)$ itself by a dispersion relation, then we have succeeded to resum the function F by a generalization of the Borel method.

The main difficulty of this program is, for a given F(x), to know whether a suitable $\rho(s)$ exists, and to understand its analytic structure. If F(x) is given by an integral like (1.1) but now *multidimensional*, we can take $\rho(s) = \int \frac{du}{s-\varphi(u)}$.

The branch points of $\rho(s)$ are again the critical values of φ (the saddle-point values), but the analytic structure of $\rho(s)$ is very hard to describe in detail [8].

Now, our main concern lies with functions F(x) that admit semi-classical (or perturbative) expansions in quantum mechanics and in field theory. Such expansions present some analogy with the previous case, as they can be generated by Feynman path integrals which formally look like (1.1), with x replaced by $\hbar^{-1}$ (or the inverse coupling constant g^{-1}), the integration variable u by the trajectories in configuration space, and $\varphi(u)$ by the classical action function. We therefore expect the previous results to hold (qualitatively at least) : there should exist an *analytic* function $\rho(s)$ on a Riemann surface $\mathcal{S}$, such that F(x) be a Laplace transform of $\rho(s)$; the branch points of $\rho(s)$ should be the stationary values of the action and should thus arise from *each classical trajectory* (real or complex) ; in principle, semi-classical methods should provide the discontinuity expansions of $\rho(s)$ at the branch points (a semi-classical description of the *global topology* of $\mathcal{S}$ would be also highly desirable, but this stands as an open question). This description of quantum mechanics in terms of analytic functions of the variable s (conjugate of $1/\hbar$) has been given in ref.[9].

Our analysis of exponentially small contributions to asymptotic expansions means here that tunneling effects (or instanton effects) should sometimes be numerically relevant even in cases when they are subdominant. Moreover they should become relatively more and more important as $\hbar$ (or g) $\to 0$, since we have seen that $|\delta/\varepsilon| \to \infty$.

A last interesting consequence of the Riemann surface structure is that the large orders of a semi-classical expansion at one branch point are governed by the leading orders of an expansion *of the same nature* at another branch point. This reciprocity property is probably universal for semi-classical expansions : it was noted in a completely different way by Dingle in the case of the WKB expansion for wave functions[3]. It provides a new insight into the semi-classical nature of large order analysis[7].

2. Semi-classical eigenvalues of the quartic oscillator.

The quartic oscillator is the one-dimensional quantum system described by the Hamiltonian :

$$\hat{H}(\hbar,g) = -\hbar^2 \frac{d^2}{dq^2} + gq^4 \qquad (g>0) \quad . \tag{2.1}$$

This simple Hamiltonian, although not (yet) exactly solvable, has been extensively studied[10-12,29]. We shall be concerned here with its eigenvalues. Because of the obvious scaling property ($\approx$ denotes unitary equivalence) :

$$\hat{H}(\hbar,g) \approx \hbar^{4/3}\, g^{1/3}\, \hat{H}(1,1) \tag{2.2}$$

(under the change of variables $q \to g^{-1/6}\,\hbar^{1/3} q$), we can restrict ourselves to the case $g=1$. But in view of a semi-classical study of (2.1), instead of letting $\hbar = 1$ immediately as suggested by Eq.(2.2) we prefer to keep it as an explicit expansion parameter for some time. The classical Hamiltonian corresponding to (2.1) with $g=1$ is $H=p^2+q^4$. Its classical trajectory for a given energy E is the solution of $\frac{dq}{dt} = 2\sqrt{E-q^4}$, namely a Jacobi elliptic function[13] :

$$q(t) = E^{1/4}\, \mathrm{cn}(2\sqrt{2}\, E^{1/4} t \mid \tfrac{1}{2}) \quad . \tag{2.3}$$

Let us first rewrite the WKB expansion for the eigenfunctions and eigenvalues of a general one-dimensional Schrödinger equation with an analytic potential V(q) :

$$\left(-\hbar^2 \frac{d^2}{dq^2} + V(q)\right)\psi(q) = E\,\psi(q) \quad . \tag{2.4}$$

The solutions of (2.4) in the complex q plane are known to have the exact form[14] :

$$\psi = u^{-1/2} \exp \frac{i}{\hbar}\int u\, dq \quad , \tag{2.5}$$

where $u(q,\hbar)$ is a solution of

$$u^2 - p^2 = \hbar^2\, (u^{-1/2})''\, u^{1/2} \quad , \tag{2.6}$$

$p(q) = \pm(E-V(q))^{1/2}$ is the classical momentum, and $' = \frac{d}{dq}$.

For each determination of the function p(q) , Eq(2.6) (which is *exact*) can be formally solved in powers of $\hbar^2$, as :

$$u(q) = u(q\,,p(q)) = p + \sum_{n=1}^{\infty} \hbar^{2n}\, u_{2n}(q,p) \quad . \tag{2.7}$$

The terms u_{2n} can be computed recursively : they are polynomial functions of $V'(q)$, $V''(q)$, ... and of p^{-1} , odd in p . For instance :

$$u_2 = \frac{V''(q)}{8p^3} + \frac{5V'(q)^2}{32p^5}$$

An alternative method uses the representation : $\psi = \exp \frac{i}{\hbar}\int v\, dq$ and $v = p + \sum_{n=1}^{\infty} (i\hbar)^n v_n$ [15,12]. This results in a somewhat simpler recursion relation

$$v_n = \frac{1}{2p}\left(v'_{n-1} - \sum_{k=1}^{n-1} v_k\, v_{n-k}\right) \quad , \quad v_o = p \tag{2.8}$$

By comparing the two expressions of ψ we see that $u_{2n} = (-1)^n v_n$, but the remarkable structure (2.5) of ψ is harder to understand by this method.

A third way of computing the expansion (2.7) is by means of a closed (i.e. non-recursive) formula[11,16].

In all methods, complexity increases rapidly with the order n . Practical calculations are best performed by computer in symbolic computing languages[17,11-12,16].

From the knowledge of the WKB solution (2.5) we can now derive an eigenvalue quantization condition correct to all orders in $\hbar$, by an argument of Wentzel and Dunham[15,18]. We take $V(q)$ to be a simple-well potential with two (real) turning points $q_- < q_+$ as in Fig.3, and we define the analytic function $p(q) = (E - V(q))^{1/2}$ in the complex q plane cut along $[q_-,q_+]$, with the values of $\varphi = \operatorname{Arg} p$ as indicated on Fig.4 (thus neglecting the effects of complex turning points).

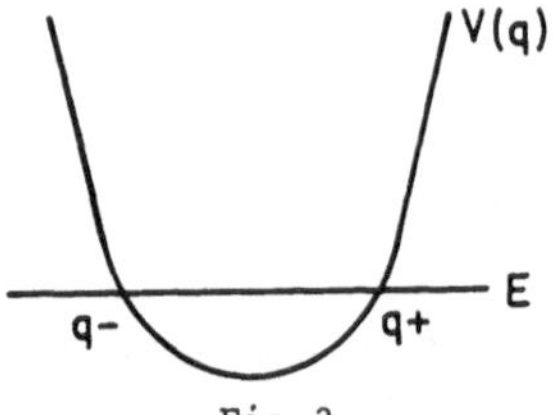

Fig.3

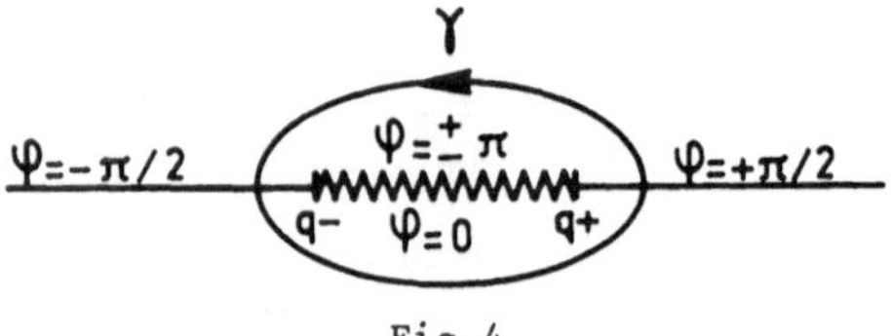

Fig.4

Then the eigenfunction ψ admits the asymptotic form :

$$\psi \sim u(q,p)^{-1/2} \exp \frac{i}{\hbar}\int_{q_o}^{q} u(q'\,,p(q'))\, dq' \tag{2.9}$$

in the whole complex q plane away from the cut, since (2.9) satisfies the correct boundary conditions for $q \to \pm\infty$. Let γ be a positive contour encircling the cut (Fig.4). The *exact* eigenfunction ψ is analytic and has an integral number k of zeros, all of them on the interval (q_-,q_+) , hence it satisfies :

$$\oint_\gamma \frac{\psi'}{\psi}\, dq = 2\pi i k \qquad .$$

The substitution of (2.9) then yields (we refer to [18,12] for details) :

$$\oint_\gamma u\, dq = \oint_\gamma p\, dq + \sum_{n=1}^{\infty} \hbar^{2n} \oint_\gamma u_{2n}\, dq = 2\pi(k+\tfrac{1}{2})\,\hbar \qquad (2.10)$$

which is an eigenvalue condition of the Bohr-Sommerfeld type, but with all corrections in powers of $\hbar$ included.

We note that the expressions (2.5) and (2.10) for the eigenfunctions and eigenvalues only differ from their usual lowest order approximations by the replacement everywhere of the classical momentum p(q) by the solution u of Eq.(2.6). Besides, the construction of u to any finite order in $\hbar$ is *purely algebraic* (and u only depends on $\hbar^2$).

We have given here a rapid derivation of the result (2.10). We do not know yet whether and when (2.10) gives a truly asymptotic expansion of the eigenvalues. Existing works only concern leading order estimates[19,20] and they suggest that this is a difficult question. But in spite of its formal derivation Eq.(2.10) appears to be the correct asymptotic expansion in all cases of interest.

We now specialize our results to the case of the potential $V(q) = q^4$, following ref.[12] but with slightly different normalizations. The recursion relation (2.8) now yields :

$$v_1 = -\frac{q^3}{p^2}$$

$$v_2\ (= -u_2) = -\left(\frac{3q^2}{2p^3} + \frac{5q^6}{2p^5}\right)$$

$$v_3 = -\left(\frac{3q}{2p^4} + \frac{27q^5}{2p^6} + \frac{15q^9}{p^8}\right)$$

$$v_4\ (= u_4) = -\left(\frac{3}{4p^5} + \frac{339q^4}{8p^7} + \frac{663q^8}{4p^9} + \frac{1105q^{12}}{8p^{11}}\right) \qquad .$$

More generally :

$$v_n = \sum_{\ell=0}^{[3n/4]} v_{n,\ell}\, q^{3n-4\ell}\, p^{-3n+2\ell+1} \qquad (2.11)$$

(the precise powers involved, and the summation upper bound [3n/4] (integer part of 3n/4) can be understood by dimensional analysis).

In the quantization condition (2.10) now written as :

$$2\pi(k+\tfrac{1}{2})\,\hbar = \sum_0^{\infty} \hbar^{2n} (-1)^n \oint_\gamma v_{2n}\, dq = \sum_0^{\infty} \hbar^{2n} \sigma_n \qquad (2.12)$$

we can evaluate the loop integrals in terms of the Euler Beta function :

$$\sigma_n = (-1)^n E^{\frac{3}{4}(1-2n)} \sum_{\ell=0}^{[3n/2]} v_{2n,\ell} B\left(-3n+\ell+\frac{3}{2}, \frac{3n}{2}-\ell+\frac{1}{4}\right) \qquad (2.13)$$

In particular, letting $c = \Gamma(1/4) = 3.62561\ldots$

$$\sigma_o = \oint_\gamma p\,dq = E^{3/4} B\left(\frac{3}{2}, \frac{1}{4}\right) = \frac{c^2}{3}\sqrt{\frac{2}{\pi}} E^{3/4} = \sigma \qquad (2.14)$$

is the classical action around the closed orbit (2.3) of energy E . In terms of σ , the quantization condition (2.12) takes the final form :

$$2\pi(k+\frac{1}{2})\hbar = \sum_{n=0}^{\infty} b_n \sigma^{1-2n} \hbar^{2n} \qquad (2.15)$$

$$b_n = \frac{(-1)^n}{2}\left(\frac{3}{c^2}\sqrt{\frac{\pi}{2}}\right)^{1-2n} \sum_{\ell=0}^{[3n/2]} v_{2n,\ell} B\left(-3n+\ell+\frac{3}{2}, \frac{3n}{2}-\ell+\frac{1}{4}\right) \qquad (2.16)$$

We give a few details about the practical computation of the b_n . The use of a formal computer algebraic system to solve the WKB recursion equations yields results in closed arithmetic form, but becomes extremely time-consuming as the order increases. With the REDUCE language[21] we found the iteration of Eq.(2.6) slightly faster than other methods, but we nevertheless stopped at b_{16} . For a more efficient evaluation of b_n for large n , we shifted to the ordinary numerical computation (in FORTRAN) of the coefficients $v_{n,\ell}$ in Eq.(2.11) recursively on n, followed by the evaluation of formula (2.16). To compensate for the errors (increasing with n) caused by huge cancellations between terms of Eq.(2.16), we worked with the multiple-precision arithmetic package MULTILONG [22]. A 30 minute IBM 360-91 computer run produced the b_n up to $n = 53$ in ordinary double-precision.

We empirically found that the sequence of signs of the b_n was + - + + - - + + - - ... and that for large n :

$$|b_n| \sim (2n-2)!\, 2^n \times C \quad , \qquad C \simeq 0.63 \qquad (2.17)$$

(these facts will be explained in section 4). We give the list (Table 1) of the computed values of $b'_n = 2^{-n} b_n/(2n-2)!$. The first exact b_n are (with $c = \Gamma(1/4)$) :

$$b_o = 1 \qquad b_1 = -\frac{\pi}{3}$$

$$b_2 = \frac{11\, c^8}{10368\, \pi^2} \qquad b_3 = \frac{4697\, c^8}{466560\, \pi}$$

$$b_4 = -\frac{390065\, c^{16}}{501645312\, \pi^4} \qquad b_5 = -\frac{53352893\, c^{16}}{1934917632\, \pi^3} .$$

It follows from Eq.(2.16) that :

$$b_{2n} = R_{2n}\, c^{8n}\, \pi^{-2n} \qquad b_{2n+1} = R_{2n+1}\, c^{8n}\, \pi^{-2n+1} .$$

where $\{R_n\}$ is a sequence of *rational* numbers, which we do not know how to generate by any simpler law.

3. An investigation of the semi-classical series.

The object of our study will be the semi-classical series (2.15). We first note that this quantization condition only involves powers of $\hbar/\sigma$, in relation with the scaling property (2.2). By formal inversion of the series (2.15), we get an eigenvalue formula in the form of an expansion for *large* k , which now depends trivially on $\hbar$:

$$\sigma_k = \frac{c^2}{3}\sqrt{\frac{2}{\pi}}\, E_k^{3/4} = 2\pi\hbar\,(k+\tfrac{1}{2})\left[1+\frac{c_1}{(k+1/2)^2}+\frac{c_2}{(k+1/2)^4}+\ldots\right]$$

Thus, in the original series (2.15) (which we found simpler to study), we can let $\hbar = 1$ and keep $1/\sigma$ as the expansion parameter. The power series

$$\sigma + \frac{b_1}{\sigma} + \frac{b_2}{\sigma^3} + \ldots = 2\pi\,(k+\tfrac{1}{2}) \tag{3.1}$$

now formally defines the quantum number as a (continuous) function of the classical action. However (3.1) certainly diverges everywhere because of the law (2.17). We are thus faced with the problem of interpreting the series (3.1) and/or resumming it.

We shall first compare the exact eigenvalues with the successive approximations defined by (3.1) : let $E_k^{(j)}$ be the solution of :

$$\sigma(E) + \frac{b_1}{\sigma(E)} + \ldots + \frac{b_j}{\sigma(E)^{2j-1}} = 2\pi(k+\tfrac{1}{2})$$

such that $\sigma(E_k^{(j)}) \sim 2\pi(k+\frac{1}{2})$ for $k \to +\infty$: $E_k^{(j)}$ is a reasonable j-th order estimate of the exact k-th eigenvalue E_k . Table 2a shows $E_k^{(j)}$ for a sample of values of k and j . For given k, $E_k^{(j)}$ exhibits the typical behaviour of the j-th partial sum of an asymptotic series : it first seems to "converge" rapidly while the successive terms of the series decrease, then it "blows up" without limit. As stated in the introduction, we take as "best" estimate E_k^* the approximation $E_k^{(j)}$ for which $|E_k^{(j+1)} - E_k^{(j)}|$ attains its minimum value ε , and we compare ε with the actual error $|E_k - E_k^*|$. From Table 2a, we find important discrepancies between the best estimates and the exact values : $|E_k - E_k^*| / \varepsilon(k)$ is much larger than 1 (and increases with k).

As in section 1, we want to interpret this phenomenon by the occurrence of subdominant corrections to the quantization condition (it is also a sign that the series (3.1) is not Borel summable in the simplest sense).

To understand the origin of subdominant corrections, we adopt the viewpoint that the Bohr-Sommerfeld rule can be formally derived from a Feynman path integral formula for the trace of the Green's function[23,24]

$$G(E) = \mathrm{Tr}(\hat{H}-E)^{-1} = \frac{i}{\hbar}\int_0^\infty dt\, e^{iEt/\hbar}\, \mathrm{Tr}\, e^{-i\hat{H}t/\hbar}$$

$$= \frac{i}{\hbar}\int_0^\infty dt\, e^{iEt/\hbar} \int_{q(t)=q(0)} \{\mathcal{D}q\}\, e^{iS\{q\}/\hbar} \qquad (3.2)$$

where $S\{q\} = \int_0^t \left[\frac{1}{4}\left(\frac{dq}{d\tau}\right)^2 - V(q(\tau))\right] d\tau$ is the classical action, and the path integral only involves paths with $q(t) = q(0)$. When (3.2) is evaluated semi-classically by the stationary phase method, only the (real) closed classical trajectories of energy E contribute, as they are the stationary points of $S\{q\} + Et$. The contributions of one such closed trajectory C_E as it is traversed 1,2,3,... times in both directions form a geometric series of ratio $\exp \frac{i}{\hbar}\oint_{C_E} p\, dq$ which sums up to :

$$G(E) \sim \frac{T(E)}{\hbar} \tan\left(\frac{1}{2\hbar}\oint_{C_E} p\, dq\right)$$

where $T(E) = \frac{d}{dE}\oint_{C_E} p\, dq$ is the primitive period of C_E. The poles of G(E) (the eigenvalues) are then given by the Bohr-Sommerfeld rule $\oint p\, dq = 2\pi(k+\frac{1}{2})\hbar$. Corrections in powers of $\hbar$ are in principle contributions to the integral (3.2) from the paths fluctuating around the (same) stationary trajectories (see ref.[24] for a review of these methods). The quantization rule (3.1) for the quartic oscillator precisely involves the action σ of the real periodic classical trajectory (2.3).

But we know, from the experience of finite-dimensional integrals of the form $\int dq\, e^{iS(q)/\hbar}$, that if S is analytic we can push the integration contour into the region $\mathrm{Im}\, S(q) > 0$ to uncover contributions from *complex* stationary points (saddle-points), each of them of the form $e^{iS_o/\hbar} \times$ [power series]. We claim that such terms give *numerically relevant* contributions, and that in particular, they account for the observed discrepancy in the series (3.1). Precisely, the complex saddle-points in the path integral (3.2) are the *closed orbits* of energy E of the classical Hamiltonian in *complex* coordinates[9]. For $H = p^2+q^4$, the complex trajectory $q_E(t) = E^{1/4}\, \mathrm{cn}(2\sqrt{2}E^{1/4}t \mid \frac{1}{2})$ has *a lattice* Λ *of periods* in the t plane[13], generated by $T_1 = (T+iT)/2$ and $T_2 = (-T+iT)/2$, where

$$T = \frac{E^{-1/4}}{2\sqrt{2}}\, 4K(1/2) = \frac{E^{-1/4}}{2\sqrt{2\pi}}\, \Gamma(1/4)^2 \qquad (3.3)$$

is the primitive real period (Fig.5).

A complex closed orbit is associated with every time path $\lambda\Big|_{t_1}^{t_2}$ with $t_2-t_1 \in \Lambda$; it can also be represented as a loop on the two-sheeted Riemann surface of the complex momentum $p = (E-q^4)^{1/2}$, whose branch points are the four classical turning points $E^{1/4}$, $iE^{1/4}$, $-E^{1/4}$, $-iE^{1/4}$ (Fig.6). In the saddle-point method, we keep the contribution of each complex period along which $\mathrm{Im}\oint p\, dq \geq 0$, instead of just the multiples of the real period as before. The actions $\oint p\, dq$ along the periods also form a lattice in the complex s plane, generated by

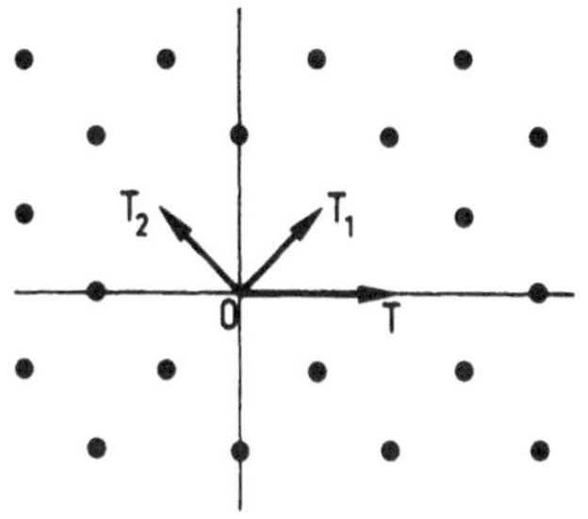

Fig.5

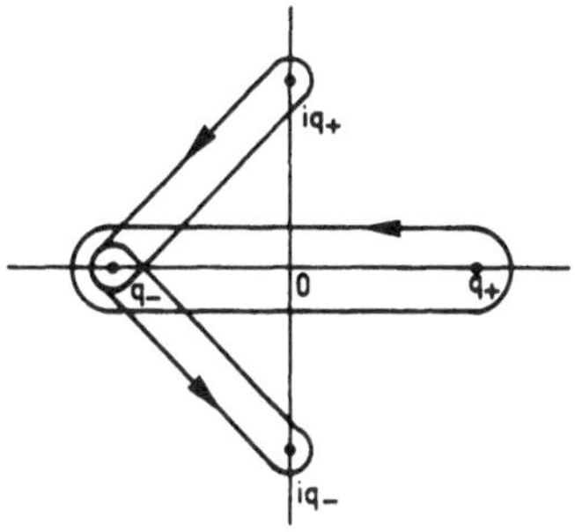

Fig.6

$$S_1 = \int_0^{T_1} p(t)\, dq(t) = \frac{\sigma+i\sigma}{2} \quad \text{and} \quad S_2 = \int_0^{T_2} p(t)\, dq(t) = \frac{-\sigma+i\sigma}{2} \tag{3.4}$$

We expect on general grounds that trajectories of periods T_1+mT ($m \in \mathbb{Z}$) will contribute terms of the order $\delta = e^{-\sigma/2\hbar}$ to the quantization rule (higher lying periods would contribute still smaller terms $e^{-\sigma/\hbar},\ldots$). But the same trajectories (and their opposites) also produce the singularities of ρ : $s_\varepsilon = \pm i(S_1+m\sigma)$ that lie closest to the dominant ones $m\sigma$ (at a distance $\sigma/\sqrt{2}$). Accordingly, the apparent accuracy of the semi-classical series is $\varepsilon \propto e^{-\sigma/\sqrt{2}\hbar} \ll \delta$. We are thus in the case where subdominant contributions *should be meaningful*. But instead of making a saddle-point expansion of (3.2), we shall rather compute them by WKB connexion formulas in the complex q plane [19,25], closely following [26].

We first place cuts in the complex q plane to define a single-valued function $p(q) = (E-q^4)^{1/2}$. For instance, we may choose the cuts and the values of $\operatorname{Arg} p = \varphi$ as in Fig.7. We then plot the *Stokes lines*, which are the lines starting from each turning point q_o, along which $\int_{q_o}^{q} p\, dq$ is real.

We initially neglect correction terms in powers of $\hbar$. In each region $\alpha = 1,2,\ldots,8$ limited by the Stokes lines, the (real) eigenfunction ψ has the asymptotic form $\psi = p^{-1/2}\, e^{-i\pi/4}\left(A_\alpha e^{i\int p\, dq} + B_\alpha e^{-i\int p\, dq}\right)$.
The WKB matching conditions relate the values of the constants A_α, B_α from one region to the adjacent one. We shall follow the changes of the constants on a distant path lying far from the cuts (dashed line on Fig.7). In region 3, because of the boundary condition for $q \to -\infty$:

$$\psi(3) = p^{-1/2}\, e^{-i\pi/4} \exp i\int_{q_-}^{q} p\, dq \quad .$$

In crossing to region 1, the turning point q_- is involved alone. The usual one

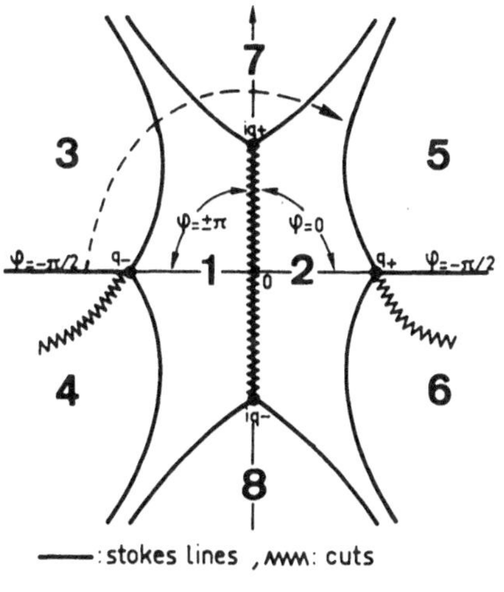

Fig.7

turning-point connexion formula implies :

$$\psi(1) = p^{-1/2}\left(e^{-i\pi/4}\ \exp\int_{q_-}^{q} p\,dq - e^{i\pi/4}\ \exp -i\int_{q_-}^{q} p\,dq\right)$$

which we rewrite as

$$\left.\begin{aligned} \psi(1) &= |p|^{-1/2}\left(A_1\ \exp i\int_{o}^{q} p\,dq + A_1^{*}\ \exp -i\int_{o}^{q} p\,dq\right) \\ \text{where}\quad A_1 &= \exp i\left(\frac{\pi}{4}-\frac{\sigma}{4}\right)\ ,\quad \frac{\sigma}{4} = -\int_{q_-}^{o} p\,dq \end{aligned}\right\} \qquad (3.5)$$

Similarly, starting from $\psi(5) = \pm p^{-1/2}\ e^{-i\pi/4}\ \exp i\int_{q_+}^{q} p\,dq$:

$$\left.\begin{aligned} \psi(2) &= |p|^{-1/2}\left(A_2\ \exp i\int_{o}^{q} p\,dq + A_2^{*}\ \exp -i\int_{o}^{q} p\,dq\right) \\ \text{with}\quad A_2 &= \pm\exp i\left(\frac{\pi}{4}-\frac{\sigma}{4}\right)\ . \end{aligned}\right\} \qquad (3.6)$$

If there were no other turning points, we could simply write $A_2 = A_1^{*}$, taking into account the fact that p(q) becomes -p(q) across the cut. But actually, to match the expressions (3.5) and (3.6), we still have to cross the system of Stokes lines associated with the turning points iq_+ and iq_- (Fig.8). The correct matching conditions must be the same as when the wave encounters an "underdense potential barrier" (Fig.9), which precisely has the turning point structure of Fig.8. Namely[26] :

$$A_2 = \sqrt{1+e^{-\sigma}}\ e^{-i\delta(\sigma)}\ A_1^{*} - ie^{-\sigma/2}\ A_1 \qquad (3.7)$$

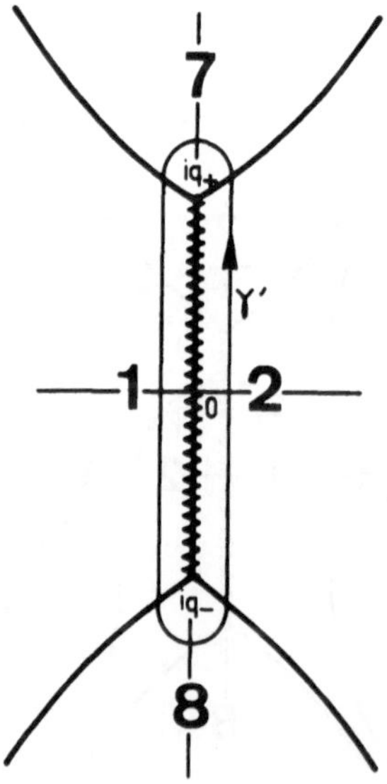

Fig.8

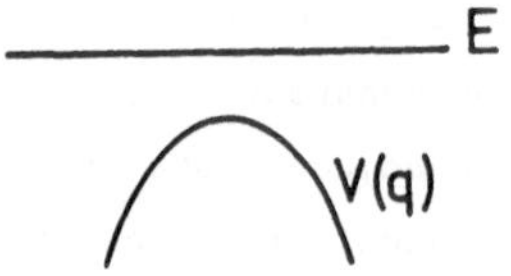

Fig.9

where the quantity $e^{-\sigma}$ arises from the contour integral $\exp i\int_{\gamma'} p\, dq$ (Fig.8), and $\delta(\sigma)$ is a phase angle which the matching procedure cannot determine in general[26]. Consistency between Eqs.(3.5), (3.6) and (3.7) leads to the quantization condition $\sqrt{1+e^{-\sigma}}\cos(\frac{\sigma}{2}-\delta(\sigma))+e^{-\sigma/2}=0$, or equivalently :

$$2\pi(k+\tfrac{1}{2}) \;=\; \sigma-\delta(\sigma)-2(-1)^k \,\mathrm{Arctg}\, e^{-\sigma/2} \tag{3.8}$$

The presence of an unknown $\delta(\sigma)$ in (3.8) seems most annoying ; fortunately we can argue that in our approach of expanding for $\sigma\to\infty$, it is consistent to let $\delta(\sigma)\equiv 0$ while incorporating power corrections in σ^{-1} into (3.8). First, in all known cases[26], $\delta(\sigma)\to 0$ as $\sigma\to\infty$, hence it can contribute only power corrections in σ^{-1} , or subdominant terms. Concerning powers of σ^{-1} , we have neglected some of them in our matching procedure, but we have already included *all of them* consistently in formula (3.1), hence $\delta(\sigma)$ is redundant in this respect. As for corrections to (3.8) in powers of $e^{-\sigma/2}$, it is reasonable to expect that in the approximation where the two turning points iq_- and iq_+ are treated as separate, the leading term (of order $e^{-\sigma/2}$) should come out correctly ; this implies that no contribution of order $e^{-\sigma/2}$ arises from $\delta(\sigma)$ in Eq.(3.8). A different argument is that for the comparison potential having no other complex turning points that the two on Fig.8 (i.e. the inverted parabolic barrier[26]), $\delta(\sigma)\equiv \mathrm{Arg}\,\Gamma(\frac{1}{2}+\frac{i\sigma}{2\pi})-\frac{\sigma}{2\pi}(\log\frac{\sigma}{2\pi}-1)$ has an asymptotic expansion $\delta\sim\frac{\pi}{12\sigma}+\frac{7\pi^3}{360\sigma^3}+\cdots$ which is known to be accurate to order $e^{-\sigma}$ when truncated at its smallest term.

Thus, in the following, we shall consistently leave out all those terms of order $e^{-\sigma}$ that might arise from $\delta(\sigma)$, or equivalently from the imperfections of our matching procedure (cooperative effects of all four turning points were overlooked). We are entitled however to include power terms in σ^{-1} . The quantization rule will then involve powers of σ^{-1} and of $e^{-\sigma/2}$ considered as independent variables.

The simplest way to write such a quantization rule is to draw the power contributions in σ^{-1} from (3.1) and the power contributions in $e^{-\sigma/2}$ from (3.8), neglecting cross terms :

$$2\pi(k+\tfrac{1}{2}) \;=\; \sigma+\frac{b_1}{\sigma}+\frac{b_2}{\sigma^3}+\ldots-2(-1)^k\,\mathrm{Arctg}\, e^{-\sigma/2} \quad . \tag{3.9}$$

As a numerical test, we consider the sequences of approximations obtained by solving Eq.(3.9) to increasing orders in $1/\sigma$ for various values of k (Table 2b). Comparison with Table 2a shows that the subdominant term in (3.9) explains indeed most of the discrepancy between the previous best estimates E_k^* and the exact values E_k .

A more consistent synthesis of Eqs.(3.1) and (3.8) should also involve terms of higher orders in $e^{-\sigma/2}$ and σ^{-1} . From the above discussion, we expect that the matching procedure, extended to higher orders in σ^{-1} , will yield safely all contributions of order $e^{-\sigma/2}\sigma^{-n}$. This is easily done if we note, as in section 2, that the analytic structure of the WKB wave functions, and hence the matching coefficients, are the same at higher orders in $\hbar$ as for the leading orders ; the only difference is that the WKB expressions are "dressed up" with power corrections in $\hbar$ (or $1/\sigma$), through the replacement everywhere of the function p(q) by the expansion u(q) of Eq.(2.7). If we now remember that in Eq.(3.8), the dominant term σ stands for the contour integral $\oint_\gamma p\,dq$ along the contour γ of Fig.4 (cf. Eq.(2.10)), whereas $e^{-\sigma/2}$ stands for $\exp i \oint_{\gamma'} p\,dq$ along the contour γ' of Fig.8, we see that the full-bodied quantization rule must be :

$$2\pi(k+\tfrac{1}{2}) = \oint_\gamma u\,dq - 2(-1)^k \,\mathrm{Arctg}\left(\exp \frac{i}{2}\oint_{\gamma'} u\,dq\right) \tag{3.10}$$

(this formula holds for any problem in which the "important" turning points have the same structure as in Fig.7). In the case of $V(q) = q^4$, we take the expressions (2.11) for $u_{2n} = (-1)^n v_{2n}$ and (2.13) for $\sigma_n = \oint_\gamma u_{2n}\,dq$, and readily find that $\oint_{\gamma'} u_{2n}\,dq = (-1)^{n+1} i\sigma_n$, hence the explicit $1/\sigma$-expansion of (3.10) reads :

$$2\pi(k+\tfrac{1}{2}) = \sigma + \frac{b_1}{\sigma} + \frac{b_2}{\sigma^3} + \frac{b_3}{\sigma^5} + \ldots - 2(-1)^k \,\mathrm{Arctg}\, e^{-\frac{1}{2}\left(\sigma - \frac{b_1}{\sigma} + \frac{b_2}{\sigma^3} - \frac{b_3}{\sigma^5} \ldots\right)} \tag{3.11}$$

The sequences of approximations obtained by solving Eq.(3.11) with both expansions truncated to the same increasing orders, are listed in Table 2c. *All discrepancies have now disappeared* : the errors on the best estimates have become of the same order as the series fluctuations. This test is sensitive enough to give a (heuristic) check for the term of order $e^{-\sigma/2}$ with its first power correction, namely $\exp\left(-\frac{1}{2}(\sigma - b_1/\sigma)\right)$ (an indirect check of the following powers will follow from the large order behaviour of the series (3.1) : see next section). However, our present computations are not accurate enough to check the validity of Eq.(3.11) to order $e^{-\sigma}$.

4. The Borel transform of the semi-classical expansion.

The following question now arises about the divergent series (3.1) : is there a natural way to sum it to a smooth function $\bar{F}(\sigma)$ such that the equation :

$$\bar{F}(\sigma_k) = 2\pi(k+\tfrac{1}{2}) \quad , \quad \sigma_k = \sigma(E_k) \tag{4.1}$$

yields the *exact* eigenvalue E_k for each k (Fig.10) ? Such an $\bar{F}(\sigma)$ would appear as an exact version of the Thomas-Fermi distribution[9]. Actually this is a very tricky question : on one hand, the existence of such a $\bar{F}(\sigma)$ is strongly suggested by the numerical accuracy of the semi-classical series, but the asymptotic series alone defines $\bar{F}(\sigma)$ only up to $\mathcal{O}(\sigma^{-\infty})$ (terms decreasing faster than any power of σ^{-1}). On the other hand, the exact relation (4.1) only defines $\bar{F}$ *at* the eigenvalues ; it does suggest an analytic relationship between energy and quantum number, but to our knowledge we cannot significantly define such a unique relationship for non-integral quantum numbers. It will thus be useful to switch to another function $F(\sigma)$, which is discontinuous but has an intrinsic meaning in the quantum theory. We define it as a "staircase" function (Fig.10) :

$$F(\sigma) = 2\pi\ \mathrm{Tr}(\sigma-\hat{I}) = 2\pi \sum_{k=0}^{\infty} \theta(\sigma-\sigma_k) \tag{4.2}$$

where θ is the Heaviside step function, $\sigma(E)$ is the action function as in Eq.(2.14) and $\hat{I} = \sigma(\hat{H})$ is the same function of the *operator* $\hat{H}$. Assuming that we have a smooth function $\bar{F}(\sigma)$ satisfying (4.1), we can recontruct $F(\sigma)$ from the following "bootstrap" relation :

$$F(\sigma) = 2\pi \sum_{k=-\infty}^{+\infty} \theta\left(\bar{F}(\sigma) - 2(k+\tfrac{1}{2})\pi\right) \tag{4.3}$$

which, by virtue of the Fourier expansion

$$2\pi \left(\sum_{k=-\infty}^{+\infty} \theta(x - 2(k+\tfrac{1}{2})\pi)\right) - x = i \sum_{\substack{m=-\infty \\ m\neq 0}}^{+\infty} \frac{(-1)^m}{m} e^{-imx} ,$$

is equivalent to :

$$F(\sigma) = \bar{F}(\sigma) + i \sum_{m\neq 0} \frac{(-1)^m}{m} e^{-im\bar{F}(\sigma)} \tag{4.4}$$

We have thus realized an explicit decomposition of $F(\sigma)$ as a sum of a smooth term $\bar{F}(\sigma)$ and of oscillating terms[9]. Moreover, the oscillating terms are completely determined by the smooth term $\bar{F}(\sigma)$.

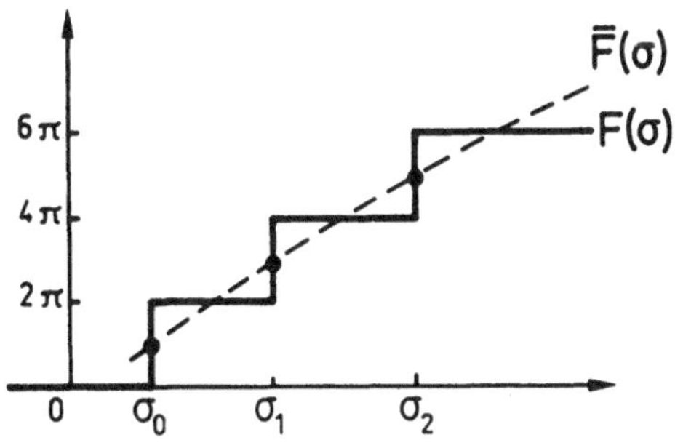

Fig.10

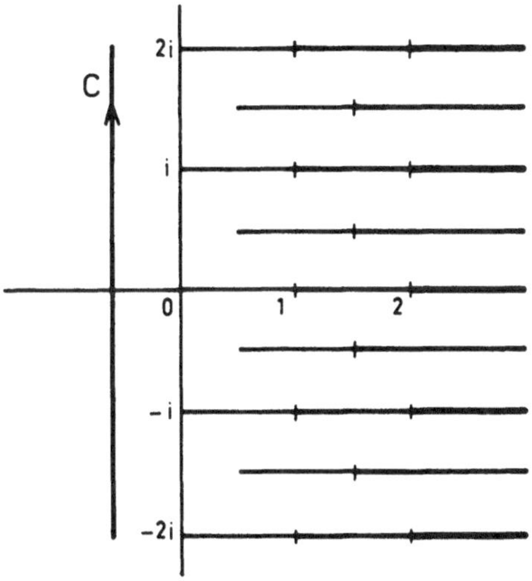

Fig.11

We now interpret this result in terms of the Laplace transform $\rho(s)$ of $F(\sigma)$ which, by Eq.(4.2), is simply :

$$\rho(s) = \frac{i}{s}\ \mathrm{Tr}\ \exp(s\hat{I}) = \frac{i}{s}\sum_{k=0}^{\infty} e^{s\sigma_k} \tag{4.5}$$

Eq.(4.5) defines ρ as a holomorphic function in the half-plane $\mathrm{Re}\, s < 0$, such that :

$$F(\sigma) = \int_C e^{-\sigma s}\ \rho(s)\ ds \tag{4.6}$$

along the contour C shown in Fig.11.

As in section 1, by pushing the contour C to the right we can express $F(\sigma)$ as a sum of contributions from the singularities s_j of $\rho(s)$, each discontinuity $\sum \rho_n^{(j)} (s-s_j)_+^n / \Gamma(n+1)$ producing a term $e^{-s_j\sigma}(\Sigma\, \rho_n^{(j)}\, \sigma^{-n-1})$.

On the other hand, by substituting into (4.4) the expansion $\bar{F}(\sigma) \sim \Sigma\, b_j \sigma^{1-2j}$ (with subdominant terms neglected), we explicitly get :

$$F(\sigma) \sim \sigma + \frac{b_1}{\sigma} + \frac{b_2}{\sigma^3} + \ldots + \sum_{m\neq 0} e^{-im\sigma}\left[\frac{(-1)^m}{m}\exp\left(-im\left(\frac{b_1}{\sigma} + \frac{b_2}{\sigma^3} + \ldots\right)\right)\right] \tag{4.7}$$

This means that $\rho(s)$ has singularities at *all integral points* on the imaginary axis. The discontinuity $\Delta_o\rho(s)$ from $s = 0$ is directly responsible for the Thomas-Fermi term $\bar{F}(\sigma)$, of which it is the Borel transform :

$$\Delta_o\rho(s) = \sum_{j=0}^{\infty} b_j\ s_+^{2j-2} / \Gamma(2j-1) \tag{4.8}$$

The discontinuities $\Delta_{im}\rho$ from the other points $s = im$ $(m \in \mathbb{Z})$ are found by identifying :

$$(-1)^m \frac{i}{m}\ \exp\left(-im\left(\frac{b_1}{\sigma} + \frac{b_2}{\sigma^3} + \ldots\right)\right) \equiv \sum \rho_\ell^{(im)}\ \sigma^{-\ell-1} \quad .$$

An equivalent expression uses the integration operator $\left(\frac{d}{ds}\right)^{-1}$ [6] :

$$\begin{aligned}\Delta_{im}\rho(im+s) &= \frac{i(-1)^m}{m}\exp\left(-im\sum_1^{\infty} b_j\left(\frac{d}{ds}\right)^{1-2j}\right)\delta(s)\\ &= (-1)^m\left(\frac{i}{m}\,\delta(s) + b_1\theta(s) - \frac{im}{2}\, b_1^2\, s_+ + \ldots\right)\end{aligned} \tag{4.9}$$

In conclusion : $\rho(s)$ has singularities distributed periodically, in order to build up the jumps of the function $F(\sigma)$; moreover, the discontinuity at $s = 0$ *generates* the other discontinuities at $s = im$ in the explicit fashion that we have just described.

We shall now see how the same thing happens with the *complex* singularities of $\rho(s)$. Actually, because of the alternative formula :

$$F(\sigma) = \frac{1}{\pi}\int_0^{\sigma} \mathrm{Im}\ \mathrm{Tr}(\hat{I} - \sigma' - i0)^{-1}\ d\sigma' \quad ,$$

we can also express $F(\sigma)$ as a path integral similar to (3.2), and deduce that the singularities of $\rho(s)$ are the actions of classical periodic orbits, multiplied by i : they form a lattice generated by $\frac{-1\pm i}{2}$ (not by $(-1\pm i)\frac{\sigma}{2}$ as in Eq.(3.4), because we now Laplace transform with respect to σ instead of $1/\hbar$). Only the singularities with positive real parts should appear on the first sheet, whose tentative structure is shown on Fig.11. The singularities at the points im ($m\in\mathbb{Z}$) arise from the real orbit traversed m times, and they were described by Eq.(4.9). We now consider the *next* row of singularities, those with $\mathrm{Re}\, s = +\frac{1}{2}$. By the same argument of contour deformation they correspond to the corrections of order $e^{-\sigma/2}$ to the function $\bar{F}(\sigma)$ in the quantization condition. Eq.(3.11) does not express directly $\bar{F}(\sigma)$ since k appears on both sides, but if we substitute once $(-1)^k = \sin\frac{\bar{F}(\sigma)}{2} + O(e^{-\sigma/2})$ in Eq.(3.11), insert the resulting modified $\bar{F}(\sigma)$ into Eq.(4.4) and collect all terms of order $e^{-\sigma/2}$, we find :

$$\begin{aligned} F(\sigma) \sim & \sum_0^\infty b_j \sigma^{1-2j} + \sum_{m\neq 0} e^{-im\sigma}\frac{(-1)^m}{m} \exp\left\{-im\sum_1^\infty b_j\,\sigma^{1-2j}\right\} \\ & + \sum_{m=-\infty}^{+\infty} e^{-i(m+\frac{1}{2})\sigma-\sigma/2}\left(-2i(-1)^m\cdot\exp\left\{-i(m+\tfrac{1}{2})\sum_1^\infty b_j\sigma^{1-2j}\right\}\cdot\exp\left\{-\tfrac{1}{2}\sum_1^\infty(-1)^j b_j\sigma^{1-2j}\right\}\right) \\ & + O(e^{-\sigma}) \end{aligned} \tag{4.10}$$

The difference between Eqs.(4.10) and (4.7) is precisely the contribution from the singularities with $\mathrm{Re}\, s = \frac{1}{2}$:

$$\sum_{m=-\infty}^{+\infty} e^{-i(m+\frac{1}{2})\sigma-\sigma/2}\int_0^\infty \Delta_{\left(\frac{1}{2}+i(m+\frac{1}{2})\right)}\,\rho\left(\tfrac{1}{2}+i(m+\tfrac{1}{2})+s\right)\, e^{-\sigma s}\, ds \tag{4.11}$$

By identification with (4.10), we find :

$$\Delta_{\left(\frac{1}{2}+i(m+\frac{1}{2})\right)}\,\rho\left(\tfrac{1}{2}+i(m+\tfrac{1}{2})+s\right) = -2i(-1)^m \exp\left\{-\sum_{j=1}^\infty\left[\frac{(-1)^j}{2}+(m+\tfrac{1}{2})i\right]b_j\left(\frac{d}{ds}\right)^{1-2j}\right\}\delta(s) \tag{4.12}$$

We thus see that those discontinuities are *also generated by the* $s=0$ *discontinuity*, as they involve the same coefficients b_j. Clearly, by further using Eq.(3.11) to all orders in $e^{-\sigma/2}$ we could similarly express all the other discontinuities of $\rho(s)$ on its first sheet. However, we cannot ascertain that Eq.(3.11) is correct to higher orders in $e^{-\sigma/2}$ as it stands, so we have not pursued that computation. We rather believe that a better understanding of all the discontinuities should involve some group action of the lattice of classical periods (real and complex) on the quantum function $\rho(s)$. An interesting consequence of this lattice structure is that it no longer makes sense to isolate $\bar{F}(\sigma)$ as "the purely non-oscillating" term in $F(\sigma)$ (the subdominant terms *are* oscillatory) and moreover, any tentative definition of $\bar{F}(\sigma)$ from $\rho(s)$ will raise problems of Borel summability, as each cut carries an infinite sequence of new branch points (Fig.11). Our analysis with Eq.(4.4) remains valid anyhow : it only involves an asymptotic representation for *some* $\bar{F}(\sigma)$ satisfying (4.1).

We now give an application (and a check) of Eq.(4.12) by considering the large order behavior of the b_j . The latter arise as coefficients in the expansion of the discontinuity $\Delta_o\rho(s)$ (Eq.(4.8)), therefore their large order behavior is controlled by the nearest singularities of $\rho(s)$, namely the four points $s_\varepsilon = \frac{(\pm 1 \pm i)}{2}$. Two of those points : $\frac{1 \pm i}{2}$, lie on the first sheet and the corresponding discontinuities are given by Eq.(4.12) with $m = 0$ and -1 . We now want to apply the large order formula (1.13). At first sight it is inadequate since $\rho(s)$ has logarithmic discontinuities whereas (1.13) involved square-root discontinuities. Fortunately, relations like (1.13) between various discontinuities of $\rho(s)$ are preserved if we multiply $F(\sigma)$ by any factor σ^α , i.e. if we apply a fractional derivative $(d/ds)^\alpha$ [6] to all discontinuities of $\rho(s)$. By selecting an (arbitrary) *half-integer* for α, we can transform $\rho(s)$ to a function $\rho'(s)$ with the desired structure (at least formally), apply (1.13) to ρ' , and express the result in terms of the original function ρ . The discontinuity $\Delta_\varepsilon \rho(s_\varepsilon + s) = \sum_{k=0}^{\infty} \rho_k^{(\varepsilon)} s_+^k / \Gamma(k+1)$ at $s = s_\varepsilon$ gives the following contribution to the large order terms in $\Delta_o \rho(s) = \sum_{k=0}^{\infty} \rho_k^{(0)} s_+^k / \Gamma(k+1)$:

$$\rho_k^{(0)} \sim (2i\pi)^{-1} s_\varepsilon^{-k} \sum_{\ell=0}^{\infty} \rho_\ell^{(\varepsilon)} s_\varepsilon^\ell \Gamma(k-\ell) \tag{4.13}$$

In our case, the contribution of the two points $s_\varepsilon = \frac{1 \pm i}{2}$ to the large order behavior of $b_j = \rho_{2j-1}^{(0)}$ has the form :

$$b_j \sim \Gamma(2j-1) \frac{2^{j+1/2}}{\pi} \cos\left(j\frac{\pi}{2} + \frac{3\pi}{4}\right) \left(1 + \frac{\alpha_1}{2j-2} + \frac{\alpha_2}{(2j-2)(2j-3)} + \dots\right)$$

We cannot directly evaluate the contribution of the two points $s_\varepsilon = \frac{-1 \pm i}{2}$ on the other sheet but, because the original Hamiltonian (2.1) is even in $\hbar$, we expect the discontinuities of $\rho(s)$ to be symmetrical under $s \to -s$; this is also necessary in order that $\rho_{2j}^{(0)} \sim 0$ at the end . In particular, the contribution of the two points $\frac{-1 \pm i}{2}$ to the large orders of b_j should be equal to that of the points $\frac{1 \pm i}{2}$ and should add to it (indeed, the empirical law (2.17) shows that the previous formula for b_j was too small by a factor 2). The final correct formula is thus

$$b_j \sim \Gamma(2j-1) \frac{2^{j+3/2}}{\pi} \cos\left(j\frac{\pi}{2} + \frac{3\pi}{4}\right) \left(1 + \frac{\alpha_1}{2j-2} + \frac{\alpha_2}{(2j-2)(2j-3)} + \dots\right) \tag{4.14}$$

where, due to Eqs.(4.12-13), the α_ℓ admit the generating function :

$$\sum_0^\infty \alpha_j t^j \equiv \exp\left(\frac{b_1 t}{2} + \frac{b_2 t^3}{2^2} - \left(\frac{b_3 t^5}{2^3} + \frac{b_4 t^7}{2^4}\right) + \frac{b_5 t^9}{2^5} + \frac{b_6 t^{11}}{2^6} - \dots\right) \tag{4.15}$$

$$\left(\alpha_1 = \frac{b_1}{2} = -\frac{\pi}{6}, \quad \alpha_2 = \frac{b_1^2}{8} = \frac{\pi^2}{72}, \quad \alpha_3 = \frac{b_1^3}{48} + \frac{b_2}{4} = -\frac{\pi^3}{1296} + \frac{11\,\Gamma(1/4)^8}{41472\,\pi^2} \dots\right)$$

Table 3 shows a good agreement between the values of α_ℓ ($\ell \le 7$) derived from Eq.(4.15) and those found by a numerical fit on the b_j' of Table 1. Moreover (4.14) predicts the correct *sign* of b_j (starting from $j = 2$). All this confirms the validity of formula (4.14), and indirectly of the correction of order $e^{-\sigma/2}$ (with power terms included) in the quantization condition (3.11).

We further note that a curious reciprocity law seems to apply to semi-classical expansions. On one hand (and on general grounds), the large order behavior of $\rho_j^{(0)}$ is governed by the first terms $\rho_j^{(\varepsilon)}$ of the nearest discontinuity (with the notations of section 1). On the other hand, and this is a peculiarity of the WKB expansion, all the discontinuities with $s \neq 0$ are generated by the one at $s = 0$. So that on the whole, the behavior of b_j $(j \to \infty)$ is governed by the first b_j themselves, as shown by Eqs.(4.14-15). It is interesting to note that a similar property was derived in a different way by Dingle, again in the case of a semi-classical expansion, this time for the wave function near a turning point (cf.[3], chap.14.3, Eq.(27)).

To conclude, let us discuss possible generalizations of the foregoing results. Apart from the particularly simple dependence of $\bar{F}(\sigma)$ on σ, the reconstruction of $F(\sigma)$ from $\bar{F}(\sigma)$ and the subsequent analysis of the singularities of $\rho(s)$ on $\mathrm{Re}\, s = 0$ are probably valid for arbitrary *one-dimensional, single-well* potentials, but the analytic continuation to $\mathrm{Re}\, s > 0$ will be very difficult except for a general *quartic* potential, in which case the classical trajectories are still elliptic functions. For multi-dimensional but *integrable* systems, our analysis can probably be transposed using a full set of action variables.

As for non-integrable systems, it follows from the general analysis of Balian and Bloch[9] that the singularities of $\rho(s)$ lie at the points $i\times\{$actions along periodic orbits$\}$ as before, but the actual structure of periodic orbits is unknown and probably quite complicated. A related family of investigations concerns the Laplace operator Δ either in a bounded cavity[27], or on a compact Riemannian manifold[28]. In both cases, the analytic function under study is $\rho(s) = \mathrm{Tr}\exp s\sqrt{-\Delta}$, $t = is$ is now the actual time variable and the singularities of ρ on $\mathrm{Re}\, s = 0$ lie at the *periods* of classical closed orbits ; the corresponding discontinuities obey explicit formulas much more complicated than Eq.(4.9)[28]. An additional difficulty in non-integrable systems is that the singularities may be dense on the imaginary axis ; even without that, the problem of analytically continuing $\rho(s)$ across $\mathrm{Re}\, s = 0$ has not been raised so far.

Acknowledgements : One of us (A.V.) has benefited from helpful discussions with C.M. Bender (who communicated his results of Ref.[12] before publication), M.V. Berry and J. Zinn-Justin. We are also grateful to J. Lascoux and T.T. Wu for stimulating discussions, to R. Schaeffer who helped us derive Eq.(3.9), and finally to Mrs. N. Tichit, Mrs. C. Verneyre and R. Conte for their assistance in the computer calculations.

References

1. G. Darboux, J. Math. 4 (1878) 5, 377.
2. H. Poincaré, Acta Math. 8 (1886) 295 ; W. Wasow, "Asymptotic Expansions for Ordinary Differential Equations" (Wiley, New York, 1965) ; F.W.J. Olver, "Asymptotics and Special Functions" (Academic Press, New York, 1974).
3. R.B. Dingle, "Asymptotic Expansions : their Derivation and Interpretation" (Academic Press, London, 1973).
4. J. Zinn-Justin, Princeton 1978 Lecture Notes (unpublished).
5. R. Balian, G. Parisi and A. Voros, Phys. Rev. Lett. 41 (1978) 1141.
6. I.M. Gelfand, G.E. Shilov, "Generalized Functions", vol.1 (Academic Press, 1968).
7. C.M. Bender, T.T. Wu, Phys. Rev. D7 (1973) 1620 ; L.H. Lipatov, JETP 72 (1977) 411 ; E. Brézin, J-C. Le Guillou and J. Zinn-Justin, Phys. Rev. D15 (1977) 1554, 1558 ; G. Parisi, Phys. Lett. 66B (1977) 167.
8. J. Leray, Bull. Soc. Math. Fr. 87 (1959) 81 ; D. Fotiadi, M. Froissart, J. Lascoux and F. Pham, Topology 4 (1965) 159.
9. R. Balian, C. Bloch, Ann. Phys. 63 (1971) 592 ; 85 (1974) 514.
10. For instance : L.I. Schiff, Phys. Rev. 92 (1953) 766 ; C. Schwartz, Ann. Phys. 32 (1965) 277 ; C.E. Reid, J. Molec. Spectrosc. 36 (1970) 183 ; P.M. Mathews, K. Eswaran, Lett. Nuovo Cimento 5 (1972) 15 ; F.T. Hioe, E.W. Montroll, J. Math. Phys. 16 (1975) 1945.
11. A. Voros, Thèse, Université Paris-Sud (Orsay, 1977).
12. C.M. Bender, K. Olaussen and P.S. Wang, Phys. Rev. D16 (1977) 1740.
13. A. Erdélyi et al., "Higher Transcendental Functions" vol.2 (Bateman Manuscript Project, McGraw Hill, New York, 1953) ; W. Magnus, F. Oberhettinger, R.P. Soni, "Formulas and Theorems for the Special Functions of Mathematical Physics" (Springer Verlag, 1966) ; M. Abramowitz, I.A. Stegun, "Handbook of Mathematical Functions"(Dover, New York).
14. For instance : A. Messiah, "Mécanique Quantique" vol.1, ch.6 (Dunod, Paris, 1959 ; English Translation : North-Holland, 1961) ; N. Fröman, Ark. för Fysik 32 (1966) 541.
15. G. Wentzel, Z. Phys. 38 (1926) 518 ; J.L. Dunham, Phys. Rev. 41 (1932) 713.
16. A. Voros, Ann. Inst. H.Poincaré 26A (1977) 343.
17. J.A. Campbell, J. Comput. Phys. 10 (1972) 308 ; 15 (1974) 413 and refs. therein.
18. N. Fröman, P.O. Fröman, J. Math. Phys. 18 (1977) 96.
19. N. Fröman, P.O. Fröman, "JWKB-Approximation, Contributions to the Theory" (North-Holland, Amsterdam, 1965).
20. E.C. Titchmarsh, "Eigenfunctions Expansions" vol.1 (Oxford Univ. Press, 1961) ; V.P. Maslov, "Théorie des Perturbations et Méthodes Asymptotiques" (Dunod, Paris, 1972) ; J.P. Eckmann, R. Sénéor, Arch. Rational Mechanics 61 (1976) 153.
21. A.C. Hearn, "REDUCE User's Manual" (University of Utah, 1973).
22. P. Bonche, M. Froissart, J-F. Renardy, "Une Chaîne de Programmes d'Arithmétique à Longueur Variable sur IBM-360" (Note CEA-N-1247, Saclay, 1970).
23. M.C. Gutzwiller, J. Math. Phys. 12 (1971) 343 ; R. Dashen, B. Hasslacher, A. Neveu, Phys. Rev. D10 (1974) 4114.
24. A. Neveu, Rep. Progr. Phys. 40 (1977) 709.
25. J. Heading, "An Introduction to Phase-Integral Methods" (Methuen, London, 1962).
26. M.V. Berry and K.E. Mount, Rep. Progr. Phys. 35 (1972) 315.
27. R. Balian, C. Bloch, Ann. Phys. 69 (1972) 76.
28. Y. Colin de Verdière, Comptes Rendus Acad. Sci. 275 (1972) 805 and 276 (1973) 1517 ; J. Chazarain, Inventiones Math. 24 (1974) 65 ; J.J. Duistermaat and V.W. Guillemin, Inventiones Math. 29 (1975) 39.
29. G. Parisi, "Trace Identities for the Schrödinger Operator and the WKB Method", Preprint LPTENS 78-9 (Ecole Normale Supérieure, Paris, March 1978).

n	b'_n	n	b'_n	n	b'_n
0	-	18	0.62690462618350	36	-0.63187742695115
1	-0.52359877559830	19	0.62743981578357	37	-0.63200858341836
2	0.40119597249644	20	-0.62792002257294	38	0.63213268306740
3	0.49832256904969	21	-0.62835082264362	39	0.63225028032241
4	-0.61771117216663	22	0.62874081905774	40	-0.63236187301556
5	-0.61443022626847	23	0.62909616407778	41	-0.63246790944398
6	0.59570714354132	24	-0.62942091956050	42	0.63256879437991
7	0.60522456343627	25	-0.62971871161799	43	0.63266489422160
8	-0.61578165645471	26	0.62999285275772	44	-0.63275654143407
9	-0.61744255110996	27	0.63024609031867	45	-0.63284403838861
10	0.61789138060000	28	-0.63048070231369	46	0.63292766069161
11	0.61995452512628	29	-0.63069865940814	47	0.63300766008082
12	-0.62185534397455	30	0.63090167915462	48	-0.63308426695388
13	-0.62300181061281	31	0.63109124785922	49	-0.63315769258223
14	0.62392636719298	32	-0.63126865722127	50	0.63322813105504
15	0.62483581881424	33	-0.63143503956354	51	0.63329576099071
16	-0.62563788089519	34	0.63159139189047	52	-0.63336074704796
17	-0.62631183052424	35	0.63173859438058	53	-0.63342324126315

Table 1 - Computed values of $b'_n = b_n \times 2^{-n} / (2n-2)!$

k = 0 $E_o = 1.060$

j	(a)	(b)	(c)
0	0.87*	1.00*	1.00*
1	0.98*	1.09*	1.08*
3	0.79	1.01	1.00
5	1.40	1.43	1.42

k = 3 $E_3 = 11.644\ 745\ 51$

j	(a)	(b)	(c)
0	11.611 525 3	11.611 501 7	11.611 501 7
1	11.644 989 5	11.644 966 4	11.644 967 0
3	11.644 765 8	11.644 742 7	11.644 743 3
5	... 768 2	... 745 2	... 745 7
7	... 767 9 *	... 744 9 *	... 745 4 *
9	... 768 1 *	... 745 0 *	... 745 6 *
11	... 767 9	... 744 8	... 745 4

k = 6 $E_6 = 26.528\ 471\ 183\ 682$

j	(a)	(b)	(c)
0	26.506 335 510 963	26.506 335 513 306	26.506 335 513 306
1	26.528 512 551 758	26.528 512 554 070	26.528 512 554 040
3	26.528 471 147 158	26.528 471 149 470	26.528 471 149 441
5	... 181 652	... 183 964	... 183 935
7	... 181 390	... 183 703	... 183 673
9	... 181 401 *	... 183 713 *	... 183 684 *
11	... 181 399 *	... 183 712 *	... 183 682 *

Table 2 - Semi-classical estimates for a sample of energy levels of the quartic oscillator, using : a) the standard WKB series (3.1) ; b) the same with an exponential term added (3.9) ; c) the same with the complete subdominant expansion added (3.11). The best estimates (up to the computed order) are shown by * . Exact levels are borrowed from Reid[10] and Ref.[12] (the latter contains a misprint for E_6).

ℓ	α_ℓ (theory)	α_ℓ (fit)
0	1	0.999 999 99
1	- 0.523 598 776	- 0.523 598 7
2	0.137 077 839	0.137 08
3	0.778 467 349	0.778 49
4	- 0.416 999 718	- 0.416 3
5	-11.850 079	-11.9
6	6.564 854	6.57 (±0.02)
7	442.947	437 (±3)

Table 3 - Coefficients α_ℓ of the large order expansion (4.14) for b_j as given by theory (Eq.(4.15)) and by a numerical fit based on the values of Table 1 . Discrepancies are compatible with computer noise.

PERTURBATION THEORY AT LARGE ORDERS

C. Itzykson*

Service de Physique Théorique
C.E.A.-Saclay, BP n°2, 91190 Gif-sur-Yvette, France

*Presented at the Conference on "Mathematical Problems in Feynman Path Integrals", Marseille June 1978

1. INTRODUCTION

Despite the tremendous successes of Quantum Field Theory in predicting electrodynamic interactions to an amazing degree of accuracy it was realized very early by Dyson[1] that the perturbative methods, the only systematic ones, were providing only asymptotic divergent series. While this is not a serious problem in this case due to the smallness of the fine structure constant α, it is clearly troublesome if one attempts to extend field theoretic models in the strong coupling regime as required by present applications to particle physics or statistical mechanics. Moreover even in the electrodynamic case it would be of interest to obtain estimates for moderately large orders in perturbation theory if only to asses the magnitude of errors without entering into detailed computations of Feynman graphs.

Dyson's argument gives the essence of the physical phenomenon underneath this mathematical property. The coupling constant measures the magnitude of the repulsive interaction between like charges. Perturbation theory however ignores the sign of α. A negative value would correspond to an unstable vacuum with respect to the creation of electron positron pairs. Consequently the type of singularity of physical quantities at zero-coupling constant is intimately connected to possible instabilities in field theory.

Jaffe gave the first proof that[2] perturbation theory diverges for super-renormalizable theories. Similar divergencies occur quantum mechanics and were investigated by Bender and Wu for the anharmonic oscillator[3]. The use of path integrals to provide heuristic means of obtaining large orders estimates was initiated by Langer[4] and Lam[5]. The type of divergences encountered, a factorial growth with the order was suggestive of Borel resumation methods[6] which were proved to be valid in the anharmonic oscillator case and in super-renormalizable theory[7].

The work of Lipatov[8] which extended these methods to renormalizable models motivated a new step forward which was developed in the work of Brézin,

Le Guillou, Zinn-Justin[9] and Parisi[10]. Many questions on the precise contributions of the subtraction scheme are still open at present. The mathematical nature of the factorial growth is easily exhibited on the generating function

$$Z(g) = \frac{1}{\sqrt{2\pi}} \int_{-\infty}^{+\infty} d\varphi \exp-\left\{ \frac{\varphi^2}{2} + g\varphi^4 \right\} \tag{1}$$

The formal expansion of Z(g) in powers of g has coefficients equal to the number Z_K of vacuum Feynmann diagrams of order k in a standard φ^4 theory

$$Z \sim \sum_0^\infty g^k Z_k$$
$$Z_K = (-4)^K \frac{\Gamma(2k+1/2)}{\sqrt{\pi}\, k!} \tag{2}$$

which behaves as k! for large k. It is obvious that the point $g=0$ is an essential singularity for Z(g) which can be exhibited as a Borel transform using the "action" $\frac{\varphi^2}{2} + g\varphi^4$ as a new variable

$$Z(g) = \int_0^\infty dt\, t^{1/2} B(gt)\, e^{-t}$$
$$B(t) = \frac{2}{\sqrt{\pi}} \sqrt{1-u(t)} \tag{3}$$
$$u(t) = \frac{\sqrt{1+16t}-1}{\sqrt{1+16t}+1}$$

The function B(t) admits a convergent power series in t

$$B(t) = \sum_0^\infty \frac{Z_K\, t^k}{\Gamma(K+3/2)} \tag{4}$$

within a circle of radius $|t| = 1/16$. Here the choice of $\Gamma(k+3/2)$ in the definition (4) in order to define a convergent series is a matter of convenience. In this example B(t) is analytic in a cut plane along the real t axis from $-\infty$ to $-1/16$ and the transformation $t \to u(t)$ maps the cut plane onto the unit circle. We are in the so-called Borel sumable case where the coefficients Z_k, i.e. the terms of the perturbation series uniquely specify the function Z(g). In mathematical terms this means that the analytic function Z(g) is bounded in a the vicinity of $g=0$ within an angle $-\frac{\pi}{2} - \varepsilon < \mathrm{Arg}\, g < +\frac{\pi}{2} + \varepsilon$.

That the number of diagrams to a given order k grows like k! is obvious from Wick's theorem applied to the Gaussian integrals

$$Z_K = \frac{(-1)^k}{k!} \sqrt{\frac{1}{2\pi}} \int_{-\infty}^{+\infty} d\varphi\, \varphi^{4k} \exp-\{\frac{\varphi^2}{2}\} \tag{5}$$

so that the dominant behavior is easily understood. But it is also instructive for further developments to rewrite Z_K as

$$Z_K = \frac{(-1)^k}{k!}\sqrt{\frac{2}{\pi}}\int_0^\infty d\varphi \exp-\left\{\frac{\varphi^2}{2} - 4k \ln\varphi\right\} \tag{6}$$

and to apply the steepest descent method to estimate it for large k. The saddle point is located at

$$\varphi_c^2 = 4k$$

and integration over quadratic deviations from φ_c yields

$$Z_K \sim \frac{(-1)^k}{k!}\sqrt{2}\exp\{2k \ln 4k - 2k\} \tag{7}$$

in agreement with (2). This method will be generalized to field theory.

Knowledge of the asymptotic behavior together with an explicit (and painful) computation of the first few orders may be combined in order to extract quantitative results for moderately large values of g. The strategy uses the asymptotic growth to find the a Borel transform together with the location of its nearest singularity. A cut plane may then be maped on a unit circle and the information on the first few terms translated in the first Taylor coefficients of this series. The latter may then be approximated via rational functions (Padé approximation) and the inverse transformation performed. This procedure yields spectacular results.

Let us also mention that generating functions to compute the number of diagrams may be easily written for more involved case than the self-coupled scalar field. In ordinary fermionic QED with j and η as the (c-number) sources for photon and electron external lines the analog of (1) incorporating Furry's theorem is

$$Z(j,\bar{\eta},\eta) = \int \frac{dA}{\sqrt{1-e^2A^2}} \exp\left\{-\frac{A^2}{2} + \bar{\eta}\ \frac{1}{1-eA}\ \eta + j\ A\right\} \tag{8}$$

The integral is meaningful for negative e^2. One finds for instance that the photon and electron propagators G and S (related to the vacuum polarization $\bar{\omega}$ and self-energy Σ) are equal

$$G = \frac{1}{(1-\bar{\omega})} = S = \frac{1}{1-\Sigma} = -2z\left(1 + \frac{d}{dz}\ln K_0(z)\right)$$

$$Z = -\frac{1}{4e^2} \tag{9}$$

Here $K_0(z)$ is the modified Bessel function whose asymptotic expansion for large z is known yielding

$$G = S = 1 + e^2 + 4e^4 + 25e^6 + 208e^8 + 2146e^{10} + 26\,368e^{12} + \dots$$

$$\bar{\omega} = \Sigma = e^2 + 3e^4 + 18e^6 + 153e^8 + 1638e^{10} + 20\,898e^{12} + \dots \tag{10}$$

Similarly the generating function for irreducible vertex diagrams Γ is

$$\Gamma = 4z(1-S)S^{-3} = 1 + e^2 + 7e^4 + 72e^6 + 891e^8 + 12\,672e^{10} + \dots \tag{11}$$

All these expansions can readily be shown to have a factorial growth similar to the one investigated above.

2. ANHARMONIC OSCILLATOR

Consider the perturbative calculation of the ground-state energy E(g) of the anharmonic oscillator with Hamiltonian

$$H = \frac{1}{2}\left(-\frac{\partial^2}{\partial\varphi^2} + \varphi^2\right) + g\,\varphi^4 \tag{12}$$

The free energy at an inverse temperature β

$$F = -\frac{1}{\beta}\ln(\mathrm{tr}\, e^{-\beta H}) \tag{13}$$

goes into the ground state energy as $\beta \to \infty$ and the partition function may be expressed as a path integral

$$Z(g,\beta) = \mathrm{tr}\, e^{-\beta H} = \int \mathcal{D}(\varphi)\, \exp - \int_{-\beta/2}^{\beta/2} dt \left(\frac{\dot{\varphi}^2+\varphi^2}{2} + g\,\varphi^4\right) \tag{14}$$

using as integrand the exponential of an Euclidean action $\int_{-\beta/2}^{\beta/2} dt\ \frac{\dot{\varphi}^2+\varphi^2}{2} + g\varphi^4$

and integrating over periodic functions $\varphi(-\beta/2) = \varphi(\beta/2)$. We have

$$E(g) - E(0) = \lim_{\beta\to\infty} -\frac{1}{\beta}\ln Z(g,\beta)/Z(0,\beta) \tag{15}$$

The expansion of $Z(g,\beta)/Z(0,\beta)$ in powers of g reads

$$Z(g,\beta)/Z(0,\beta) = \sum_0^\infty Z_K\, g^k$$
$$Z_K = \frac{(-1)^k}{k!}\frac{1}{Z(0)}\int \mathcal{D}(\varphi)\, \exp - \left\{ \int_{-\beta/2}^{\beta/2} \frac{\dot{\varphi}^2+\varphi^2}{2} - k\ln \int_{-\beta/2}^{\beta/2} dt\ \varphi^4 \right\} \tag{16}$$

For large k we look for a saddle point $\varphi_c(t)$ of the effective action in the exponential. With

$$\varphi_c(t) = \left(\frac{4k}{\int_{-\beta/2}^{\beta/2} dt\ \tilde{\varphi}^4(t)}\right)^{1/2} \tilde{\varphi}(t) \tag{17}$$

it satisfies the equation

$$\ddot{\tilde{\varphi}} - \tilde{\varphi} + \tilde{\varphi}^3 = 0 \tag{18}$$

invariant under time translation and symmetry $\tilde{\varphi} \to -\tilde{\varphi}$.

In the limit $\beta \to \infty$ the solution corresponding to the highest saddle point is readily found to be

$$\tilde{\varphi}(t) = \frac{\sqrt{2}}{\cosh t} \tag{19}$$

up to a sign or a translation in time. This continuous degeneracy is the origin of a technical difficulty in evaluating the contribution of quadratic deviations from the "instanton" (19). One has to separate explicitly the zero mode corresponding to time translation yielding a factor β upon integration. Setting :

$$\varphi = \varphi_c + \chi \qquad \psi = \frac{\dot{\tilde{\varphi}}}{\|\dot{\tilde{\varphi}}\|} \qquad u = \frac{\sqrt{3}}{2}\tilde{\varphi}^3 \tag{20}$$

expanding the effective action up to second order in χ one finds

$$Z_k \sim \beta \frac{(-1)^k}{k\,!} \; 2\,k^{1/2} \; \exp\{k \ln 3\, k^2 - 2k\}$$

$$\times \frac{\int \mathcal{D}\chi \; \delta(\langle\psi|\chi\rangle) \exp\left\{-\frac{1}{2}\langle\chi|(K+1+|u\rangle\langle u|)|\chi\rangle\right\}}{\int \mathcal{D}\chi \; \exp\left\{-\frac{1}{2}\langle\chi|(K_0+1)|\chi\rangle\right\}} \tag{21}$$

with

$$K = -\frac{d^2}{dt^2} - \frac{6}{\cosh^2 t} \qquad K_0 = -\frac{d^2}{dt^2} \tag{22}$$

The spectrum of (K+1) contains a discrete zero eigenvalue with eigen function ψ and an other discrete negative eigenvalue with eigenvalue u. The result of the integration (21) is the inverse square root of a Fredholm determinant in the subspace orthogonal to ψ which can be evaluated using various techniques[9] including the exact scattering solutions corresponding to the operator (22). The result reads

$$Z_K = \beta\, \Gamma\left(k + \frac{1}{2}\right)(-3)^k \left(\frac{6}{\pi^3}\right)^{1/2}\left[1 + 0\left(\frac{1}{K}\right)\right]$$

for large β. Due to this factorial increase of Z_K the corresponding coefficients in the expansion of E(g) given by (15)

$$E(g) = \frac{1}{2} + \sum_1^\infty g^k E_k$$

are found to be

$$E_K = (-1)^{k+1}\, \Gamma\left(k + \frac{1}{2}\right) 3^k \left(\frac{6}{\pi^3}\right)^{1/2}\left[1 + 0\left(\frac{1}{K}\right)\right] \tag{23}$$

in agreement with the expression given by Bender and Wu[3]. The corrective terms already found by these authors can also be derived using perturbative methods[11] by keeping higher order terms in the χ-expansion of the effective action and the bracket is found equal to $\left[1 - \frac{95}{72}\frac{1}{k} + \ldots\right]$.

Similar methods readily extend to other quantum mechanical systems. The existence of euclidean instantons translates in mathematical terms the existence of a tunnelling amplitude computed in a semi-classical fashion out of an unstable vacuum here for unphysical (negative) values of the coupling constant.

3. BOSONIC FIELD THEORIES

An example of a similar treatment in field theory is given by a self interacting field φ in d-dimensional space time. The generating functional for Euclidean Green's functions is

$$Z(j) = \int \mathcal{D}(\varphi) \exp\{- S_0(\varphi) - g\, S_1(\varphi) - \int d^d x\, j\, \varphi\} \tag{24}$$

A normalization factor is included in the measure and $S_0(S_1)$ denotes the free (interacting) part of the action. The source term involving j is always understood in the sense of a power series. The perturbative expansion reads

$$\begin{aligned} Z(j) &= \Sigma(-g)^k\, Z_K(j) \\ Z_K(j) &= \frac{1}{k!} \int \mathcal{D}(\varphi)\, S_1(\varphi)^k \exp\{-S_0(\varphi) - \int d^d x\, j\, \varphi\} \end{aligned} \tag{25}$$

and for any Green's function will be expressed as a sum of contributions of Feynman diagrams of order k. For the time being we shall assum a suitable regularization implying some ultraviolet cutoff Λ and we study the limit of large but fixed Λ and k going to infinity.

Classical stability requires $S_1(\varphi)$ to be bounded from below so that we may assume it to be non-negative. Under these circumstances we can interpret the saddle point method as applied to Eq. (24) in the following way in terms of purely real quantities. One looks for the set of classical fields φ for which $S_1(\varphi)$ is maximal for a given value of $S_0(\varphi)$. For simplicity we consider a normalizable case where

$$\begin{aligned} S_0(\varphi) &= \int d^d x \left[\frac{1}{2} (\partial\varphi)^2 + \frac{m_0^2}{2} \varphi^2\right] \\ S_1(\varphi) &= \int d^d x\, (\varphi^2)^N \end{aligned} \tag{26}$$

The condition of renormalizability is

$$N = \frac{d}{d-2} \tag{27}$$

so that we find the φ^4 theory in four dimensions for instance. In his original work Lipatov studied in particular the limit $N \to \infty$, $d \to 2$.

The answer to the maximum problem raised above is afforded by the following inequality due to Sobolev[12]

$$[S_1(\varphi)]^{1/N} \leq \frac{8}{d(d-2)} \sigma_d^{-2/d} S_0(\varphi) \tag{28}$$

where σ_d is the area of the unit sphere in $d+1$ dimensions

$$\sigma_d = 2 \frac{\pi^{\frac{d+1}{2}}}{\Gamma\left(\frac{d+1}{2}\right)} \tag{29}$$

This inequality controls the potential part of the action in terms of its kinetic part. For $m_0 > 0$ it cannot be saturated (eventhough the constant cannot be improved) while for $m_0 = 0$ it can on the conformal invariant set of classical fields

$$\varphi_{a,\lambda}(x) = \left[\frac{2\lambda}{\lambda^2+(x-a)^2}\right]^{(d-2)/2} \tag{30}$$

These can be seen to be the solutions to the corresponding variational problem in the massless case i.e. they correspond to the saddle points discussed previously.

The Sobolev inequality suggests to perform a Borel transformation by introducing the function

$$B(t) = \sum_k Z_K(-t)^k \frac{k!}{Nk!} = \sum_{p=0}^{N-1} \int \mathcal{D}(\varphi) \exp\left\{-S_0(\varphi) - \exp\left\{\frac{2i\pi p}{N t}\right\} S_1^{1/N}(\varphi) - \int d^dx\, j\varphi\right\} \tag{31}$$

From (28) this should converge in a circle of radius

$$|t| < \frac{1}{8} d(d-2)\sigma_d^{2/d} \tag{32}$$

with a singularity arising in the massless case from integrating in the vicinity of the instantons (30). This yields for the positive Z_K the estimate

$$Z_K \sim \frac{(\mu k)!}{k!} \left(\frac{8}{d(d-2)} \sigma_d^{-2/d}\right)^{kN} \tag{33}$$

To go beyond this leading estimate requires Gaussian integration over quadratic fluctuations around the instantons. Again zero modes have to be treated apart. The resulting integration restores translational invariance on one hand while the one over dilatations requires the introduction of a renormalization scale. It is noteworthy that this renormalization procedure seems consistent in the following sense. The renormalization counterterms required to go beyond the leading approximation in k are of lower order in k than the leading terms in the action of the saddle point. We refer for details to the authors of references [8] and [9].

A very impressive application of these techniques has been given by Le Guillou and Zinn-Justin in the case of the φ^4 theory in three dimensions[13] in their computation of critical indices.

It is of course interesting to extend this program to QED where we face problems of statistics and gauge invariance. First we investigate scalar electrodynamics and neglect the mass of the charged scalar field[14]. The generating functional reads in that case

$$Z = \int \mathcal{D}(A,\varphi,\varphi^*) \exp - \int d^4x \left\{ \frac{F^2}{4} + \frac{\lambda}{2} (\partial . A)^2 + (\partial + ie A) \varphi^* . (\partial - ie A)\varphi + j\varphi^* + j^* \varphi + J . A \right\} \tag{34}$$

Here A_μ is the potential $F_{\mu\nu} = \partial_\mu A_\nu - \partial_\nu A_\mu$ the field intensity and λ is a gauge fixing parameter. Using the Sobolev inequalities it is possible to show a priori that the (regularized) perturbation theory coefficients cannot grow faster than a factorial of the order. In this case however the use of simultaneous inequalities forbids to find by these means a common saturating set of fields and one has to resort to a numerical investigation of the saddle point equations. It is convenient to use a combined integration over fields and coupling constant to extract the k-th order term from a Cauchy integral over e. In this way we look for solutions of the classical equations of motion with finite (extremal) action.

A saddle point involving real fields requires finding a solution to the following set of coupled equations

$$\begin{aligned} &(\partial^2 - e^2 A^2)\varphi = 0 \\ &(\partial A + A . \partial)\varphi = 0 \\ &-\partial^2 A_\mu + (1-\lambda)\partial_\mu\ \partial . A + 2\ e^2 A_\mu\ \varphi^2 = 0 \end{aligned} \tag{35}$$

complemented by a stationarity condition on the coupling constant

$$k + e^2 \int d^4x\ A^2 \varphi^2 = 0 \tag{36}$$

The latter shows, in agreement with Dyson's argument, that the saddle point value of e^2 is negative and small as $\frac{1}{k}$. A complete study of the non-linear set of coupled equations (35) is a formidable task. Under the prejudice that the most symmetrical situations corresponds to the lowest saddle point we look for an ansatz of the form

$$\begin{aligned} &(-e_c^2)^{1/2}\ A_\mu = M_{\mu\nu}\ x^\nu\ a(x^2)/x^2 \\ &(-e_c^2)^{1/2}\ \varphi = (2/x^2)^{1/2}\ f(x^2) \\ &-e_c^2 = \frac{2\pi^2}{k} \int_0^\infty \frac{dx^2}{x^2}\ a^2\ f^2 \end{aligned} \tag{37}$$

with $M_{\mu\nu}$ a fixed antisymmetric matrix in such a way that the conditions

$$\partial . A = 0 \qquad\qquad x . A = 0 \tag{38}$$

are automatically fulfilled. They insure that the gauge fixing parameter λ will disappear from the solution and the second equation in (35) verified. In reference[14] we originally assumed that $M_{\mu\nu}$ was a non singular matrix fulfilling therefore

$$M = -M^T \qquad\qquad M^2 = -I \tag{39}$$

This leads to a reduced one dimensional problem in the variable $t = \ln x^2$ corresponding to the equations

$$\ddot{a} = a(1-f^2) \qquad\qquad \ddot{f} = \frac{1}{4} f(1-a^2) \tag{40}$$

with the extremal of the effective action given by

$$\begin{aligned}&\int d^4x \left[\frac{1}{4} F^2 + \frac{\lambda}{2} (\partial . A)^2 + (\partial + ie A)\varphi^* . (\partial - ie A)\varphi\right] \\ &= -\frac{2\pi^2}{e_c^2} \int_{-\infty}^{+\infty} dt \{\dot{a}^2 + a^2 + 4\dot{f}^2 + f^2 - a^2 f^2\}\end{aligned} \tag{41}$$

It was required to find the unique (up to symmetries) solution such that the integrals

$$I = \int_{-\infty}^{+\infty} dt\; a^2 f^2 = \int_{-\infty}^{+\infty} dt(\dot{a}^2 + a^2) = \int_{-\infty}^{+\infty} dt(4\dot{f} + f^2) \tag{42}$$

were finite. A numerical solution was found leading to a behavior of k-th order of one particle irreducible Euclidean Green's functions of the form

$$\begin{aligned}\Gamma_n &= \sum_k \Gamma_n^{(k)} \left(-\frac{e^2}{4\pi}\right)^k \\ \Gamma_K &= k!\, A^k\, K^{b_n}\, C_n \left[1 + O\left(\frac{1}{k}\right)\right]\end{aligned} \tag{43}$$

with a constant A given by

$$A = 0.0808 \tag{44}$$

Lipatov and Bogomolny[15] have reinvestigated the same problem with M a **degenerate matrix** . They succeeded in improving the above estimate by a tenpercent decrease of the constant A. No complete treatment of the system (35), (36) is yet available but it seems unlikely that any very significant improvement is to be expected.

4. FERMIONIC QED

A new ingredient is required in the fermionic case to cope with Pauli's principle. The latter introduces relative signs among diagrams which might lead to cancellations. On the other hand the use of semi-classical methods for the study of anticommuting path integrals as required in this case needs some elaboration. The original development of a relativistic Thomas-Fermi method was initiated by Parisi[16] in the case of a Yukawa coupling and was developed for QED in reference[17] .

In the Euclidean domain we define the Green's functions by averaging a string of A's, ψ's and $\bar{\psi}$'s over paths weighted by the exponential of minus the action

$$S = S_A + S_F$$

$$S_A = \int d^4x \ \frac{1}{4} F^2 + \frac{\lambda}{2} (\partial.A)^2 \tag{45}$$

$$S_F = \int d^4x \ \bar{\psi}(i\,\gamma.\partial - e\,\gamma.A - M)\,\psi$$

The fermionic field variables ψ and $\bar{\psi}$ belong to an anticommuting Grassmann algebra and the antihermitian γ matrices fulfill

$$\{\gamma^\mu,\gamma^\nu\} = -2\,\delta^{\mu\nu} \qquad \gamma^{\mu +} = -\gamma^\mu \tag{46}$$

Let us concentrate on the set of Green's functions without external charged lines. We can integrate over Fermi fields so that an n-photon amplitude will read

$$G_{\mu_1\ldots\mu_n}(x_1,\ldots,x_n) = \int \mathcal{D}(A) A_{\mu_1}(x_1)\ldots A_{\mu_n}(x_n)\ \frac{\Delta(eA)}{\Delta(0)}\ e^{-S_A} \tag{47}$$

where $\Delta(eA)$ is a suitably subtracted Fredholm determinant

$$\Delta(eA) = \det[(i\,\bar{\partial} - e\bar{A})(i\,\underline{\partial} - e\underline{A}) + M^2] \tag{48}$$

The notations $\bar{n}$ and $\underline{n}$ stand for the quaternionic representations

$$\bar{n} = n^0 + i\vec{n}.\vec{\sigma} \qquad \underline{n} = n_0 - i\,\vec{n}.\vec{\sigma} \tag{49}$$

if $n \equiv (n_0,\vec{n})$ is an arbitrary four-vector and $\vec{\sigma}$ the Pauli matrices.

Expanding $\ln \frac{\Delta(eA)}{\Delta(0)}$ in even powers of e according to Furry's it is expressed in terms of the one (charged) loop diagrams with photon internal lines and requires therefore a renormalization to order e^2 but not to order e^4 according to gauge invariance. Moreover $\frac{\Delta(eA)}{\Delta(0)}$ is real and is an entire function of e^2.

For a regular real A let $e_n(A)$ be the discrete eigenvalues of the Dirac equation in the potential A_μ corresponding to zero eigen-modes. We find from the previous definitions that $\Sigma\, e_n^{-2}(A)$ diverges but $\Sigma\, e_n^{-4}(A)$ converges it is therefore tempting to use the standard Hadamard representation to write

$$\frac{\Delta(eA)}{\Delta(0)} = \exp[-e^2\bar{\omega}(A)]\prod_n \left(1 - \frac{e^2}{e_n^2(A)}\right) \exp \frac{e^2}{e_n^2(A)} \tag{50}$$

where e_n runs over the eigenvalues with positive imaginary part say and $\bar{\omega}(A)$ is the subtracted vacuum polarization contribution. We shall see however that (50) is misleading and that the order of this entire function is more likely to be four rather than two as implied by (50).

Using Sobolev's inequality it is possible to show that

$$|e_n(A)|^2 \geq \frac{16\pi^2}{3} \frac{1}{\int d^4x \, \frac{1}{4} F^2} \tag{51}$$

which insures that it if $\sum_n |e_n(A)|^{-p}$ converges for some p, $\ln \frac{\Delta(eA)}{\Delta(0)}$ will be analytic in the vicinity of $e = 0$. In reference[17] various examples of external fields A were considered and the eigenvalues $e_n(A)$ computed. They all exhibited the following behaviour $\sum e_n^{-4}(A) < \infty$, $\sum_n |e_n(A)|^{-4} = \infty$, $\sum_n |e_n(A)|^{-4-\varepsilon} < \infty$ for positive ε. This indicated that $\frac{\Delta(eA)}{\Delta(0)}$ is of order four and a thorough investigation was initiated in the second reference[17] to prove this statement crucial to the large order estimates. Eventhough we were not able to obtain a mathematical theorem, I am reasonably confident that the result is true as already suggested by Adler[18]. What is somehow paradoxical is that the behaviour of the perturbative series, i.e. the behaviour of the fully quantized theory in the vicinity of zero coupling is dominated by the question of the large e (i.e. large field) behaviour of $\frac{\Delta(eA)}{\Delta(0)}$, or stated differently by the large external field problem (strong coupling) of the Dirac equation . This looks at first like a tractable question since it involves large quantum numbers and should be resolved using semi-classical methods. The difficulty lies in the question of gauge invariance. Indeed were we studying the Yukawa case

$$(i\gamma.\partial + M + g\sigma)\psi = 0 \tag{52}$$

we would find that for large g (large σ) it is possible to neglect the relative variation of σ and therefore one obtains that $\frac{\Delta(g\sigma)}{\Delta(0)}$ suitably subtracted cannot grow faster than exp cst $\int d^dx |g\sigma(x)|^d$ in dimension d, indicating an order four in four dimensions. Unfortunately this does not directly apply to the electromagnetic case since a locally constant A can be gauged away. Nevertheless we have found that it is very likely that

$$\left|\frac{\Delta(eA)}{\Delta(0)}\right| < \exp \mathrm{cst} |e|^4 \int d^4x \, |A_T(x)|^4 \tag{53}$$

where $A_T(x)$ means $A_\mu(x)$ in the transverse gauge $\partial.A = 0$.

If this is granted, applying the same ideas as those indicated in the preceding paragraphs we find that any (regularized) amplitude behaves for large order k as

$$G = \sum G^{(k)}(e^2)^k \qquad G^k \sim \Gamma\left(\frac{k}{2}\right) \tag{54}$$

This indicates an enormous cancelation.

A way to avoid the difficult question of the order of $\Delta(eA)$ is to restrict one self to a subset of diagrams involving one charged particle loop. Typical of it is the study of the vacuum polarization $\bar{\omega}_{\mu\nu}(k)$ in this approximation. We then also escape the problems of internal renormalization due to the unique overall logarithmic ultraviolet divergence. With α standing for the fine structure constant the coefficient $F(\alpha)$ of this divergence introduced by Baker Johnson and Willey [19] is known up to order six

$$F(\alpha) = \frac{2}{3}\left(\frac{\alpha}{2\pi}\right) + \left(\frac{\alpha}{2\pi}\right)^2 - \frac{1}{4}\left(\frac{\alpha}{2\pi}\right)^3 + \ldots \tag{55}$$

It may be obtained by replacing $\frac{\Delta(eA)}{\Delta(0)}$ by $\ln \frac{\Delta(eA)}{\Delta(0)}$ in Eq. (47). The large order can then be estimated by retaining the contribution of the lowest eigen-mode in the two point function and solving a coupled classical set of equations with the result

$$\begin{aligned} F(\alpha) &= \sum_{k=1}^{\infty} \left(\frac{\alpha}{2\pi}\right)^k F_K \\ F_K &\sim (-1)^k\, k!\left(\frac{1}{I}\right)^k k^3 f \end{aligned} \tag{56}$$

with f a constant and

$$I = 5.9201 \tag{57}$$

5. CONCLUSIONS

Apart from stressing a number of spectacular successes we could not find enough space to discuss many of the very interesting problems open in this field. We will conclude by mentioning two of them.

The first has to do with situations analogous to the one encountered for non abelian gauge fields. Here the perturbative vacuum is unstable for real values of the coupling constant as manifested by the corresponding existence of instantons. A simpler model is afforded in quantum mechanics by the degenerate double well potential when perturbation is carried around one of the minima. The series cannot be summed using Borel methods and one has to introduce some (but which) extra information. The physical relevance of this problem to quantum chromodynamics is clear and no neat answer is known to this day.

The second set of open problems is related to the role of renormalization in introducing new singularities in the Borel transformed plane as recognized by 't Hooft [20]. This does not affect results such as the one just discussed for $F(\alpha)$ where only overall divergences occur but may arise in the case of strings of internal subtractions. A simple example was discussed by Lautrup [21] where he considered a vertex diagram in which a string of lowest order electron-positron bubbles is inserted in a photon line. This gives a diagram which grows like $+k!$ thus spoiling eventually the Borel summability. This is readily shown to be related to the difficulties associated to the Landau ghost, an unphysical singularity in the photon propagator, which in turn is related to the bad ultraviolet behaviour of QED. An analysis of these questions has been given recently by Parisi [22]. More work is clearly needed to provide a complete picture. Nevertheless large order estimates of perturbation theory have opened an exciting new field of investigation involving deep questions in quantum field theory and path integral methods have been among the major tools.

REFERENCES

[1] F.J. Dyson, *Phys. Rev.* 85, 631 (1952)

[2] A.M. Jaffe, *Comm. Math. Phys.* 1,127 (1965)

[3] C.M. Bender, T.T. Wu, *Phys. Rev.* 184,1231 (1969)
Phys. Rev. Lett. 27, 461 (1971), *Phys. Rev.* D7, 1620 (1972)

[4] J.S. Langer, *Ann. Phys.* 41, 108 (1967)

[5] C.S. Lam, *Nuovo Cimento* 55, 258 (1968)

[6] E. Borel, Leçons sur les séries divergentes, Paris (1928). For the anharmonic oscillator see

[7] S. Graffi, V. Grecchi and B. Simon, *Phys. Lett.* 32B,631 (1970). In the case of field theory see J.P. Eckmann, J. Magnen and R. Seneor, *Commun. Math. Phys.* 39, 251 (1975), J.S. Feldman and K. Osterwalder, *Ann. of Phys.* (N.Y) 97,80 (1976)

[8] L.N. Lipatov, JETP Lett 24,179 (1976), 25, 116 (1977)

[9] E. Brézin, J.C. Le Guillou, J. Zinn-Justin, *Phys. Rev.* D15, 1544, 1558 (1977)

[10] G. Parisi, *Phys. Lett.* 66B, 167 (1977)

[11] J.C. Collins and D.E. Soper,University Preprint (1977)
G. Auberson, G. Mennessier and G. Mahoux, to be published. R. Seznec, Thèse de 3ème cycle, Saclay

[12] See for instance E.M. Stein, Singular integrals and differentiability properties of functions - Princeton University Press (1970)

[13] J.C. Le Guillou and J. Zinn-Justin, *Phys. Rev. Lett.* 39, 95 (1977)

[14] C. Itzykson, G. Parisi and J.B. Zuber, *Phys. Rev. Lett.* 38, 306 (1977)

[15] L.N. Lipatov, Bogomolny , Leningrad Preprint (1978)

[16] G. Parisi, *Phys. Lett.* 66B, 382 (1977)

[17] C. Itzykson, G. Parisi and J.B. Zuber, *Phys. Rev.* D16, 996 (1977)

[18] S. Adler, *Phys. Rev.* D10 2399 (1974)

[19] K. Johnson, R. Willey, and M. Baker, *Phys. Rev.* 163, 1699 (1967)

[20] G.'t Hooft, Lectures of the Banff summer school (1977)

[21] B. Lautrup, *Phys. Lett.* 69B 109 (1977)

[22] G. Parisi, Ecole Normale Preprint LPTENS 78/8

DIAS-TP-78-35

Anomalous Behaviour of the Effective Potential*

by

L. O'Raifeartaigh

and

G. Parravicini

Dublin Institute for Advanced Studies

Dublin 4, Ireland

---

*Talk given at Colloquium on "Mathematical Problems in Feynman Path Integrals" at Marseille, May 22 - 26, 1978.

1. Introduction

It has been known for some time[1] that in quantum field theory the effective potential $E(\phi_c)$ corresponding to a classical Lagrangian $\mathcal{L}(\phi_c)$, where ϕ_c is a set of classical fields, is equal to the sum of all connected one-particle irreducible[2] (CIPI) vacuum graphs for a 'shifted' Lagrangian, that is,

$$E(\phi_c) = W(\hat{\mathcal{L}}(\phi,\phi_c)), \tag{1.1}$$

where W denotes the sum of all CIPI vacuum graphs, and $\hat{\mathcal{L}}(\phi,\phi_c)$ is the 'shifted' Lagrangian, defined as

$$\hat{\mathcal{L}}(\phi,\phi_c) = \mathcal{L}(\phi+\phi_c) - \mathcal{L}(\phi_c) - \phi\cdot\frac{\delta\mathcal{L}(\phi_c)}{\delta\phi_c}, \tag{1.2}$$

where ϕ is a set of quantum fields. For simplicity, we shall assume that the fields ϕ and ϕ_c are Lorentz scalar fields. The first purpose of this talk will be to simplify the usual functional integral proof of the vacuum-graph formula (1.1) so that the result becomes intuitively evident. For polynomial Lagrangians the advantage of the formulae (1.1) and (1.2) is that $\hat{\mathcal{L}}(\phi,\phi_c)$ does not have a new functional form but is the same as $\mathcal{L}(\phi_c)$ except that ϕ_c is replaced by ϕ and the mass and coupling parameters, which we shall denote generically by m and f, are replaced by functions of ϕ_c. Thus

$$\hat{\mathcal{L}}(m,f,\phi,\phi_c) = \mathcal{L}(m(\phi_c), f(\phi_c), \phi), \tag{1.3}$$

and hence (1.1) reduces to

$$E(\phi_c) = W(\mathcal{L}(m(\phi_c), f(\phi_c), \phi), \tag{1.4}$$

which expresses $E(\phi_c)$ as a function of the vacuum graphs of the <u>original</u>

Lagrangian. To convince oneself of (1.3) consider the single-field renormalizable Lagrangian

$$\mathcal{L}(\phi_c) = \frac{1}{2}(\partial_\mu\phi)^2 - \frac{1}{2}\phi_c^2(m^2 + 2f\phi_c + g^2\phi_c^2). \tag{1.5}$$

From (1.2) one easily calculates that

$$\hat{\mathcal{L}}(\phi,\phi_c) = \frac{1}{2}(\partial_\mu\phi)^2 - \frac{1}{2}\phi^2(m^2(\phi_c) + 2f(\phi_c)\phi + g^2\phi^2), \tag{1.6}$$

where

$$f(\phi_c) = f + 2g^2\phi_c \quad \text{and} \quad m^2(\phi_c) = m^2 + 6f\phi_c + 6g^2\phi_c^2. \tag{1.7}$$

It is clear that a similar result will hold for any polynomial potential $V(\phi_c)$. Using (1.4) it is very tempting to go one step further and write

$$E(\phi_c) = \left\{ W(\mathcal{L}(m,f,\phi)) \right\}_{\substack{m \to m(\phi_c) \\ f \to f(\phi_c)}}, \tag{1.8}$$

which would incorporate the simple rule: "to calculate $E(\phi_c)$ simply compute the vacuum graphs for the <u>original</u> Lagrangian $\mathcal{L}(m,f,\phi)$ and then change the parameters m, f to functions of ϕ_c according to (1.7), or the corresponding expressions for general $V(\phi_c)$ ". And indeed in many cases this simple rule is correct and has been used in the literature. However, the second purpose of this talk is to point out that the rule is not always correct, and that the exceptions include such well-known Lagrangians as ones which are supersymmetric and/or spontaneously broken. In view of the importance of these exceptions it may be desirable to have an alternative rule which covers these cases, and a further purpose of the talk is to propose such a rule. The rule proposed leads to effective potentials which have discontinuities in their first derivatives, and therefore incorporate a sort of phase transition.

Much of the talk is contained in two recently published papers by the authors,[3] but the presentation here is somewhat different. Where possible, technical details are omitted and the reader is referred to the above references.

2. Definition of the Effective Potential

Let $G(J)$ be the generating functional for the Greens functions in quantum field theory defined (up to a J-independent normalization constant) in the usual way[1] by a functional integral

$$e^{iG(J)} = \int d\mu(\phi)\, e^{i\int(\mathcal{L}(\phi)+J\phi)dx} \tag{2.1}$$

Then the effective potential $E(\phi_c)$ is defined to be the Legendre transform of $G(J)$, that is

$$E(\phi_c) = G(J) - \int J\phi_c\, dx\,, \tag{2.2}$$

where ϕ_c, the so-called classical field, is defined as

$$\phi_c = \frac{\delta G(J)}{\delta J} = \int \phi\, d\mu(\phi)\, e^{i\int(\mathcal{L}+J\phi)dx}\,, \tag{2.3}$$

and hence is the average of ϕ with respect to functional integral. Note that the Legendre transform (2.2) (2.3) is just the analogue of the transformation which changes any Lagrangian into (minus) its Hamiltonian

$$-H = \mathcal{L} - \pi\dot{\phi} \qquad \text{where} \qquad \pi = \frac{\delta\mathcal{L}}{\delta\dot{\phi}} \tag{2.4}$$

and just as the second equation in (2.4) has the inverse $\dot{\phi} = \frac{\delta H}{\delta\pi}$ so (2.3) has the inverse

$$J = -\frac{\delta E(\phi_c)}{\delta \phi_c} \quad . \tag{2.5}$$

The analogy between $E(\phi_c)$ and the Hamiltonian partly explains the name effective potential, but the precise meaning of $E(\phi_c)$ is that it is a generating functional for the connected one-particle-irreducible (CIPI) Feynman graphs of the theory

$$E(\phi_c) = \sum_{CIPI} \frac{1}{n!} \Gamma(x_1 \ldots x_n)\, \phi_c(x_1) \ldots \phi_c(x_n) \ . \tag{2.6}$$

This property was first noted explicitly by Jona-Lasinio[4] and a simple inductive proof has been given by Abers and Lee[5]. Here we are interested only in the result (2.6), in particular in the very special case

$$E(0) = \Gamma_{CIPI}(0) = \overline{W}(\mathcal{L}(\phi)), \tag{2.7}$$

where $\overline{W}$ denotes the sum of all CIPI vacuum graphs, except for the zero-loop ones. The result (2.7) will play a crucial role in the sequel.

3. <u>Proof of the Vacuum Graph Formula.</u>

To establish the vacuum graph formula (1.1) we note first that from definitions of $G(J)$ and $E(\phi_c)$ in (2.1) and (2.2) respectively, we have

$$e^{iE(\phi_c)} = \int d\mu(\phi)\, e^{i\int(\mathcal{L}(\phi) + J(\phi - \phi_c))dx} \ . \tag{3.1}$$

Hence, using the assumed translational invariance of $d\mu(\phi)$ we have

$$e^{iE(\phi_c)} = \int d\mu(\phi)\, e^{i\int(\mathcal{L}(\phi+\phi_c) + J\phi)dx} \ . \tag{3.2}$$

Equation (3.2) already contains the main part of the shifted Lagrangian

The remaining terms of zero and first order in ϕ enter when we use the inverse formula (2.5) for J and subtract from $E(\phi_c)$ its zero loop, or classical part - $-\int \mathcal{L}(\phi_c)dx$. A short computation gives

$$e^{i\Delta E(\phi_c)} = \int d\mu(\phi)\, e^{i\int\left(\hat{\mathcal{L}}(\phi,\phi_c)+K\phi\right)dx} , \tag{3.3}$$

where

$$\Delta E(\phi_c) = E(\phi_c) + \int \mathcal{L}(\phi_c)dx \quad \text{and} \quad K = \frac{-\delta\Delta E(\phi_c)}{\delta\phi_c} , \tag{3.4}$$

and $\hat{\mathcal{L}}(\phi,\phi_c)$ is just the shifted Lagrangian defined in (1.2). But now if we compare (3.2) with (2.1) we see that $\Delta E(\phi_c)$ is just the generating functional for the Greens functions of the shifted Lagrangian, evaluated at the value of K given by (3.4),

$$\Delta E(\phi_c) = \left\{ G(\hat{\mathcal{L}},K)\right\}_{K=\frac{-\delta\Delta E(\phi_c)}{\delta\phi_c}} . \tag{3.5}$$

This leads us to consider in turn the effective potential $\tilde{E}(\tilde{\phi}_c)$ corresponding to $G(\hat{\mathcal{L}},K)$, which is

$$\tilde{E}(\tilde{\phi}_c) = G(\hat{\mathcal{L}},K) - \int K\tilde{\phi}_c , \tag{3.6}$$

where

$$\tilde{\phi}_c = \frac{\delta G(\hat{\mathcal{L}},K)}{\delta K} = \int \phi\, d\mu(\phi)\, e^{i\int(\hat{\mathcal{L}}-K\phi)dx} . \tag{3.7}$$

Now comes the crucial point. Since K in (3.5) is fixed, so is $\tilde{\phi}_c$ in (3.7). Furthermore we can compute it by using the last expression in (3.7) and reversing the steps from (3.1) to (3.3). Since the steps are the same except that the massive $d\mu(\phi)$ is replaced by $\phi d\mu(\phi)$, and $\phi \to \phi - \phi_c$ in the translation from

(3.2) to (3.1) we obtain by inspection

$$\tilde{\phi}_c = e^{i\int(\mathcal{L}(\phi_c)+J\phi_c)dx} \left\{ \int(\phi-\phi_c)d\mu(\phi)\, e^{i\int(\mathcal{L}(\phi)+J\phi)dx} \right\}. \tag{3.8}$$

But the integral in brackets is zero from the definition of ϕ_c ! Thus the new classical field is zero

$$\tilde{\phi}_c = 0. \tag{3.9}$$

Inserting this result in (3.6) and (3.5) we obtain at once

$$\Delta E(\phi_c) = G(\hat{\mathcal{L}}, K) = \tilde{E}(\hat{\mathcal{L}}, 0), \tag{3.10}$$

and since from (2.7) $\tilde{E}(\hat{\mathcal{L}}, 0)$ is just the sum of all the CIPI vacuum diagrams for $\hat{\mathcal{L}}$, except the zero-order ones, we have

$$E(\phi_c) = \tilde{E}(\hat{\mathcal{L}},0) + \int\mathcal{L}(\phi_c)dx = \overline{W}(\hat{\mathcal{L}}) + \int\mathcal{L}(\phi_c)dx = W(\hat{\mathcal{L}}), \tag{3.11}$$

where $W(\hat{\mathcal{L}})$ is the sum of all the CIPI vacuum diagrams for $\hat{\mathcal{L}}(\phi,\phi_c)$, including the zero order ones. This is the required result.

4. Anomalous Cases

As pointed out in the Introduction, for polynomial Lagrangians, the formula (3.11) for the effective potential can be written as

$$E(\phi_c) = V(\mathcal{L}(m(\phi_c)\ f(\phi_c)\ \phi), \tag{4.1}$$

and in many cases this can be further reduced to

$$E(\phi_c) = \left\{ V(\mathcal{L}(m,f,\phi)) \right\}_{\substack{m\to m(\phi_c)\\ f\to f(\phi_c)}} \tag{4.2}$$

where $V(\mathcal{L}(m,f,\phi))$ is the sum of all vacuum graphs for the original

Lagrangian. In these cases $E(\phi_c)$ can evidently be obtained by direct substitution from the vacuum graphs of the original Lagrangian. The usefulness of such a formula may be illustrated by considering the one-loop contribution. It is well-known[4] that for any Lagrangian in (3+1) dimensions the one-loop contribution to the vacuum graphs are

$$\mathrm{tr}\, M^4 \log(M^2/\mu^2) \,, \tag{4.3}$$

where M is the mass-matrix and μ an arbitrary renormalization parameter. Then according to (4.2) the one-loop contribution to the effective potential is just

$$\overset{(1)}{E}(\phi_c) = \mathrm{tr}\, M^4(\phi_c) \log(M^2(\phi_c)/\mu^2) \,, \tag{4.4}$$

where $M(\phi_c)$ is the mass-matrix for the shifted Lagrangian $\hat{\mathcal{L}}(\phi,\phi_c)$. From the definition of $\hat{\mathcal{L}}(\phi,\phi_c)$ in (1.2) it is easy to see that $M^2(\phi_c)$ is given by

$$M^2(\phi_c) = \frac{\partial^2 V(\phi_c)}{\partial \phi_c^2} \,, \tag{4.5}$$

where $V(\phi_c)$ is the potential in the original Lagrangian $\mathcal{L}(\phi_c)$. Thus we see that quite generally the effective potential corresponding to any classical potential $V(\phi_c)$ is just

$$\overset{(1)}{E}(\phi_c) = \mathrm{tr}\,(V''(\phi_c))^2 \log(V''(\phi_c)/\mu^2) \,. \tag{4.6}$$

For example for a ϕ^4 potential, the expression (4.6) is proportional to $\phi^4 \log(\phi^2/\mu^2)$. In general for ϕ^{2k} it is proportional to $\phi^{4k-4} \log(\phi^{2k-2}/\mu^2)$, which is a polynomial increase on ϕ^{2k} for $k > 2$. Note that the term $(-\phi^{4k-4} \log \mu^2)$ exhibits explicitly the non-renormalizability of the theory for $k > 2$.

Unfortunately the formula (4.6) cannot be correct for all potentials, since there exist potentials for which $V''(\phi_c)$ is not positive, and hence $\overset{(1)}{E}(\phi_c)$ is not real, for all ϕ_c. Examples of such potentials are given below. Since potentials exist for which the one-loop contribution to (4.2) is not valid one must conclude that in such cases, and perhaps other cases, the formula (4.2) itself is not valid. In that case one must try to locate where the step from (4.1) to (4.2) can go wrong, and find a means to correct it. This will be done in the next section. But first we list some examples of potentials for which $V''(\phi_c)$ is not positive for all ϕ_c.

Example 1. Let $V(\phi_c)$ be the potential of the example given in the introduction,

$$V(\phi_c) = \frac{1}{2}\phi_c^2(m^2+2f\phi_c+g^2\phi_c^2) \quad \text{and} \quad V''(\phi_c) = m^2+6f\phi_c+6g^2\phi_c^2 , \tag{4.7}$$

where $m^2, g^2, m^2g^2-f^2 > 0$. Then $V''(\phi_c)$ is negative for

$$3(f+2g^2\phi_c)^2 > (3f^2-2m^2g^2) > 0 . \tag{4.8}$$

Example 2. Let $V(\phi_c)$ be a spontaneously symmetry breaking potential for a single scalar field,

$$V(\phi_c) = \frac{m^2}{2}(\phi_c^2-a^2)^2 \quad \text{and} \quad V''(\phi_c) = 2m^2(3\phi_c^2-a^2) \tag{4.9}$$

where m^2, $a^2 > 0$. Then $V''(\phi_c)$ is negative for $\phi_c^2 < a^2/3$.

Example 3. Let $V(\phi_c)$ be a spontaneous symmetry breaking potential for a set of SO(N) vector fields,

$$V(\phi_c) = \frac{m^2}{2}(\phi_c^r\phi_c^r-a^2)^2 \quad \text{and} \quad V_{rs}(\phi_c) = 2m^2(\phi_c^t\phi_c^t-a^2)\delta_{rs} + 4m^2\phi_c^r\phi_c^s , \tag{4.10}$$

where r, s, t = 1 ... N. Then $V''(\phi_c)$ has N-1 negative eigenvalues for $\phi_c^t\phi_c^t < a^2$ and N negative eigenvalues for $\phi_c^t\phi_c^t < a^2/3$.

Example 4. Let V(A,B) be the supersymmetric Wess-Zumino potential,[6]

$$V(A,B) = m^2(A^2+B^2) + 2mgA(A^2+B^2) + g^2(A^2+B^2)^2 , \tag{4.11}$$

and

$$V''(A,0) = \text{diag}\{3(m+2gA)^2 - m^2 , (m+2gA)^2 + m^2\} . \tag{4.12}$$

Hence one of the eigenvalues of $V''(A,B)$ is negative for $(m+2gA)^2 < m^2/3$ and B = 0.

Thus, for spontaneous symmetry breaking and supersymmetry, the anomalous case would seem to be the rule rather than the exception.

5. Origin of the Anomaly.

The fact that the mass-squared term $m^2(\phi_c) = V''(\phi_c)$ becomes negative in the anomalous examples of the last section already indicates the origin of these anomalies, because, for any Lagrangian, the mass-squared term can be negative only if $\phi = 0$ is not a minimum of the potential. For example, the minima of the potential

$$V(\phi_c) = -m^2\phi_c^2 + g^2\phi_c^4 , \qquad m^2, g^2 > 0, \tag{5.1}$$

are at $\phi_c^2 = m^2/2g^2$, not at $\phi = 0$. The anomalous examples therefore show that the potential minimum for $\mathcal{L}(m(\phi_c), f(\phi_c), \phi)$ is not necessarily at $\phi = 0$ for all ϕ_c, even if the potential minimum for $\mathcal{L}(m, f, \phi)$ is at that point. More generally the examples show that the absolute potential minima (or vacua) of $\mathcal{L}(m(\phi_c)\ f(\phi_c)\ \phi)$ and $\mathcal{L}(m, f, \phi)$ do not necessarily coincide. With this possibility in mind it is easy to understand why the assumed equality

$$W(\mathcal{L}(m(\phi_c)\ f(\phi_c)\ \phi)) = \{W(\mathcal{L}(m, f, \phi)\}_{\substack{m \to m(\phi_c) \\ f \to f(\phi_c)}} , \tag{5.2}$$

can fail, namely, because the vacuum graphs $W(\mathcal{L}(m, f, \phi)$ are calculated at the potential minimum of $\mathcal{L}(m, f, \phi)$, whereas the vacuum graphs $W(\mathcal{L}(m(\phi_c)\ f(\phi_c)\ \phi)$ are calculated at the potential minimum of $\mathcal{L}(m(\phi_c)\ f(\phi_c)\ \phi)$. Furthermore, one should expect the rule (5.2) to fail if, and only if, the potential minima of $\mathcal{L}(m,f,\phi)$ and $\mathcal{L}(m(\phi_c)\, f(\phi_c)\, \phi)$ differ. This means that so long as the potential minima are the same the rule should work. On the other hand, once the potential minima differ it should no longer work, even if the one-loop contribution to $E(\phi_c)$ remains real. For example, for the potential (1.5) of the introduction, the absolute minimum of $V(m,f,\phi)$ is at $\phi = 0$, but that of $V(m(\phi_c), f(\phi_c), \phi)$ is at $\phi = 0$ only if $(f+2g\phi_c)^2 \geq (3f^2-2m^2g^2)$. Hence we expect the rule to fail for the values $(f+2g\phi_c)^2 < (3f^2-2m^2g^2)$ of ϕ_c. But the one-loop contribution to becomes negative only for $(f+2g\phi_c)^2 < (3f^2-2m^2g^2)/3$. Thus for $3f^2 > 2m^2g^2$ there is a non-trivial range of values of ϕ_c, namely $(3f^2-2m^2g^2)/3 < (f+2g\phi_c)^2 < (3f^2-2m^2g^2)$ for which the rule fails but the formal value obtained for the one-loop contribution to is still real (but wrong). The complexity of the one-loop contribution is therefore only a signal that we are deep inside the region where the rule fails. A detailed investigation of the different regions for the anomalous examples of the last section are given in references 3 . the reason for the failure of the rule (5.2) is understood, it is not difficult to suggest a remedy, and such a remedy is suggested in the next section.

6. Proposed Modification to Treat Anomalous Cases.

In the previous section we have seen that the rule (5.2) fails when the potential minima of $\mathcal{L}(m(\phi_c), f(\phi_c), \phi)$ and $\mathcal{L}(m,f,\phi)$ do not coincide. Since the relation

$$E(\phi_c) = W(\mathcal{L}(m(\phi_c), f(\phi_c), \phi)), \tag{6.1}$$

is still valid, we could, of course, still calculate the vacuum graphs directly from $\mathcal{L}(m(\phi_c), f(\phi_c), \phi)$. However, the question is whether there is a substitution formula analogous to (5.2), which allows one to calculate with the original Lagrangian $\mathcal{L}(m,f,\phi)$ and afterwards make a substitution analogous to (5.2). In this section we wish to show that there does indeed exist such a formula and that it is

$$V(\mathcal{L}(m(\phi_c), f(\phi_c), \phi)) = \left\{ V(\mathcal{L}(m, f, \phi)) \right\}_{\substack{m \to m(\phi_e) \\ f \to f(\phi_e)}} \tag{6.2}$$

where $m(\phi_e)$ and $f(\phi_e)$ are the same functions of ϕ_e as $m(\phi_c)$ and $f(\phi_c)$ are of ϕ_c, but ϕ_e is a new field which we call the <u>effective</u> field, and is defined as

$$\phi_e = \phi_c + \overset{\circ}{\phi}(\phi_c) , \tag{6.3}$$

where $\overset{\circ}{\phi}(\phi_c)$ is the value of ϕ at the potential minimum of $\mathcal{L}(m(\phi_c), f(\phi_c), \phi)$. (It is tacitly assumed that ϕ has been chosen so that the potential minimum of $\mathcal{L}(m, f, \phi)$ is at $\phi = 0$.) Thus when the minima of $\mathcal{L}(m, f, \phi)$ and $\mathcal{L}(m(\phi_c), f(\phi_c), \phi)$ coincide, so that $\overset{\circ}{\phi}(\phi_c)$ is zero, we recover the old formula (5.2) but when they no longer coincide the formula is modified by shifting ϕ_c by a further amount $\overset{\circ}{\phi}(\phi_c)$. Before going on to the proof of the prescription (6.2) (6.3) we illustrate its content by means of an example, namely the potential $g^2(\phi_c^2 - a^2)/2$ for a single scalar field. From the definition (1.2) (with the kinetic terms omitted) one finds at once that

$$V(m(\phi_c), f(\phi_c), \phi) = \hat{V}(\phi, \phi_c) = \tfrac{1}{2} g^2 \phi^2 \left\{ (\phi + 2\phi_c)^2 + 2(\phi_c^2 - a^2) \right\}. \tag{6.4}$$

A little algebra shows that the absolute minimum of the expression (6.4) is at the points

$$\phi = 0 \quad \text{for} \quad \phi_c^2 \geq a^2 \quad \text{and} \quad \phi = -\tfrac{1}{2}\left\{ 3\phi_c + \sqrt{4a^2 - 3\phi_c^2} \right\} \quad \text{for} \quad \phi_c^2 < a^2 . \tag{6.5}$$

Thus by definition

$$\phi_e = \phi_c \quad \text{for} \quad \phi_c^2 \geq a^2 \quad \text{and} \quad \phi_e = -\tfrac{1}{2}\left\{ \phi_c + \sqrt{4a^2 - 3\phi_c^2} \right\} \quad \text{for} \quad \phi_c^2 < a^2 . \tag{6.6}$$

Note that ϕ_e is discontinuous, and that ϕ_e^2 is continuous, but has a discontinuous first derivative, at $\phi_c^2 = a^2$. If we define $m^2(\phi_c)$ and $f(\phi_c)$ to be the coefficients of $\phi^2/2$ and $\phi^3/2$ in (6.4) in the usual way we see that

$$m^2(\phi_c) = g^2(6\phi_c^2 - 2a^2) \qquad \text{and} \qquad f(\phi_c) = 4g^2\phi_c \,. \tag{6.7}$$

Hence by definition

$$m^2(\phi_e) = g^2(6\phi_e^2 - 2a^2) = \begin{cases} g^2(6\phi_c^2 - 2a^2) & \text{for} \quad \phi_c^2 \geq a^2 \\ g^2\left(4a^2 - 3\phi_c^2 + 3\phi_c\sqrt{4a^2 - 3\phi_c^2}\right) & \text{for} \quad \phi_c^2 \leq a^2 \end{cases} \tag{6.8}$$

and

$$f(\phi_e) = 4g^2\phi_e = \begin{cases} 4g^2\phi_c & \text{for} \quad \phi_c^2 \geq a^2 \,, \\ -2g^2\left(\phi_c + \sqrt{4a^2 - 3\phi_c^2}\right) & \text{for} \quad \phi_c^2 < a^2 \end{cases} \tag{6.9}$$

Note that $m^2(\phi_e)$ and $f^2(\phi_e)$ are continuous, but have discontinuous first derivatives at $\phi_c^2 = a^2$. Since the sum of the vacuum graphs $V(\mathcal{L}(m,f,\phi))$ is a continuous function of m^2 and f^2 we therefore expect it to be continuous at $\phi_c^2 = a^2$ but to have discontinuous first derivatives. This turns out to be the case. For example the one-loop contribution

$$m^4(\phi_e) \log\left(m^2(\phi_e)/\mu^2\right) , \tag{6.10}$$

is easily seen to have these properties. In ref. 3 it is shown that this pattern of discontinuity in the first derivatives holds for all four of the anomalous cases discussed in section 4. A particularly pleasant consequence of the prescription (6.2) is that in the supersymmetric case the one-loop contribution (6.10) is not only real, but if μ^2 is chosen to be the original mass m^2, it is also positive, as a supersymmetric potential should be.

It remains to justify the prescription (6.2). The justification rests on the fact that in $\mathcal{L}(m(\phi_c), f(\phi_c), \phi)$ a translation $\phi \to \phi + \varepsilon$ of the field can be absorbed by a similar translation $\phi_c \to \phi_c + \varepsilon$ of the classical field. More precisely, since by definition,

$$\mathcal{L}(m(\phi_c), f(\phi_c), \phi) = \mathcal{L}(\phi + \phi_c) - \mathcal{L}(\phi_c) - \phi\, \mathcal{L}'(\phi_c) \,, \tag{6.11}$$

we have

$$\mathcal{L}(m(\phi_c), f(\phi_c), \phi + \varepsilon) = \mathcal{L}(\phi + \phi_c + \varepsilon) - \mathcal{L}(\phi_c) - (\phi + \varepsilon)\, \mathcal{L}'(\phi_c) \,, \tag{6.12}$$

$$\mathcal{L}(m(\phi_c + \varepsilon), f(\phi_c + \varepsilon), \phi) = \mathcal{L}(\phi + \phi_c + \varepsilon) - \mathcal{L}(\phi_c + \varepsilon) - \phi\, \mathcal{L}'(\phi_c + \varepsilon) \,, \tag{6.13}$$

and on differentiation of (6.12) at $\phi = \varepsilon$,

$$\mathcal{L}'(m(\phi_c), f(\phi_c), \varepsilon) = \mathcal{L}'(\phi_c + \varepsilon) - \mathcal{L}'(\phi_c) \,. \tag{6.14}$$

Hence subtracting (6.13) and ϕ times (6.14) from (6.12) we obtain

$$\mathcal{L}(m(\phi_c), f(\phi_c), \phi + \varepsilon) = \mathcal{L}(m(\phi_c + \varepsilon), f(\phi_c + \varepsilon), \phi) + \mathcal{L}(m(\phi_c), f(\phi_c), \varepsilon) + \phi\, \mathcal{L}'(m(\phi_c), f(\phi_c), \varepsilon) \,, \tag{6.15}$$

which shows that $\phi \to \phi + \varepsilon$ has been absorbed by $\phi_c \to \phi_c + \varepsilon$ at the expense of adding the two extra terms shown.

Suppose now that $\overset{\circ}{\phi}$ is the potential minimum of $\mathcal{L}(m(\phi_c), f(\phi_c), \phi)$. Then by definition

$$\mathcal{L}'(m(\phi_c), f(\phi_c), \overset{\circ}{\phi}) = 0 \,. \tag{6.16}$$

Hence for $\varepsilon = \overset{\circ}{\phi}$ eq. (6.15) reduces to

$$\mathcal{L}(m(\phi_c), f(\phi_c), \phi + \overset{\circ}{\phi}) = \mathcal{L}(m(\phi_e), f(\phi_e), \phi) + \mathcal{L}(m(\phi_c), f(\phi_c), \overset{\circ}{\phi}) \,, \tag{6.17}$$

where ϕ_e is defined as $\phi_c + \overset{c}{\phi}$. Since the last term in (6.17) is constant with respect to ϕ it plays no role in the evaluation of vacuum graphs and hence

$$W(\mathcal{L}(m(\phi_e), f(\phi_e), \phi)) = W(\mathcal{L}(m(\phi_c), f(\phi_c), \phi + \overset{\circ}{\phi}), \tag{6.18}$$

the potential minimum for both sides being just $\phi = 0$. But then

$$\begin{aligned} E(\phi_c) &= W(\mathcal{L}(m(\phi_c), f(\phi_c), \phi))_{\phi = \overset{\circ}{\phi}} \\ &= W(\mathcal{L}(m(\phi_c), f(\phi_c), \phi + \overset{\circ}{\phi}))_{\phi = 0} \\ &= W(\mathcal{L}(m(\phi_e), f(\phi_e), \phi))_{\phi = 0} \end{aligned} \tag{6.19}$$

where the subscript denotes the value of ϕ at the potential minimum in each case. Since $\mathcal{L}(m(\phi_e), f(\phi_e), \phi)$ and $\mathcal{L}(m, f, \phi)$ both have potential minimum $\phi = 0$, we then have

$$E(\phi_c) = W(\mathcal{L}(m(\phi_e), f(\phi_e), \phi)_{\phi=0} = \Big\{ W(\mathcal{L}(m, f, \phi))_{\phi=0} \Big\}_{\substack{m \to m(\phi_e) \\ f \to f(\phi_e)}} \tag{6.20}$$

which is the required justification for the prescription (6.2).

REFERENCES

(1) R. Jackiw, Phys. Rev. D9, 1686 (1974).

(2) J. Schwinger, Proc. Nat. Acad. Sci. US 37, 452 (1951).

J. Iliopoulos, C. Itzykson and A. Martin, Rev. Mod. Phys. 47, 165 (1975).

B. Zumino, Brandeis University Summer Institute Lectures (MIT Press, London 1971).

S. Coleman, Proc. Ettore Majorana Summer School, Erice 1973.

S. Coleman and E. Weinberg, Phys. Rev. D7, 1888 (1973).

(3) L. O'Raifeartaigh and G. Parravicini, Nucl. Phys. B111, 501, 516 (1976).

(4) G. Jona-Lasinio, Nuovo Cim. 34, 1790 (1964).

(5) E. Abers and B. Lee, Physics Reports 9C (1973).

(6) J. Wess and B. Zumino, Phys. Letters 49B, 52 (1974).

- SHORT COMMUNICATIONS -

- SECTION VII -

NON-AFFINE PATH ALGORITHM IN THE FUNCTIONAL

INTEGRAL CALCULUS OF SCHRÖDINGER KERNELS

J. Bertrand and M. Ginocchio

Université Paris VII

Taking Feynman procedure to quantify a classical system, we want to show how a change in the time measure and/or in the definition of paths in phase space can lead to a large family of quantization rules.

I.- Construction of a fractional kernel.

Let $h_\tau(q,p)$ be a time dependent classical hamiltonian. We characterize the corresponding quantum system by its propagator $\mathcal{K}_{t_0 t}(q_0,q)$ defined by a functional integral in the following manner. Let $\Pi \equiv [t_0,\dots,t_m,t]$ be a partition of the time interval $[t_0,t]$ with $t_0 < t_1 < \dots < t_{m+1} \equiv t$. We set:

(1) $$\mathcal{K}_{t_0 t} = \lim_{\Pi \nearrow} \mathcal{K}_{\Pi}$$

where the formal limit is taken when the partition modulus goes to zero and the fractional kernel $\mathcal{K}_\Pi$ is written in terms of the hamiltonian action S ,viz.

(2) $$\mathcal{K}_\Pi(q_0,q) \equiv \int \frac{dp_0}{h} \int \prod_{i=1}^{m} \frac{dq_i\,dp_i}{h} \exp\left\{\frac{1}{\alpha}\sum_{j=0}^{m} S_{t_j t_{j+1}}(q_j,q_{j+1},p_j)\right\}$$

with

(3) $$\alpha = h/2\pi i \ , \ q_{m+1} \equiv q \ ; \ p_j, q_j \in \mathbb{R}^N$$

(4) $$S_{ut}(x,q,p) = \int_u^t \left[p\,\frac{d\xi}{d\tau} - h_\tau(\xi,p)\right]\delta\tau$$

In this expression, p is taken constant in the time interval $[u, t]$ as usual[1]. As for the time measure element $\delta\tau$ and the configuration space path $\xi(\tau)$ going from (x, u) to (q, t), they are still to be specified. It is this freedom that will allow for different quantization rules. The only restrictions come from the fact that we want $\mathcal{K}$ obtained from (1) and (2) to be a propagator, i.e. to satisfy:

(5) $$\mathcal{K}_{t_0 t_0}(q_0, q) = \delta(q - q_0)$$

(6) $$\mathcal{K}_{t_0 t'} \circ \mathcal{K}_{t' t}(q_0, q) \equiv \int dq' \, \mathcal{K}_{t_0 t'}(q_0, q') \mathcal{K}_{t' t}(q', q) = \mathcal{K}_{t_0 t}(q_0, q)$$

Let us define the "mean hamiltonian function" $\bar{h}_{ut}$ by:

(7) $$\bar{h}_{ut}(x, q, p) \equiv \frac{1}{t-u} \int_u^t h_\tau(\xi, p) \, \delta\tau$$

For hamiltonians not depending explicitly on time, we require that $\bar{h}_{ut}$ be independent of u and t whenever the path $\xi(\tau)$ is a straight line. This will ensure that (6) goes into the composition law for an additive semi-group in time. We are thus led necessarily to $(t-u)^{-1} \delta\tau = \rho(\tau) d\tau$ where ρ is a distribution such that

(8) $$\begin{cases} \int_u^t \rho(\tau) d\tau = \varpi_0 = 1 \\ \int_u^t \left(\frac{\tau - u}{t - u}\right)^n \rho(\tau) d\tau = \varpi_n \\ \text{with } \varpi_n \text{ constant with respect to } u \text{ and } t \end{cases}$$

This implies that ρ has the form

$$\rho(\tau) = \sum_{j=0}^{n+1} \beta_j \delta(\tau - \tau_j) + \beta \frac{\chi_{[u,t]}(\tau)}{t-u}$$

where :

$[u, \tau_1, \ldots \tau_n, t]$ is a partition of the time interval $[u, t]$ with $u \equiv \tau_0 < \tau_1 < \ldots < \tau_{n+1} \equiv t$, $\tau_j = u + \theta_j \times (t-u)$ $0 \leq \theta_j \leq 1$

β_j and β are constants and $\sum_{j=0}^{n+1} \beta_j + \beta = 1$,

$\chi_{[u,t]}$ is the characteristic function of $[u, t]$.

Setting $\theta \equiv \frac{\tau - u}{t - u}$, we can write

$$\varpi_n = \int_0^1 \theta^n \, \delta\theta \, , n \geq 0$$

with

$$\delta\theta \equiv (t-u) \rho[u + \theta(t-u)] d\theta = \left[\sum_{j=0}^{n+1} \beta_j \delta(\theta - \theta_j) + \beta \chi_{[0,1]}(\theta)\right] d\theta$$

In configuration space, we choose curves defined by a function $\gamma_\theta(x,q)$ which is analytical in θ , $0 \leq \theta \leq 1$, for x,q belonging to an open set in $\mathbb{R}^N$ and subject to the conditions:

(i) $\xi(\tau) = \gamma_\theta(x,q)$ with $\gamma_0(x,q) = x$, $\gamma_1(x,q) = q$,

(ii) equation (5) is satisfied ,i.e. r defined by

$$r(x,q) \equiv \int_u^t \frac{d\xi}{d\tau}\,\delta\tau = \int_0^1 \frac{\partial\gamma_\theta}{\partial\theta}\,\delta\theta \tag{9}$$

must verify

$$\left.\frac{\partial r}{\partial x}\right|_{x=q} = 1$$

We note that $r(x,q)$ will reduce to the usual $q-x$ in two typical cases:

1) polygonal paths in configuration space

$$\xi = x + \theta\,(q-x)$$

2) $\delta\tau = d\tau$

II. - The evolution operator.

We shall now look for the formal expression of the hamiltonian operator appearing in the associated Schrödinger equation.

Rearranging terms in (2), we can write:

$$\mathcal{K}_\pi = \mathcal{K}_{[t_0,t_1]} \circ \cdots \circ \mathcal{K}_{[t_m,t]} \tag{10}$$

where

$$\mathcal{K}_{[u,t]}(x,q) = \int \frac{dp}{h}\, e^{\frac{p}{\alpha} r(x,q)}\, e^{-\frac{t-u}{\alpha}\bar{h}_{ut}(x,q,p)} \tag{11}$$

$\mathcal{K}_\pi$ is the kernel of an operator U_π which acts on a suitable test function φ as :

$$(12)\quad (U_\pi \varphi)(q) = (\varphi \circ \mathcal{K}_\pi)(q) \equiv \int dq' \varphi(q') \mathcal{K}_\pi(q', q)$$

whence

$$(13)\quad U_\pi = U_{[t_m, t]} \cdots U_{[t_0, t_1]}$$

The only values of $\bar{h}$ needed next are :

$$\bar{h}_{uu}(x, q, p) = \lim_{t' \to u} \frac{1}{t'-u} \int_u^{t'} h_\tau(\xi, p) \delta\tau$$

which becomes, if we assume that $h_\tau(q,p)$ is C^∞ in q :

$$(14)\quad \bar{h}_{uu}(x, q, p) = \int_0^1 h_u(\gamma_\theta(x, q), p)\, \delta\theta$$

Expanding the second exponential in (11), we can write :

$$(15)\quad U_{[u,t]} = 1 - \frac{t-u}{\alpha} H_u + \text{terms of order higher than } (t-u) ,$$

where

$$(16)\quad (H_u \varphi)(q) \equiv \int \frac{dx\,dp}{h}\, e^{\frac{p}{\alpha} r(x,q)}\, \bar{h}_{uu}(x, q, p)\, \varphi(x)$$

Inserting (15) into (13), we obtain the evolution operator by [2] :

$$(17)\quad U_{t_0 t} \equiv \lim_{\pi \nearrow} U_\pi = \exp\left\{ -\frac{1}{\alpha} \int_{t_0}^{t} H_u\, du \right\}$$

III. - Formulation of associated correspondence rules

What we have just seen is that, given a measure $\delta\theta$ and a function γ_θ satisfying suitable conditions, we obtain the form of a correspondence rule between the classical and quantum hamiltonians : $h_u(q,p) \longrightarrow H_u$ by using equations (14) and (16). The same rule can then be applied to any other analytical function in phase space.

a) Family of curves defined by : $\gamma_\theta(x, q) = x + \Gamma_\theta(x - q)$

where

$$(18)\quad \begin{cases} \Gamma_0(x-q) = 0 \\ \Gamma_1(x-q) = q-x \end{cases}$$

Γ_θ must be such that (5) is verified, which allows to write, according to (9)

$$(19)\quad r(x,q) = q - x + s(x-q)$$

where s is a series without terms of degree one, and zero.

Computation of (16) then gives :

$$(20)\quad H_u = \chi\left[R\left(\tfrac{p}{\alpha}, \alpha\partial_p, \partial_q\right) h(q,p)\right]$$

where χ performs the operation : $p \longrightarrow \alpha\partial_q$ and q before ∂_q .

Here we use a space of partially ordered series in X, Y, Z ("X before Y") and R is defined by the following internal product :

$$(21)\quad R(X,Y,Z) = S(X+\partial_{Y'}\,;Y)\,T(Z,Y')\Big|_{Y'\to Y} = T(Z,Y)\,e^{\overleftarrow{\partial}_Y\overrightarrow{\partial}_X}\,S(X;Y)$$

where

$$(22)\quad S(X+Z\,;Y) = \sum_{n\geq 0}\ \sum_{i+j+k=n} \frac{1}{i!\,j!\,k!}\, X^i\, s^{n,j}(Y)\, Z^k$$

$$= \sum_{n\geq 0} \frac{1}{n!}\left(X+Z+\partial_Y\right)^n s^n(Y)$$

$$(23)\quad T(Z,Y) = \int_0^1 \delta\theta\; e^{Z\Gamma_\theta(Y) + ZY}$$

In general, R consists of a series in p, ∂_p and ∂_q . Yet, when applied to q and p , it performs a simple quantization :

$$(24)\quad q \longrightarrow q + \int_0^1 \delta\theta\, \Gamma_\theta(0)\ ,\quad p \longrightarrow \alpha\partial_q$$

<u>Rules defined by a series in ∂_p, ∂_q .</u>

Such rules are obtained from (20) - (23) if and only if

$$r(x,q) = q - x$$

They are given by :

(25) $$H_\mu = \chi\left[T(\partial_q, \alpha\partial_p)\, h_\mu(q,p)\right]$$

They now belong to Cohen's [3] family if $\int_0^1 \delta\theta\, \Gamma_\theta^n(0) = 0 \;, \forall n \geq 1.$

Specializing still more to $\Gamma_\theta(Y) = \gamma(\theta)\, Y$, we get

(26) $$T(Z,Y) = \int_0^1 \delta\theta\; e^{[1+\gamma(\theta)]YZ}$$

Classical rules are recovered by choosing

1) $\delta\theta = d\theta \;,\; \gamma(\theta) = -\theta$

hence $T(Z,Y) = \sum_{n\geq 0} \frac{(YZ)^n}{(n+1)!}$ (Born-Jordan)

2) $\delta\theta = \delta(\theta - \lambda)\, d\theta \;,\; 0 \leq \lambda \leq 1 \;,\; \gamma(\theta) = -\theta$

hence $T(Z,Y) = e^{(1-\lambda)YZ}$ (Weyl if $\lambda = \frac{1}{2}$)

3) $\delta\theta = \frac{1}{2}\left[\delta(\theta) + \delta(\theta - 1)\right] d\theta \;,\; \gamma(\theta) = -\theta$

(symmetrization rule).

Conversely, given a definite correspondence principle expressed in terms of $T(Z,Y)$, we can invert formula (23). We then find, when $\delta\theta = d\theta$, the function Γ_θ characterizing the family of configuration space curves. In other cases, Γ_θ is not uniquely determined, which leads to some freedom in interpolation.

On the other hand, if h_μ and H_μ are given, T can be explicitely determined [4]; but it will depend on h_μ and H_μ and cannot define a linear application between spaces of classical functions and operators.

b) Family of curves defined by an implicit equation :

$\xi = x + \theta(q - x)\, \frac{g(\xi)}{g(q)}$ where g is a C^∞ function on $\mathbb{R}$.

Locally the solution of this equation is

$$\xi = \gamma_\theta(x,q) = x + \sum_{n\geq 1} \frac{\theta^n}{n!} \left(\frac{q-x}{g(q)}\right)^n \frac{d^{n-1}}{dx^{n-1}} \left[g^n(x)\right]$$

and conditions (i), (ii) are automatically satisfied.

In the case where $r(x,q) = q - x$, (16) gives

$$H_u = \chi\Big\{h_u(q,p) + \sum_{k\geq 1} \sum_{j=0}^{k-1} \frac{\alpha^k (k-1)!}{j!\,(k-j-1)!} \sum_{m=0}^{k} \frac{(-1)^m \varpi_m\, g^{m,j}(q)}{m!\,(k-m)!\,g^m(q)}\, \partial_q^{k-j}\, \partial_p^k\, h_u(q,p)\Big\}$$

Here the new feature with respects to Cohen's rule is the q dependence of the coefficients in the formal series in ∂_p, ∂_q . Yet it leads to the quantization rules : $q \longrightarrow q$ and $p \longrightarrow \alpha\partial_q$.

In the case where $r(x,q) \neq q - x$, there are tractable examples where both p and q are transformed into various expressions in q and ∂_q .

Finally we may remark that in the formal series defining the above rules only terms in $\partial_p^m \partial_q^n$, $n < m$ occur. This is related to the fact that we always keep step functions in p in the definition of the functional integral.

B I B L I O G R A P H Y

(1) C. Garrod : Rev. mod. Phys. 38 483 (1966)

(2) J. Bertrand and M. Irac : in this Colloquium and references quoted there.

(3) L. Cohen : Journ. Math. Phys. 7 781 (1966)

(4) M. Mizrahi : Preprint 1978

NON-UNIQUENESS IN WRITING SCHRODINGER KERNEL

AS A FUNCTIONAL INTEGRAL

J. Bertrand and M. Irac.

Université PARIS VII
Laboratoire de Physique théorique et mathématique.
Tour 33-43 - 2, Place Jussieu - 75221 PARIS CEDEX 05

....

Feynman integrals [1] have become a privileged tool for quantizing a classical system; this is mainly because the functional integral form of a Schrödinger kernel involves directly a classical function, be it lagrangian or hamiltonian. Yet the actual meaning of these integrals, the nature of the quantum systems thus set up and the relation of configuration space to phase space expressions are still a matter of discussion [2]-[4]. Therefore, we think it worthwhile to point out a few results that can be obtained in a consistent though formal study of these subjects.

I. Functional integral in phase space.

Let us consider a quantum system characterized by a Schrödinger equation

$$(1)\qquad i\hbar \frac{\partial \psi}{\partial t}(q,t) = H\left(\frac{\partial}{\partial q}, q, t\right) \psi(q,t)$$

where $q = (q^\alpha) \in \mathbb{R}^N$ and H is a given differential operator. We tentatively write the propagator $G_{q_0 q}(t_0,t)$ in terms of an arbitrary function h as:

$$(2)\qquad G_{q_0 q}(t_0,t) = \lim_{n\to\infty} G^n_{q_0 q}(t_0,t)$$

where

$$(3)\qquad G^n_{q_0 q}(t_0,t) = \int_{\mathbb{R}^{2nN}} \prod_{i=1}^{n} dq_i \frac{dp_i}{2\pi\hbar} \delta^N(q_n - q) \exp \frac{i}{\hbar} \sum_{i=1}^{n} \left[p_i(q_i - q_{i-1}) - h(p_i, q_i, q_{i-1}, t_i, t_{i-1}) \frac{t - t_0}{n} \right]$$

Here $\frac{dp_i}{2\pi\hbar}$ stands for $\frac{dp_i^1}{2\pi\hbar} \cdots \frac{dp_i^N}{2\pi\hbar}$, $0 \leq t_0 \leq t_i \leq t$, $\forall i$.

[This is a standard [5]-[9] expression if h is a member of the one parameter family $\{h_\lambda\}$ obtained from a classical hamiltonian $\mathcal{H}(p,q,t)$ by:

$$(4)\qquad h_\lambda(p_i, q_i, q_{i-1}, t_i, t_{i-1}) = \mathcal{H}(p_i, \lambda q_i + (1-\lambda) q_{i-1}, t_i) \quad , 0 \leq \lambda \leq 1 \;].$$

We then prove that (2) has all the properties of a Schrödinger kernel and we shall presently write the corresponding H.

Indeed, we compute:

(5) $G_{q_0 q}(t,t) = G^n_{q_0 q}(t,t) = \delta^N (q_0 - q)$

This result is independent of the order of integrations. Now, if t_0 and t' are fixed instants of time such that $0 \leqslant t_0 \leqslant t'$, the following equality holds for all t belonging to $E_{t_0 t'} = \{ t \mid t = t' + \frac{m}{n}(t' - t_0)$, m, n positive integers$\}$

(6) $\int dq' \, G^{kn}_{q_0 q'}(t_0, t') \, G^{km}_{q' q}(t', t) = G^{k(n+m)}_{q_0 q}(t_0, t)$, k positive integer,

provided we can perform the q' integral first. In the limit $k \to \infty$, (6) gives the composition law:

(7) $\int dq' \, G_{q_0 q'}(t_0, t') \, G_{q' q}(t', t) = G_{q_0 q}(t_0, t)$

for all $t \in E_{t_0 t'}$; since $E_{t_0 t'}$ is densely contained in $\mathbb{R}^+$, this result can be extended to the latter under suitable continuity hypotheses for G. From (5) and (7) we can infer that the matrix elements of the operator H in a basis of position eigenstates $|q\rangle$ are:

$$(8) \quad \langle q | H(t) | q_0 \rangle = \lim_{n \to \infty} i\hbar \frac{d}{dt} G^n_{q_0 q}(t_0, t) \Big|_{t = t_0} = \int \frac{dp}{2\pi\hbar} h(p, q, q_0, t, t) \, e^{\frac{i}{\hbar} p (q - q_0)}$$

provided the t - derivative of h (p, q, q_0, t, t) exists. This formula implies that a hermitian H is obtained from functions h satisfying:

$$\int dp \; \exp \frac{i}{\hbar} p (q - q_0) \left[h(p, q, q_0, t, t) - h^*(p, q_0, q, t, t) \right] = 0$$

From now on we shall restrict to functions h depending on one time variable since the only values of h that come into the expression of H (t) are h (p, q, q_0, t, t).

Finally (8) leads to the more explicit form:

$$(9) \quad H(P, Q, t) = \int \frac{dp' \, dq' \, dr \, ds}{4\pi^2 \hbar^2} h\left(p', q' - \frac{s}{2}, q' + \frac{s}{2}, t\right) \exp \frac{i}{\hbar} \left[(q' - Q) r + (p' - P) s \right]$$

Where Q is multiplication by q and $P = -i\hbar \frac{\partial}{\partial q}$

Now, if we return to our original problem, namely if we are given an operator H (P, Q, t), we shall be able to write the propagator of the associated Schrödinger equation as a functional integral as soon as we have found a funtion h(p, q, q_0, t) satisfying (8). This h is not uniquely determined by H and we can use this freedom to write the most tractable expression of the functional integral in each case [10] .

II. Relation to correspondence rules.

So far, no correspondence rule has been needed; yet, if we want to "quantize" a classical system characterized by a hamiltonian $\mathcal{H}(p,q,t)$, we have to find out a way to determine h and this amounts to choosing a quantization rule [11] .

Let us study in more detail the case where $h = h_\lambda$ is given by (4); it leads to a hamiltonian H_λ obtained from (9) by:

(10) $$H_\lambda = \exp - i\hbar\lambda \frac{\partial}{\partial p}\frac{\partial}{\partial q}\mathcal{H}(p,q,t)\Big|_{\substack{q \text{ before } p \\ q \to Q \\ p \to P}}$$

Thus, when λ varies, propagators written in terms of $\mathcal{H}$ using (2), (3) and (4) are solutions of different quantum problems, except in special cases. For example, if $\mathcal{H}$ is of the form:

(11) $$\mathcal{H}(p,q,t) = \sum_{P=(i_1,\dots i_N)\in\{\text{Permutations}\}} f_P(p^{i_1}\cdots p^{i_\ell}, q^{i_{\ell+1}},\dots q^{i_N})$$

with f_p arbitrary, H is independent of λ .

If N = 1, one can show that (11) gives all the hamiltonians possessing this property. If N ≠ 1, there are in addition hamiltonians whose Fourier transform with respects to p is an arbitrary function of $(q^i - q_0^i)$, $((q^1-q_0^1)q^i - (q^i-q_0^i)q^1)$ and t.

We want to emphasize once more that the choice of a definition for the functional integral is somewhat arbitrary and does not deliver us from the problem of quantization.

On the other hand, looking for a solution of a given Schrödinger equation, Mizrahi [7] and Cohen [5] have proposed a device to compute $\langle q|H|q_0\rangle$; they choose h of the form:

(12) $$h(p,q,q_0) = \int \mathcal{H}_f(p,x)\, f\left(\theta, \frac{q-q_0}{\hbar}\right)\exp - i\theta\left[x - \frac{q+q_0}{2}\right]\, dx\,\frac{d\theta}{2\pi}$$

where $\mathcal{H}_f$ is the classical function obtained from H by the dequantization rule corresponding to f. Taking (12) back into (3), we find different expressions of the same propagator, thus recovering an explicit form of Cohen's result.

III. Configuration space integrals.

Still another expression of the functional Schrödinger propagator may be obtained if we perform the p integration and write the result in terms of the Legendre transform L of the function h:

(13) $$L(y,q_0,q,t) \equiv py - h(p,q_0,q,t)$$

and $y^{\alpha} = \frac{\partial h}{\partial p_{\alpha}}(p, q_0, q, t)$, $\alpha = 1, \dots N$, yields $p(y, q_o, q, t)$.

This gives

(14) $$G_{q_0 q}(t_0, t) = \lim_{n \to \infty} \int \prod_{i=1}^{n} dq_i \, \delta(q_n - q) \, S\left(q_i, q_{i-1}, t_i, \frac{t-t_0}{n}\right)$$

where

$$S\left(q_i, q_{i-1}, t_i, \frac{t-t_0}{n}\right) = \int \frac{dy_i}{2\pi\hbar} \left| \det \frac{\partial^2 L}{\partial y_i^{\alpha} \partial y_i^{\beta}} \right| \exp \frac{i}{\hbar} \frac{t-t_0}{n} \left[L(y_i, q_i, q_{i-1}, t_i) + (q_i - q_{i-1} - y_i) \frac{\partial L}{\partial y_i} \right]$$

To compare this expression with usual Feynman [1] integrals in configuration space, we choose $h = h_{\lambda}$ for $\lambda = 1$. There L is the classical lagrangian; the corresponding Feynman integral is:

(15) $$F_{q_0 q}(t_0, t) = \int_{q_n = q} \prod_{i=1}^{n-1} dq_i \prod_{i=1}^{n} B\left(q_i, t_i, \frac{t-t_0}{n}\right) \exp \frac{i}{\hbar} \sum_{i=1}^{n} L\left(\frac{q_i - q_{i-1}}{t - t_0} n, q_i, t_i\right) \frac{t-t_0}{n}$$

where the explicit form of B depends on L. It can be seen that (14) and (15) coincide, i.e. that a function B depending only upon q_i, t_i and $\frac{t - t_o}{n}$ can be determined from (14), if $\mathcal{H}$ has the form:

(16) $$\mathcal{H}(p, q, t) = \lambda(q, t) p^2 + \mu(q, t) p + \nu(q, t)$$

with arbitrary functions λ, μ and ν. For N = 1, we can show that there are no other hamiltonians for which this property of the integrals holds.

It is a pleasure to acknowledge fruitful discussions with E. Amar, M. Ginocchio and J.C. Houard.

References.

1. R.P. Feynman and A.R. Hibbs: Quantum Mechanics and Path Integrals (Mac Graw-Hill, 1965)
   R.P. Feynman: Phys. Rev. 84 10 (1951)
   C. Garrod: Rev. mod. Phys. 38 483 (1966)
   D.I. Blokhintsev and B.M. Barbashov: Sov. Phys. Usp. 15 193 (1972)
2. F.A. Berezin: Teor. Mat. Fiz. 6 194 (1971)
3. See for example:
   W. Garczynski: Rep. Math. Phys. 4 21 (1973)
   A. Truman: J. Math. Phys. 17 1852 (1976)
   Ph. Combe, G. Rideau, R. Rodriguez and M. Sirugue-Collin: preprint CPT (Marseilles) 1976
   M. M. Mizrahi: J. Math. Phys. 19 298 (1978)
   and references cited there
4. C. de Witt-Morette et al.: Gen. Rel. Grav. 8 581 (1977)
5. L. Cohen: J. Math. Phys. 17 597 (1976)

6. F.J. Testa: J. Math. Phys. 12 1471 (1971)
   J.S. Dowker: J. Math. Phys. 17 1873 (1975)
   M. Sato: Prog. Theor. Phys. 58 1262 (1977)
   H. Leschke and M. Schmutz: Z. Physik B 27 85 (1977)
7. M.M. Mizrahi: J. Math. Phys. 16 2201 (1975)
8. R. Fanelli: J. Math. Phys. 17 490 (1976)
9. I.W. Mayes and J.S. Dowker: J. Math. Phys. 14 434 (1973)
10. U. Brandt, H. Leschke and J. Stolze: Preprint Dortmund (1978)
11. J. Bertrand and M. Ginocchio: Preprint Paris VII (1978)

ABOUT THE CONFORMAL PROPERTIES OF YANG - MILLS FIELDS

G. BURDET, C. MARTIN, M. PERRIN
Laboratoire de Physique Mathématique
Universite de Dijon
21000 - DIJON (France)

I - Let $\{A_\mu(x)\}$ be a solution of Yang-Mills equations, without matter field, defined on a (pseudo) Riemannian manifold M with gauge group G (semi-simple compact Lie group) and let $F^{\mu\nu}(x)$ be the corresponding Yang-Mills fields. It is well known that $\{A_\mu(x)\}$ may be interpreted as a connection 1-form A on a principal fibre bundle P(M,G) over the base M with structure group G ; $\{F^{\mu\nu}(x)\}$ is then the curvature form F which defines the field strength of the gauge field A.

A gauge transformation is a diffeomorphism f of P(M,G) into P(M,G) such that :

- $f(p.g) = f(p).g \qquad p\in P \;, \; g\in G$
- f preserves any fibre.

Let G_∞ be the group of all gauge transformations ; G_∞ consists of the sections of the associated fibre bundle $P \times_G G$, G acting on itself by conjugation. $f \in G_\infty$ may be locally represented as a G-valued function on M. The action of $f\in G_\infty$ on the 1-form A is given by :

$$A \to f^{-1} A f + f^{-1} df$$

and on the 2-form F by :

$$F \to f^{-1} F f$$

Let us suppose now there exists a group K of transformations which acts upon the base M. To study the behaviour of the Yang-Mills theory under the action of this group requires to lift its action from the base space M to the principal fibre bundle P(M,G), i.e. to construct a homomorphism of K into the group D_G of diffeomorphisms of P(M,G) which commute with the action of G .

Let us recall that under the action of an element $f \in D_G$ a 1-form connection A in P(M,G) transforms into a 1-form connection $\hat{A}$ defined by [1] :

(1) $$\hat{A}(V) = A(f_*(V))$$

where V belongs to the tangent space to P(M,G) at u : $T_u(P)$ and f_* is the differential of f which sends $T_u(P)$ into $T_{f(u)}(P)$. Let us suppose that K is a group of C^∞ transformations acting on the base M

$$k : x \to k.x \qquad k\in K,\; x\in M \text{ with } k.(k'.x) = (kk').x$$

We shall say that the Yang-Mills equation solution A transforms covariantly under the action of K if :

i) there exists a homomorphism S of K into D_G such that

(2) $\forall k \in K \qquad \dot{S}_k = k.x \qquad \forall x \in M$

where $\dot{S}_k$ is the 1-1 map induced on M from $S_k \in D_G$ by projection.

ii) The 1-form connection $\hat{A}^{(k)}$ transformed of A by S_k is solution of Yang-Mills equations written with the new variables.

Moreover A will be said invariant under the action of a subgroup $K' \subset K$ if

$$\hat{A}^{(k)} = A \qquad \forall k \in K'$$

- Existence of S :

P(M,G) is defined by a family $\{\mathcal{U}_\alpha\}$ of open sets covering M and by G-valued transition functions $\psi_{\alpha\beta}(x)$ defined on $\mathcal{U}_\alpha \cap \mathcal{U}_\beta$.

Let σ_α be a local section above $\mathcal{U}_\alpha$ such that $\forall\alpha$:

$$\sigma_\beta(x) = \sigma_\alpha(x)\ \psi_{\alpha\beta}(x)$$

$\forall k \in K$ we define S_k on $\sigma_\alpha(\mathcal{U}_\alpha)$, $\forall\alpha$, in the following way :

Let $x \in \mathcal{U}_\alpha$, then there exists β such that $k.x \in \mathcal{U}_\beta$, and we set

$$S_k[\sigma_\alpha(x)] = \sigma_\beta(k.x)\ \lambda_k^{\alpha\beta}(x) \qquad \text{with} \qquad \lambda_k^{\alpha\beta}(x) \in G.$$

Then S_k is extended to $\pi^{-1}(\mathcal{U}_\alpha)$ (π being the projection of P onto M) by using the invariance under G as follows

$$S_k(u) = S_k(\sigma_\alpha[\pi(u)])a$$

with $\quad u = \sigma_\alpha[\pi(u)]a \quad , a \in G, \quad \forall u \in \pi^{-1}(\mathcal{U}_\alpha)$

Finally S_k is defined on P(M,G) by setting :

(3) $\qquad S_k(u) = \sigma_\beta[k.\pi(u)] \qquad \lambda_k^{\alpha\beta}[\pi(u)]a$

$\forall u \in P$ with $\pi(u) \in \mathcal{U}_\alpha$ and $k.\pi(u) \in \mathcal{U}_\beta$.

The image of a point of P(M,G) will be uniquely defined if and only if :

(4) $\quad \lambda_k^{\beta\delta}(x) = \psi_{\delta\gamma}(k.x)\ \lambda_k^{\alpha\gamma}(x)\ \psi_{\alpha\beta}(x) \qquad$ if $x \in \mathcal{U}_\alpha \cap \mathcal{U}_\beta$, $kx \in \mathcal{U}_\gamma \cap \mathcal{U}_\delta$

If such λ's exist, they are C^∞ mappings on their domain.

By writting that S is an homomorphism we get

(5) $\quad \lambda_{kk'}^{\alpha\gamma}(x) = \lambda_k^{\beta\gamma}(k'.x)\ \lambda_{k'}^{\alpha\beta}(x) \qquad$ if $x \in \mathcal{U}_\alpha$, $k'.x \in \mathcal{U}_\beta$ and $(kk').x \in \mathcal{U}_\gamma$

It is easy to verify that (5) is consistent with (4).

Remark : If we suppose that the mapping $k \to k.x$ is continuous, $\forall x$ there exists a neighbourhood $\mathcal{N}$ of the identity in K such that for $k,k' \in \mathcal{N}$, $k'.x$ and $(kk').x \in \mathcal{U}_\alpha$.

Then from rel.(5) it is seen that $\lambda_k^{\alpha\alpha}$ is a G-valued local coboundary.

In summary, if a family $\{\lambda_k^{\alpha\beta}(x),\ \forall k \in K\}$ satisfying (4) and (5) can be found, the mapping S_k from P to P defined by (3) is an automorphism of P(M,G) satisfying (2). So, the 1-form connection $\hat{A}$ defined by (1) is now explicitely given by a family $\{\hat{A}_\beta\}$ of 1-forms $\hat{A}_\beta$ defined on $\mathcal{U}_\beta$ by $\hat{A}_\beta = \sigma_\beta^* \hat{A}$ (σ_β^* denoting the dual of the differential of the mapping σ_β) obtained from the family $\{A_\alpha\}$ as follows :

$$(6) \qquad k^*.\hat{A}_\beta = \lambda_k^{\alpha\beta}(x)\ A_\alpha(x)\ \lambda_k^{\alpha\beta}(x)^{-1} + \lambda_k^{\alpha\beta}(x)\ d\lambda_k^{\alpha\beta}(x)^{-1} \qquad x \in \mathcal{U}_\alpha \quad k.x \in \mathcal{U}_\beta$$

where k^* is the dual of the tangent map associated to k.

II - Application to instanton solutions

For the sequel we specialize to SU(2)-bundles over S^4. The dual $^*F$ of a curvature 2-form F is defined by using the Hodge operator [2] and F is said to be self-dual (respectively anti self-dual) if

$$^*F = F \qquad \text{(resp. } ^*F = -F\text{)},$$

accordingly the corresponding connexion is said to be self-dual (resp. anti self-dual).

Any fibre bundle $P(S^4,SU(2))$ is characterized by the 2nd Chern class $C_2(P) \in H^4(S^4,\mathbb{Z})$. One needs also the 1^{rst} Pontryagin class $p_1(P)$ which is defined by

$$p_1(P) = -2\,C_2(P) = -\frac{1}{4\pi^2}\,\mathrm{Tr}(F \wedge F)$$

By integrating over the fundamental cycle of S^4 a topological invariant n is obtained which is called the Pontryagin index or winding number. n is a positive integer if the connection is self-dual and a negative integer if it is anti self-dual.

By definition a $|n|$-instanton solution of Yang-Mills equations (more exactly a gauge equivalence class of solutions or modulus) will be any (anti)self-dual connection form on a SU(2)-bundle over S^4 the Pontryagin index of which is n.

Now it is interesting to note that the Hodge operator is conformally invariant on 2-forms :

$$^*(k^{-1*}.\ F_\alpha) = k^{-1*}.\ ^*F_\alpha \qquad \forall k \in SL(2,\mathbb{H})$$

If the construction described in Part I has been explicitely carried out for K being the conformal group $SL(2,\mathbb{H})$ the curvature $\hat{F}$ associated to $\hat{A}$ given by (6) , can be evaluated from :

$$\hat{F}_\beta(\hat{x}) = \lambda_k^{\alpha\beta}(k^{-1}\hat{x})(k^{-1*}.F_\alpha)\ \lambda_k^{\alpha\beta}(k^{-1}x)^{-1}\ , \quad k^{-1}\hat{x} \in \mathcal{U}_\alpha,\ \hat{x} \in \mathcal{U}_\beta$$

Then the conformal invariance of the Hodge operator implies that if F_α is (anti) self-dual, $k^{-1*}.F_\alpha$ too is (anti) self-dual as well as $\hat{F}_\beta(\hat{x})$ which will be a $|n|$-instanton Yang-Mills field. Therefore according to the definition given in Part I the instanton solutions will transform covariantly under the conformal group if its action can be lifted from S^4 to the SU(2)-bundles. We shall treat explicitely the cases $n=\pm 1$. The notations used in what follows are defined in the Appendix.

The (anti) self-dual 1-instanton solutions correspond respectively to connections on the following principal fibre bundles [3 a,b] :

$$P^{s(a)}(S^4, SU(2)_{s(a)}) = (Sp(2)/_{SU(2)_{a(s)}} = S^7) \to S^4$$

The base $S^4 (= Sp(2)/Sp(1) \times Sp(1))$ can be described by using two open sets

$$\mathcal{U}_1 = \{\Lambda_1\} \quad \text{and} \quad \mathcal{U}_2 = \{\Lambda_2\} \quad , \ S^4 = \mathcal{U}_1 \cup \mathcal{U}_2$$

The fibre bundles $P^{s(a)}$ are then described by two local trivializations which need two charts :

Let $u \in P^{s(a)}$ be a representative of the class of $\begin{pmatrix} A & B \\ C & D \end{pmatrix} \in Sp(2)$, if $u \in \pi^{-1}(\mathcal{U}_\alpha)$, $\alpha = 1,2$, one considers the charts $u \to (\Lambda_\alpha, U_\alpha)$ $((\Lambda_\alpha, V_\alpha)$ resp.).

The corresponding transition functions are given by :

$$\psi_{21}^s = -\frac{X_1^+}{|X_1|} = -\frac{X_2}{|X_2|} \quad \text{and} \quad \psi_{21}^a = \frac{X_1}{|X_1|} = \frac{X_2^+}{|X_2|}$$

Up to a gauge transformation any 1-instanton solution is given by :

$$\text{(7a)}\ n = 1 \ : \quad A_\alpha^s(X_\alpha) = \frac{|W_\alpha|^2}{\mu_\alpha^2 + |W_\alpha|^2}\ \frac{W_\alpha}{|W_\alpha|}\ d\ \frac{W_\alpha^+}{|W_\alpha|}$$

$$\text{(7b)}\ n = -1 \ : \quad A_\alpha^a(X_\alpha) = \frac{|W_\alpha|^2}{\mu_\alpha^2 + |W_\alpha|^2}\ \frac{W_\alpha^+}{|W_\alpha|}\ d\ \frac{W_\alpha}{|W_\alpha|} \qquad \alpha = 1,2$$

with $W_\alpha = X_\alpha - X_\alpha^o$ where X_α^o and μ_α are the "position" and the "size" of the instanton respectively.

Now, let $k = \begin{pmatrix} A & B \\ C & D \end{pmatrix}$ and $k^{-1} = \begin{pmatrix} A' & B' \\ C' & D' \end{pmatrix}$ belong to $SL(2,\mathbb{H})$; k acts on S^4 as follows :

- in U_1 : $X_1 \to \hat{X}_1 = (A\,X_1 + B)\ (C\,X_1 + D)^{-1}$
- in U_2 : $X_2 \to \hat{X}_2 = (C + D\,X_2)\ (A + B\,X_2)^{-1}$

Let us note that $X_1 = X_2^{-1}$ and $\hat{X}_1 = \hat{X}_2^{-1}$ in $\mathcal{U}_1 \cap \mathcal{U}_2$.

It is then easy to verify that the following quantities satisfy relations (4) and (5) :

$$\text{(8a)} \quad \lambda_k^{11}(X_1) = \frac{A'^+ - C'^+ X_1^+}{|A'^+ - C'^+ X_1^+|} \quad , \quad \lambda_k^{22}(X_2) = \frac{D'^+ - B'^+ X_2^+}{|D'^+ - B'^+ X_2^+|} \quad , \quad \lambda_k^{12}(X_1) = \frac{B'^+ - D'^+ X_1^+}{|B'^+ - D'^+ X_1^+|}$$

in the self-dual case and

$$\text{(8b)} \quad \lambda_k^{11}(X_1) = \frac{CX_1 + D}{|CX_1 + D|} \quad , \quad \lambda_k^{22}(X_2) = \frac{A + BX_2}{|A + BX_2|} \quad , \quad \lambda_k^{12}(X_1) = \frac{AX_1 + B}{|AX_1 + B|}$$

in the anti self-dual one.

Under the action of k, according to rel.(6) the 1-instanton solutions $A_\alpha^s(X_\alpha)$ and $A_\alpha^a(X_\alpha)$ turn into $\hat{A}_\alpha^s(\hat{X}_\alpha)$ and $\hat{A}_\alpha^a(\hat{X}_\alpha)$ respectively. One establishes that $\hat{A}_\alpha^s(\hat{X}_\alpha)$ and $\hat{A}_\alpha^a(\hat{X}_\alpha)$ appear under the same form as r.h.s. of (7a) and (7b) respectively in terms of the new variables $\hat{W}_\alpha = \hat{X}_\alpha - \hat{X}_\alpha^o$ with the new "positions" and "sizes" given by :

$$\text{(9a)} \quad (\mu_1, X_1^o) \to (\hat{\mu}_1, \hat{X}_1^o) = \left(\frac{\mu_1}{|CX_1^o + D|^2 + \mu_1^2 |C|^2} , \frac{(AX_1^o + B)(CX_1^o + D)^+ + \mu_1^2 AC^+}{|CX_1^o + D|^2 + \mu_1^2 |C|^2} \right)$$

$$\text{(9b)} \quad (\mu_2, X_2^o) \to (\hat{\mu}_2, \hat{X}_2^o) = \left(\frac{\mu_2}{|A + BX_2^o|^2 + \mu_2^2 |B|^2} , \frac{(C + DX_2^o)(A + BX_2^o)^+ + \mu_2^2 DB^+}{|A + BX_2^o|^2 + \mu_2^2 |B|^2} \right)$$

Moreover it is easy to verify that $\hat{A}_\alpha^{s(a)} = A_\alpha^{s(a)}$ if k belongs to Sp(2) embedded into SL(2,$\mathbb{H}$), which makes clear the SO(5) invariance of the 1-instanton solutions [4]. The dimension of the space of moduli for $|n| = 1$ being 5 [5] it follows that the manifold of 1-instanton solutions is a single 5-dimensional orbit under the action of SL(2,$\mathbb{H}$), any point of which being stabilized by Sp(2), i.e. SL(2,$\mathbb{H}$)/Sp(2) [3b] .

To end we want to translate the above results in terms of the linear formalism described in Ref.[6a,b] because this one permits a complete construction for all instanton-solutions which allows to hope a geometric characterization of the n-instanton solution manifold as a family of SL(2,$\mathbb{H}$)-orbits. Briefly let us recall that this approach requires the construction of a quaternionic $(n+1) \times n$-matrix M with some constraints and whose elements are linear in $P^1(\mathbb{H})$ isomorphic to S^4.
Then the linear equivalence which consists in replacing M by M' = PMR with $P \in Sp(n+1)$ and $R \in Gl(n,\mathbb{R})$ corresponds to a SU(2) equivalence of the 1-form connection. Explicitely for n = 1 let us set

$$M = \begin{pmatrix} \ell_1 \\ \ell_2 \end{pmatrix} \text{ , then } M' = \begin{pmatrix} A & B \\ C & D \end{pmatrix} \begin{pmatrix} \ell_1 \\ \ell_2 \end{pmatrix} r, \text{ where } \begin{pmatrix} A & B \\ C & D \end{pmatrix} \in Sp(2) \text{ and } r \in \mathbb{R}.$$

Then the gauge potential corresponding to M' deduces from the one corresponding to M by the gauge transformation

$$\frac{A|\ell_2|^2 - B\ell_2 \ell_1^+}{|A|\ell_2|^2 - B\ell_2\ell_1^+|} \in Sp(1)$$

The solution $A_\alpha^s(X_\alpha)$ given in (7a) corresponds to $M_\alpha = \begin{pmatrix} X_\alpha - X_\alpha^o \\ \mu_\alpha \end{pmatrix}$ in the linear formalism, while $\hat{A}_\alpha^s(\hat{X}_\alpha)$ corresponds to $\hat{M}_\alpha = \begin{pmatrix} \hat{X}_\alpha - \hat{X}_\alpha^o \\ \hat{\mu}_\alpha \end{pmatrix}$

Then one can show that there exist pairs (P_α, R_α) belonging to $(Sp(2), GL(1,\mathbb{R}))$ and given by

$$P_1 = (|D+CX_1^o|^2 + \mu_1^2|C|^2)^{-1/2} \begin{pmatrix} A'^+ - C'^+X_1^{o+} & \mu_1 C'^+ \\ \mu_1 C & D+CX_1^o \end{pmatrix}, \quad R_1 = (|D+CX_1^o|^2+\mu_1^2|C|^2)^{-1/2}$$

$$P_2 = (|A+BX_2^o|^2 + \mu_2^2|B|^2)^{-1/2} \begin{pmatrix} D'^+ - B'^+X_2^{o+} & \mu_2 B'^+ \\ \mu_2 B & A+BX_2^o \end{pmatrix}, \quad R_2 = (|A+BX_2^o|^2+\mu_2^2|B|^2)^{-1/2}$$

which are such that $\hat{M}_\alpha = P_\alpha M_\alpha R_\alpha$. So, in the linear formalism the pairs (P_α, R_α) play the role of the $\lambda_k^{\alpha\alpha}(X_\alpha)$ given by (8a).

We want to thank J.L. RICHARD for fruitful discussions.

References

1. S. KOBAYASHY, K. NOMIZU : "Foundations of differential geometry", Interscience, New York (1969).

2. M.F. ATIYAH, N.J. HITCHIN, I.M. SINGER : "Self-duality in four-dimensional Riemannian Geometry", Preprint.

3a. A. TRAUTMAN : International Journal of Theoretical Physics 16, 561 (1977).

3b. R. STORA : "Yang Mills Instantons, Geometrical Aspects",Erice (1977).CPT 77/P.943
R. DURIEUX : Thèse de 3e Cycle, Marseille (1978).

3c. A. BELAVIN, A. POLYAKOV, A. SCHWARZ, Yu. TYUPKIN : Phys. Lett. B5 9, 85-87 (1975).

4. R. JACKIW, C. REBBI : Phys. Rev. D 14, 517 (1976).

5. M.F. ATIYAH, N.J. HITCHIN, I.M. SINGER : Proc.Nat.Acad.Sci.U.S.A. 74,2662 (1977).

6a. M.F. ATIYAH, N.J. HITCHIN, V.G. DRINFELD, Yu. I. MANIN : Phys.Lett. 65A, 185 (1978).

6b. J. MADORE, J.L. RICHARD, R. STORA :"An Introduction to the Twistor Programme", Les Houches (1978).CPT 78/P.1005.

APPENDIX

Let $GL(2,\mathbb{H})$ be the set of all regular quaternionic 2x2 matrices. $GL(2,\mathbb{H})$ is isomorphic to the group of regular linear transformations on $\mathbb{H}^2$. ($\mathbb{H}$ being the quaternionic field). By excluding the real dilatations on $\mathbb{H}^2$, we get $SL(2,\mathbb{H})$ which is connected and isomorphic to $SU^*(4)$. Moreover $SL(2,\mathbb{H})/Z_2$ is isomorphic to $SO_0(5,1)$, one has

$$SL(2,\mathbb{H}) = \{ \begin{pmatrix} A & B \\ C & D \end{pmatrix} \in GL(2,\mathbb{H}) \mid \Delta = |A|^2 |D|^2 + |B|^2 |C|^2 - AC^+ DB^+ - BD^+ CA^+ = 1 \}$$

Let us introduce the symplectic group $Sp(2)$ which keeps invariant the quaternionic product and is isomorphic to $SL(2,\mathbb{H}) \cap U_4$:

$$Sp(2) = \{K \in SL(2,\mathbb{H}) \ , \ K^+K = 1 \}$$

One has the following isomorphism $Sp(2)/Z_2 \approx SO(5)$.

We use the following decompositions of an element $\begin{pmatrix} A & B \\ C & D \end{pmatrix} \in Sp(2)$:

$$\begin{pmatrix} A & B \\ C & D \end{pmatrix} = \Lambda_\alpha \begin{pmatrix} U_\alpha & 0 \\ 0 & V_\alpha \end{pmatrix} \text{ with } \begin{cases} D \neq 0 : \Lambda_1 = \dfrac{1}{\sqrt{1+|X_1|^2}} \begin{pmatrix} 1 & X_1 \\ -X_1^+ & 1 \end{pmatrix}, \ U_1 = \dfrac{A}{|A|}, \ V_1 = \dfrac{D}{|D|}, X_1 = BD^{-1} \\ \\ B \neq 0 : \Lambda_2 = \dfrac{1}{\sqrt{1+|X_2|^2}} \begin{pmatrix} -X_2^+ & 1 \\ 1 & X_2 \end{pmatrix}, \ U_2 = \dfrac{C}{|C|}, \ V_2 = \dfrac{B}{|B|}, X_2 = DB^{-1} \end{cases}$$

$\alpha = 1,2$.

The set of matrices $\begin{pmatrix} U & 0 \\ 0 & V \end{pmatrix}$ is the direct product $Sp(1) \otimes Sp(1)$ isomorphic to $SU(2) \otimes SU(2)$ and we denote $\{U\} = SU(2)_s$ and $\{V\} = SU(2)_a$.

It must also be noticed that $SO(5)/SO(4) \approx S^4$ and $Sp(2)/Sp(1) \otimes Sp(1) \approx P^1(\mathbb{H})$, so that by using the previously quoted isomorphisms one sees that S^4 is isomorphic to $P^1(\mathbb{H})$.

Added Note : To be more complete, we have to mention that, in the context of gauge theories, homomorphisms of a given group into the diffeomorphisms of a principal fibre bundle have also been considered to construct a priori invariant gauge fields. See for instance : A.S. SCHWARZ , Commun.math.Phys. 56, 79 (1977) ; J.P. HARNAD, S. SCHNIDER, L. VINET, Phys. Lett. B76, 589 (1978) , C.R.M. 774 and 792 Preprints.

Infrared Problem and Zero-mass Limit in a Model of Non-abelian Gauge Theory[†]

G. Curci

CERN, Geneva, Switzerland

and

R. Ferrari

Istituto di Fisica, Università di Pisa, Pisa, Italy
and Istituto Nazionale di Fisica Nucleare, Sezione di Pisa

Abstract. Infrared divergences of Yang Mills theory are regularized by an explicit mass term for the vector meson. By using the invariance properties under Becchi, Rouet and Stora transformations we demonstrate that the theory violates physical unitarity also in the limit of zero mass.

[†] Seminar presented by R. Ferrari at the conference "Colloquium on Mathematical Problems in Feynman Path Integral" Marseille, May 22-26, 1978.

Various attempts have been made recently, to evaluate predictions in Quantum Chromodynamics by using standard perturbation theory. The fundamental interaction of Quantum Chromodynamics is given by vector mesons coupled to the quarks according to a Yang Mills field theory. The perturbative approach to this field theory encounters difficulties for the physical interpretation. To be definite, if one uses the covariant quantization, the theory is renormalizable according to the power counting theorem, however the unphysical modes do not decouple from the physical states, i.e. the probability is not conserved.

The way out for this difficulty was indicated by Feynman[1] and fully developed by Faddeev and Popov[2], 't Hooft and Lee and Zinn-Justin[3]. They showed that the contribution of the unphysical modes of the vector mesons to transition probabilities is cancelled, in an enlarged theory, by other unphysical modes given in terms of a complex scalar fermi field (Faddeev Popov ghost) and, in some case, of auxiliary scalar fields (e.g. Higgs fields). The physical unitarity is satisfactory established if the vector mesons acquire a mass by some mechanism (e.g. Higgs mechanism or violation of minimal coupling principle[4]). For massless theory the proof is formal and, in our opinion, unreliable.

We will discuss the problem of physical unitarity by using an infrared regulator in the lagrangian: an explicit mass term for the vector meson[5]†:

$$\mathcal{L} = -\frac{Z_g}{4g^2}[G_{\mu\nu}]^2 + \frac{\alpha Z_\alpha}{g^2}\Big[-\frac{1}{2}(\partial^\mu A_\mu)^2 + \partial^\mu\varphi^*\partial_\mu\varphi - \frac{1}{2}(\partial_\mu\varphi^*\times\varphi - \varphi^*\times\partial_\mu\varphi)A^\mu + \frac{1}{8}(\varphi^*\times\varphi)^2\Big] + \frac{\mu^2 Z_\mu}{g^2}\Big[\frac{A^2}{2} - \varphi^*\varphi\Big] + \mathcal{L}_{matter} \qquad (I)$$

This lagrangian differs from the one proposed by Faddeev and Popov even for $\mu^2=0$. The reason of our choice lies on the invariance of the action under Becchi Rouet Stora (BRS) transformations[6] (each term in the square braket has this property):

$$\begin{aligned} [F, A_\mu] &= D_\mu\varphi \,, \\ \{F, \varphi\} &= -\frac{1}{2}\varphi\times\varphi \\ \{F, \varphi^*\} &= -\Big(\partial^\mu A_\mu + \frac{1}{2}\varphi^*\times\varphi\Big) \end{aligned} \qquad (2)$$

whereF is the generator of the BRS transformations (for simplicity we neglect the presence of the matter fields).

The perturbative series is defined by dimensional regularization and BRS inva-

---

† Where the group indices are suppressed and the products are given by the Kroneker δ and the structure constants of the group, e.g. $\varphi^*\varphi = \varphi^{*i}\varphi^j\delta^{ij}$, $(\varphi^*\times\varphi)^i = f^{ijk}\varphi^{*j}\varphi^k$

riance guarantees that the normalization conditions are simply obtained by finite renormalization constants (Z_g, Z_α, Z_μ)[7]. We choose on-shell conditions by using the inverse of the propagator of the vector meson

$$\Gamma_{\mu\nu}(k) = A(k^2)\, g_{\mu\nu} + B(k^2)\, k_\mu k_\nu \tag{3}$$

i.e.

$$\begin{aligned} A(\mu^2) &= 0 \\ A'(\mu^2) &= g^{-2} \\ B(\mu^2) &= 0 \end{aligned} \tag{4}$$

This is the Feynman gauge, since the transverse and the longitudinal part of $\Gamma_{\mu\nu}$ have the same mass.

The problem of physical unitarity can be formulated in the following way. The physical vectors form a subspace G' of the space G of the asymptotic states. G' must have reasonable properties, e.g. positive metric, helicity states according to the unitary representations of the Poincaré group etc. Since the lagrangian (I) is hermitian the S-matrix is unitary, i.e.

$$\sum_{\{n\}} \langle \Phi' | S^* | n \rangle \langle n | S | \Phi \rangle = \langle \Phi' | \Phi \rangle \tag{5}$$

where {n} is an orthonormal basis for G. However G contains unphysical states, therefore a physical interpretation is possible only if physical unitarity is valid, i.e. if for any Ψ' and Ψ of G'

$$\sum_{\{n\}}{}' \langle \Psi' | S^* | n \rangle \langle n | S | \Psi \rangle = \langle \Psi' | \Psi \rangle \tag{6}$$

where Σ' is the sum over an orthonormal basis of G' (conservation of probability).

For $\mu^2 \neq o$ it easy to see that the S-matrix is not physically unitary[8]. Here we want to investigate whether, in the limit $\mu^2 = o$, physical unitarity is recovered.

The choice of the *physical space* is crucial and non-trivial because of the following problems: first, for $\mu^2 \neq o$ the vector meson is a spin-one mode with three polarization states but for $\mu^2 = o$ the vector meson should have only two helicity states and second, in Yang Mills theory, unlike QED, the scalar part described by $\partial^\sigma A_\sigma$ does not decouple from the transverse states.

For $\mu^2 \neq o$ the asymptotic field describes a spin-one particle and a scalar mode with negative metric. We choose the following basis for a momentum k $(k^2 = \mu^2)$

$$\varepsilon^{(i)} = (0; \underline{\rho}^{(i)}) \qquad i = 1, 2 \tag{7}$$

with $\underline{\varsigma}^{(i)} \cdot \underline{k} = \underline{\varsigma}^{(1)} \cdot \underline{\varsigma}^{(2)} = 0 \qquad \underline{\varsigma}^{(i)2} = 1$

and

$$\varepsilon_l = \mu^{-1}\left(k\,;\ \underline{k}\,\frac{k_0}{k}\right)$$
$$\varepsilon_s = \mu^{-1}(k_0\,;\ \underline{k}) \qquad (8)$$

For $\mu^2 = o$ the helicity-one states are described by equivalence classes. We assume the following prescription: in order to evaluate the zero-mass limit from the $\mu^2 \neq o$ theory, we take as representative of the equivalence classes the vectors $\varepsilon_\mu^{(i)}$, i=I,2, in eq. (7). Therefore for any given momentum and group index the contribution of the polarization states ε_1 and ε_s should cancel the Faddeev Popov part in the unitarity eq. (5) when the vector states Ψ and Ψ' are physical and the limit $\mu^2 = o$ is performed†.

We shall prove that this cancellation does not occur in the limit of zero mass. This fact demonstrates that the formal proof for the massless theory is unreliable, moreover it indicates that our choice of physical space is not the right one. Our conjecture is that the theory can be interpreted physically if the finite resolution in energy *and* direction of momentum is taken into account.

In order to prove our statement we use extensively the properties of the generator F of the BRS transformations[11]. Since the action is invariant under BRS transformations we have

$$[F, S] = 0 \qquad (9)$$

Moreover the perturbative solution in terms of Green functions satisfies the Slavnov Taylor identities[12] associated to the BRS invariance. Therefore the symmetry is not spontaneously broken, i.e.

$$F\,|0\rangle = 0 \qquad (10)$$

For $\mu^2 \neq o$ the asymptotic field can be splitted in a unique way into a spin-one and spin-zero part. The spin-one part is transverse; therefore, by using the reduction formulae in the first eq. (2), F commutes with the creation- and annihilation-operator associated to the transverse part of $A_\mu^{in,out}$. Using this fact and eq. (10) we see that

$$F\,|\Psi\rangle = 0 \qquad (11)$$

if Ψ contains no FP ghosts or ε_s-polarization states.

The asymptotic states are normalized in such a way that $\langle \Phi'^{in} | S | \Phi^{in} \rangle$ is

† A similar prescription has been used by Faddeev and Slavnov[9]. However they used the Landau gauge and therefore the infrared divergences are not fully regularized. On the contrary van Dam and Veltman[10] do not consider unphysical the polarization state ε_1 and therefore the violation of physical unitarity occurs as for $\mu^2 \neq o$.

the relevant amputated Green function evaluated on the mass-shell. Then from eq.(4) we get the norm of the one-particle states

$$\begin{aligned} \langle a'_T | a_T \rangle &= 2 k_0 \, g^{-2} \, \delta(\underline{k}' - \underline{k}) \\ \langle a'_s | a_s \rangle &= -2 k_0 \, r^{-2} \, \delta(\underline{k}' - \underline{k}) \\ \langle c' | c \rangle &= -\langle \bar{c}' | \bar{c} \rangle = 2 k_0 \, s^{-2} \, \delta(\underline{k}' - \underline{k}) \end{aligned} \tag{12}$$

where a_T is any one of the transverse polarization modes, a_s is the scalar part of the vector meson and c, $\bar{c}$ are the ghost and anti-ghost. r^2 and s^2 are the residua of the longitudinal part of the vector meson propagator and of the Faddeev Popov propagator. From eq. (4) we get

$$r^{-2} = g^{-2} + h(1) \tag{13}$$

where h is given by the relation

$$B(k^2, \mu^2) = h(k^2/\mu^2)\,(k^2/\mu^2 - 1) \tag{14}$$

The unitarity equation (5) can be written in the following form

$$\langle \Phi' | \Phi \rangle = \langle \Phi' | S^* \,\mathcal{U}[\mathcal{E}(\lambda)]\, S | \Phi \rangle \tag{15}$$

where

$$\mathcal{E}(\lambda) = \exp\left\{ \int \frac{d^3k}{2k_0} \left[\lambda(\underline{k}) \left(\mathcal{A}(\underline{k}) + g^2 a_\ell^*(\underline{k}) a_\ell(\underline{k}) \right) + g^2 \left(a_1^*(\underline{k}) a_1(\underline{k}) + a_2^*(\underline{k}) a_2(\underline{k}) \right) \right] \right\} \tag{16}$$

$$\mathcal{A}(\underline{k}) = -r^2 a_s^*(\underline{k}) a_s(\underline{k}) + s^2 \left(c^*(\underline{k}) c(\underline{k}) - \bar{c}^*(\underline{k}) \bar{c}(\underline{k}) \right) \tag{17}$$

and the sum over the group indices is understood. $\mathcal{U}$ is a product defined on monomials of creation- and annihilation-operators, where the projector on the vacuum state is inserted between the N-ordered creation- and annihilation-operators[11]. Let us now consider states Ψ and Ψ' where no FP ghosts or ε_s-polarization states are present. By using the properties of the $\mathcal{U}$-product, we get

$$\begin{aligned} &2 k_0 \frac{\delta}{\delta \lambda(\underline{k})} \langle \Psi' | S^* \,\mathcal{U}[\mathcal{E}(\lambda)]\, S | \Psi \rangle = \\ &= \langle \Psi' | S^* \,\mathcal{U}\left[\left(\mathcal{A}(\underline{k}) + g^2 a_\ell^*(\underline{k}) a_\ell(\underline{k}) \right) \mathcal{E}(\lambda) \right] S | \Psi \rangle \end{aligned} \tag{18}$$

We consider now the contribution of the longitudinal states to the RHS of eq.(18). According to our normalization of the asymptotic states, this is given by the index contraction of the truncated Green functions by the tensor

$$\varepsilon_{\ell\mu}\varepsilon_{\ell\nu} = -\frac{k_0^2}{k^2}\varepsilon_{s\mu}\varepsilon_{s\nu} + \frac{k_0}{k}(\varepsilon_{s\mu}\varepsilon_{\ell\nu} + \varepsilon_{s\nu}\varepsilon_{\ell\mu}) + \frac{\mu^2}{k^2} g_{\mu 0} g_{\nu 0} \tag{19}$$

For *fixed* momentum k, the last term can be neglected in the limit of zero-mass†. Therefore the longitudinal term in eq. (18) can be replaced by

$$a_\ell^* a_\ell \longrightarrow -\frac{k_0^2}{k^2} a_s^* a_s + \frac{k_0}{k}(a_\ell^* a_s + a_s^* a_\ell) = -a_s^* a_s + (a_\ell^* a_s + a_s^* a_\ell) + O(\mu^2) \tag{20}$$

We evaluate now the (anti-) commutator of F with the creation- and annihilation -operators. We need the following amplitudes

$$\frac{\delta^2 \Gamma}{\delta\varphi\,\delta K^\mu} = -i\, I\, p_\mu \tag{21}$$

$$\frac{\delta^2 \Gamma}{\delta A^\mu\,\delta R} = -i\, J\, p_\mu \tag{22}$$

where Γ is the generating functional of the one-particle irreducible amplitudes and K_μ and R are the external sources associated to $D_\mu\phi$ and $\partial^\mu A_\mu + \frac{1}{2}\phi^*\times\phi$. By using the Slavnov Taylor identities[12] for Γ we get[5]

$$-\frac{\delta^2\Gamma}{\delta A_\mu(x)\,\delta A^\rho}\,\frac{\delta^2\Gamma}{\delta\varphi(y)\,\delta K_\rho} + \frac{\delta^2\Gamma}{\delta A_\mu(x)\,\delta R}\,\frac{\delta^2\Gamma}{\delta\varphi(y)\,\delta\varphi^*} = 0 \tag{23}$$

(where summation over the indeces of the dummy sources is understood), i.e.

$$J\, r^2 = I\, s^2 \tag{24}$$

Finally we can use the reduction formulae for the fields in eq. (2) and by using the Jacobi identity and eq. (24) we get

$$\begin{aligned} &[F, a_s] = \mu J c && [F, a_s^*] = \mu J \bar{c}^* \\ &\{F, c\} = 0 && \{F, c^*\} = \mu I a_s^* \\ &\{F, \bar{c}\} = \mu I a_s && \{F, \bar{c}^*\} = 0 \end{aligned} \tag{25}$$

and

$$[F, \mathcal{A}] = -r^2\mu J\,\bar{c}^* a_s - r^2\mu J\, a_s^* c + \mu J s^2\, a_s^* c + \mu I s^2\,\bar{c}^* a_s = 0 \tag{26}$$

By using the properties of F given by eq. (9), (11), (25) and (26), we see that

† For soft vector mesons the last term does contribute. The same effect appears in QED. See for instance the footnotes in J.M. Jauch and F. Rohrlich, "The Theory of Photons and Electrons", Springer-Verlag, New York Heidelberg Berlin 1976, p.337.

the 1-s terms in eq.(20) do not contribute to the RHS of eq. (18). The remaining terms give

$$2k_0 \frac{\delta}{\delta\lambda(\underline{k})} \langle \Psi' | S^* \mho[\mathcal{E}(\lambda)] S | \Psi \rangle =$$

$$= \langle \Psi' | S^* \mho[(\mathcal{A} - g^2(-a_s^* a_s + \frac{1}{\mu I}\{F, c^*\} a_s + \frac{1}{\mu I} a_s^* \{F, \bar{c}\}))\mathcal{E}(\lambda)] S | \Psi \rangle + O(\mu^2)$$

$$= \langle \Psi' | S^* \mho[(\mathcal{A} - g^2(-a_s^* a_s + \frac{J}{I} c^* c - \frac{J}{I} \bar{c}^* \bar{c}))\mathcal{E}(\lambda)] S | \Psi \rangle + O(\mu^2) \qquad (27)$$

and by eq. (24)

$$= \langle \Psi' | S^* \mho[(1 - g^2/\Gamma^2)\mathcal{A}\,\mathcal{E}(\lambda)] S | \Psi \rangle + O(\mu^2)$$

$$= \langle \Psi' | S^* \mho[-g^2 h(1)\mathcal{A}\,\mathcal{E}(\lambda)] S | \Psi \rangle + O(\mu^2) \qquad (28)$$

Eq. (28) can be now integrated. We introduce the minimal observable energy ΔE of the quanta, in order to cut off the soft modes. By letting λ = 1 we get

$$\langle \Psi' | S^* S | \Psi \rangle = \langle \Psi' | S^* \mho[\mathcal{E}_{IR} \exp(\int_{k_0 > \Delta E} d^3k/2k_0 \; g^2 (a_2^*(\underline{k}) a_1(\underline{k}) + a_2^*(\underline{k}) a_2(\underline{k})))$$

$$\exp(-g^2 h \int_{k_0 > \Delta E} d^3k/2k_0 \, \mathcal{A}(\underline{k}))] S | \Psi \rangle + O(\mu^2) \qquad (29)$$

where $\mathcal{E}_{IR}$ is given by the sum over *all* unobservable quanta. The violation of physical unitarity is given by the last factor, since it gives the non-zero contribution of the unphysical modes to the transition probabilities.

We conclude our discussion with the following two comments. From the point of view of practical computation, the violation of physical unitarity appears at high order in the loop expansion. In the series expansion of the exponential containing $\mathcal{A}$, the linear term does not contribute since S-matrix elements containig only one F-P ghost in the initial or in the final state are zero (conservation of the F-P charge). By using eq. (9), (11) and (26) the same can be shown for the scalar part of the vector meson. The violation starts from the second power of $\mathcal{A}$, therefore at the two-loop *correction* of the lowest non-zero amplitude, because of the factor $g^2 \cdot h(1)$.

Our analysis shows that the physical interpretation is very crucial for the choice of the physical space. An example is given by the presence of the energy cut-off ΔE in eq. (29). This indicates that we should be more consistent in our definition of the physical space and take into account also the finite accuracy in the angle separation of two or more massless quanta[13]. In our opinion, neglecting this important aspect is the origin of the difficulties with physical unitarity for Yang Mills theories.

References

1 - R.P. Feynman, Acta Phys. Polon. 24, 697 (1963).

2 - L.D. Faddeev and V.N. Popov, Phys. Lett. 25B, 29 (1967).

3 - G. 't Hooft, Nucl. Phys. B33, 173 (1971) and B35, 167 (1971).
B.W. Lee and J. Zinn-Justin, Phys. Rev. D5, 3121, 3137, 3155 (1972).

4 - C. Becchi, A. Rouet and A. Stora, Ann. Phys. (New York) 98, 287 (1976).
G. Curci and R. Ferrari, Nuovo Cimento 35A, 474 (1976).

5 - G. Curci and R. Ferrari, Nuovo Cimento 32A, 151 (1976).

6 - C. Becchi, A. Rouet and A. Stora, Phys. Lett. 52B, 344 (1974).

7 - G. 't Hooft and M. Veltman, Nucl. Phys. B44, 189 (1972).
C.G. Bollini and J.J. Giambiagi, Nuovo Cimento 12B,20 (1972).
G.M. Cicuta and E. Montaldi, Lett. Nuovo Cimento 4, 329 (1972).
P. Breitenlohner and D. Maison, Comm. math. Phys. 52, 11,39,55 (1977).

8 - G. Curci and R. Ferrari, Nuovo Cimento 35A, 1 (1976).

9 - A.A. Slavnov and L.D. Faddeev, Theor. and Math. Phys. 3, 321 (1970).
A.A. Slavnov, Theor. and Math. Phys. 8, 201 (1972).

10 - H. van Dam and M. Veltman, Nucl. Phys. B22, 397 (1970).

11 - G. Curci and R. Ferrari, Nuovo Cimento 35A, 273 (1976).

12 - A.A. Slavnov, Theor. and Math. Phys. 10, 99 (1972).
J.C. Taylor, Nucl. Phys. B33, 436 (1971).

13 - The same suggestion comes from the analysis of gauge invariance of transition probabilities: G. Curci and E. d'Emilio, "On Gauge Invariance and Perturbative Calculations in Quantum Chromodynamics", TH.2582-CERN.

UNITARITY RESTRICTIONS ON SEMI-CLASSICAL APPROXIMATIONS TO CERTAIN FUNCTIONAL INTEGRALS†

H. M. Fried
Physics Department
Brown University
Providence, Rhode Island, U. S. A. 02912

ABSTRACT

Estimates of physically relevant quantities (cross sections, multiplicities, correlations) in eikonal models of high-energy field theories lead to a class of semi-classical solutions to non-linear Euler equations. The requirement of unitarity restricts these possible solutions to a unique function of energy and impact parameter.

†Submitted to the Meeting in Marseille "Mathematical Problems in Feynman Path Integrals", May 1978. Supported in part by the U. S. Department of Energy.

Stationary phase or saddle-point approximations to functional integrals lead naturally to non-linear Euler equations, with the degree of non-linearity depending on the specific problem. For solutions to such equations there is a type of quasi-topological invariant familiar from complex variable theory. This invariant does not depend on the strength of singularities, but only on their location. This invariant is an integer, long familiar from the work of Cauchy and Riemann, and is used to denote the branch, n, of the particular multi-sheeted function under consideration. Thus n may be thought of as that integer which characterizes a solution as long as appropriate boundaries of singularities are not crossed.

The question then arises: How is one to specify the appropriate branch of a particular solution to the Euler equations? The purpose of these remarks is to suggest, for physical problems involving the scattering and production of particles and/or resonances, that the requirement of Unitarity may be used to determine the branch; and to illustrate this possibility with a simple and concrete example drawn from a problem in high-energy eikonal physics.[1]

In essence, this problem may be formulated as a semi-classical approximation to that functional integral representing the total cross section (as well as multiplicities, etc.) for the scattering of a pair of very high-energy particles, with the elastic scattering amplitude defined essentially as the shadow of absorption corresponding to inelastic particle production. The calculation is set up within the context of an eikonal pionization approximation, with the scattering amplitude T(s,t) given in terms of an eikonal function $\chi(b,s)$,

$$T(s,t) = i\,\frac{s}{2}\int d^2b\; e^{i\vec{q}\cdot\vec{b}}\,[1 - e^{i\chi(b,s)}] \qquad (1)$$

where $q^2 = -t \ll s$. This eikonal is to correspond to a pair of nucleons (heavy horizontal lines) that scatter by the exchange of an increasing number of virtual vector mesons (vertical lines) which themselves contain the exchanges of all possible numbers of virtual scalar mesons (light horizontal lines). At asymptotic energies, ladder graphs provide the dominant contribution, and only these are included.

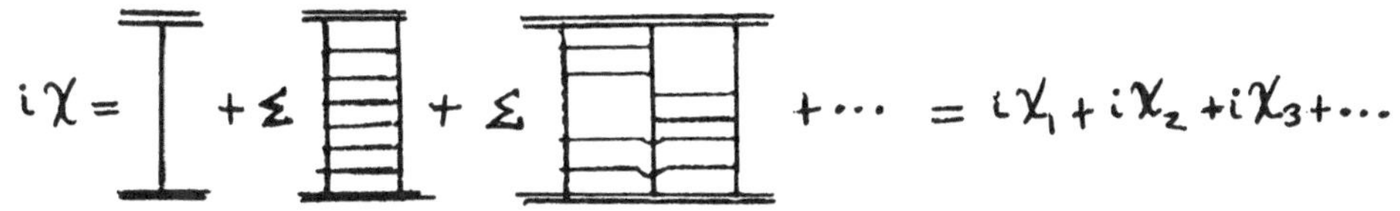

The retention of only $i\chi_1 + i\chi_2$ generates an eikonal with absorptive behavior[2] $i\chi \sim -\, s^{\alpha}(\ell n s)^{-1}\exp[-\beta b]$ or[3] $\sim s^{\alpha'}(\ell n s)^{-1}\exp[-\beta' b^2/\ell n s]$, from which follows the physically observed effect of rising total cross sections,

$$\sigma_{TOT} = 2\ \mathrm{Re}\int d^2b\ [1 - e^{i\chi}] \qquad (2)$$

which in this case yields

$$\sigma_{TOT} \sim \ell n^2 s \quad . \tag{3}$$

Estimates of the remaining χ_n are difficult to obtain, but have been attempted.[4] There, one found suggestions of very strong cancellations, generated by all remaining $i\chi_n$, such that the result (3) is replaced by a σ_{TOT} which vanishes as a power of s. One difficulty with these estimates is that they essentially assume an expansion can be made in the nucleon-vector meson coupling constant to yield a specific $i\chi_n$; and then, at fixed n, the $s \to \infty$ limit is effectively taken, after which the summation on all n is performed. This method may well be suspect if the coefficient of g^{2n}, in the nth order expansion, increases sharply with energy.

A different method of estimation has subsequently been given,[1] in which functional integral representations were written for the eikonal and all relevant cross sections, and then approximated in a semi-classical way. From that analysis there was obtained a non-linear Euler equation for a semi-classical field $\tilde{\phi}_o(k)$, in terms of impact parameter b, mass m of the vertical mesons, mass μ of the horizontal mesons, coupling constants g and λ, and the rapidity $Y \sim \ell ns$:

$$\tilde{\phi}_o(k) = -\frac{g^2\lambda}{(2\pi)^2}\frac{1}{k^2+\mu^2}\int\frac{d^2q\; e^{i\vec{q}\cdot\vec{b}}}{q^2+m^2}\cdot\frac{e^{R(q/m)}}{(q-k_\perp)^2+m^2} \quad , \tag{4}$$

with

$$R(q/m) = \lambda\int d^4p\;\tilde{\phi}_o(p)[m^2+(q-p_\perp)^2]^{-1} \quad ,$$

a somewhat complicated pair of equations.

However, for the case of physical interest corresponding to large impact parameter, mb>>1, this non-linear functional relationship may be approximated by a single non-linear algebraic equation. One expands $R(q/m) \simeq R_o + i\vec{q}\cdot\vec{R}_1 - q_i q_j R_2^{ij}$, and assumes $q_{max} << m$, to obtain

$$R_1^i = 0,\; R_2^{ij} = R_2\delta_{ij},\; R_2 = R_o/6m^2 \quad ,$$

and

$$R_o^2 = -\,i\xi\,\exp[R_o - \eta/R_o] \tag{5}$$

with

$$\xi = 3(g\lambda)^2\pi Y/2m^2 \quad , \quad \eta = \frac{3}{2}(mb)^2 \quad .$$

If solutions to (5) can be found which have the property $Re(R_o)>0$, necessary for the convergence of the q-integral of (4), $\tilde{\phi}_o$ can then be written as

$$\tilde{\phi}_o(k) \simeq -i\left(\frac{m^2R_o}{\lambda\pi^2Y}\right)(k^2+\mu^2)^{-1}\,(k_\perp^2+m^2)^{-1} \quad . \tag{6}$$

The eikonal function corresponding to this semi-classical approximation is then given by

$$i\chi = \left(\frac{m}{\pi\lambda}\right)^2 \frac{R_o}{Y} - \left(\frac{m}{\pi\lambda}\right)^2 \frac{R_o^2}{2Y} - \frac{1}{2}\,\ell n(1-R_o) - \frac{1}{2} R_o \quad .$$

For such a non-linear relation as (5), one may expect dependence upon the branch of the possible solutions. With $R_o \equiv \rho \exp[i(\theta+2\pi n)]$, where ρ, θ, and n denote magnitude, phase $(-\pi \leq \theta \leq +\pi)$ and branch, respectively, substitution into (5) yields the pair of relations

$$(\rho - \eta/\rho) \cos\theta = \ell n(\rho^2/\xi) \tag{7}$$

$$(\rho + \eta/\rho) \cos\theta = 4\pi n + 2\theta - \pi/2 \tag{8}$$

The solutions to (7) and (8) depend upon n, and they are qualitatively different for different values of n.

The requirement of Unitarity, easily given in this eikonal context, is now invoked to provide one more (non-linear) relation, in order to uniquely fix n, or rather to fix that value of n which should be used in different b and Y regions. In the notation of (7) and (8), Unitarity requires that

$$\rho^2 = \rho^2 \cos^2\theta - \rho \cos\theta + aY \left[\ell n(1+\rho^2-2\rho\cos\theta) + 2\rho\cos\theta\right], \tag{9}$$

where $a = (\lambda\pi/2m)^2$. As Y, and b, increase it follows that the necessary n values must also increase; and in the limit of exceedingly large Y, one can, in effect, treat n essentially as a continuous variable. One thus finds

$$\rho^2 \sim aY\ell nY \ , \quad \theta \sim \pi/2 \ , \quad n \sim \frac{1}{4\pi}[aY\ell nY]^{\frac{1}{2}}$$

in the region of relatively small impact parameter, $\eta < Y\ell nY$; but that

$$\rho \sim \alpha Y + \beta \ell nY \ , \quad \theta \to \theta_o < \pi/2 \ , \quad n \sim Y$$

with $\alpha = 2a\cos\theta_o \cdot \csc^2\theta_o$, $\beta^{-1} = 2\cos\theta_o$ in the region of larger impact parameter $\eta \leq \eta_{max} = (\alpha Y)^2 \lesssim \rho^2$. The maximum η corresponds to $mb_{max} \sim Y$, that impact parameter at which one expects the Froissart bound to become operative. As η exceeds η_{max}, there is a discontinuous change as ρ falls sharply below $\xi \sim Y$, exhibiting $\rho \sim O(1)$ and $\theta \sim 0$ for any n. Finally, for the largest impact parameters, $\eta > \xi > \rho^2$, ρ falls off very rapidly with increasing Y, $\rho \sim Y^{\frac{1}{2}} \exp[-\frac{a}{4\pi}\eta Y]$ while $n \sim \eta/\rho$ and $\theta \sim \pi/2$.

The result of this analysis is to reproduce, in a somewhat cumbersome way, the intuitive Froissart distribution in terms of an absorptive eikonal function which effectively vanishes for values of impact parameter which exceed Y/m; and from this there follows (among other physical predictions) an increase of cross sections of the form (3). However, the calculated behavior of σ_{TOT} is really not the point of this exercise, for hadronic physics is far more complicated than sketched here. What is (to our knowledge) unique, and what may be important, is the use of Unitarity as a restriction on the class of acceptable semi-classical solutions needed in connection with a particular functional integral. It may be hoped that this technique will find other applications in related problems.

REFERENCES

1. Hector Moreno and H. M. Fried, Phys. Rev. D12, 2031 (1975); and D17, 352 (1978).
2. H. Cheng and T. T. Wu, Phys. Rev. Lett. 24, 1455 (1970).
3. S-J Chang and T-M Yan, Phys. Rev. D4, 537 (1971).
4. R. Aviv, R. L. Sugar, and R. Blankenbecler, Phys. Rev. D5, 3252 (1972); and D6, 2216 (1972). R. Blankenbecler and H. M. Fried, Phys. Rev. D8, 678 (1973).

ON THE FOKKER-PLANCK LAGRANGIAN

L. Garrido and M. San Miguel
Dpto. de Física Teórica, Universidad de Barcelona
Diagonal 647, Barcelona-28, Spain

1. INTRODUCTION

The path integral representation of the Fokker-Planck equation (F.P.E.) has become increasingly popular [1,2]. One of the reasons for this is the hope that the Lagrangian featuring in the path integral approach could be the key quantity for developing a consistent theory of non equilibrium thermodynamics. In such a way, path integral methods have come back to their original field in physics, namely stochastic processes.

As it happens in the path integral formulation of Quantum Mechanics, ordering ambiguities appear giving rise to different Lagrangian related to different correspondence rules or different definitions of the path integral [3,4]. These ordering ambiguities are related to the commutation relations of the operator formulation [5] of F.P.E. In this communication we show how the Lagrangian for some n-dimensional, non constant diffusion processes can be singled out by the requirement that its Euler-Lagrange equations define the most probable path of the process. This is done by reducing the problem to a constant diffusion one. We also analize the properties under time reversal of the obtained Lagrangian and of the most probable path as a first step towards a thermodynamic interpretation. This Lagrangian defines a classical canonical problem and the stochastic description amounts to the introduction of fluctuations around the most probable path either by the path integral or by impossing commutation relations that can be derived from the path integral formulation [6].

2. LAGRANGIAN

Almost all paths of a diffusion process are nowhere differentiable and furthermore, the probability of a single path vanishes. For these two reasons the concept of most probable path has often been criticized. Nevertheless a sensible mathematical meaning can be given to a differentiable most probable path by considering the differentia ble path in whose neighborhood lie the actual paths most likely to occur [7]. It is in this sense that we will use the concept of most probable path.

Let us consider a general FPE with constant diffusion $D_{\mu\nu}$,

$$\frac{\partial P(q,t)}{\partial t} = -\frac{\partial}{\partial q_\mu} v_\mu(q) P(q,t) + \frac{\partial^2}{\partial q_\mu \partial q_\nu} D_{\mu\nu} P(q,t) \tag{1}$$

The transition probablity $\alpha(q,t;q',t')$ can be written as a path integral in configuration space

$$\alpha(q,t;q',t') = \int \delta^n q(\tau)\, e^{-\int_{t'}^{t} d\tau\, \mathcal{L}(q(\tau),\dot{q}(\tau))} \tag{2}$$

We here choose the Stratonovich-Graham Lagrangian [2]

$$\mathcal{L}(q,\dot{q}) = \frac{1}{4} D^{-1}_{\mu\nu}(\dot{q}_\mu - v_\mu)(\dot{q}_\nu - v_\nu) + \frac{1}{2}\frac{\partial v_\mu}{\partial q_\mu} \tag{3}$$

This Lagrangian is known to correspond to a mid point discretization or to Weyl's correspondence rule [3,4] and to the usual Stratonovich rules of calculus. Therefore, the path maximizing the action in (2) is obtained from (3) by the usual Euler-Lagrange equations as explicitly proved in one dimension in Ref. 7 . Assuming this last property for (3) we now derive the Lagrangian for non constant diffusion with the same property. So, let $D_{\mu\nu}$ depend on q in equation (1). Being $D_{\mu\nu}(q)$ a symmetric positive definite nxn matrix, there exists a symmetric, real, nxn matrix $g_{\mu\nu}(q)$ such that

$$g_{\mu\alpha}(q)\, g_{\alpha\nu}(q) = D_{\mu\nu}(q) \tag{4}$$

Let us introduce a new set of n variables Q_μ by

$$dQ_\mu = g^{-1}_{\mu\nu}\, dq_\nu \tag{5}$$

The existence of such a set implies that the following condition should be fulfilled

$$\frac{\partial g^{-1}_{\alpha\mu}}{\partial q_\nu} = \frac{\partial g^{-1}_{\alpha\nu}}{\partial q_\mu} \tag{6}$$

In a covariant formulation [8] the assumption (6) means that we are restricting ourselves to an euclidean space and thus, there exists a set of coordinates in which $D_{\mu\nu} = \delta_{\mu\nu}$.

Denoting $P(q(Q),t)$ by $\mathcal{P}(Q,t)$ we have

$$\langle f(q(t))\rangle = \int d^n q \, P(q,t) f(q) = \int d^n Q \, \left\| \frac{\partial q_\mu}{\partial Q_\nu} \right\| \mathcal{P}(Q,t) f(q(Q)) \tag{7}$$

so that the Fokker-Planck distribution function in the Q variables is

$$P'(Q,t) = \mathcal{P}(Q,t)\sqrt{D} \tag{8}$$

where $D = \| D_{\mu\nu} \|$

From equation (8) and (1) and making appropiate use of (6) the FPE for $P'(Q,t)$ can be derived [6]. It reads

$$\frac{\partial P'(Q,t)}{\partial t} = -\frac{\partial}{\partial Q_\mu} V'_\mu P'(Q,t) + \frac{\partial^2}{\partial Q_\mu^2} P'(Q,t) \tag{9}$$

where

$$V'_\mu = g^{-1}_{\mu\nu} V_\nu + g^{-1}_{\alpha\mu} \frac{\partial g_{\beta\alpha}}{\partial Q_\beta} \tag{10}$$

Once we have succeded in transforming the original FPE to another FPE with constant diffusion, the Lagrangian for the latter is according to (3)

$$\mathcal{L}(Q,\dot{Q}) = \frac{1}{4}(\dot{Q}_\mu - V'_\mu)^2 + \frac{1}{2}\frac{\partial V'_\mu}{\partial Q_\mu} \tag{11}$$

Since the Euler-Lagrange equations associated to (11) define a most probable path that should be independent of the set of coordinates used, the Lagrangian in the primitive q coordinates is derived by requiring invariance of that equations: $\mathcal{L}(q,q')$ is obtained from (11) by performing the point canonical transformation (5)

$$\mathcal{L}(q,\dot{q}) = \frac{1}{4} D^{-1}_{\mu\nu}(\dot{q}_\mu - W_\mu)(\dot{q}_\nu - W_\nu) + \frac{1}{2}\sqrt{D}\frac{\partial}{\partial q_\mu}\frac{W_\mu}{\sqrt{D}} \tag{12}$$

where

$$W_\mu = V_\mu - \sqrt{D}\frac{\partial}{\partial q_\nu}\frac{D_{\mu\nu}}{\sqrt{D}} \tag{13}$$

The Lagrangian (12) is the one deduced from the path integral solution to the FPE (1) for the case of euclidean spaces by Gra ham [2] and it <u>does not</u> correspond to the mid point discretization or Weyl's correspondence rule that yields the Lagrangian (3) for constant diffusion [3,4]. The discrepancy existing on the rules needed to derive the Lagrangian (3) and (12) from the formal path integral can therefore be understood by the above derivation of one from the other as a requirement on the way the most probable path is evaluated.

3. TIME REVERSAL PROPERTIES

For the one dimensional linear problem, Onsager and Machlup [9] already noted that the most probable path is invariant under time reversal, the irreversibility of the process being introduced by the different weight given by the Lagrangian to the two independent solutions of the Euler-Lagrange equation. We show here that this result also holds for the general process considered in section 2 as long as Detailed Balance holds [10].

The time reversal operation is defined by

$$
T: \quad \begin{aligned} t &\longrightarrow -t \\ q_\mu &\longrightarrow q_\mu \\ V^R_\mu &\longrightarrow -V^R_\mu \\ V^I_\mu &\longrightarrow V^I_\mu \\ D_{\mu\nu} &\longrightarrow D_{\mu\nu} \end{aligned} \qquad (14)
$$

where we have assumed even variables and where V_μ^R and V_μ^I are respectively the reversible and irreversible parts of the drift V_μ. The irreversible drift V_μ^I and $D_{\mu\nu}$ are the two sources of irreversibility.

The most probable path associated to (12) is given by

$$
D^{-1}_{\alpha\mu}\left(\ddot q_\mu - \frac{\partial W_\mu}{\partial q_\nu}\dot q_\nu\right) + (\dot q_\mu - W_\mu)\left[\frac{\partial D^{-1}_{\alpha\mu}}{\partial q_\nu}\dot q_\nu + D^{-1}_{\nu\mu}\frac{\partial W_\nu}{\partial q_\alpha} - \frac{1}{2}\frac{\partial D^{-1}_{\nu\mu}}{\partial q_\alpha}(\dot q_\nu - W_\nu)\right] -
$$
$$
- \frac{\partial}{\partial q_\alpha}\left(\sqrt{D}\frac{\partial}{\partial q_\nu}\frac{W_\nu}{\sqrt{D}}\right) = 0 \qquad (15)
$$

This differential equation is invariant under the operation (14) whenever

$$
\dot q_\mu\left[\frac{\partial}{\partial q_\alpha}(D^{-1}_{\nu\mu}W^I_\nu) - \frac{\partial}{\partial q_\mu}(D^{-1}_{\alpha\nu}W^I_\nu)\right] - D^{-1}_{\nu\mu}\left(\frac{\partial W^R_\nu}{\partial q_\alpha}W^I_\mu + \frac{\partial W^I_\nu}{\partial q_\alpha}W^R_\mu\right) -
$$
$$
- \frac{1}{2}\frac{\partial D^{-1}_{\nu\mu}}{\partial q_\alpha}(W^R_\mu W^I_\nu + W^I_\mu W^R_\nu) - \frac{\partial}{\partial q_\alpha}\left(\sqrt{D}\frac{\partial}{\partial q_\nu}\frac{W^R_\nu}{\sqrt{D}}\right) = 0 \qquad (16)
$$

where, of course

$$W_\mu^R = V_\mu^R \tag{17}$$

$$W_\mu^I = V_\mu^I - \sqrt{D}\,\frac{\partial}{\partial q_\nu}\,\frac{D_{\mu\nu}}{\sqrt{D}} \tag{18}$$

The assumption of Detailed Balance is equivalent to the fulfillment of the potential conditions [11]

$$V_\mu^I = \frac{\partial D_{\mu\nu}}{\partial q_\nu} - D_{\mu\nu}\,\frac{\partial \phi_{st}}{\partial q_\nu} \tag{19}$$

$$\frac{\partial}{\partial q_\mu}(V_\mu^R P_{st}) = 0 \tag{20}$$

where

$$\phi_{st} = -\log P_{st}(q) = -\log\left(\lim_{t\to\infty} P(q,t)\right) \tag{21}$$

It is a straightforward calculation to check that (19) and (20) imply the vanishing of the coefficient of $\dot{q}_\mu$ and of the remaining part in (16), so that (15) becomes invariant under time reversal.

In spite of this symmetry the Lagrangian itself is not invariant under time reversal. In fact, relying on the potential conditions (19) and (20), it can be decomposed as

$$\mathcal{L}(q,\dot{q}) = \mathcal{L}^{inv}(q,\dot{q}) + \frac{1}{2}\frac{d}{dt}(\phi_{st} - \log\sqrt{D}) \tag{22}$$

where

$$\mathcal{L}^{inv}(q,\dot{q}) = \frac{1}{4}D_{\mu\nu}^{-1}[(\dot{q}_\mu - W_\mu^R)(\dot{q}_\nu - W_\nu^R) + W_\mu^I W_\nu^I] + \frac{1}{2}\sqrt{D}\frac{\partial}{\partial q_\nu}\frac{W_\nu^I}{\sqrt{D}} \tag{23}$$

is a Lagrangian invariant under time reversal that differs from $\mathcal{L}(q,\dot{q})$ in a total derivative with respect to time and therefore defines the same Euler-Lagrange equations.

If the decomposition (22) is introduced in (2) it becomes clear that $\mathcal{L}^{inv}$ gives a reversible contribution, while $\frac{1}{2}(\phi_{st}-\log\sqrt{D})$ is a "potential" such that the irreversible contribution is given by the difference of its value at the final and initial state.

Summarizing, the condition of detailed balance guarantees that the classical canonical problem associated to the Lagrangian (12) ("most probable path") is invariant under time reversal. This symmetry is broken when fluctuations around that path are introduced by means of a path integral. If fluctuation are introduced by means of an operator formalism [5], the breaking of the symmetry under time reversal happens to be a consequence of the commutation relations imposed [10].

It is interesting to point out that if $W_\mu^R = 0$, the Lagrangian has no ordering problems, i.e., the canonical associated Hamiltonian does not contain any term including products of coordinates q_μ and momenta $p_\mu = \frac{\partial \mathcal{L}}{\partial \dot{q}_\mu}$. Therefore, in this case the decomposition (22) leads to a prescription independent path integral. This property has already been used [4] in a different context for a one dimensional problem. The conditions of Detailed Balance and vanishing reversible drift are automatically fulfilled in one dimension if natural boundary conditions are assumed.

The authors have profited from helpful conversations with H. Leschke, F. Langouche and D. Roekaerts during the conference.

REFERENCES

1. H. Haken, Z. Physik B24, 321(1976)
2. R. Graham, Z. Physik B26, 281(1977)
3. H. Leschke and M.Schmutz , Z. Physik B27, 85(1977)
4. F. Langouche, D. Roekaerts and E. Tirapegui, KU Leuven Preprints, KUL-TF-77/203(1977); 78/003, 78/007, 78/015 (1978).
5. L. Garrido and M. San Miguel, Progr. Theor. Phys. 59,40(1978), 59, 55(1978).
6. L. Garrido, D. Luriè and M. San Miguel, Phys. Lett. A (to appear).
7. D. Dürr and A. Bach, Comm. Math. Phys. 60, 153(1978).
8. R. Graham, Z. Physik B26, 397(1977).
9. L. Onsager and S. Machlup, Phys. Rev 91, 1505(1953)
10. L. Garrido, D. Luriè and M. San Miguel (unpublished)
11. R. Graham and H. Haken, Z. Physik 243, 289(1971).

DISTRIBUTION DEFINITION OF PATH INTEGRALS

W. KERLER

Fachbereich Physik, Universität Marburg, D355o Marburg

By starting from quantum mechanics it turns out that a rather general definition of quantum functional integrals can be given which is based on distribution theory. It applies also to curved space and provides clear rules for non-linear transformations. The refinements necessary in usual definitions of path integrals are pointed out. Since the quantum nature requires special care with time sequences, it is not the classical phase space which occurs in the phase-space form of the path integral. Feynman's configuration-space form only applies to a highly specialized situation, and therefore is not a very advantageous starting point for general investigations. It is shown that the commonly used substitutions of variables do not properly account for quantum effects. The relation to the traditional ordering problem is clarified. The distribution formulation has allowed to treat constrained systems directly at the quantum level, to complete the path integral formulation of the equivalence theorem, and to define functional integrals also for space translation after the transition to fields.

Path integrals have become very important now in modern quantum field theory, in particular for the formulation of gauge theories and for the expansion around solutions of non-linear classical equations as solitons and instantons. However, in order to be more than a heuristic tool and to allow really to go beyond perturbation theory their definition must properly account for quantum effects also in the case of non-linear substitutions of variables. Furthermore, it is advantageous not to be fixed to the case of quadratic momenta.

A suitable method to find an adequate definition[1)] is to start from quantum mechanics. For the matrix element of time translations

$$U(q',q;t'-t) = \langle q'| \exp(-\frac{i}{\hbar}H(t'-t)) |q\rangle$$

with states normalized as $\langle q'| q\rangle = \delta(q'-q)$ one may write

$$U(q_b,q_a\ ;t_b-t_a) = \lim_{n\to\infty} \int dq_1\ldots dq_{n-1}\ U(q_b,q_{n-1};t_b-t_{n-1}) \times U(q_{n-1},q_{n-2};t_{n-1}-t_{n-2})\ldots U(q_1,q_a;t_1-t_a). \tag{1}$$

If one requires $t_{\nu+1}-t_\nu > o$ in (1) for all ν, the interdependence of the numbering and of t is strictly monotonic. Then introducing continuous functions q(t) one can identify $q_\nu = q(t_\nu)$. With the abbreviation

$$K(q(t'),q(t)) = U(q(t'),q(t);t'-t) \text{ for } t'> t$$

and writing $H(q',q) = \langle q'| \hat{H} |q\rangle$, (1) can be replaced by

$$K(q(t_b),q(t_a)) = \lim_{n\to\infty} \int dq(t_{n-1})dq(t_{n-2})\ldots dq(t_1) \prod_{\nu=0}^{n-1} \Big(\delta(q(t_{\nu+1})-q(t_\nu)) - \frac{i}{\hbar}(t_{\nu+1}-t_\nu)H(q(t_{\nu+1}),q(t_\nu))\Big) \tag{2}$$

where $t_b \equiv t_n > t_{n-1} > \ldots > t_1 > t_o \equiv t_a$, and $\max(t_{\nu+1}-t_\nu)\to o$.

Of course, (2) is understood to act on a suitable test function space. The distribution H(q',q) has the general form

$$H(q',q) = \sum_{k=0}^{m} c_k(v)\left(\frac{\hbar}{i}\right)^k \delta^{(k)}(u) \tag{3}$$

where u = q'-q, and with v(q',q) such that the correspondence between q',q and u,v is one-to-one. For the standard example $\hat{H} = \frac{1}{2}\hat{p}^2+V(\hat{q})$ one has m=2, $c_2=\frac{1}{2}$, $c_1=o$, and $c_o(v) = V(v)$ for $v=v_\lambda \equiv \lambda q'+(1-\lambda)q$. More general functions v(q',q) can be readily introduced since with the relation

$v_I = v_I(u,v_{II})$ one gets in this simple case already an equivalent description by using $c_o^{II}(v^{II}) = c_o^I(v^I(o,v^{II}))$. However, if the c_k for $k > o$ are not constant (as in curved space problems) from c_k^I one gets contributions to all $c_o^{II},\dots,c_k^{II}$. This comes about since one has to use the formula

$$f(u)\,\delta^{(k)}(u) = \sum_{\nu=0}^{k}\binom{k}{\nu}(-1)^\nu f^{(\nu)}(o)\,\delta^{(k-\nu)}(u) \tag{4}$$

which for the determination of the coefficients of an equivalent description gives the relation

$$c_k^I(v_I)\,\delta^{(k)}(u) = \sum_{\nu=0}^{k}\binom{k}{\nu}(-1)^\nu\left[\partial^\nu c_k^I(v_I(u,v_{II}))/\partial u^\nu\right]_{u=o}\delta^{(k-\nu)}(u). \tag{5}$$

For linear v the transition to equivalent descriptions can be shown to correspond in the operator formulation to simple reordering by using the commutation relation. The distribution formulation allows, however, also to go easily to complicated non-linear v, which is important for the consistency of non-linear transformations of variables in path integrals[1].

If the Fourier representation of the delta function and of its derivatives is used, (3) can be cast into the form[1]

$$H(q',q) = \frac{1}{2\pi\hbar}\int dp\; e^{\frac{i}{\hbar}pu}\quad h(p,v) \tag{6}$$

where

$$h(p,v) = \sum_{k=0}^{m} c_k(v)p^k. \tag{7}$$

To compare with the traditional ordering rule mappings, considered as mappings of one quantum theory, it is convenient to use the general representation[2]

$$\hat{H} = \left(\frac{1}{2\pi\hbar}\right)^2\int dp\,dq\,d\xi\,d\eta\,\exp\left\{\frac{i}{\hbar}\left((\hat{p}-p)\xi + (\hat{q}-q)\eta\right)\right\} \times c(\xi,\eta)\,H_c(p,q) \tag{8}$$

where $c(o,\eta) = c(\xi,o) = 1$. By calculating $H(q',q)$ from (8) and comparing with (6) it follows that

$$h(p,v(q',q)) = \frac{1}{2\pi\hbar}\int dq''d\eta\,\exp\left\{\frac{i}{\hbar}\left(\frac{q'+q}{2} - q''\right)\eta\right\} \times c(-(q'-q),\eta)\,H_c(p,q''). \tag{9}$$

For the family of orderings given by

$$c(\xi,\eta) = \exp\left\{\frac{i}{\hbar}\left(\frac{1}{2}-\lambda\right)\xi\eta\right\} \tag{10}$$

one obtains from (9) $h(p,v(q',q)) = H_c^{\lambda}(p,\lambda q'+(1-\lambda)q)$.

This means that for $v = v_\lambda \equiv \lambda q'+(1-\lambda)q$ the relation to a classical Hamiltonian is given by $h^\lambda(p,v^\lambda)=H_c^\lambda(p,v^\lambda)$, which includes Weyl, qp and pq ordering for $\lambda=\frac{1}{2}$, 1 and o respectively. From (9) it is seen that in cases not described by (1o) there is no simple connection between h and H_c.

Inserting (6) into (2) and introducing also continuous functions p(t) we obtain the form[1)]

$$K(q(t_b),q(t_a)) = \lim_{n\to\infty}\int dp(t_{2n-1})dq(t_{2n-2})\ldots dq(t_2)dp(t_1)$$

$$\left(\frac{1}{2\pi\hbar}\right)^n \exp\left\{\frac{i}{\hbar}\sum_{\alpha=0}^{n-1}(p(t_{2\alpha+1})(q(t_{2\alpha+2})-q(t_{2\alpha})) -h(p(t_{2\alpha+1}),v(q(t_{2\alpha+2}),q(t_{2\alpha})))(t_{2\alpha+2}-t_{2\alpha}))\right\} \qquad (11)$$

where $t_b \equiv t_{2n} > t_{2n-1} > \ldots > t_2 > t_1 > t_o \equiv t_a$ and $\max(t_{\nu+1}-t_\nu)\to o$.

With (11) we have the generalization of what is usually called the phase space form of the path integral. It is, however, to be realized that it is not the classical phase space but a complicated limit of alternate p and q integrations which occurs. The quantum nature requires that for products p(t)q(t') in the integrand it is always specified which factor is earlier. The more compact form of (11), where in the limit the integrand is replaced by an exponential of an ordinary integral, is somewhat formal in view of the underlying product of distributions.

The generalization of Feynman's configuration space form occurs for m = 2 (i.e. for h quadratic in p). It follows from (11) by a suitably defined Gaussian integration. It can be shown[1)] that the highest order coefficient in (3) resp. (7) is real. Then Gaussian integration makes sense if a suitable small imaginary term proportional to p^2 is added to h, which is supposed to be put equal to zero in the final result. For the example $h = \frac{1}{2}g(v)p^2+V(v)$ with $g>o$ we give the additional term for convenience the form $-i\eta\, gp^2/2$ with $\eta>0$. Then integrating out the $p(t_\nu)$ in (11) we obtain

$$K(q(t_b),q(t_a)) = \lim_{n\to\infty}\int dq(t_{n-1})dq(t_{n-2})\ldots dq(t_1)$$

$$\times\,(\pi i\hbar 2(1-i\eta))^{-\frac{1}{2}n}\left(\prod_{\nu=0}^{n-1}((t_{\nu+1}-t_\nu)g(v(q(t_{\nu+1}),q(t_\nu))))^{-\frac{1}{2}}\right)$$

$$\times \exp\left\{\frac{i}{\hbar}\sum_{\nu=0}^{n-1}((t_{\nu+1}-t_\nu)(\frac{1}{2}\left(\frac{q(t_{\nu+1})-q(t_\nu)}{t_{\nu+1}-t_\nu}\right)^2\left(g(v(q(t_{\nu+1}),q(t_\nu)))(1-i\eta)\right)^{-1}\right. \tag{12}$$
$$\left.-V(v(q(t_{\nu+1}),q(t_\nu)))))\right\}$$

where $t_b \equiv t_n > t_{n-1} > \cdots > t_1 > t_o \equiv t_a$, and where $\max(t_{\nu+1}-t_\nu)\rightarrow 0$.

It is seen that the $i\eta$ prescription appears in (12) just in the manner that, when going further to the generating functional of Green's functions, the Feynman prescription for propagators comes out without any additional assumption. The more compact form of (12), where in the limit the integrand is replaced by an exponential of an ordinary integral, is again rather formal due to the underlying product of distributions. This has, for example, the consequence that the then appearing derivative of q(t) has not the ordinary properties under non-linear transformations, as has first been observed[3] by using the short-time expansion of the action, and which gets its full explanation from the distribution formulation. Therefore, if one starts from the Lagrangian (even for $g \equiv 1$) it is conceivable that one runs into some difficulties with defining path integrals.

In the treatment[1] of non-linear transformations $q = F(Q)$ one has to replace $u = q'-q$ by $u_o = Q'-Q$ in (3) which can be done by using $\delta^{(k)}(a(x)x) = (a(x))^{-k-1}\,\delta^{(k)}(x)$. Since the new coefficients must not depend on u_o, one has to apply (4) to the factors in front of the $\delta^{(k)}(u_o)$. For each $\delta^{(k)}$ term one thus gets $\delta^{(l)}$ terms with $k \geq l \geq o$. To absorb the effect of the integration volume we put $dF/dQ = \alpha(Q)\,\beta(Q)$ and within one time interval apply (4) to $\beta(Q')\,\delta^{(l)}(u_o)\alpha(Q)$ which due to the u_o dependence of Q and Q' again produces lower order terms. We thus obtain

$$\beta(Q')\;\delta(u)\,\alpha(Q) = \delta(u_o)$$

which corresponds to naive expectations, and

$$\beta(Q')\;H(q',q)\,\alpha(Q) = \tilde{H}(Q',Q)$$

where the derivatives of the delta function produce additional terms such that from (3) one gets

$$\tilde{H}(Q',Q) = \sum_{k=o}^{m}\sum_{k'=o}^{k} c_k(v)\left(\frac{\hbar}{i}\right)^k M_{kk'}(v)\,\delta^{(k')}(u_o).$$

The path integral now transforms as

$$K(q(t_b),q(t_a)) = \tilde{K}(Q(t_b),Q(t_a)) \quad (\beta(Q(t_b))\,\alpha(Q(t_a)))^{-1}$$

where $\tilde{K}$ has the form (2) with variable Q and generating distribution $\tilde{H}$. The usual naive substitution of variables neglects the additional terms which arise from non-linear transformations.

Our formulation can be readily extended to N degrees of freedom[1]. Also the calculation for non-linear transformations can be generalized[4]. This gives the possibility to treat constrained systems directly at the quantum level[4], in contrast to Faddeev's approach[5] which uses classical phase space considerations and classical transformations. The extension to fields[1,6] is described by a similar limit as the one of the path integral definition. Therefore functional integrals can be as well defined for space translations[1]. Furthermore, our formalism allows to show[6] that the up to now used path integral formulation of the equivalence theorem[7] is incomplete.

References

1) W. Kerler, Nucl. Phys. B 138 (1978)
2) L. Cohen, J. Math. Phys. 7 (1966) 781;
G.S. Agarwal and E. Wolf, Phys. Rev. D2 (197o) 2161
3) S.F. Edwards and Y.V. Gulyaev, Proc. Roy. Soc. A 279 (1964) 229;
J.-L. Gervais and A. Jevicki, Nucl. Phys. B 110 (1976) 93
4) W. Kerler, Phys. Lett. 76 B (1978) 423
5) L.D. Faddeev, Theor. and Math. Phys. 1 (197o) 1
6) W. Kerler, preprint Marburg April 1978
7) A. Salam and J. Strathdee, Phys. Rev. D 2 (197o) 2869;
Y.-M.P. Lam, Phys. Rev. D7 (1973) 2943, in particular Sec. V

FUNCTIONAL INTEGRAL REPRESENTATIONS AND INEQUALITIES FOR BOSE PARTITION FUNCTIONS

Hajo Leschke
Institut für Physik der Universität
D-4600 Dortmund 50, Germany

Abstract

Different phase space descriptions of quantum mechanics are used to construct different functional integral representations for the partition function of a general system where Bose degrees of freedom are involved. The "static approximation" to these functional integrals yields (pseudo-)classical partition functions providing either a lower or an upper bound on the original partition function. The upper bound generalizes a result due to HEPP and LIEB.

1. Introduction

It is known[1] that the factor-ordering multiplicity of noncommuting operators corresponds to the existence of different "discretization" or "sequential limit" prescriptions, which can be used to define a path or functional integral. This is in accord with the fact that a given quantum theory allows for different but equivalent c-number representations or, conversely, that a given classical theory can be quantized in various inequivalent ways. There is nothing strange about a multiplicity or so-called ambiguity inherent in functional integration, because mathematics offers a variety of one-to-one mappings between a noncommutative and a commutative structure. What seems to puzzle some people is simply that mathematics cannot tell them which mapping they like to prefer or, conversely, which nature prefers. This situation is of course not only true for quantum mechanics but also for quantum statistical mechanics, which is nowadays also called Euclidean quantum mechanics. In the present contribution I want to make the following two points:

(i) The partition function of a given boson Hamiltonian can be represented by different functional integral expressions, and

(ii) Although these expressions often do not help to calculate the partition function explicitly they may help to find calculable bounds on it.

For the sake of notational transparency the methods will be demonstrated only for a system consisting of a single boson mode representing physically e.g. a photon or phonon mode.
The generalization to several boson modes is straightforward.

2. Normal and antinormal symbols

Let me describe the boson mode in terms of an annihilation operator a and its adjoint a^+, the creation operator, which obey the canonical commutation relation

$$[a, a^+] = 1 \tag{1}$$

The dynamics is generated by the Hamiltonian $H = H(a, a^+)$ which is some given self-adjoint operator function of a and a^+. Since H may be written with the help of (1) in various equivalent forms, there can be associated with H various (real valued) functions $h(\alpha, \alpha^*)$ of a complex number α and its conjugate α^*. These functions are known in the literature[2] as (pseudo-)classical phase space functions or symbols corresponding to the quantum Hamiltonian. Here I restrict my attention to two such symbols, the normal symbol $h_+(\alpha, \alpha^*)$ and the antinormal symbol $h_-(\alpha, \alpha^*)$. They are defined by substituting α for a and α^* for a^+ in the normal-ordered and antinormal-ordered form of H, respectively. The normal-ordered (resp. antinormal-ordered) form of H results by bringing with the help of (1), all a^+'s to the left (resp. right) of all a's. The following relations hold

$$h_+(\alpha', \alpha^*) = \langle\alpha|H|\alpha'\rangle / \langle\alpha|\alpha'\rangle \tag{2}$$

$$H = \int \frac{d^2\alpha}{\pi} h_-(\alpha,\alpha^*) \; |\alpha\rangle \langle\alpha| \tag{3}$$

because the coherent states

$$|\alpha\rangle := \exp(\alpha a^+ - \alpha^* a) \; |0\rangle \tag{4}$$

enjoy the properties[3)]

$$a|\alpha\rangle = \alpha|\alpha\rangle \tag{5}$$

$$\int \frac{d^2\alpha}{\pi} |\alpha\rangle \langle\alpha| = 1 \tag{6}$$

$$\langle\alpha|\alpha'\rangle = \exp\left(-\frac{|\alpha|^2}{2} - \frac{|\alpha'|^2}{2} + \alpha^* \alpha'\right) \tag{7}$$

Here $|0\rangle$ denotes the normalized state which is annihilated by a, and $\int d^2\alpha$ means $\int_{-\infty}^{\infty} d\mathrm{Re}\,\alpha \int_{-\infty}^{\infty} d\mathrm{Im}\,\alpha$.

As an example I exhibit the normal and the antinormal symbol

$$h_\pm(\alpha,\alpha^*) = \omega(\alpha^*\alpha \pm \tfrac{1}{2}) + \lambda \left[(\alpha + \alpha^*)^4 \pm 6(\alpha+\alpha^*)^2 + 3 \right] \tag{8}$$

of the anharmonic oscillator

$$H = \omega\,(a^+ a + \tfrac{1}{2}) + \lambda\,(a + a^+)^4 \tag{9}$$

3. Normal and antinormal functional integration

The quantity of basic interest in (equilibrium) quantum statistical mechanics is the partition function

$$Z := \mathrm{tr}\; e^{-\beta H} \tag{10}$$

Here "tr" denotes the trace over the (Fock-) Hilbert space of the boson mode and the positive parameter β has the physical meaning of the inverse temperature. According to (2) - (7) one can write

$$Z = \int \frac{d^2\alpha}{\pi} w_+(\alpha,\alpha^*) = \int \frac{d^2\alpha}{\pi} w_-(\alpha,\alpha^*) \tag{11}$$

where w_+ (resp. w_-) denotes the normal (resp. antinormal) symbol of the "density" operator $W := e^{-\beta H}$. A functional integral representation of Z may be viewed as an attempt to express w_+ (resp. w_-) in terms of h_+ (resp. h_-). The resulting functional representations may be called the normal and the antinormal functional integral representation, respectively.

Let me begin with the construction of the normal representation. Employing (6), one has for each natural number n

$$Z = \int \frac{d^2\alpha_1}{\pi} \cdots \int \frac{d^2\alpha_n}{\pi} \prod_{j=1}^{n} \langle\alpha_j| e^{-\frac{\beta}{n}H} |\alpha_{j-1}\rangle \tag{12}$$

where I have set

$$\alpha_0 := \alpha_n \tag{13}$$

In the limit $n \to \infty$ it is sufficient to compute each of the n factors occurring in (12) accurately up to order 1/n. One can therefore use the following "high-temperature approximation"

$$\langle\alpha|e^{-\tau H}|\alpha'\rangle \approx \langle\alpha|\alpha'\rangle\, e^{-\tau h_+(\alpha',\alpha^*)} \tag{14}$$

which is accurate up to order τ, due to (2). Combining (12), (14), and (7) one arrives at

$$Z = \lim_{n\to\infty} \int \frac{d^2\alpha_1}{\pi} \cdots \int \frac{d^2\alpha_n}{\pi}\, e^{S_+^{(n)}(\alpha_1, \ldots, \alpha_n)} \tag{15}$$

if one uses the abbreviation

$$S_+^{(n)}(\alpha_1, \ldots, \alpha_n) := \sum_{j=1}^{n} \frac{\beta}{n} \left(\alpha_{j-1} \frac{\alpha_j^* - \alpha_{j-1}^*}{2(\beta/n)} - \alpha_j^* \frac{\alpha_j - \alpha_{j-1}}{2(\beta/n)} - h_+(\alpha_{j-1}, \alpha_j^*)\right) \tag{16}$$

Because $S_+^{(n)}$ appears formally as a Riemann sum approximating the normal action functional

$$S_+[\alpha] := \int_0^\beta d\tau \left(\frac{1}{2}\alpha(\tau) \frac{d}{d\tau} \alpha^*(\tau) - \frac{1}{2} \alpha^*(\tau) \frac{d}{d\tau} \alpha(\tau) - h_+(\alpha(\tau),\alpha^*(\tau))\right) \tag{17}$$

evaluated at a differentiable function $\alpha(\tau)$, it is customary and mnemonicly convenient to use a continuum notation for the r.h.s. of (15). The notation which I propose here is

$$\oint_{+} \delta\alpha \; e^{S_{+}[\alpha]} \tag{18}$$

This expression may (formally) be interpreted as a functional integral over all closed paths or loops (i.e. $\alpha(\beta) = \alpha(0)$, compare (13)) in $\mathbb{C} \cong \mathbb{R}^2$, but it should be noted that it is in fact defined (!) as the r.h.s. of (15), i.e. as the limit of a definite sequence of finite dimensional integrals. In particular, the subscript "+" attached to the functional integration sign $\oint_{+}\delta\alpha$ indicates that the value $h(\alpha(\tau), \alpha^*(\tau))$ of the phase space function, or symbol, occurring in the action functional must appear in its discretized version precisely as $h(\alpha_{j-1}, \alpha_j^*)$ and e.g. not as $h(\alpha_j, \alpha_j^*)$ or anything else (compare (16)). Let me call the functional integration concept defined by this discretization prescription the <u>normal functional integration</u>.

In order to construct the antinormal functional integral representation of the partition function I start from the operator identity

$$e^{-\beta H} = \lim_{n\to\infty} \left(\int \frac{d^2\alpha}{\pi} \; e^{-\frac{\beta}{n} h_{-}(\alpha,\alpha^*)} |\alpha\rangle \langle\alpha| \right)^n \tag{19}$$

valid due to (3) and because the exponential on the r.h.s. contributes only up to the first order in $1/n$. Taking the trace of (19) yields the desired antinormal representation

$$\begin{aligned} Z &= \lim_{n\to\infty} \int \frac{d^2\alpha_1}{\pi} \cdots \int \frac{d^2\alpha_n}{\pi} \; e^{S_{-}^{(n)}(\alpha_1, \ldots, \alpha_n)} \\ &=: \oint_{-} \delta\alpha \; e^{S_{-}[\alpha]} \end{aligned} \tag{20}$$

Here $S_{-}^{(n)}$ is defined by the r.h.s. of (16), with $h_{+}(\alpha_{j-1},\alpha_j^*)$ replaced by $h_{-}(\alpha_j, \alpha_j^*)$, and the <u>antinormal action functional</u> $S_{-}[\alpha]$ is defined by the r.h.s. of (17), with $h_{+}(\alpha(\tau), \alpha^*(\tau))$ replaced by $h_{-}(\alpha(\tau), \alpha^*(\tau))$. Accordingly, the discretization prescription defining the <u>antinormal functional integration</u> $\oint_{-}\delta\alpha$ differs from that of the normal one in the respect that $h(\alpha(\tau), \alpha^*(\tau))$ must be discretized as $h(\alpha_j, \alpha_j^*)$, and not as $h(\alpha_{j-1}, \alpha_j^*)$.

Summarizing, we see that each phase space description (or ordering scheme) must be combined with the appropriate discretization (or lattice, or sequential limit) prescription of the functional integral in order to guarantee that the same (!) partition function of a given quantum Hamiltonian is recovered. This can be nicely illustrated by explicit calculations for the harmonic oscillator ($\lambda = 0$ in (9)).

4. Pseudo-classical partition functions as bounds on the quantum partition function

The pseudo-classical Hamiltonians h_+ and h_- associated with the quantum Hamiltonian H give rise to the definition of corresponding pseudo-classical partition functions

$$Z_\pm := \int \frac{d^2\alpha}{\pi} \; e^{-\beta h_\pm(\alpha,\alpha^*)} \tag{21}$$

Heuristically they may be viewed as resulting from the functional integral representations (18) and (20) for the true partition function Z by the neglect of all "dynamic fluctuations" in the paths to be integrated over. This restriction to the constant paths in functional integrals is also called the "static approximation". Since it leads to a (prescription independent) ordinary integration over different integrands one cannot expect Z_+ and Z_- to be identical. Instead one finds that Z_+ provides a lower and Z_- an upper bound on Z, i.e.

$$Z_+ \le Z \le Z_- \tag{22}$$

These inequalities are due to Hepp and Lieb[4]. One interesting point about them is their generality. They hold essentially for every Hamiltonian, for which all three Z's make sense. One can prove (22) by making use of the functional integral representations of Z. In fact, this was done by Hepp and Lieb for the upper bound. And it will be done for the generalized upper bound to be presented in the next section. But the original bounds in (22) can be proved more directly and simply as I now proceed to demonstrate. The essential tool is Jensen's inequality[5].

According to (2) and (7) one has

$$Z_+ = \int \frac{d^2\alpha}{\pi} \; e^{-\beta \langle\alpha|H|\alpha\rangle} \tag{23}$$

Employing the (Jensen-)Peierls-Bogoliubov inequality

$$e^{\langle\psi|A|\psi\rangle} \le \langle\psi|e^{A}|\psi\rangle \tag{24}$$

valid for any self-adjoint operator A and any normalized state vector $|\psi\rangle$, equation (23) yields

$$Z_+ \le \int \frac{d^2\alpha}{\pi} \; \langle\alpha|\; e^{-\beta H}|\alpha\rangle \tag{25}$$

which according to (10) and (6) completes the proof of the lower bound. Denoting by $\{|n\rangle\}$ the complete orthonormal system of eigenstates of H one can write as the kernel of an operator K in the Hilbert space $\mathcal{L} := L^2(\mathbb{R}^2)$ one can rewrite (33) as

$$Z = \lim_{n\to\infty} \mathrm{tr}_{\mathcal{L}}\, K^n_{\beta/n} \tag{35}$$

For the proof of (30) it is therefore sufficient to show

$$|\mathrm{tr}_{\mathcal{L}}\, K^n_{\beta/n}| \leq \mathrm{tr}_{\mathcal{L}}\, K_\beta \qquad \text{for } n = 2^m,\ m \in \mathbb{N} \tag{36}$$

To this end observe

$$K_\tau = P_\tau\, Q_\tau \tag{37}$$

where the self-adjoint operators P_τ and Q_τ are defined through the kernels

$$\begin{aligned} P_\tau(\alpha,\alpha') &:= \frac{1}{\pi} \langle\alpha|\, e^{-\tau A}\, |\alpha'\rangle \\ Q_\tau(\alpha,\alpha') &:= \delta^{(2)}(\alpha-\alpha')\, e^{-\tau b_-(\alpha',\alpha'^*)} \end{aligned} \tag{38}$$

and enjoy the (semi)group properties

$$P_\tau P_{\tau'} = P_{\tau+\tau'} \quad , \qquad Q_\tau Q_{\tau'} = Q_{\tau+\tau'} \tag{39}$$

By using (37), an inequality due to Golden and Thompson[6], and an iterated version of (39), one finds for $n = 2^m$

$$\begin{aligned} |\mathrm{tr}_{\mathcal{L}}\, K^n_{\beta/n}| &= |\,\mathrm{tr}_{\mathcal{L}}\, (P_{\beta/n}\, Q_{\beta/n})^n| \leq \mathrm{tr}_{\mathcal{L}}\, P^n_{\beta/n}\, Q^n_{\beta/n} \\ &= \mathrm{tr}_{\mathcal{L}}\, P_\beta\, Q_\beta = \mathrm{tr}_{\mathcal{L}}\, K_\beta \end{aligned} \tag{40}$$

which completes the proof of (36) and therefore of (30). It is tempting to conjecture that the upper bound $\tilde{Z}_-$ as defined in (29) is less accurate than another upper bound

$$\tilde{Z} := \mathrm{tr}(e^{-\beta A}\, e^{-\beta B}) \tag{41}$$

derived by Golden and Thompson[6] along analogous lines. This is because $\tilde{Z}_-$ may be viewed as resulting from $\tilde{Z}$ by the substitution

$$e^{-\beta B} \to \int \frac{d^2\alpha}{\pi}\, e^{-\beta b_-(\alpha,\alpha^*)}\, |\alpha\rangle\langle\alpha| \tag{42}$$

which for small β always implies an increasement. If this is true for all β I do not know yet. In any case I want to stress that $\tilde{Z}_-$ is in general easier to compute than $\tilde{Z}$.

6. Additional remarks

Let me add that many of the above conclusions may be generalized from boson coherent states to other (e.g. spin) coherent states[4, 7]. Furthermore, they may be generalized to cases where other degrees of freedom are coupled to the boson mode(s).

For applications of the inequalities (22) in these cases the reader is referred e.g. to Ref.[4, 8].

$$Z = \sum_n e^{-\beta <n|H|n>} = \sum_n e^{-\beta \int \frac{d^2\alpha}{\pi} |<\alpha|n>|^2 h_-(\alpha,\alpha^*)} \tag{26}$$

by making use of (3). Since $\frac{1}{\pi}|<\alpha|n>|^2$ may be interpreted for every n as a probability density in $\mathbb{R}^2$, another version of Jensen's inequality yields

$$e^{-\beta \int \frac{d^2\alpha}{\pi} |<\alpha|n>|^2 \; h_-(\alpha,\alpha^*)} \leq \int \frac{d^2\alpha}{\pi} |<\alpha|n>|^2 \; e^{-\beta h_-(\alpha,\alpha^*)} \tag{27}$$

A proof of the upper bound in (22) now follows from a combination of (26), (27), and

$$\sum_n |<\alpha|n>|^2 = <\alpha|\alpha> = 1 \tag{28}$$

5. Generalized upper bound

The upper bound in (22) can be generalized to

$$\tilde{Z}_- := \int \frac{d^2\alpha}{\pi} <\alpha| \; e^{-\beta A} |\alpha> \; e^{-\beta b_-(\alpha,\alpha^*)} \tag{29}$$

when the Hamiltonian H = A+B is divided into two self-adjoint operators A and B. The antinormal symbol of B is denoted as b_-. The generalization of $\tilde{Z}_-$ with respect to Z_- lies in the amount of freedom to choose some clever divison of H. For A = 0 and B = H one recovers Z_-. For A = H and B = 0 the true partition function Z is reproduced. The general proof of

$$Z \leq \tilde{Z}_- \tag{30}$$

is in close analogy to the proof given by Hepp and Lieb for their upper bound Z_-, except that it starts from the Lie-Trotter formula

$$e^{-\beta H} = \lim_{n\to\infty} \left(e^{-\frac{\beta}{n} A} \; e^{-\frac{\beta}{n} B} \right)^n \tag{31}$$

which can be rewritten as

$$e^{-\beta H} = \lim_{n\to\infty} \left(e^{-\frac{\beta}{n} A} \int \frac{d^2\alpha}{\pi} \; e^{-\frac{\beta}{n} b_-(\alpha,\alpha^*)} |\alpha> <\alpha|\right)^n \tag{32}$$

analogously to (19). The trace yields

$$Z = \lim_{n\to\infty} \int d^2\alpha_1 \; \dots \int d^2\alpha_n \prod_{j=1}^{n} K_{\beta/n} \, (\alpha_j, \alpha_{j-1}) \tag{33}$$

which reduces to (2o) for A = 0. Interpreting the "transfer kernel"

$$K_\tau(\alpha,\alpha') := \frac{1}{\pi} \; <\alpha| \; e^{-\tau A} \; |\alpha'> \; e^{-\tau b_-(\alpha',\alpha'^*)} \tag{34}$$

Acknowledgements

I am grateful to Uwe Brandt for stimulating discussions, to Joachim Stolze for a helpful calculation, and to Allen C. Hirshfeld for a critical reading of the manuscript.

References

1) See, e.g.
J.S. Dowker, J. Math. Phys. 17, 1873 (1976);
H. Leschke and M. Schmutz, Z. Physik B27, 85 (1977) and references quoted there

2) L. Cohen, J. Math. Phys. 7, 781 (1966)
K.E. Cahill and R.J. Glauber, Phys. Rev. 177, 1857 (1969);
G.S. Agarwal and E. Wolf, Phys. Rev. D 2, 2161 (1970)
F.A. Berezin, Theor. Math. Phys. 6, 141 (1971)

3) J.R. Klauder, Ann. of Physics 11, 123 (1960)
R.J. Glauber, Phys. Rev. 131, 2766 (1963)

4) E.H. Lieb, Commun. Math. Phys. 31, 327 (1973)
K. Hepp and E.H. Lieb, Phys. Rev. A8, 2517 (1973)

5) See, e.g.
G.H. Hardy, J.E. Littlewood, and G. Pólya "Inequalities",
At the University Press, Cambridge 1967

6) S. Golden, Phys. Rev. 137, B 1127 (1965);
C.J. Thompson, J. Math. Phys. 6, 1812 (1965)

7) A. M. Perelomov, Sov. Phys. Usp. 20, 703 (1977)

8) U. Brandt and H. Leschke, Z. Physik 271, 295 (1974)

RENORMALIZATION OF YANG-MILLS THEORY
DEVELOPED AROUND AN INSTANTON

A. ROUET

C.N.R.S. - LUMINY - Case 907
Centre de Physique Théorique
F-13288 MARSEILLE CEDEX 2 (France)

It is now a common belief that gauge theories may provide reasonable models for strong interactions. However, usual perturbation theory is certainly too poor for working out such a model, and non perturbative contributions have to be taken into account. At least we need to consider non perturbative contributions at the classical level, and to compute perturbative developments around these classical configurations.

Crucial classical configurations are the instantons [1], and, in this note, we shall restrict ourself to this case. A calculation of perturbation at zeroth order around one instanton has been carried out by 't Hooft [2]. It consists in a saddle point approximation of the Feynman path integral together with a collective coordinates calculation for treating the zero mode problem. A proper definition of the perturbation theory at any order has been worked out by Amati and the present author [3]. Even in the Feynman path formalism the problem is quite more tricky at upper orders than at zeroth order. One cannot neglect any more the dependence in the integration variable of the zero mode jacobian. It makes this determinant no more diagonal, coupling the contributions of the gauge zero modes (usual Faddeev Popov determinant), of the translation zero modes and of the dilatation one. To solve this problem we have been led to consider the three types of zero modes on the same footing [4]. We have then introduced Faddeev-Popov like -but space time independent- ghosts together with the usual ones to deal with this hudge jacobian mixing the translation and dilatation zero modes contribution to the usual gauge one. Let me point out that introducing these extra ghosts is not a fancy way of doing a calculation we could do otherwise : it is the only way we know to define a perturbation theory at upper orders.

More specifically, we want to give a meaning to the ill defined quantity $\frac{Z(j_\mu)}{Z(0)}$, with

$$Z(j_\mu) = \int \mathcal{D} Q_\mu \, \exp\left[\int \mathcal{L}(A_\mu^{cl} + Q_\mu) + j_\mu Q^\mu \right]$$

$$\mathcal{L}(\mathcal{A}_\mu) = -\frac{1}{4}\left(\partial_\mu \mathcal{A}_\nu - \partial_\nu \mathcal{A}_\mu + g\, \mathcal{A}_\mu \times \mathcal{A}_\nu\right)^2 \qquad (1)$$

$$A_{\mu,i}^{cl} = \frac{2}{g} \frac{\eta_{i\mu\nu} x_\nu}{x^2 + 1}$$

where the notations are the usual ones [2,3]. A covariant basis for the zero modes consists in $D_\mu^{ij}(x)\,\delta(x-y)$, $F_{\mu\nu}^i(x)$, $x_\nu F_{\mu\nu}^i(x)$, where D_μ^{ij} and $F_{\mu\nu}^i$ are the covariant derivative and field strength associated to $A_{\mu,i}^{cl}$. The gauge zero modes $D_\mu^{ij}(x)\,\delta(x-y)$ are parametrized by the isospin index j and the space time variable y, the translation zero modes $F_{\mu\nu}^i(x)$ by the Lorentz index ν. The dilatation zero mode $x_\nu F_{\mu\nu}^i(x)$ is unique. Dealing with these zero modes, we are able to properly define $\frac{Z(j_\mu)}{Z(0)}$ by

$$Z(j_\mu) = \int \mathcal{D}\omega^j \, da_\nu \, d\lambda \int \mathcal{D} Q_\mu \, \mathcal{D} C \, \mathcal{D} \bar{C} \, d\psi_\nu \, d\bar{\psi}_\nu \, d\psi \, d\bar{\psi} \times \exp\left[\int L + j_\mu Q^\mu \right]$$

$$\begin{aligned} L = & -\frac{1}{4}\left[\partial_\mu (A_\nu^{cl} + Q_\nu) - \partial_\nu (A_\mu^{cl} + Q_\mu) + (A_\mu^{cl} + Q_\mu) \times (A_\nu^{cl} + Q_\nu) \right]^2 \\ & - \frac{1}{2\alpha}(D_\mu Q^\mu)^2 - \frac{\lambda^2}{2\beta}\left(\int F_{\mu\nu} Q_\mu\right)^2 - \frac{\lambda^4}{2\gamma}\left(\int x_\nu F_{\mu\nu} Q_\mu\right)^2 \\ & + C D_\mu U^\mu + \psi_\nu F_{\mu\nu} U^\mu + x_\nu \psi F_{\mu\nu} U^\mu \end{aligned}$$

$$U^\mu = D_\mu \bar{C} + Q_\mu \times \bar{C} + (\bar{\psi}_\nu + x_\nu \bar{\psi})(D_\nu Q_\mu - F_{\mu\nu}) + Q_\mu \bar{\psi} \qquad (2)$$

C and $\bar{C}$ are the usual Faddeev Popov ghosts associated to the gauges, ψ_ν and $\bar{\psi}_\nu$ are the x independent ghosts associated to the translations, ψ and $\bar{\psi}$ those associated to the dilatations. $\omega^j(y)$, a_ν, λ, are the corresponding gauge, translation, dilatation parameters. The "gauge fixing terms" $(D_\mu Q^\mu)^2$, $(\int F_{\mu\nu} Q^\mu)^2$, $(\int x_\nu F_{\mu\nu} Q^\mu)^2$, cure the zero mode problem respectively for the gauge, translation and dilatation zero modes. They raise the degeneracy of the quadratic form in Q_μ, and provide a unique definition of the propagator.

Lagrangian (2) possesses a generalized BRS symmetry [5,3]:

$$\begin{aligned}
&\delta A_\mu^{cl} = 0 \\
&\delta Q_\mu = \hat{\imath}\, U_\mu \\
&\delta C = \hat{\imath}\, \frac{1}{\alpha}\, D_\mu Q_\mu \\
&\delta \psi_\nu = \hat{\imath}\, \frac{\lambda^2}{\beta^2} \int F_{\mu\nu} Q_\mu \\
&\delta \psi = \hat{\imath}\, \frac{\lambda^4}{\gamma} \int x_\nu F_{\mu\nu} Q_\mu \\
&\delta \bar{C} = \hat{\imath} \left[-\frac{1}{2} \bar{C} \times \bar{C} + (\bar{\psi}_\nu + x_\nu \bar{\psi}) D_\nu \bar{C} - \frac{1}{2} (\bar{\psi}_\nu + x_\nu \bar{\psi})(\bar{\psi}_\rho + x_\rho \bar{\psi}) F_{\nu\rho} \right] \\
&\delta \bar{\psi}_\nu = \hat{\imath}\, \bar{\psi}_\nu \bar{\psi} \\
&\delta \bar{\psi} = 0
\end{aligned} \tag{3}$$

where $\hat{\imath}$ is an infinitesimal anticommuting parameter. This symmetry leads to a Slavnov type identity. The theory can be renormalized using the BPHZL [6] formalism. In this procedure we add a mass term proportional to $(1-s)^2$. Then, together with the usual ultraviolet BPHZ [7] substractions with respect to the external momenta, we perform infrared substractions with respect to the variable s. Let me point out that this does not correspond to introduce a mass in an ad hoc way : it is one of the standard procedures to renormalize massless theories. Relevant Green functions are to be calculated at $s = 1$. In this framework we have proved that it is possible to choose the counterterms in such a way that the Slavnov identity is satisfied at any order, up to terms proportional to $(1-s)^2$ -i.e. is satisfied for $s = 1$. The perturbation theory is thus completely defined, up to a finite number of normalization conditions. It is possible to prove, using standard methods [5,3], that the Green functions of gauge fields transverse to the translation and dilatation zero modes do not depend on the parameters β , γ , related to the translation and dilatation zero modes.
Let me notice that adding fermions to the theory (quantum chromodynamics) has been done, and does not generate any extra difficulty.

Until now I have discussed the perturbative development around one instanton A_μ^{cl} , solution of the classical equation of motion. Actually we can develop around $r A_\mu^{cl}$, where r is a parameter varying from zero to one. Obviously $r A_\mu^{cl}$ is not a classical solution for r not equal to zero or one. In the Lagrangian, this brings a term $({}^r D_\nu\, {}^r F_{\mu\nu}) Q_\mu$, linear in Q_μ , arising from the development of $\left[\partial_\mu (A_\nu^{cl} + Q_\nu) - \partial_\nu (A_\mu^{cl} + Q_\mu) + (A_\mu^{cl} + Q_\mu) \times (A_\nu^{cl} + Q_\nu) \right]^2$ where ${}^r D_\mu$ and ${}^r F_{\mu\nu}$ denote the covariant derivative and the field strength associated to $r A_\mu^{cl}$. We can trivially modify the theory by substracting this linear term. Let me notice that changing L to $\mathcal{L} - ({}^r D_\nu\, {}^r F_{\mu\nu}) Q_\mu$ results in changing the one irreducible generating functional Γ to

$\Gamma_{-}(^{r}D_{\nu}\,{}^{r}F_{\mu\nu})\,Q_{\mu}$. The procedure sketched above for the development around A_{μ}^{cl} applies here [8] and provides us with a theory parametrized by r . This theory interpolates Yang Mills theory developed around zero (r = 0) and Yang Mills theory developed around one instanton (r = 1). Let me however underline that for r not equal to zero or one this is not the Yang Mills theory developed around $r\,A_{\mu}^{cl}$ because of the substraction of the linear term. Using the fact that A_{μ}^{cl} is rapidly decreasing for large euclidean momenta, we can prove that the counterterms which are dominant in this regime do not depend on r , and thus are equal for r = 0 and r = 1. As a result the $\beta(g)$ Callan-Symanzik function is the same either computed in the perturbation theory around zero or around the instanton [8]. This is a crucial result to justify such methods as those developed by Callan, Dashen, Gross [9].

Very recently, using similar methods, I have proved that the knowledge of the counterterms in the zeroth sector provides (in a non trivial way) all the counterterms -and not only the dominant ones- in the instanton sector. I have also obtained a formula giving the one particle irreducible Green functional in the instanton sector in terms of this functional in the zero sector. This point will be developed in a forthcoming publication.

REFERENCES

1 A.A. BELAVIN, A.M. POLYAKOV, A.S. SCHWARTZ, Yu.S. TYUPKIN, Phys. Letters 59B, 85 (1975).

2 G. 't HOOFT, Phys. Rev. D14, 3432 (1976).

3 D. AMATI, A. ROUET, CERN Preprint TH 2468.

4 D. AMATI, A. ROUET, Phys. Letters 73B, 39 (1978).

5 C. BECCHI, A. ROUET, R. STORA, Ann. Phys. 98, 2 (1976).

6 J.H. LOWENSTEIN, W. ZIMMERMANN, Nuclear Phys. B86, 77 (1975).
J.H. LOWENSTEIN, Comm. math. Phys. 47, 53 (1976).

7 W. ZIMMERMANN, Ann. Phys. 77, 536 (1973) ; 77, 570 (1973).

8 A. ROUET, Phys. Letters, To be published.

9 C. CALLAN, R. DASHEN, D. GROSS, Phys. Lett. 66B, 375 (1977).

- CLOSING ADDRESS -

J'ai accepté de clore ce congrès parce que cela me fait très plaisir de remercier ceux qui nous ont offert ces journées intéressantes, fructueuses et agréables. Je suis sûre que l'un ou l'autre des membres de l'équipe qui forme le comité d'organisation a particulièrement donné de sa personne. Mais je pense aussi qu'ils aimeront que je respecte la personnalité que constitue l'équipe de Marseille et, au nom de tous les participants, je remercie Sergio Albeverio, Philippe Combe, Raphael Høegh-Krohn, Madeleine Sirugue-Collin, Michel Sirugue, Guy Rideau, Raymond Stora, Maryse Cohen-Solal et Geneviève Niard.

Aucun participant ne s'étonnera que j'ai inclus le nom de Michel Sirugue dans la liste du comité d'organisation et terminé par le nom de celle à qui l'on adresse la correspondance ("except the hotel reservation form") et le nom de celle qui est venue à la rescousse lorsque le flot des physiciens et mathématiciens attirés par la conférence a dépassé les prévisions les plus larges.

L'accueil professionnel et personnel de l'équipe de Marseille, le cadre du CNRS et les services qu'il offre à ses chercheurs et à ses invités ont fait de ce congrès une rencontre des plus sympathiques. Mais cette rencontre n'aurait pas eu lieu si l'Université de Provence (Aix-Marseille I), l'Université d'Aix-Marseille II, l'Université de Paris VII et le Centre de Physique Théorique du CNRS n'avaient pas accordé les crédits nécessaires.

The scientific closing remarks, on the other hand, are not easy to make. I can neither summarize in a few minutes, nor select the high points of this conference. An authority higher than mine is needed to pass judgement on our works and I shall quote the comment Dirac made at the occasion of his 75th birthday, "A physicist needs that his equations be mathematically sound". As a physicist I have used expressions which are mathematically unsound - but each time I listened to what a mathematician had to say about my transgressions, a new aspect of the problem appeared and progress followed in a direction which I had not suspected.

But mathematicians cannot provide the concepts and the tools needed by physicists if physicists cannot say what they are studying. The labels "physicist" and "mathematician" need not be put on different persons. One and the same person can be both - admittedly with stronger leanings one way on the other when tackling a problem.

Choosing the mathematical problems in Feynman path integral as the core of the conference was forward looking. It brings together good physics and good mathematics and the marriage can be expected to be very fruitful. Feynman began by analyzing some quantum phenomena. As a physicist he said what he was studying : quantum interferences. He then introduced an expression to describe the basic features of quantum phenomena. He used path integration in its very crude form to construct the celebrated diagrams and give the rules for computing them. But diagrams are only a perturbative computational method.

Path integration is much more. Path integration was not conceived as a solution of the Schrödinger equation with Cauchy data. Path integration provides a description of nature which is different from partial differential equations.

The World is global and stochastic, and physical laws are local and deterministic. Path integration is global and stochastic, partial differential equations are local and deterministic. The beautiful thing is that their complementary descriptions can be brought together : path integrals are solutions of partial differential equations.

Some 35 years have elapsed since the first path integral was written. Path integrals have been a source of frustration : they seem to be so powerful but how does one unleash this power. This conference is the first one, I believe, which takes the bull by the horns and ask "What are they ?". It is hoped that from the definitions which have been presented here an increasingly versatile path integration technology will follow and path integration will give both qualitative and quantitative answers. We now have several formulation of quantum physics and it was good to have in the same conference other formulations so that the relationship between the various formulations can be investigated. It was good also to have a presentation of some of the problems that path integration should ultimately be able to handle qualitatively and quantitatively.

Because path integration entered the scene when partial differential equations were so well developed, it has often been underestimated and misunderstood - even by Feynman himself. The organizers of this conference should be congratulated for having recognized its importance and all of us should be encouraged to pursue the good works presented during the Marseille conference.

Cécile De Witt-Morette
26 Mai 1978

Selected Issues from

Lecture Notes in Mathematics

Vol. 594: Singular Perturbations and Boundary Layer Theory, Lyon 1976. Edited by C. M. Brauner, B. Gay, and J. Mathieu. VIII, 539 pages. 1977.

Vol. 596: K. Deimling, Ordinary Differential Equations in Banach Spaces. VI, 137 pages. 1977.

Vol. 605: Sario et al., Classification Theory of Riemannian Manifolds. XX, 498 pages. 1977.

Vol. 606: Mathematical Aspects of Finite Element Methods. Proceedings 1975. Edited by I. Galligani and E. Magenes. VI, 362 pages. 1977.

Vol. 607: M. Métivier, Reelle und Vektorwertige Quasimartingale und die Theorie der Stochastischen Integration. X, 310 Seiten. 1977.

Vol. 615: Turbulence Seminar, Proceedings 1976/77. Edited by P. Bernard and T. Ratiu. VI, 155 pages. 1977.

Vol. 618: I. I. Hirschman, Jr. and D. E. Hughes, Extreme Eigen Values of Toeplitz Operators. VI, 145 pages. 1977.

Vol. 623: I. Erdelyi and R. Lange, Spectral Decompositions on Banach Spaces. VIII, 122 pages. 1977.

Vol. 628: H. J. Baues, Obstruction Theory on the Homotopy Classification of Maps. XII, 387 pages. 1977.

Vol. 629: W.A. Coppel, Dichotomies in Stability Theory. VI, 98 pages. 1978.

Vol. 630: Numerical Analysis, Proceedings, Biennial Conference, Dundee 1977. Edited by G. A. Watson. XII, 199 pages. 1978.

Vol. 636: Journées de Statistique des Processus Stochastiques, Grenoble 1977, Proceedings. Edité par Didier Dacunha-Castelle et Bernard Van Cutsem. VII, 202 pages. 1978.

Vol. 638: P. Shanahan, The Atiyah-Singer Index Theorem, An Introduction. V, 224 pages. 1978.

Vol. 648: Nonlinear Partial Differential Equations and Applications, Proceedings, Indiana 1976–1977. Edited by J. M. Chadam. VI, 206 pages. 1978.

Vol. 650: C*-Algebras and Applications to Physics. Proceedings 1977. Edited by R. V. Kadison. V, 192 pages. 1978.

Vol. 656: Probability Theory on Vector Spaces. Proceedings, 1977. Edited by A. Weron. VIII, 274 pages. 1978.

Vol. 662: Akin, The Metric Theory of Banach Manifolds. XIX, 306 pages. 1978.

Vol. 665: Journées d'Analyse Non Linéaire. Proceedings, 1977. Edité par P. Bénilan et J. Robert. VIII, 256 pages. 1978.

Vol. 667: J. Gilewicz, Approximants de Padé. XIV, 511 pages. 1978.

Vol. 668: The Structure of Attractors in Dynamical Systems. Proceedings, 1977. Edited by J. C. Martin, N. G. Markley and W. Perrizo. VI, 264 pages. 1978.

Vol. 675: J. Galambos and S. Kotz, Characterizations of Probability Distributions. VIII, 169 pages. 1978.

Vol. 676: Differential Geometrical Methods in Mathematical Physics II, Proceedings, 1977. Edited by K. Bleuler, H. R. Petry and A. Reetz. VI, 626 pages. 1978.

Vol. 678: D. Dacunha-Castelle, H. Heyer et B. Roynette. Ecole d'Eté de Probabilités de Saint-Flour. VII-1977. Edité par P. L. Hennequin. IX, 379 pages. 1978.

Vol. 679: Numerical Treatment of Differential Equations in Applications, Proceedings, 1977. Edited by R. Ansorge and W. Törnig. IX, 163 pages. 1978.

Vol. 681: Séminaire de Théorie du Potentiel Paris, No. 3, Directeurs: M. Brelot, G. Choquet et J. Deny. Rédacteurs: F. Hirsch et G. Mokobodzki. VII, 294 pages. 1978.

Vol. 682: G. D. James, The Representation Theory of the Symmetric Groups. V, 156 pages. 1978.

Vol. 684: E. E. Rosinger, Distributions and Nonlinear Partial Differential Equations. XI, 146 pages. 1978.

Vol. 690: W. J. J. Rey, Robust Statistical Methods. VI, 128 pages. 1978.

Vol. 691: G. Viennot, Algèbres de Lie Libres et Monoïdes Libres. III, 124 pages. 1978.

Vol. 693: Hilbert Space Operators, Proceedings, 1977. Edited by J. M. Bachar Jr. and D. W. Hadwin. VIII, 184 pages. 1978.

Vol. 696: P. J. Feinsilver, Special Functions, Probability Semigroups, and Hamiltonian Flows. VI, 112 pages. 1978.

Vol. 702: Yuri N. Bibikov, Local Theory of Nonlinear Analytic Ordinary Differential Equations. IX, 147 pages. 1979.

Vol. 704: Computing Methods in Applied Sciences and Engineering, 1977, I. Proceedings, 1977. Edited by R. Glowinski and J. L. Lions. VI, 391 pages. 1979.

Vol. 710: Séminaire Bourbaki vol. 1977/78, Exposés 507–524. IV, 328 pages. 1979.

Vol. 711: Asymptotic Analysis. Edited by F. Verhulst. V, 240 pages. 1979.

Vol. 712: Equations Différentielles et Systèmes de Pfaff dans le Champ Complexe. Edité par R. Gérard et J.-P. Ramis. V, 364 pages. 1979.

Vol. 716: M. A. Scheunert, The Theory of Lie Superalgebras. X, 271 pages. 1979.

Vol. 720: E. Dubinsky, The Structure of Nuclear Fréchet Spaces. V, 187 pages. 1979.

Vol. 724: D. Griffeath, Additive and Cancellative Interacting Particle Systems. V, 108 pages. 1979.

Vol. 725: Algèbres d'Opérateurs. Proceedings, 1978. Edité par P. de la Harpe. VII, 309 pages. 1979.

Vol. 726: Y.-C. Wong, Schwartz Spaces, Nuclear Spaces and Tensor Products. VI, 418 pages. 1979.

Vol. 727: Y. Saito, Spectral Representations for Schrödinger Operators With Long-Range Potentials. V, 149 pages. 1979.

Vol. 728: Non-Commutative Harmonic Analysis. Proceedings, 1978. Edited by J. Carmona and M. Vergne. V, 244 pages. 1979.

Vol. 729: Ergodic Theory. Proceedings, 1978. Edited by M. Denker and K. Jacobs. XII, 209 pages. 1979.

Vol. 730: Functional Differential Equations and Approximation of Fixed Points. Proceedings, 1978. Edited by H.-O. Peitgen and H.-O. Walther. XV, 503 pages. 1979.

Lecture Notes in Physics

Vol. 68: Y. V. Venkatesh, Energy Methods in Time-Varying System Stability and Instability Analyses. XII, 256 pages. 1977.

Vol. 69: K. Rohlfs, Lectures on Density Wave Theory. VI, 184 pages. 1977.

Vol. 70: Wave Propagation and Underwater Acoustics. Edited by J. Keller and J. Papadakis. VIII. 287 pages. 1977.

Vol. 71: Problems of Stellar Convection. Proceedings 1976. Edited by E. A. Spiegel and J. P. Zahn. VIII, 363 pages. 1977.

Vol. 72: Les instabilités hydrodynamiques en convection libre forcée et mixte. Edité par J. C. Legros et J. K. Platten. X, 202 pages. 1978.

Vol. 73: Invariant Wave Equations. Proceedings 1977. Edited by G. Velo and A. S. Wightman. VI, 416 pages. 1978.

Vol. 74: P. Collet and J.-P. Eckmann, A Renormalization Group Analysis of the Hierarchical Model in Statistical Mechanics. IV, 199 pages. 1978.

Vol. 75: Structure and Mechanisms of Turbulence I. Proceedings 1977. Edited by H. Fiedler. XX, 295 pages. 1978.

Vol. 76: Structure and Mechanisms of Turbulence II. Proceedings 1977. Edited by H. Fiedler. XX, 406 pages. 1978.

Vol. 77: Topics in Quantum Field Theory and Gauge Theories. Proceedings, Salamanca 1977. Edited by J. A. de Azcárraga. X, 378 pages 1978.

Vol. 78: Böhm, The Rigged Hilbert Space and Quantum Mechanics. IX, 70 pages. 1978.

Vol. 79: Group Theoretical Methods in Physics. Proceedings, 1977. Edited by P. Kramer and A. Rieckers. XVIII, 546 pages. 1978.

Vol. 80: Mathematical Problems in Theoretical Physics. Proceedings, 1977. Edited by G. Dell'Antonio, S. Doplicher and G. Jona-Lasinio. VI, 438 pages. 1978.

Vol. 81: MacGregor, The Nature of the Elementary Particle. XXII, 482 pages. 1978.

Vol. 82: Few Body Systems and Nuclear Forces I. Proceedings, 1978. Edited by H. Zingl, M. Haftel and H. Zankel. XIX, 442 pages. 1978.

Vol. 83: Experimental Methods in Heavy Ion Physics. Edited by K. Bethge. V, 251 pages. 1978.

Vol. 84: Stochastic Processes in Nonequilibrium Systems, Proceedings, 1978. Edited by L. Garrido, P. Seglar and P. J. Shepherd. XI, 355 pages. 1978

Vol. 85: Applied Inverse Problems. Edited by P. C. Sabatier. V, 425 pages. 1978.

Vol. 86: Few Body Systems and Electromagnetic Interaction. Proceedings 1978. Edited by C. Ciofi degli Atti and E. De Sanctis. VI, 352 pages. 1978.

Vol. 87: Few Body Systems and Nuclear Forces II, Proceedings, 1978. Edited by H. Zingl, M. Haftel, and H. Zankel. X, 545 pages. 1978.

Vol. 88: K. Hutter and A. A. F. van de Ven, Field Matter Interactions in Thermoelastic Solids. VIII, 231 pages. 1978.

Vol. 89: Microscopic Optical Potentials, Proceedings, 1978. Edited by H. V. von Geramb. XI, 481 pages. 1979.

Vol. 90: Sixth International Conference on Numerical Methods in Fluid Dynamics. Proceedings, 1978. Edited by H. Cabannes, M. Holt and V. Rusanov. VIII, 620 pages. 1979.

Vol. 91: Computing Methods in Applied Sciences and Engineering, 1977, II. Proceedings, 1977. Edited by R. Glowinski and J. L. Lions. VI, 359 pages. 1979.

Vol. 92: Nuclear Interactions. Proceedings, 1978. Edited by B. A. Robson. XXIV, 507 pages. 1979.

Vol. 93: Stochastic Behavior in Classical and Quantum Hamiltonian Systems. Proceedings, 1977. Edited by G. Casati and J. Ford. VI, 375 pages. 1979.

Vol. 94: Group Theoretical Methods in Physics. Proceedings, 1978. Edited by W. Beiglböck, A. Böhm and E. Takasugi. XIII, 540 pages. 1979.

Vol. 95: Quasi One-Dimensional Conductors I. Proceedings, 1978. Edited by S. Barišić, A. Bjeliš, J. R. Cooper and B. Leontić. X, 371 pages. 1979.

Vol. 96: Quasi One-Dimensional Conductors II. Proceedings 1978. Edited by S. Barišić, A. Bjeliš, J. R. Cooper and B. Leontić. XII, 461 pages. 1979.

Vol. 97: Hughston, Twistors and Particles. VIII, 153 pages. 1979.

Vol. 98: Nonlinear Problems in Theoretical Physics. Proceedings, 1978. Edited by A. F. Rañada. X, 216 pages. 1979.

Vol. 99: M. Drieschner, Voraussage – Wahrscheinlichkeit – Objekt. XI, 308 Seiten. 1979.

Vol. 101: A. Martin-Löf, Statistical Mechanics and the Foundations of Thermodynamics. V, 120 pages. 1979.

Vol. 102: H. Hora, Nonlinear Plasma Dynamics at Laser Irradiation. VIII, 242 pages. 1979.

Vol. 103: P. A. Martin, Modèles en Mécanique Statistique des Processus Irréversibles. IV, 134 pages. 1979.

Vol. 104: Dynamical Critical Phenomena and Related Topics. Proceedings, 1979. Edited by Ch. P. Enz. XII, 390 pages. 1979.

Vol. 105: Dynamics and Instability of Fluid Interfaces. Proceedings, 1978. Edited by T. S. Sørensen. V, 315 pages. 1979.

Vol. 106: Feynman Path Integrals, Proceedings, 1978. Edited by S. Albeverio et al. XI, 451 pages. 1979.